THIN FILM TECHNOLOGY. *By* Robert W. Berry, Peter M. Hall and Murray T. Harris.

SIGNALS AND NOISE IN COMMUNICATION SYSTEMS. *By* Harrison E. Rowe.

PROBABILITY AND ITS ENGINEERING USES, Second Edition. *By* Thornton C. Fry.

PRINCIPLES OF ELECTRON TUBES. *By* James W. Gewartowski and Hugh A. Watson.

APPLIED MATHEMATICS FOR ENGINEERS AND SCIENTISTS, Second Edition. *By* S. A. Schelkunoff.

THE PROPERTIES, PHYSICS AND DESIGN OF SEMICONDUCTOR DEVICES. *By* John N. Shive.

TRANSISTOR TECHNOLOGY, Vol. I. *Edited by* H. E. Bridgers, J. H. Scaff and J. N. Shive.

TRANSISTOR TECHNOLOGY, Vol II. *Edited by* F. J. Biondi.

TRANSISTOR TECHNOLOGY, Vol. III. *Edited by* F. J. Biondi.

PHYSICAL ACOUSTICS AND THE PROPERTIES OF SOLIDS. *By* Warren P. Mason.

SWITCHING RELAY DESIGN. *By* R. L. Peek, Jr. and H. N. Wagar.

THEORY AND DESIGN OF ELECTRON BEAMS, Second Edition. *By* J. R. Pierce.

SPEECH AND HEARING, Second Edition. *By* Harvey Fletcher.

MODULATION THEORY. *By* Harold S. Black.

FERROMAGNETISM. *By* Richard M. Bozorth.

DESIGN OF SWITCHING CIRCUITS. *By* William Keister, Alistair E. Ritchie and Seth H. Washburn.

PIEZOELECTRIC CRYSTALS AND THEIR APPLICATION TO ULTRASONICS. *By* Warren P. Mason.

PRINCIPLES AND APPLICATIONS OF WAVEGUIDE TRANSMISSION. *By* George C. Southworth.

ELECTRONS AND HOLES IN SEMICONDUCTORS. *By* William Shockley.

ELECTROMECHANICAL TRANSDUCERS AND WAVE FILTERS, Second Edition. *By* Warren P. Mason.

FOURIER INTEGRALS FOR PRACTICAL APPLICATIONS. *By* the late George A. Campbell and Ronald M. Foster.

CAPACITORS—THEIR USE IN ELECTRONIC CIRCUITS. *By* M. Brotherton.

NETWORK ANALYSIS AND FEEDBACK AMPLIFIER DESIGN. *By* Hendrik W. Bode.

ELECTROMAGNETIC WAVES. *By* S. A. Schelkunoff.

POISSON'S EXPONENTIAL BINOMIAL LIMIT. *By* the late E. C. Molina.

ECONOMIC CONTROL OF QUALITY OF MANUFACTURED PRODUCT. *By* the late W. A. Shewhart.

TRANSMISSION NETWORKS AND WAVE FILTERS. *By* T. E. Shea.

MICROWAVE ELECTRONICS. *By* John C. Slater.

TRAVELING-WAVE TUBES. *By* J. R. Pierce.

Thin Film Technology

Prepared with the collaboration of the following members of the Technical Staff of the Bell Telephone Laboratories and members of the Engineering Staff of the Western Electric Company:

A. COUCOULAS	W. H. ORR
R. E. KERWIN	L. A. PRIOLO
G. KLEINEDLER	C. C. SHIFLETT
E. A. LaCHAPELLE	D. K. THOMSON

THIN FILM TECHNOLOGY

by

ROBERT W. BERRY
PETER M. HALL
MURRAY T. HARRIS

Members of the Technical Staff,
Bell Telephone Laboratories, Inc.

W. MATSUKADO WA7JBE

D. VAN NOSTRAND COMPANY, INC.

Princeton, New Jersey

Toronto   London   Melbourne

Van Nostrand Regional Office: *New York, Chicago, San Francisco*

D. Van Nostrand Company, Ltd., *London*

D. Van Nostrand Company (Canada), Ltd., *Toronto*

D. Van Nostrand Australia Pty. Ltd., *Melbourne*

Library of Congress Catalog Card No. 68-25817

PRINTED IN THE UNITED STATES OF AMERICA

"All of nature is to be found in the smallest things."

—*Latin proverb*

Foreword

About ten years ago, a small group of chemists led by D. A. McLean started to explore sputtered films of tantalum for use as circuit elements in integrated networks. Early interest focused on resistor networks that, as H. Basseches showed, could be readily adjusted to high initial precision by trim anodization. D. Gerstenberg's subsequent discovery of the dramatic improvement in tantalum film stability, which results from the use of nitrogen during sputtering, extended this high initial precision to long-term performance. R. W. Berry's invention of the thin film capacitor broadened the use of tantalum thin films to another basic circuit element, and made possible the design of filters in a single materials system without the use of inductance. More recently, R. W. Wyndrum utilized tantalum thin film technology in a functional approach to network design. In his approach, individual circuit elements are no longer discernible, but are instead distributed in resistive areas of appropriate geometry or in multilayered areas of resistive, dielectric, and conductive films.

This rapid growth and diversification of tantalum thin film technology has been further accelerated by the advent of silicon integrated circuit technology. Tantalum thin film technology complements silicon technology. It provides integration and specialized performance characteristics for circuit elements that silicon cannot provide. Silicon technology is best suited for integrating transistors, diodes, and low valued, noncritical resistors. Tantalum technology excels in integrating precision resistors and capacitors over a wide range of values, as well as in completing interconnections among silicon integrated circuits.

Tantalum thin film technology and beam-lead sealed-junction technology are topologically and metallurgically compatible. Together, these two technologies form the dual foundation of a flexible approach to integrated electronics that is applicable to all types of communication needs and to any level of system integration. Together, they offer the circuit designer two batch-fabricated and batch-interconnected materials systems with a flexible interface. The designer can shift this interface smoothly toward one system or the other, depending upon the required function. Future improvements, which will come with the maturing of each of these materials technologies, will likewise shift this interface.

McLean's original choices of a materials system based on tantalum and a film deposition process based on sputtering have stood the test of time. Sputtered films of tantalum have demonstrated unmatched stability, reliability, and the opportunity for custom tailoring to specific performance needs. The anodically grown oxide of tantalum has provided adjustability and environmental protection for resistors as well as strong, uniform, and stable dielectrics for thin film capacitors.

Having been widely accepted by designers of station equipment, transmission systems, and switching systems, tantalum thin film technology has reached high-volume manufacture. The successful transition of this technology from the laboratory to the field could not have occurred without essential contributions by manufacturing engineers. Collaborating closely with Bell Laboratories, Western Electric teams led by J. F. Paulsen at Allentown, K. A. Rahlfs at the Princeton Engineering Research Center, and J. D. Schiller at North Carolina made important advances in manufacturing methods. It is through such innovations as in-line sputtering and automated high-speed trim anodizing that tantalum film technology can today offer low cost in addition to outstanding performance and versatility.

This book contains a comprehensive treatment of all major aspects of tantalum thin film technology, namely, materials, processes, structures, design, and manufacture. Written by specialists in these disciplines, it is broadly based on years of experience, not only in research and development, but also in manufacturing.

Tantalum thin film technology is still evolving and growing. The accomplishments to date, however, are sufficiently concrete to make this publication timely and relevant.

March 1968   E. D. REED

Preface

The commercial use of thin films has been growing at an increasingly rapid rate for the past several years, particularly in the electronics industry. When the Western Electric Company recently established an internal, nine-month course intended to afford comprehensive coverage of integrated circuits, it clearly felt the absence of a reference book that could also satisfy specialized course needs. The material presented in this book was compiled in order to provide a textbook for that portion of the Western Electric course dealing with thin films.

THIN FILM TECHNOLOGY has as its objective the provision of scientific bases for the methods and materials used in thin film electronics. Additionally, it provides sufficient detail in the areas of application to permit an understanding of that aspect of the subject that might still be termed "art."

The book has three broad divisions: first, a discussion of the various methods of film formation; second, properties of materials as they relate to the electronic uses of thin films; and third, information necessary for the actual application of thin films to integrated circuits.

Thin films made from the tantalum system currently have a special importance in the field of microelectronics, and the authors have had more personal experience with this system than with any other. Accordingly, in the section devoted to properties of materials, tantalum has been covered in much greater detail than other thin film material systems. Nevertheless, in the interest of thoroughness, every effort has been made to include all the materials currently in use.

The chapters were originally written by the following contributors:

    R. W. Berry, Chapter 4
    P. M. Hall, Chapters 2, 6, and 7
    M. T. Harris, Chapters 1, 5, 8, and 9
    E. A. LaChapelle, A. Coucoulas, and
      G. Kleinedler, Chapter 12
    W. H. Orr, Chapter 11
    L. A. Priolo, Chapter 13
    C. C. Shiflett, Chapter 3
    D. K. Thomson and R. E. Kerwin, Chapter 10

These contributors are not further credited with authorship of their respective chapters, however. Because of the considerable changes from the original texts, the editors alone must accept full responsibility for the content of the present work.

In any work of this kind, acknowledgment is merited by many more people than practical considerations permit acknowledging. Many colleagues on the staffs of the Bell Telephone Laboratories and the Western Electric Company have provided information and helpful discussions throughout the progressive stages of the book's development; we acknowledge our indebtedness to them collectively. We must acknowledge specifically E. D. Reed of the Bell Telephone Laboratories and A. E. Anderson of the Western Electric Company for their encouragement and support of this work. The critical comments of D. A. McLean of the Bell Telephone Laboratories contributed significantly to the text, and the efforts of C. E. Pounds, Peggy Wyatt, and Nancy Davis in editing proof and preparing figures were invaluable.

R. W. BERRY
P. M. HALL
M. T. HARRIS

Contents

CHAPTER   PAGE

Foreword   ix

Preface   xi

1 INTRODUCTION   1

1.1   General   2
1.2   Physical Properties of Thin Films   2
1.3   Thin Film Deposition   3
      Electrodeposition   3
      Chemical Reduction   5
      Electroless Plating   5
      Vapor Plating   6
      Evaporation   7
      Sputtering   8
      Anodization   9
      Polymerization   10
1.4   Applications of Thin Films   11
      Optical   11
      Magnetic   12
      Electronic   13

2 VACUUM TECHNOLOGY   19

2.1   Kinetic Theory of Gases   19
      Ideal Gas Laws   20
      Molecular Motion   22
      Mean Free Path   23
      Collisions of Molecules with a Surface   24
      Pressure   25
      Units of Pressure   29
      Electrical Conductivity   30
      Thermal Conductivity   30

2.2   Gas Flow   31
      Quantity of Gas and Gas Flow Rates   31
      Conductance and Pump Speed   32
      Regions of Gas Flow   35
      Viscous Flow Conductances   36
      Molecular Flow Conductances   36

2.3   Fore- and Roughing-Pumps   38
      Rotary Oil Pumps   38
      Roots Blower Pumps   40
      Sorption Pumps   41

2.4 High Vacuum Pumps   44
      Diffusion Pumps   44
      Diffusion Pump Fluids   46
      Diffusion Pump Performance   47
      Diffusion Pump Operation   52
      Getter and Ion Pumps   53
      Advantages and Disadvantages of Sputter-Ion Pumps   54
      Turbomolecular Pumps   56

2.5 Vacuum Systems   60
      Gas Flow Analysis   61
      Pumpdown   65
      Selection of Pump Sizes   67
      Operation of a Bell Jar System   69
      Maintenance and Leak Detection   71
      Continuous-Feed Vacuum System   74

2.6 Vacuum Gauges   78
      Mechanical Gauges   78
      Hydrostatic Gauges   79
      Thermal Conductivity Gauges   82
      Thermionic Ionization Gauges   85
      Cold-Cathode Ionization Gauges   91
      Gauge Locations   93

2.7 Vacuum Valves   94
      High Vacuum Valves   94
      Foreline and Roughing Valves   96
      Ultrahigh Vacuum Valves   97
      Leak Valves   99

2.8 Vacuum Materials and Techniques   101
      Elastomers   103
      O-Ring Seals   106
      Metal Joining Techniques   108

3 EVAPORATED FILMS   113

3.1 The Environment for Evaporation of Films   113

3.2 The Vaporization Process   117

3.3 Growth, Structure, and Adhesion of Evaporated Films   123
      Adsorption Phenomena   124
      Nucleation and Initial Growth   126
      Factors Affecting Film Growth and Film Properties   130
      Adhesion of the Film   133

3.4 Types of Evaporation Sources   135
     Resistance Heated Sources   135
     Electron Bombarded Evaporation Sources   139
     Miscellaneous Sources   142

3.5 Evaporation of Specific Types of Materials, Special Techniques   145
     Metals   146
     Metal Alloys   149
     Insulators and Dielectrics   151
     Evaporation in the Presence of a Glow Discharge   156

3.6 Evaporated Film Thickness and Thickness Distribution   158
     Theoretical Film Thickness Distribution   158
     Emissive Characteristics of Practical Sources   161
     Methods of Producing Uniform Deposits   162
     Film Thickness Measurement Techniques   164
     In-Process Deposition Monitors   183

4 SPUTTERED FILMS   191

4.1 Theories of Sputtering   193
     Distribution of Sputtered Atoms   194
     Sputtering Yield   195
     Energy of Sputtered Atoms   198

4.2 Sputtering in a Glow Discharge   199
     Glow Discharge Characteristics   199
     Normal and Abnormal Glow Discharge   202

4.3 Film Deposition in a Glow Discharge   204
     Deposition Rate   205
     Contamination During Deposition   208
     Reactive Sputtering   210
     Deposition Parameters and Their Influence   211
     Alloy Deposition   212
     Film Adherence   213

4.4 Sputtering with an Artificially Supported Discharge   214
     Thermionic Electron Emission   214
     Radio-Frequency Excitation   216
     Direct Sputtering of Dielectrics   217

4.5 Effects of Sputtering Conditions on Film Properties   219
     Sputtering Atmosphere   220
     Deposition Parameters   225
     Substrate Influences   235

4.6 Sputtering Technology   235
     Laboratory or Bell Jar Sputtering Facility   236
     Other Designs   248
     Film Monitors During Deposition   252

5 CHEMICAL METHODS OF FILM FORMATION   255

5.1 Electroplating   255
     Deposition Parameters   257
     Plating Technology   258
     Plating on Thin Film Circuits   262

5.2 Electroless Plating   263
     Deposition Process   263
     Deposition Parameters   265
     Properties of Electroless Plated Films   265

5.3 Chemical Vapor Plating   266
     Film Deposition Mechanism   267
     Deposition Parameters   267
     Film Properties   268
     Film Deposition Technique   268

5.4 Anodization   271
     Anodization Mechanism   272
     Anodization Kinetics   274
     Anodization Parameters   278
     Properties of Anodic Oxide Films   281

6 ELECTRICAL CONDUCTION IN METALS   289

6.1 Conduction in Bulk Metals   289
     Conduction Electron Energy Distribution   291
     Mean Free Path   294
     Phonon Scattering   296
     Impurity and Defect Scattering   303
     Effect of Stress   304
     Alloys   306
     Magnetic Alloys   308

6.2 Conduction in Thin Films   310
     Sondheimer Theory   310
     Temperature Dependence   314
     Stress Effects in Thin Films   317
     Real Films   323

7 RESISTOR AND CONDUCTOR MATERIALS   329

   7.1 Thin Film Resistor Parameters   329
      Sheet Resistance   329
      Temperature Coefficient of Resistance   333
      Voltage Coefficient of Resistance   335
      Noise Coefficient   335
      High-Frequency Performance   336
      Stability   336
      Power Density   336

   7.2 Thin Film Resistor Materials   337
      Chromium   337
      Nickel-Chromium Alloys   337
      Tin Dioxide   339
      Chromium-Silicon Monoxide   339
      Palladium-Silver Glaze   340

   7.3 Tantalum-Based Thin Film Resistors   341
      Adjustment of Resistor Value   344
      Aging and Stabilization   349

   7.4 Thin Film Conductors   364
      Adherence   365
      Sheet Resistance   366
      Joinability   366
      Heat Treatment Resistance   367

8 CAPACITOR MATERIALS   371

   8.1 Fundamental Capacitor Properties   371
      Capacitance   371
      Energy Stored   372
      Leakage Current   372
      Dielectric Breakdown   373
      Maximum Working Stress   373
      Dielectric Loss   376

   8.2 Thin Film Capacitor Materials   377
      Silicon Monoxide Capacitors   378
      Tantalum Oxide Capacitors   380
      Manganese-Oxide Tantalum-Oxide Capacitors   387
      Silicon-Monoxide Tantalum-Oxide Duplex Capacitors   388
      Organic Polymer Capacitors   389

9 SUBSTRATES   395

   9.1 Substrate Manufacture   396
      Glass   397
      Ceramics   397
      Synthetic Sapphire   398

9.2 Material Properties   398
     Surface Smoothness   398
     Flatness   400
     Porosity   401
     Mechanical Strength   402
     Thermal Expansion   402
     Thermal Conductivity   402
     Resistance to Thermal Shock   405
     Thermal Stability   406
     Chemical Stability   406
     Electrical Conductivity   411
     Cost   411

9.3 Substrate Cleaning   413
     Testing for Cleanness   416

10 PATTERN GENERATION   419

10.1 General Example   419

10.2 Photolithographic Processing   421
      Photomasks   422
      Photographic Techniques   425
      Metal-on-Glass Masks   444
      Photoresists   445
      Etching   451

10.3 Mechanical Masking   451

10.4 Other Methods of Pattern Generation   452

10.5 Manufacture of Photoshaping Tools   455
      Generating the Art Master   455
      Making the Master Mask   456
      Manufacturing Aperture Masks   462

11 COMPONENT AND CIRCUIT DESIGN   467

11.1 Film Resistors   473
      Choice of Materials   473
      Resistor Properties   474
      Stability, Power, and Size   474
      Film Thickness   480
      Resistor Line Width   482
      Number of Squares   486
      Counting Squares   487
      Resistor Parasitics   494

11.2 Film Capacitors   510
      Capacitor Properties and Materials   510
      Dielectric Thickness   511
      Capacitor Yield   514
      Capacitor Losses   515
      Capacitor Patterns   519
      Capacitor Parasitics   523

11.3 Conducting Films   525
      Uses and Materials   525
      Conducting Paths and Land Areas   526
      Crossovers   530
      Film Inductors   531

11.4 Distributed Film Components   533
      Area-Resistor Networks   533
      Distributed RC Structure   535

11.5 Film Circuit Layout   543
      Examples of Film Circuit Layouts   544

12 LEAD ATTACHMENT   559

12.1 Metallurgical Background   561
      Atomic Theory   561
      Crystal Structure Theory   562
      Metallic Solidification   566
      Alloying and Intermediate Phases   567
      Diffusion   568
      Wetting   569

12.2 Metallic Surfaces   570
      Ideal Surfaces   570
      Real Surfaces   572

12.3 Real Metal Cohesion   575
      Electron Interaction   575
      The Grain Boundary   576

12.4 Factors Affecting Joining   580
      Surface Contamination   580
      Interface Contact   581
      Binding Activation Energy   582
      Connection Stability   583

12.5 Thin Film Joining   583
      Thin Film Adhesion   584
      Thin Film Joining Constraints   587

12.6 Alloy Bonding   589
         Fundamentals of Soldering   589
         Fluxes   590
         Solders   591
         Soldering Equipment and Methods   595
         Soldering to Thin Films   599

12.7 Solid-Phase Bonding   602
         Thermocompression Bonding   604
         Ultrasonic Bonding   612

12.8 Fusion Welding   618
         Laser Welding   618
         Electron Beam Welding   621
         Parallel-Gap Resistance Welding   622

12.9 Connection Evaluation   622
         Destructive Testing   623
         Nondestructive Testing   627

13 TECHNIQUES AND FACILITIES FOR LARGE
      SCALE MANUFACTURE   633

13.1 Substrates   633

13.2 Sputtering   638
         Continuous Sputtering Machine   638
         Cleaning   642
         Cleanness Evaluation   642
         Environment   644
         Control of Film Properties   644
         Bell Jar Versus In-Line Machine   645

13.3 Evaporation   647

13.4 Pattern Generation   653
         Photolithography   655
         Negative Versus Positive Resists   656
         Resist Coating Methods   656
         Exposure   657
         Photolithography Versus Mechanical Masks   661
         Thermal Stabilization   661

13.5 Trim Anodization   661
         Trim-Anodization Sets   661
         Fixturing   665

13.6 Circuit Separation   666
        Glass Scribing Machines   666
        Scribe Line and Diamond Details   669
        Breaking   670
        Ceramic Substrate Separation   671

13.7 Lead and Component Attachment   673
        Soldering to Glass Substrates   673
        Soldering to Ceramic Substrates   676
        Soldering Problems   677

13.8 Testing   679
        Tantalum Nitride Film Evaluation   679
        Resistance Measurements   682
        Overload Pulse Test   682
        Circuit Performance Test   685
        Accelerated-Aging Test   685

Index   689

Thin Film Technology

Chapter 1

Introduction

This book is pointed primarily toward thin films for electronic applications. There is, however, much common ground with other uses and with other studies of thin films, especially in relation to general methods and general properties. Hence a major part of this book is devoted to general film techniques and to the general theories and practice of thin films. The recent impact of thin films in the electronic field is based upon two features.

(1) *Compactness* results from the very thinness. A resistor film, for example, which is 400 A thick and 1 mil wide has 500 times the resistance per unit length as a 1 mil wire of the same material. For capacitors, the area required for a given capacitance value is proportional to the dielectric film thickness.

(2) *Integration* is accomplished by fabricating many interconnected components on each supporting substrate to form a "thin film integrated circuit." In many cases other components are bonded to the same substrate, especially transistors, diodes, or "silicon integrated circuits." Such combinations are referred to as "hybrid circuits." Integration has many desirable results; low cost, reproducibility and high performance, and especially reliability.

There are other only slightly less compelling reasons for interest in thin films for electronics. These include precise and stable components, the possibility of prescribed temperature coefficients, and in situ functional adjustment of completed networks. Also, thin films are naturally adaptable for distributed parameter networks.

The authors and editors of this book have participated in the development of tantalum films for integrated circuitry; because of their familiarity with this field, tantalum films are covered more completely than films of other materials, especially in the applied sections.

1

1.1   GENERAL

Thin films of the noble metals have been used for decorating glass and ceramics for over a thousand years. As early as the seventh century, it was known how to paint a suspension of a silver salt onto a glass medicine vial and then heat it to convert the salt to metallic silver. Another very old method of forming thin films is the beating of gold to form "gold leaf." The thinnest films made this way are about four or five millionths of an inch thick. Too much stress applied to this gold leaf leads to rupture, thus imposing a limitation on its thinness. Thickness of films is usually discussed in terms of angstrom units (A). This dimension is equivalent to four billionths of an inch and is of the same order of magnitude as the dimensions of a single atom.

Much more recently, thin films have been used in the study of the relationship between the structure of solids and their physical properties. Practical applications include electrical circuits, optical instruments, and magnetic information-storage devices. These more modern films are formed by depositing material onto a clean supporting substrate to build up film thickness, rather than by thinning down bulk material. Thin films for electronic circuit applications range from a few hundred angstroms to tens of thousands of angstroms in thickness. A typical film might be 1000 A thick, while the glass substrate supporting it would be 10,000 times as thick. Films vary in structure from amorphous films such as anodic oxide dielectrics to single crystal films such as expitaxially grown silicon; most films, however, fall between these extremes and are polycrystalline, composed of many small crystals, called crystallites or grains, fitted together with more or less random orientation. For very thin films, crystallite size is observed to be a function of thickness, but for thicker films, it is usually independent of thickness. Sizes range from about 50 A to 2000 A. Crystallite size also depends on deposition parameters such as substrate temperature and deposition rate, and after deposition may be altered by annealing, with the larger crystallites growing at the expense of the smaller ones.

1.2   PHYSICAL PROPERTIES OF THIN FILMS

The mechanical properties of thin films are quite different from those of the bulk material. For example, the strengths exhibited by some films appear to be as much as 200 times as great as for well annealed bulk samples, and are usually several times as great as the strengths of severely cold worked bulk material. This behavior has been explained in two ways; first, the polycrystalline film structure is more disordered (smaller crystallites or grains) than can be achieved by cold working, and second, if the films are sufficiently thin, the dislocations could extend throughout the entire thickness of the film and thus be locked in place, giving no mechanism for yielding.

An electronic effect of thinness is observed in very thin dielectric films as an abrupt change in the conductance at constant field as the thickness is reduced below some finite dimension. With dielectric films thicker than about 100 A, the field required to cause a given current to flow is generally independent of film thickness. Thinner films, however, exhibit a large increase in current density due to "tunneling." This is because the probability of an electron with a given energy penetrating the potential barrier presented by the dielectric increases exponentially with decreasing thickness.

Thinness also results in a marked change in the conductivity of metallic films as film thickness becomes of the same order as, or smaller than the electron mean free path. When this occurs, scattering of electrons at the film surface becomes a significant factor, and the effective conductivity is reduced. Observations of film resistivity as a function of thickness thus provide one means of estimating the electron mean free path. This method has been found difficult in practice, since films of a high degree of perfection are required. The resistivity of metal films may be separated into two portions, one caused by electrons being scattered by atoms in thermal motion, and another caused by scattering at lattice imperfections. The imperfections in deposited metal films result in resistivities higher than those observed for bulk metals. The effect of temperature on the two modes of scattering is not the same, and this results in differences between the temperature coefficients of resistance of bulk metals and those of polycrystalline metallic films. The magnitude of this difference may be used to infer the degree of imperfection in films.

1.3   THIN FILM DEPOSITION

Electrodeposition

In the early 19th century, gold electroplating was discovered, and shortly thereafter other metals were produced in film form by electrodeposition. The deposition of decorative and protective films was soon followed by the use of electrodeposition as a means of refining gold and copper. Electroplated copper found further application in "electroforming" and as an undercoating for the chromium plating of automotive "bright work." Thicknesses of up to several thousandths of an inch may be plated so that this process may also be used to alter the dimensions of an object. Films may be plated onto non-conductors if they are first coated with a conducting material.

Electrodeposition of metals involves the movement of metallic ions in solution toward a cathode in an applied electric field. The ions accept electrons and are deposited on the cathode as metal atoms. The characteristics of the deposited film depend on current density, agitation and temperature of solution, diffusion velocity of the metallic ions, and the shape and structure of the electrodes. Impurities in the electrolyte solution may also be adsorbed or deposited with the metal and may affect film properties. Metal deposition on the cathode surface does not proceed by successively laying down atoms from one side to the other, as in building a brick wall. Deposition starts at a few favored sites (apparently at defects in the surface), and the metal ions landing at these sites may then move along the surface to kinks, edges, steps, or other discontinuities before becoming incorporated in the metal lattice. The crystals are thus built up layer by layer, normal to the surface and also laterally, until the growing faces meet each other. Neighboring crystals have different orientations, and thus a polycrystalline structure results. If the concentration of the electrolyte is reduced near a growing nucleus, outward growth is encouraged more than lateral growth. On the other hand, high ionic concentration over the surface favors lateral growth.

A simple example of a plating cell consists of two copper electrodes immersed in an aqueous solution of copper sulfate. The electrodes are connected to an external source of direct current, and copper ions in the copper sulfate solution are attracted to the negative copper electrode or cathode. When a copper ion receives two electrons and is deposited on the cathode, a copper atom of the anode gives up two electrons and dissolves in the solution as a positively charged copper ion. The acceptance of electrons by a copper ion to form an uncharged copper atom is called chemical reduction, while the loss of electrons by the copper atom at the anode is called oxidation. The physical law describing this process, first formulated by Michael Faraday in 1833, remains unmodified by newer discoveries and stands as an important part of the foundation of physical science. Faraday's law states that 96,500 coulombs of electricity will reduce one gram equivalent weight of material at the cathode and will oxidize one gram equivalent weight at the anode.

The modern plating bath is a complex system with special additives to control the bath acidity and film brightness, to aid in plating into holes and crevices, and to produce a uniformly thick film. In thin film circuit applications, electroplating is used in the fabrication of combination metal masks and for increasing the conductance of interconnections. Magnetic films have been electrodeposited for use in computer memories. For most optical uses, the complex nature of the plating bath provides control difficulties which preclude the use of plated films.

Chemical Reduction

At about the same time that electrodeposition was discovered, Justus von Liebig, one of the founding fathers of chemistry, discovered another film deposition technique which is still used today, primarily for silvering mirrors. In his method, the silver is in an ammoniacal solution of silver nitrate. The silver metal is subsequently precipitated by the addition of a reducing agent. Sugar, Rochelle's salt (sodium potassium tartrate), or formaldehyde is commonly used as the reducing agent. Similar techniques have also been used to deposit copper, platinum, and lead sulfide. Lead sulfide films were suggested for use in the fabrication of resistors in 1910.

Electroless Plating

A film deposition method similar to "chemical reduction" but of more recent discovery is electroless plating. This process was invented in 1946 by A. Brenner and G. E. Riddell, who were attempting to electrodeposit a nickel-tungsten alloy on the inside of a tube. They had added a hypophosphite to the bath to decrease the extent of anodic oxidation and had placed the anode inside the tube. This combination of events led to the plating of the tube both inside and out. Brenner next made the surprising observation that the same result ensued without the use of any electric current. Apparently, the hypophosphite ion is capable of reducing nickel salts in contact with some metal surfaces. It was then found that materials which do not normally accept an electroless plating do so if first washed with a solution of palladium chloride. This technique then permits plating on selected areas and inside holes where electroplating is accomplished with difficulty. The process resembles electroplating in that it may be carried on continuously to build up a thick coating. In addition to nickel, palladium and gold films have also been electrolessly plated. Some firms are making use of electrolessly deposited nickel films on ceramic as a resistor material.

Vapor Plating

A material may be plated from a vapor if it forms a volatile compound which can readily be dissociated or reduced at some temperature below the melting point of the material. This volatile compound must also be sufficiently stable to prevent decomposition or reduction before reaching the deposition surface. The vapor plating technique then consists of reducing or decomposing this volatile compound upon a heated surface. An example of this method is the coating of an object with titanium metal by the hydrogen reduction of titanium tetrabromide at 1100° to 1400°C. Titanium tetrabromide, a solid at room temperature, is heated to 50° to 100°C and hydrogen gas is passed over it to yield the vapor plating gas mixture. This gas mixture is then passed over the heated specimen where it reacts at the heated surface to produce an adherent coating of the non-volatile titanium metal. The remaining hydrogen and hydrogen bromide, together with unreacted titanium tetrabromide, are then passed out of the plating vessel and may be recovered.

An advantage of the vapor plating technique is that refractory metals may be coated on objects for high-temperature protective coatings in thicknesses much greater than those obtained with other methods. Plating of massive deposits on fine wires permits the purification of refractory metals such as zirconium and titanium to sufficient purity to make them workable. High-melting compounds such as silicon carbide and tantalum carbide may also be deposited. Thicknesses from a few hundredths of a mil up to one-half inch are possible, while the average film thickness is from one to ten mils. A major limitation of the method is that the object to be coated must be heated to a high temperature. Nevertheless, thin carbon film resistors formed by pyrolytic decomposition are made in quantities of millions annually. Epitaxially grown single crystal silicon, however, is prepared by this technique by using silicon tetrachloride and hydrogen gas.

A similar technique is used to produce tin oxide. Tin chloride solution, usually with a few percent antimony chloride added, is sprayed onto a heated substrate where it is hydrolized to the oxide. The resulting resistive films are used in both discrete resistors and in thin film integrated circuits.

Evaporation

Vacuum evaporation requires somewhat more sophisticated equipment than do most of the methods already discussed; thus, its development had to await the necessary advances in vacuum technology. The four most important developments which tended to popularize vacuum evaporation were the improvements in vacuum pumps in the early 20th century, the practice of heating the evaporation charge with bare tungsten wire, the adaption of the process to the deposition of nonmetals, and J. Strong's development of an economical method of evaporating aluminum films for large telescope mirrors.

Evaporation is accomplished by using vacuum pumps to reduce the pressure inside a deposition chamber to a millionth of atmospheric pressure or less, and then heating the material to be evaporated in a filament or boat made of a high-melting-point material. The heat is supplied by resistance heating, radio-frequency induction, or by electron bombardment. Both metals and thermally stable compounds such as metal oxides may be deposited. Films produced by evaporation are relatively pure and thus are of interest from a theoretical standpoint. Films of very pure alkali metals are useful in the study of photoelectric phenomena and electrical conductivity, while heavy-metal films are used for experiments on surface properties and on the growth mechanism of single crystal films. The pressure in an electron microscope is sufficiently low to allow for evaporation studies in the microscope, and this fact has been of major importance in the study of growth mechanisms of films produced by the evaporation technique.

Some of the early research using thin-metal films was concerned with their optical properties. These studies were directed toward learning the refractive indices and absorption of bulk metals by the measurement of these properties in thin films. Structural differences between evaporated films and bulk material, however, made impossible the extrapolation of the film properties to bulk metals. Advances in vacuum technology and the development of electron-optical methods of examination have allowed the anomalous behavior of deposited films to be explained.

The properties of evaporated films, just as with other films, are dependent on film structure and also on the interaction of the film with its substrate. Metal evaporated onto other than single crystal substrates results in more or less randomly oriented polycrystalline films. Sometimes, however, there is a preference for a particular lattice plane of the film to be oriented parallel to the substrate surface. By depositing onto single crystals, it has been possible to produce films with a limited number of well defined grain orientations and even single crystal films. Film growth during deposition involves the nucleation of small islands which move about and grow together with a mobility that gives them a liquid appearance when viewed with an electron microscope. The final crystallite size is a function of the substrate temperature, deposition rate, and subsequent annealing. Various imperfections occur in the crystallites, such as missing atoms, impurity atoms, misplaced atoms, and misfits of whole rows or planes of atoms. These defects are a result of the growth mechanism and have marked effects on properties. Electrical conductivity, for example, is a property which is a function of the degree of perfection of the crystal lattice.

Both metal and dielectric films have been evaporated and used in thin film electrical circuits. Metal films are used for conductors, resistors, inductors, and capacitor electrodes, while dielectrics are used in making capacitors, crossovers, and in field-effect transistors. These films are also used in optical devices such as reflectors, filters, protective coatings for hygroscopic materials, and in the prevention of electrostatic charging of fluorescent screens in cathode ray tubes. Magnetic films for computer information storage have been prepared by evaporation, but the largest single use is undoubtedly in the manufacture of decorative items such as the plastic trim parts used by the automotive industry.

Sputtering

Cathode sputtering was discovered in the middle of the 19th century but did not experience very great popularity until quite recently. A familiar application is its use in phonograph record manufacture, wherein a gold film is sputtered onto the wax "master" disc to make it conductive (thus, the term "gold sputtered master"). The object of this process is to enable the grooved surface to be copied by electroforming a copper facsimile, which requires a conductive surface.

Sputtering is similar to evaporation in that reduced pressures are required for each. The principal difference is that while thermal energy is used in evaporating the coating material, ion bombardment of the material, causing ejection of atoms, is used for sputtering. Thus, thin films of refractory materials may be deposited by sputtering without high source temperatures, such as required by evaporation. The ions are formed when a high electric field is applied to a low-pressure gas such as argon, creating a glow discharge in the deposition chamber. The positively charged argon ions are accelerated through the field to strike a cathode, made of the material to be sputtered. The atoms of a cathode surface are torn loose by this ion bombardment and condense on the surroundings. The actual gas pressure during sputtering is of the order of a ten thousandth of an atmosphere. Nevertheless, if films of high purity are desired, it is necessary to evacuate the sputtering chamber to less than a millionth of an atmosphere prior to sputtering so as to control the contamination of the deposited film. This is especially important with the refractory metals normally deposited by sputtering since they are noted for their "gettering" ability. This ability has also been put to advantage to produce films which are purposely "doped" by the addition of reactive gases to the system during sputtering. This technique is known as reactive sputtering.

An advantage of sputtering is the capability of depositing refractory metals onto relatively cool substrates at reasonable rates, and the capability of depositing compounds of these metals by addition of reactive gases. Control of the deposition parameters is somewhat more tenuous than that of evaporation due to the higher complexity of conditions in sputtering. Sputtered films are used primarily in thin film circuitry to form resistors, capacitor electrodes, and the starting material for fabrication of metal oxide capacitor dielectrics. These films of refractory metals are very stable chemically and thus produce stable, long-lived components.

A large amount of work has been done on sputtering methods in the past several years. New developments include "triode sputtering," in which much lower pressures may be used, "bias sputtering," for controlling film properties, and "radio frequency sputtering," which is especially useful for depositing dielectric films.

Anodization

Although this process of forming oxide coatings on metals was known from the middle of the 19th century, the first relatively thick protective coating was developed in 1924. This application was for the protection of a "duraluminum" seaplane and made use of chromic acid as the anodizing electrolyte. A few years later, a sulfuric acid bath was applied to the same process. The technique of anodization is related to that of electrodeposition. The apparatus used is much the same for both methods but, as the name implies, the thin film is formed at the anode or positive electrode rather than at the cathode. In the example given for the electrodeposition of a copper film, it was noted that oxidation of the copper anode occurred, resulting in copper atoms from the anode being dissolved in the electrolyte solution as copper ions. Some metals, such as tantalum and aluminum, behave differently under these conditions and the metal ions, rather than dissolving, react with oxygen ions from the solution to form an adherent metal oxide film on the anode. At constant cell potential, the growth of this high-resistance anodic oxide film causes a continuous decrease in current. The growth rate of the film decreases with the current; thus, the film thickness is essentially limited by the voltage applied to the cell. The field strength necessary to cause ionic current flow through these oxide films is of the order of 10^7 volts per centimeter. Thus, they are suitable dielectrics for high-working-field capacitors, and both tantalum and aluminum anodic oxides are used commercially in electrolytic capacitors. Anodization is used both to form dielectric films and to adjust the resistance value of tantalum film resistors by uniformly converting conducting tantalum to nonconducting tantalum oxide.

The maximum thickness of anodic films is limited during anodization by film breakdown, characterized by sparking and oxygen evolution. This breakdown occurs at different thicknesses for different metal oxides and also varies with film structure, substrate structure, and electrolyte composition. Anodic oxide films on tantalum show very intense interference colors, and these have been used to estimate the thickness of the dielectric films. Interference colors from aluminum oxide films are not very intense due to the high reflectivity of the underlying aluminum metal. The brilliant colors commonly associated with decorative anodized aluminum are the result of dyeing the oxide film.

Polymerization

Another approach to forming thin dielectric films is the spreading of a solution of a polymer material, with subsequent evaporation of the solvent, and this technique has been successfully utilized in the manufacture of "lacquer film" capacitors. Care must be exercised to prevent the formation of pinholes during solvent evaporation. These films are often porous, and thus they are not used as protective coatings for use against reactive gases. This technique is limited to polymers soluble in volatile solvents, which seriously restricts the available materials. An improved technique is to spread a film of the monomeric material without using a solvent and then to polymerize *in situ*. The monomer must be stable in the liquid state to enable spreading, or it may be a stable gas for coating by evaporation. Polymerization may be initiated thermally or by irradiation with ultraviolet light or an electron beam. The avoidance of solvent evaporation results in fewer pinholes in the film. Film porosity may still permit the diffusion of vapors, but the absence of pinholes greatly reduces the number of electrical shorts when such films are used as capacitor dielectrics. Some materials which have been polymerized by irradiation are butadiene, methyl-methacrylate, acrolein, and divinyl benzene.

Thermal polymerization has been used to produce thin polymer films which are suitable for capacitor and cryotron fabrication. A poly para-xylylene film called "parylene" was recently introduced by the Union Carbide Company. Their film deposition process consists of evaporating the dimer of para-xylylene and then pyrolyzing it to produce an intermediate which is stable in the gas phase. When this intermediate condenses onto a relatively cooler substrate, it polymerizes, covering the substrate with a film of polymer. Films have been produced which were made into capacitors with capacitance densities of from 250 to 25,000 picofarads per square centimeter. This polymer film has a dielectric constant of 2.7. It is also being investigated as a possible "crossover" dielectric.

1.4 APPLICATIONS OF THIN FILMS

Optical

The use of thin films as coatings on optics began just before 1900 when it was observed that "tarnished" lenses had a higher transmittance than clear ones. In the 1930's, it was deduced that this effect was due to an interference phenomenon, and within a few years practical coatings were developed. These antireflection coatings have several optical uses. The first is the prevention of light loss, as already mentioned, and the second one is the reduction of unwanted light in an optical image, resulting from multiple reflections between the surfaces of optical elements. A third use was developed during the second world war to remove sky reflections from the faces of airplane instruments, making them more readable. Some of the materials used for these antireflection films are magnesium fluoride, silicon oxide, and aluminum oxide. These and several others are all deposited by vacuum evaporation.

Another very early use of thin films in optical coatings was the making of reflectors. Improvements in reflecting films were important in the development of the interferometer, which in turn enabled an impressive accuracy in spectroscopic work. In addition to producing better reflectors, thin films were also used to protect these mirrors from corrosion and loss of reflectivity. Some reflectors have been made by electrodeposition, but better control is obtained using vacuum evaporation. More recent optical films have resulted in a variety of filters and include films for controlling the temperature of satellites by adjustment of the properties of solar absorbency and thermal emittance. Current development in optical films is mainly concerned with multilayer films for use in filters, and includes both work on fabrication methods and work aimed toward an explanation of the film properties.

Magnetic

Magnetic thin films have been prepared by a variety of methods, with emphasis on electrodeposition and vacuum evaporation. The necessity for a conducting film to be used as a primary layer in the electroplating process is something of a disadvantage. This two-step technique results in a rougher film than is obtained from evaporation onto a smooth glass substrate. Vacuum evaporation provides better control of product than does plating, in addition to avoiding the unwanted roughness. The first magnetic film was prepared in 1951 by Drigo, using electroplating, while the first successful evaporated magnetic film was prepared by Blois in 1955.

The need for magnetic films arose with the development of high-speed computers and their need for large memories. Earlier computer memories involved the use of magnetic drums for information storage. These machines were relatively slow and had very limited information storage ability. The magnetic core devices which followed used a ferrite toroid or "memory core" for storage. These cores are magnetized in one of two stable magnetic directions by induction from electrical pulses appearing simultaneously at the cores. These pulses arrive on wires threaded through the cores from two orthogonal directions. When the magnetic direction is reversed, a pulse is generated in a third "sensing" wire threaded through the core; thus, the sense of the magnetization direction is retrieved. Thin magnetic films may have several advantages over cores to provide similar functions.

It was correctly predicted by Kittel that a critical thickness exists, below which a thin magnetic film becomes a single domain. As a consequence, the magnetic direction reversal time becomes less than

10^{-9} seconds. Not all thin magnetic films possess this property, but some of the permalloys do. These films contain about 80 percent nickel and 20 percent iron. In addition to fast magnetization reversal, the geometry of the magnetic films permits large-scale inexpensive computer memory construction. Current research in this field is aimed at obtaining fast switching with a minimum input pulse. At the same time, a maximum change in magnetic induction is desired to produce a high signal-to-noise ratio in readout.

Electronic

Early "film" resistors were formed from a resistive paint composed of carbon and a binder. In 1925, H. Pender built a machine which drew glass rods through this resistive paint, cured the film, measured the resistance, and automatically cut off a resistor of the proper length. Somewhat improved stability was obtained when the need for the binder was obviated. This was accomplished by depositing carbon films by the pyrolysis of hydrocarbon gases as mentioned above. The large negative temperature coefficient of resistance of these pyrolytic carbon films was reduced by the addition of boron to the film, and the addition of silicon and oxygen was found to increase the maximum usable resistance values, the film hardness, and the maximum operating temperature. Resistance to humidity, however, was usually reduced by these additions.

An early application of metal films to making resistors occurred in Germany in 1926 when Sigmund Loewe first applied the method of spraying a platinum solution onto glass rods, with subsequent firing to form a metallic platinum film resistor. Not long after this, a solution of tin oxide was atomized onto glass insulators in order to reduce their corona effect. Modification of this solution, by the addition of antimony oxide, resulted in coatings with stable electrical properties and with a temperature coefficient of resistance which could be made either positive or negative. Organic materials also present in the solution act to reduce some of the tin oxide to produce an oxygen deficient "metal-oxide" film which behaves as an n-type semiconductor. The resistance and temperature coefficients are affected by the presence of donor or acceptor atoms added to the solution. Loewe also invented a hermetically sealed evaporated film resistor in 1924 by placing an insulating rod inside a metal cylinder and sealing them in an evacuated glass tube. Conducting wires led through the tube to the insulating rod inside. The metal cylinder was then partially evaporated so as to coat the insulating rod and produce a hermetically sealed thin film resistor.

Both chromium and Nichrome® films have been vacuum evaporated onto insulating substrates to yield precision resistors. The resistance values of these resistors are increased and adjusted by forming them on cylindrical cores and then cutting a helical pattern in the films. This "spiraling" technique was invented by Kruger in 1919 for adjusting the resistance of a sputtered resistor and has also been used for increasing the resistance of carbon film resistors. On flat substrates, the resistance path length, and thus the resistance of deposited films, is controlled either by depositing through a mask or by etching a pattern in the film after deposition. The Nichrome® film resistors provide a wide range of resistance values with small temperature coefficients of resistance and a variously quoted precision from ±0.5 to ±0.05 percent. A disadvantage of this type of resistor is the corrosion of the film in humid atmospheres, particularly under a d.c. load. Sealing the resistors in glass removes this moisture instability to yield resistors comparable in quality to the "wire wound" variety, but cheaper and smaller in size.

The electrical resistance of thin films is dependent on both their geometry and structure. Films may be made thinner in order to decrease the substrate area required for resistors, but this results in reduced stability. A second technique for conserving substrate area is to increase film resistivity by altering its composition and structure. Several metal-dielectric combinations have been simultaneously evaporated in order to obtain higher resistivities. Of these "cermet" films, the combination of chromium and silicon monoxide has received the most attention. A mixture of powdered chromium and silicon monoxide is "flash" evaporated to produce the films, which are then annealed to adjust their resistance values. Annealing may either increase or decrease sheet resistance, depending upon the initial value. Thus, films which have a composition such that their initial sheet resistance is low show a resistance decrease on annealing, while films with a higher initial sheet resistance show an increase in resistance when annealed. These films are said to show a stability at high temperature greater than that of the evaporated Nichrome® resistors.

One of the very earliest metal film resistors was formed by sputtering. The process was slow and difficult to control and was little used for this purpose until recently. In the 1950's at Georgia Tech, Belser reported on the resistivities and temperature coefficients of resistance of a large variety of sputtered metal films, including tantalum. The inclusion of nitrogen in tantalum films, by Bell Telephone Laboratories, greatly improved the stability, and sputtered tantalum nitride films yield very stable resistors which require no encapsulation and are capable of extremely precise resistance value adjustment.

Aluminum and zinc are deposited onto dielectrics such as paper and polymer films to produce thin metal electrodes for capacitors. Capacitors constructed from metallized paper occupy only one third the volume of their counterparts made from paper and aluminum foil. In addition to this size reduction, these capacitors have a self-healing property. This is due to vaporization and removal of the thin metal electrode in the vicinity of a dielectric breakdown.

More recently, thin film capacitor dielectrics have also been deposited. Thinner dielectrics are attainable this way; thus, for material of a given dielectric constant there is an increase in capacitance per unit area. Among evaporated films, silicon monoxide is most often used as a thin film capacitor dielectric. Because practical limits exist on usable thinness, further increases in capacitance density are sought in materials of higher dielectric constant. Anodically formed tantalum oxide is superior in this respect.

Organic polymers are also used as thin film dielectrics. Their low dielectric constants permit the obtaining of more precise capacitance values when low-valued capacitors are required. These films also suffer less damage from mechanical and thermal shock than do the inorganic dielectrics. The organic films possess good insulation resistance, low dielectric loss, and good chemical stability. Silicon monoxide, for example, has a much higher moisture sensitivity than does parylene. The organic films generally tend to decompose at high temperatures and thus are inferior to the inorganic materials under such conditions.

The fabrication of completely thin film circuits would appear to be an obvious advantage, and the development of compatible thin film active devices is therefore of interest. One type of thin film transistor consists of a narrow, evaporated semiconductor film between two thin film metal electrodes. A thin dielectric film over the semiconductor insulates it from a metallic "gate" conductor above the dielectric film. Current flow between the "source" and "drain" electrodes is regulated by applying a bias potential to the gate electrode. Cadmium sulfide, cadmium selenide, and tellurium have been used as semiconductor films in making this field-effect transistor. Both thin film amplifiers and "flip-flop" circuits have been successfully made in thin film form using these transistors. However, the availability of inexpensive silicon devices (transistors and integrated circuits) which are compatible with thin film structures tends to obviate the need for these thin film devices.

The increasing complexity of modern electronic systems produces an unprecedented demand for reliability. The extraordinary terrestrial and interplanetary environments created by the "space age" further increase the severity of the problems associated with reliability. In some instances, the excessive cost of maintenance is overshadowed by the virtual impossibility of making repairs, so that component and interconnection failure must be reduced just to provide acceptable performance. In addition to requiring improved performance from conventional components, functional devices must be created to replace individual components and thus provide the impetus for the development of electronic systems. These devices will obviate the need for the complex interdependence of a nearly infinite number of individual components.

Tantalum thin film circuitry is one of the newest and fastest growing techniques available for meeting any of the stability, precision, or size reduction requirements of modern complex electronic systems. At room temperature and slightly above, tantalum is almost as immune to direct chemical attack as either gold or platinum; yet, it has a higher resistivity than the noble metals. The refractory nature of tantalum implies a high recrystallization temperature, which minimizes annealing of its structural features at the normal operating temperature of resistors. This book provides a considerable amount of information and detail on tantalum-based films.

REFERENCES

1. *Thin Films*. American Society for Metals. Metals Park, Ohio, 1964.
2. Hass, G., editor. *Physics of Thin Films*. New York: Academic Press; vol. 1, 1963; vol. 2, 1964; vol. 3, 1966.
3. Neugebauer, C. A., J. B. Newkirk, and D. A. Vermilyea, editors. *Structure and Properties of Thin Films*. New York: John Wiley and Sons, 1959.
4. Holland, L. *Vacuum Deposition of Thin Films*. London: Chapman Hall, Ltd., 1963.
5. Lowenbein, F. A. *Modern Electroplating*. New York: John Wiley and Sons, 1962.
6. Young, L. *Anodic Oxide Films*. London: Academic Press, 1961.
7. Powell, C. F., I. E. Campbell, and B. W. Gonser. *Vapor Plating*. New York: John Wiley and Sons, 1955.
8. Prutton, M. *Thin Ferromagnetic Films*. London: Butterworths, 1964.
9. Heavens, O. S. *Optical Properties of Thin Film Solids*. London: Butterworths, 1955.

10. Dummer, G. W. A. *Modern Electronic Components*. New York: Philosophical Library, 1959.
11. Holland, L., editor. *Thin Film Microelectronics*. New York: John Wiley and Sons, 1965.
12. Schwartz, N., and R. W. Berry. "Thin Film Components and Circuits," *The Physics of Thin Films*, 2, 363-425 (1964).
13. McLean, D. A., N. Schwartz, and E. D. Tidd. "Tantalum Film Technology," *Proceedings of the IEEE*, 52, 1450-1462 (1964).

Chapter 2
Vacuum Technology

As we have already seen in Chap. 1, thin films can be prepared by a number of methods. Two of the most important of these methods are vacuum evaporation and glow discharge sputtering, both of which rely heavily on vacuum technology. In fact, the fields of vacuum technology and thin film technology have evolved together, each assisting the other in its development. Because of this close association, a familiarity with vacuum technique is important for anyone involved in thin films. This chapter provides background material to serve as an introduction to this subject. More complete works on vacuum technology are listed at the end of the chapter [1,2,3]. The specific interactions of the vacuum environment with films and their properties will be discussed in detail in Chaps. 3 and 4.

This introduction to vacuum technology is not intended to be complete. There are many topics of interest (e.g., ultrahigh vacuum technology) that are omitted, and some of the available vacuum components are not discussed. The purpose of this chapter is to introduce basic principles and discuss some of the standard components and common techniques that have served well in the past, especially in the production of thin films. Vacuum technology is advancing rapidly, however, with important new developments occurring each year. The current literature must be consulted regularly to keep abreast of the field. The publications and meetings of the American Vacuum Society are especially recommended for this purpose.

2.1   KINETIC THEORY OF GASES

In general, the kinetic theory of gases provides an adequate framework for understanding behavior of gases at low pressures. This theory fairly accurately describes many phenomena, although it is based on the following simple assumptions:

1. Gases consist of molecules; all molecules of a given gas are alike.
2. The molecules of a gas are in constant motion, and this motion is related to the temperature of the gas.
3. The molecules undergo perfectly elastic collisions with each other. These collisions obey the fundamental laws of classical mechanics (conservation of momentum and energy).

Some of the important conclusions of the kinetic theory which are useful in vacuum work are summarized below.

Ideal Gas Laws

Pressure, volume, temperature, and the quantity of an ideal gas can be related by

$$pV = NkT \tag{2-1}$$

where p = pressure, V = volume, N = total number of molecules, k = Boltzmann's constant = 1.3805×10^{-16} erg per °K, and T = temperature in °K (°K = °C + 273.16). Real gases behave more and more like ideal gases as the pressure is lowered. Therefore, Eq. (2-1) may be considered valid for most vacuum work. Equation (2-1) applies to noncondensable gases, but not to vapors which can be liquefied at the temperature in question. Equation (2-1) is a combination of Charles' law (which states that at a constant pressure the volume of a given mass of gas varies directly as the absolute temperature) and Boyle's law (which states that at a constant temperature the product of pressure and volume is a constant, for a given mass of gas). Thus,

$$\frac{pV}{T} = \text{constant} = n_M R_o \tag{2-2}$$

where R_o is the so-called gas constant and n_M is the number of moles of gas (one mole of a gas contains M grams, where M is the molecular weight).

Avogadro's law states that equal volumes of all gases at any given temperature and pressure contain equal numbers of molecules. The number of molecules in one mole of gas is a constant, called Avogadro's number, $N_A = 6.023 \times 10^{23}$ molecules per mole. A mole of gas at 0° C and a pressure of 760 torr* occupies a volume of 22.415 liters. From Eqs. (2-1) and (2-2),

$$Nk = n_M R_o$$

but

$$n_M = \frac{N}{N_A}$$

Therefore,

$$Nk = \left(\frac{N}{N_A}\right) R_o$$

or

$$R_o = N_A k$$

$$= 8.31 \text{ joules per } °K \text{ per mole}$$

$$= 82.06 \text{ atm cm}^3 \text{ per } °K \text{ per mole}$$

Equation (2-1) may be used to calculate n, the number of molecules per unit volume:

$$n = \frac{N}{V} = \frac{p}{kT} \tag{2-3}$$

When p is in torr and n is the number of molecules per cm^3, Eq. (2-3) becomes

$$n = 9.656 \times 10^{18} \frac{p}{T}$$

Even at low pressures there are still large numbers of molecules present; e.g., at 25°C and 10^{-8} torr,

$$n = \frac{(9.656 \times 10^{18}) \ (10^{-8})}{298.16}$$

$$= 3.24 \times 10^8 \qquad \text{molecules per cm}^3$$

*1 torr = 1 mm of Hg. Pressure units are defined later in this section.

Figure 2-1 gives values for n at several other pressures.

Pressure (torr)	Molecules per cm³, n	Mean free path, L (cm)	Average time between collisions, τ (sec)	Monolayer* time (sec)
760 = atm	2.5×10^{19}	6.7×10^{-6} = 670 A	1.4×10^{-10}	2.9×10^{-9}
1	3.2×10^{16}	5.1×10^{-3}	1.8×10^{-7}	2.2×10^{-6}
10^{-2}	3.2×10^{14}	5.1×10^{-1}	1.8×10^{-5}	2.2×10^{-4}
10^{-4}	3.2×10^{12}	5.1×10 = 20 inches	1.8×10^{-3}	2.2×10^{-2}
10^{-6}	3.2×10^{10}	5.1×10^{3}	1.8×10^{-1}	2.2
10^{-8}	3.2×10^{8}	5.1×10^{5} = 3 miles	18	3.6 minutes
10^{-10}	3.2×10^{6}	5.1×10^{7} = 300 miles	30 minutes	6 hours

*Time required for a monomolecular layer of air to form on a surface, assuming that all molecules stick.

FIG. 2-1. Some properties of air at 25°C.

Molecular Motion

The kinetic theory of gases postulates that gas molecules are in constant motion and that their velocities are temperature dependent. The distribution of velocities is described by the Maxwell-Boltzmann distribution law, which states that for a given speed v,

$$\frac{1}{N} \frac{dN}{dv} = \frac{4}{\sqrt{\pi}} \left(\frac{m}{2kT} \right)^{3/2} v^2 \exp \left(\frac{-mv^2}{2kT} \right) \qquad (2\text{-}4)$$

where m is the mass of a molecule. This says that if there are N molecules in the volume of gas under consideration (at a temperature T), there will be dN molecules having speeds between v and $v + dv$.

The speed v_m which maximizes Eq. (2-4) is the most probable speed. Hence,

$$v_m = \left(\frac{2kT}{m} \right)^{1/2} = 1.289 \times 10^4 \left(\frac{T}{M} \right)^{1/2} \qquad \text{cm/sec} \qquad (2\text{-}5)$$

where M is the molecular weight of the gas. The average speed, v_a, and the root-mean-square speed, v_r, are related to v_m by

$$v_a = \left(\frac{2}{\sqrt{\pi}}\right) v_m = 1.455 \times 10^4 \left(\frac{T}{M}\right)^{1/2} \qquad \text{cm/sec} \qquad (2\text{-}6)$$

and

$$v_r{}^2 = \frac{3 v_m{}^2}{2}$$

The root-mean-square speed should be used in energy calculations. It should be noted that these are rather high speeds; for air with $M = 29$ at $25°C$ $(298°K)$, we get

$$(v_a)_{\text{air}} = 0.47 \text{ kilometer/sec} = 0.29 \text{ mile/sec} = 1040 \text{ miles/hr}$$

The average speed is thus about 30 percent greater than the speed of sound propagation in the gas. Yet, when a sample of ammonia is released at one side of a room, minutes elapse before the odor is apparent on the other side of the room. The reason for this discrepancy is the long zig-zag path taken by the ammonia molecules because of many intermolecular collisions, a phenomenon known as gaseous diffusion.

Mean Free Path

The numerous collisions that the gas molecules undergo lead to the notion of a mean free path. The mean free path is defined as the average distance that a molecule travels between successive collisions with other molecules in the gas phase. The mean free path enters into many aspects of vacuum technology, particularly the design of vacuum plumbing. The mean free path, L, is related to the average molecular speed by

$$L = v_a \tau$$

where τ is the average time for a given molecule between collisions. The mean free path depends on the density and diameter of the molecules:

$$L = (\pi \sqrt{2} \, n d_0{}^2)^{-1} = \frac{0.225}{n d_0{}^2} \qquad (2\text{-}7)$$

where d_0 is the equivalent molecular diameter (assuming hard elastic spheres). Values of d_0 for several common gases are listed below.

	d_0		d_0
He	2.2×10^{-8} cm	Air	3.7×10^{-8} cm
Ne	2.6×10^{-8} cm	N_2	3.8×10^{-8} cm
H_2	2.7×10^{-8} cm	Hg	4.3×10^{-8} cm
Ar	3.6×10^{-8} cm	CO_2	4.6×10^{-8} cm
O_2	3.6×10^{-8} cm	H_2O	4.7×10^{-8} cm

For air at room temperature, Eq. (2-7) reduces to

$$L = \frac{5 \times 10^{-3}}{p} \qquad \text{cm} \tag{2-8}$$

where p is the gas pressure in torr. Figure 2-1 gives values for L and τ for air at 25°C, and various pressures.

Collisions of Molecules with a Surface

Another useful quantity is the molecular arrival rate Z, defined as the rate of molecular collisions per unit area on the wall of a container. It can be shown that

$$Z = \frac{nv_a}{4}$$

Substituting from Eqs. (2-3), (2-5), and (2-6) gives

$$Z = \frac{1}{4} \left(\frac{p}{kT} \right) \left(\frac{2}{\sqrt{\pi}} \right) \left(\frac{2kT}{m} \right)^{1/2}$$

$$= p \left(\frac{1}{2\pi mkT} \right)^{1/2} \tag{2-9}$$

When p is in torr, Eq. (2-9) becomes

$$Z = 3.51 \times 10^{22} \, p \left(\frac{1}{TM} \right)^{1/2} \qquad \text{molecules/cm}^2 \text{ sec} \tag{2-10}$$

Equation (2-10) can also be used to calculate how much gas passes through small holes in thin plates at low pressures.

Knudsen [4] has shown that when a molecule collides with a container wall, the angle at which it leaves the container wall is independent of the angle at which it arrived. The distribution of molecules leaving the wall is a function of θ, where θ is the angle between a given trajectory and the normal to the surface. Knudsen showed that

$$dN = \frac{nv_a \cos \theta \, d\phi}{4\pi}$$

where dN is the number of molecules leaving a unit area of the wall per unit time with trajectories included in the solid angle $d\phi$. In the case shown in Fig. 2-2, $d\phi$ is the solid angle between the two cones. It is given by

$$d\phi = 2\pi \sin \theta \, d\theta$$

Note that integration over the entire hemisphere gives

$$\int dN = \frac{nv_a}{4} = Z$$

Pressure

The pressure on the wall of a container is the force per unit area. According to the kinetic theory of gases, the force is due to impulses imparted to the wall by molecules striking it or being emitted from it. As shown above, the reflection of these molecules is nonspecular. Instead, an incoming molecule has an equal probability of coming off the surface in any direction, regardless of its incoming direction. This is due to two causes, one being the microscopic roughness of the surface which is always present, and the other being long "dwell times" on the surface, which effectively erase all memory of the incoming direction. If the wall temperature is the same as the temperature in the body of the gas, the distribution of velocities of molecules striking the wall is the same as the distribution of velocities of molecules leaving the wall.*

*Molecular reflection from a surface has been discussed in greater detail by Knudsen [4].

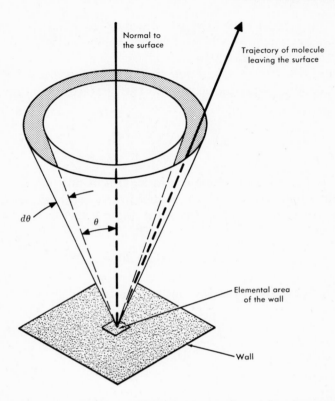

FIG. 2-2. Knudsen's cosine law. Molecules leaving the surface have a distribution which varies as cos θ.

To calculate the pressure on a wall, let the x axis be perpendicular to the wall, and directed in toward the wall. Now consider only those molecules with a given velocity, $\mathbf{v}$; that is, the velocity components are between v_x and $v_x + dv_x$, v_y and $v_y + dv_y$, and v_z and $v_z + dv_z$. (Note that v_x must be positive in order for the molecule to strike the wall.) The impulse imparted to the wall by the arrival of one of these molecules is mv_x. The incident flux of molecules with velocity $\mathbf{v}$ impinging on a unit area per unit time is v_x times the number of such molecules per unit volume. Let the fraction of such molecules be $f(v_x, v_y, v_z)\, dv_x dv_y dv_z$.* The pressure due to arrival of molecules is the impulse per molecule times the incident flux of molecules.

*For a Maxwellian distribution,

$$f(v_x, v_y, v_z) = \left(\frac{m}{2\pi kT}\right)^{3/2} \exp\left[\frac{-m(v_x^2 + v_y^2 + v_z^2)}{2kT}\right]$$

Thus, the pressure due to the arrival of these molecules is

$$(mv_x) \ [v_x n f(v_x,v_y,v_z) \ dv_x dv_y dv_z]$$

The total pressure due to incoming molecules is found by integrating over the entire velocity distribution (except for negative v_x):

$$p_{\text{incoming}} = nm \int_{-\infty}^{\infty} \int_{-\infty}^{\infty} \int_{0}^{\infty} v_x{}^2 f(v_x,v_y,v_z) \ dv_x dv_y dv_z$$

The function $f(v_x,v_y,v_z)$ is symmetrical in v_x, so this can be rewritten as

$$p_{\text{incoming}} = nm \ \frac{\overline{v_x{}^2}}{2}$$

where $\overline{v_x{}^2}$ is the average of $v_x{}^2$ over the entire velocity distribution.

Now, if the temperature of the wall is the same as the temperature of the gas, the distribution of velocities will be the same for the incoming as for the outgoing molecules. Therefore, the pressure due to outgoing molecules will be equal to p_{incoming}, and the total pressure will be twice p_{incoming}. Thus,

$$p = nm\overline{v_x{}^2}$$

In the body of the gas, there are no preferred directions, so we can say, and it can be shown, that the kinetic energy associated with any one of the three orthogonal directions is one-third of the total kinetic energy; that is,

$$m\overline{v_x{}^2} = \frac{mv_r{}^2}{3}$$

where v_r is the root-mean-square speed as defined above. Or,

$$\overline{v_x{}^2} = \frac{v_r{}^2}{3}$$

and the pressure is

$$p = \frac{nmv_r{}^2}{3} \tag{2-11}$$

Now the average kinetic energy of a molecule is

$$\overline{E} = \frac{mv_r^2}{2}$$

Thus, we have the important result that

$$p = \frac{2n\overline{E}}{3}$$

or, the pressure is equal to two-thirds of the total kinetic energy per unit volume.

Referring back to Eqs. (2-5) and (2-6), we recall that

$$v_r{}^2 = \frac{3v_m{}^2}{2} = \frac{3kT}{m}$$

Thus, we see that

$$p = nkT$$

where n is the number of molecules per unit volume. Thus, the ideal gas law [Eq. (2-1)] has been derived using the kinetic theory. Note that the entire development of expressions for the pressure has been for an equilibrium system, in contrast to the situation where there is a net flow of molecules or energy.

Consider two chambers, both containing the same species of gas at the same temperature, pressure, and density. Let them be separated by a thin but impervious membrane. The pressure and the distribution of velocities will be the same on both sides of the membrane. Thus, if the membrane is removed, no change will occur in the velocity distributions. It is convenient to consider that the pressure still exists, with the gas from one side exerting pressure on the gas from the other side. The pressure in the body of a gas at equilibrium will then be the same in all directions, and will be equal to the pressure on the walls. In vacuum applications, however, the term "pressure" usually means the pressure far away from the walls. In fact, most vacuum gauges measure some other property associated with the density of molecules, because the actual forces in the high vacuum region are too small to be measured directly.

In a mixture of two or more gases, each gas exerts its pressure independently of the others. The pressure of each gas is the pressure it would have if it occupied the chamber by itself, and the total pressure is the sum of these partial pressures.

Units of Pressure

In the cgs system, the unit of force is the dyne and the unit of pressure is the microbar:

$$1 \text{ microbar} = 1 \text{ dyne/cm}^2$$

$$1 \text{ bar} = 10^6 \text{ dyne/cm}^2$$

In the mks system, the unit of pressure is the newton/meter^2:

$$1 \text{ newton/meter}^2 = 10 \text{ microbars}$$

Other units of pressure more commonly used in vacuum technology are discussed below.

A standard atmosphere is defined by the National Bureau of Standards as

$$1 \text{ atm} = 1.01325 \text{ bars}$$

The suggested international standard term to replace the term millimeter of mercury (mm of Hg) is torr. The torr is defined as 1/760 of a standard atmosphere. Hence (except for a very small correction),

$$1 \text{ torr} = 1 \text{ mm Hg}$$

$$1 \text{ atm} = 760 \text{ mm Hg} = 760 \text{ torr}$$

Also,

$$1 \text{ micron} = 10^{-6} \text{ meters Hg} = 10^{-3} \text{ mm Hg} = 1 \text{ millitorr}$$

$$1 \text{ bar} = 750.06 \text{ mm Hg}$$

Electrical Conductivity

Under certain conditions, a gas may become ionized, and thereby become a conductor of electricity. One way by which a gas may be ionized is by the actions of free energetic electrons bombarding the molecules of the gas. At low pressures, the number of ions created depends on the number of molecules available for a given number of free electrons. Under these conditions, the "conductivity" of a gas is proportional to its pressure. This property is used in some types of pressure gauges. Ionized gases are also used in the process of cathodic sputtering, which is discussed in Chap. 4.

Thermal Conductivity

At high pressures (atmospheric and above), cooling of a body by the gas surrounding it is primarily by convection. In this region there are many molecules and they are so close together that reduction in number (i.e., pressure) has very little effect on the thermal conductivity of the gas. Down to pressures of approximately 0.1 torr, the thermal conductivity is almost independent of pressure.

The process of heat conduction at very low pressures can be illustrated by a very simplified discussion. Consider two parallel plates separated by a distance that is small compared to the planar dimensions of the plates. If the intermediate space contains a gas with a mean free path that is large compared to the plate separation, a molecule will have a very small probability of colliding with another molecule on its path from one plate to the other. Assume it will leave one plate with energy $3kT_1/2$ (on the average), where T_1 is the temperature of the plate it left; it can release energy $3k(T_1 - T_2)/2$ to the other plate (at temperature T_2). Thus, energy is indeed transferred from the one plate to the other, and the heat flow is proportional to $T_1 - T_2$. There is, however, no temperature gradient in the body of the gas; that is, the velocity distribution is the same at all points in the gas. Also, the heat flow is independent of the plate separation. Thus, one cannot speak of thermal conductivity in the usual sense (ratio of heat flux to temperature gradient) for gases at such low pressures. The heat flow at these pressures, however, is proportional to the density of molecules in the gas, and thus is proportional to the pressure.

This discussion has been considerably simplified in the interests of clarity. Actually, the molecules may be reemitted before reaching the temperature of the wall. Also, polyatomic molecules can release more energy than monatomic molecules because of the additional degrees of freedom. The essential conclusions, however, remain unaltered; the heat flow by conduction is proportional to the pressure and to the temperature difference, and it is independent of the plate separation.

Many pressure gauges have been designed using this dependence of heat flow on pressure. They are generally useful in the range from about 10^{-3} to 10^{-1} torr. The amount of heat transferred by conduction also depends on the type of gas; e.g., hydrogen transfers more heat than carbon dioxide under identical conditions. Thus, the type of gas present affects the readings of gauges which estimate the pressure by measuring the thermal conductivity of the gas.

2.2 GAS FLOW

Most vacuum systems are dynamic systems, and an understanding of the flow of gases is essential for proper design and use of these systems. The vacuum plumbing and pumps are of special interest since they determine the rate at which gas can be removed from a system.

Quantity of Gas and Gas Flow Rates

In dealing with problems where the total number of molecules is an important factor (such as the evacuation of a vessel), it is useful to think in terms of the quantity of gas, q. If a certain amount of gas is contained in a volume V, at pressure p, the quantity of gas is defined by

$$q = pV$$

The quantity of gas is also a function of temperature, but discussions of vacuum processes are usually based on a fixed temperature of 25°C. Commonly used units of q are:

1 micron liter (1 μl) $= 10^{-3}$ torr liters
$=$ quantity of gas at a pressure of 1 micron (10^{-3} torr), in a volume of 1 liter, at $T = 25°C$
$= 3.24 \times 10^{16}$ molecules for an ideal gas (at $T = 25°C$)
1 micron cu ft $= 2.832 \times 10^{-2}$ torr liters

A closely related concept is the gas flow rate Q, also called the throughput. It is a measure of the quantity of gas at a specified temperature which flows across an area per unit time; i.e.,

$$Q = \frac{d}{dt}(pV) \tag{2-12}$$

that is, Q is the time derivative of pV. Convenient units for Q are torr liters per second. Also used is the standard cc per second:*

> 1 std cc/sec = 0.76 torr liter/sec
> 1 "lusec" = 1 micron liter/sec = 10^{-3} torr liters/sec

Gas flow rates are especially convenient for characterizing leaks and pumps. In practice, gas flow rates are determined by measuring the gas pressure at a given point and obtaining the gas volume passing that point per unit time by other means. From the ideal gas law, $pV = NkT$, and for a constant temperature Eq. (2-12) becomes

$$Q = \frac{d}{dt}(pV) = kT\frac{dN}{dt} \tag{2-13}$$

Thus, Q is directly proportional to a molecular current, and in a closed pumping line under steady state conditions, Q must be the same at all points which are at the same temperature. Note that Q is defined for gas flow across an area. If all of the molecules that cross this area per unit time, dN/dt, come from a fixed volume, then Eq. (2-13) becomes

$$Q = kT\frac{dN}{dt} = V\frac{dp}{dt}$$

for a constant volume system, with no gas sources such as leaks.

Conductance and Pump Speed

In order to maintain a gas flow rate, Q, through a pipe or aperture, a certain pressure drop must be maintained across it. The gas flow rate per unit pressure differential is called conductance, and is expressed as

$$C = \frac{Q}{p_1 - p_2} \tag{2-14}$$

*This quantity is sometimes referred to as an atm cc per sec, a cc per sec, or a standard cc.

VACUUM TECHNOLOGY 33

where C is the conductance, and p_1 is the pressure at the entrance and p_2 the pressure at the exit of the pipe (or aperture). The quantity $1/C$ may be treated as an impedance or resistance. The units of C are volume per unit time (usually liters per second). Conductance can be a function of both pressure and geometry.

The usefulness of the concept of conductance is shown in the following example. Consider two pipes, A and B, connected in series. The pressures are as follows: p_1 at the inlet of A, p_2 at the exit of A, p_2 at the inlet of B, and p_3 at the exit of B. From Eq. (2-14), the conductances are

$$C_A = \frac{Q}{p_1 - p_2}$$

$$C_B = \frac{Q}{p_2 - p_3}$$

$$C = \frac{Q}{p_1 - p_3}$$

where C is the total conductance. Rearranging, we get

$$p_1 - p_2 = \frac{Q}{C_A}$$

$$p_2 - p_3 = \frac{Q}{C_B}$$

$$p_1 - p_3 = \frac{Q}{C}$$

Combining these three equations gives

$$\frac{Q}{C} = \frac{Q}{C_A} + \frac{Q}{C_B}$$

Thus,

$$\frac{1}{C} = \frac{1}{C_A} + \frac{1}{C_B}$$

In general, for conductances in series (the most common situation), we get

$$\frac{1}{C} = \frac{1}{C_1} + \frac{1}{C_2} + \frac{1}{C_3} + \cdots$$

For conductances in parallel, the total conductance is

$$C = C_1 + C_2 + C_3 + \cdots$$

Conductance has the same dimensions as another quantity used in vacuum technology, namely, pump speed S_p, which is defined by

$$S_p = \frac{Q}{p} \qquad (2\text{-}15)$$

where Q is the net* throughput of gas and p is the pressure at the inlet of a pump. The concept of pump speed can be extended to apply not only to the gas flow at the intake of a pump, but also to the gas flow at any other point in the system, where it is defined by

$$S = \frac{Q}{p} = \frac{1}{p} \frac{d}{dt} (pV)$$

where p is the pressure at the point where S is determined.

The concept of conductance can also be used to calculate the net speed S_E (called speed of exhaust) available at the outlet of the vacuum chamber. Let p_1 and p_2 be the pressures at two different points in the system. The corresponding pump speeds S_1 and S_2 are

$$S_1 = \frac{Q}{p_1}$$

$$S_2 = \frac{Q}{p_2}$$

The conductance between the two points is

$$C = \frac{Q}{p_1 - p_2} = \frac{Q}{(Q/S_1) - (Q/S_2)}$$

That is,

$$\frac{1}{C} = \frac{1}{S_1} - \frac{1}{S_2}$$

Transposing, we get

$$\frac{1}{S_1} = \frac{1}{C} + \frac{1}{S_2} \qquad (2\text{-}16)$$

*In some cases, some molecules can flow through a pump in the reverse direction.

Now, let S_2 be the pump speed S_p, and let C be the conductance between the pump and the vacuum chamber. Then S_1 is the speed of exhaust S_E, and Eq. (2-16) becomes

$$\frac{1}{S_E} = \frac{1}{C} + \frac{1}{S_p} \qquad (2\text{-}17)$$

Equation (2-17) points out the importance of the conductance of connecting tubing in a vacuum system as it affects the pumping speed available in the system. For example, if a 50 liter-per-second pump is connected to a vacuum chamber by tubing having a conductance of only 10 liters per second, the net speed is 8.3 liters per second. Using Eq. (2-16), the pump speed at any point of the system can be determined if the pump speed is known at some other point, and if the conductance between the two points can be calculated or measured.

Regions of Gas Flow

When a vacuum pump is first applied to a pipe line at atmospheric pressure, a pressure gradient is produced and the flow of gas in the vicinity of the pump is quite turbulent. The flow lines are very irregular and distorted, as eddies and vortices appear and disappear. Turbulent flow normally exists only for the first few seconds of pumpdown. When the turbulence disappears, the system enters the *viscous flow* region in which gas flows like a viscous liquid. The normal chaotic motion of the molecules has a drift velocity superimposed on it. This drift velocity is very small near the walls of the pipe, and increases to a maximum at the middle of the cross section. The flow is essentially laminar, and the flow rate is limited by the viscous drag on the molecules. The flow can be considered laminar only when the mean free path between molecular collisions is much less than the pipe diameter. This is true for most of the range of fore- and roughing-pumps.

When the pressure is so low that the mean free path exceeds the pipe diameter, the result is *molecular flow*. Under this condition, the molecules move around independently of each other and collide only with the walls. High vacuum pumps operate in the molecular flow region.

There is a smooth transition between molecular and viscous flow. This transition is characterized by the ratio L/d (the Knudsen number), where L is the mean free path and d is the characteristic dimension (e.g., diameter) of the vessel through which the gas flows.

When L/d is less than 0.01, the flow is viscous; when L/d is greater than 1.0, the flow is molecular. Between these values lies the transition range.

Viscous Flow Conductances

The calculation of conductances in the region of viscous flow is difficult because it depends not only on the shape of the pipe and the type of gas, but may also depend on the gas pressure. However, some equations have been developed for this region. For air at 25°C, the conductance of an aperture in the viscous range is given by

$$C = 20A \qquad \text{liters/sec}$$

where A is the area of the aperture (cm²). This equation is applicable only when the pressure drop across the aperture is at least a factor of 10. For a long cylindrical pipe, the conductance in the viscous range is

$$C = \frac{180\, d^4 \bar{p}}{l} \qquad \text{liters/sec}$$

where d is the diameter (cm), $\bar{p}$ is the average pressure in the pipe (torr), and l is the pipe length (cm).

Molecular Flow Conductances

For high vacuum engineering, the region of molecular flow is more important than the viscous flow region. In the molecular flow region ($L >> d$), conductances are not pressure-dependent. The molecular flow conductance of an aperture between two large vessels is given by

$$C = 11.7A \qquad \text{liters/sec} \qquad (2\text{-}18)$$

where A is the aperture area (cm²) and the conductance is for air at 25°C. A long cylindrical pipe will have a conductance in the molecular flow region of

$$C = 12.2\frac{d^3}{l} \qquad \text{liters/sec} \qquad (2\text{-}19)$$

where d is the pipe diameter (cm) and l is its length (cm). For "fairly short" pipes ($0.1d < l < 10d$), Dushman [1] suggests a combination of Eqs. (2-18) and (2-19):

$$C = \frac{12.2\, d^3}{l\left(1 + \dfrac{4d}{3l}\right)} \qquad \text{liters/sec}$$

A correction can be applied for the pressure range intermediate between molecular and viscous conductance:

$$C = \alpha C_{\text{molecular}} \qquad (2\text{-}20)$$

For a cylindrical pipe, the correction factor α is a function of pd. It is shown in Fig. 2-3, where p is in torr and d is in centimeters. Figure 2-3 shows that conductances in the viscous region are much higher than in the molecular flow region.

All the above formulas for molecular flow conductance are valid for air at 25°C, but they can be generalized for other conditions by the relation

$$C = \beta C_{\text{air, 25° C}}$$

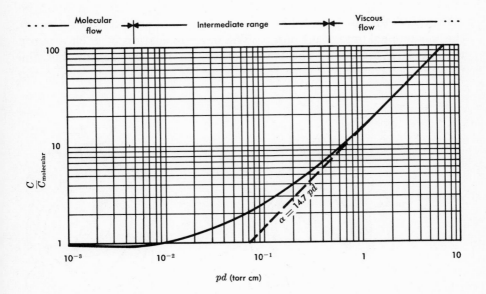

FIG. 2-3. Intermediate range correction factor $\alpha = \dfrac{C}{C_{\text{molecular}}}$

where

$$\beta = 0.31 \sqrt{\frac{T}{M}}$$

T is the absolute temperature (°K) and M is the molecular weight of the gas. Calculations of conductance have been made for many other geometries and conditions. These are reviewed in books on vacuum technology such as Dushman [1].

2.3 FORE- AND ROUGHING-PUMPS

The vacuum engineer is usually faced with the problem of pumping over a pressure range so large that no single pump is efficient over the entire range. Most pumps can be classified as operational either in the viscous flow region (fore- and roughing-pumps) or in the molecular flow region (high vacuum pumps). Fore- and roughing-pumps will be considered first.

Rotary Oil Pumps

For applications involving continuous operation over the pressure range from atmosphere to about 10^{-3} torr, where the gas must be discharged to the atmosphere, the modern industrial workhorse is the mechanical rotary oil pump. The basic operation of such a pump is shown in Fig. 2-4. The rotor rotates on its own center, but the rotor is not concentric with the housing. It makes a seal at the top

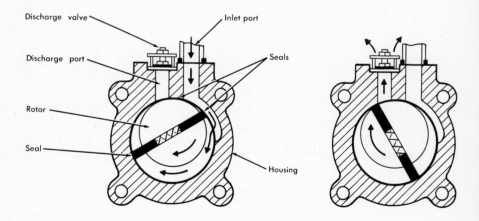

Fig. 2-4. Rotary oil pump. (From A. Guthrie, *Vacuum Technology*, 1963; courtesy John Wiley and Sons, Inc.)

between the inlet and outlet ports. Two other seals are provided by spring-loaded vanes. A film of oil effects the seal at all sealing points. As the rotor rotates, the vanes trap a certain volume of gas, compress it, and release it into the outlet port. The seal at the top between the rotor and housing prevents the gas from getting back into the inlet port. Typical rotor speeds are between 200 and 500 rpm, and these pumps can be designed to operate with very little noise or vibration. Many pumps of this type are designed with two or more stages operated in series. Such an arrangement can extend the useful range to much lower pressures than those obtained with a single-stage pump.

Some typical pumping speed curves are shown in Fig. 2-5. The pumping speed is seen to be fairly constant over a wide pressure range, but it eventually drops off and goes to zero at the "ultimate pressure." These pressures are usually measured with a McLeod gauge, which measures only noncondensable gases; therefore, production experience may indicate ultimate pressures as much as an order of magnitude higher than manufacturers' specifications. It is clear from Fig. 2-5 that two-stage pumps (dashed curves) have the lowest ultimate pressures.

If all three of the seals on a rotary oil pump were perfect, it would pump without loss of speed down to the vapor pressure of the oil used (about 10^{-5} torr). These seals are not perfect, however, and some gas can leak past them. Also, contaminants in the oil, such as

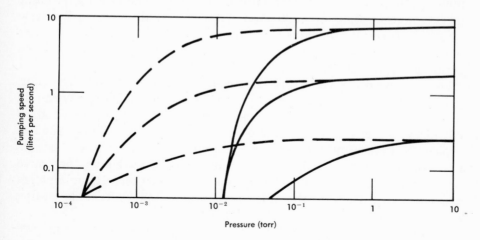

FIG. 2-5. Pumping speed of some typical rotary oil pumps. The speed becomes very small at the ultimate pressure, which is much lower for two-stage pumps (dashed lines) than for single-stage pumps (solid lines). (From Dresser Vacuum Co.)

dissolved gases, can be reemitted on the inlet side to limit the vacuum. For thin film operations, where the oil can be kept free of contaminants, these pumps are extremely reliable and require very little maintenance. Some contaminants may cause gumming and corrosion (e.g., chemically active vapors, mercury, and fine grit). Other contaminants may limit the ultimate pressure (e.g., the condensable vapors, such as water and solvent vapors). At room temperature, the vapor pressure of water is about 20 torr. Water in the oil can limit the ultimate pressure of a rotary oil pump to between 1 and 10 torr. If large amounts of condensable vapors are anticipated, provision can be made for "gas ballasting." This is the practice of letting some air into the pump before the compression stage in order to limit the actual compression ratio. The condensables then are not compressed as much, and may be prevented from condensing inside the pump at all. In a two-stage pump, the ballasting is inserted between the two stages. Gas ballasting, however, does increase the ultimate pressure. When the oil does become contaminated, it is easily changed, although several flushings may be necessary before the contamination is effectively removed. The oil should not be left exposed to the air, as it will absorb water.

Roots Blower Pumps

Another pump that operates in the viscous flow region is the Roots blower pump, shown schematically in Fig. 2-6. The two figure-eight lobes counter-rotate at 4000 rpm without actually touching. These lobes are carefully machined and aligned such that they are always within a few thousandths of an inch of each other. Since they never actually touch, lubrication is unnecessary. As the lobes rotate, they trap and compress gas from the inlet and discharge it after compression. Because of their limited compression, Roots pumps normally discharge into a mechanical backing pump. Their region of high efficiency is limited to the range of 10^{-3} to 10 torr. This range is especially important for sputtering applications; thus, these pumps may become important for thin film applications, although they have not been extensively used in the past for this purpose. Roots blowers are currently used on the inlet and outlet stages of some in-line sputtering systems for thin film deposition.

Typical pumping speed curves for a two-stage Roots pump are shown in Fig. 2-7. One of the major limitations at low pressures is the finite clearance between the lobes, which allows some gas to leak back into the chamber being pumped. Another problem is leakage

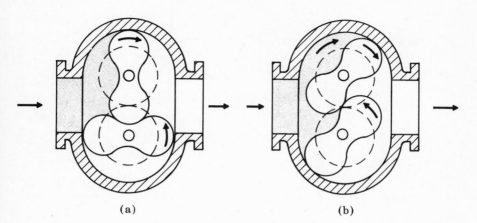

(a) (b)

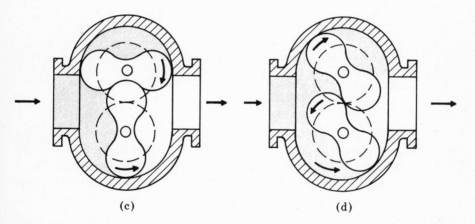

(c) (d)

FIG. 2-6. Operation of a Roots pump. (From A. Guthrie, *Vacuum Technology*, 1963; courtesy John Wiley and Sons, Inc.)

at the seal where the drive-shaft enters the vacuum; some versions of this pump have the motor mounted inside the housing to avoid this problem. Another source of difficulty is that heat generated from gas compression must be removed. The temperature rise can be limited by water cooling and by limiting the rotation speed during the time that the pump is operating at the high end of its pressure range.

Sorption Pumps

The simplest type of roughing pump is the sorption pump. A sorbent material (sometimes called a molecular sieve) is placed in

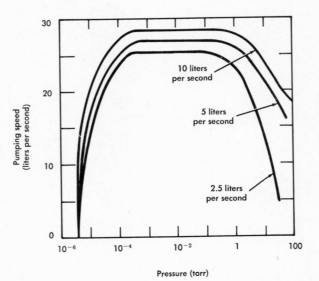

FIG. 2-7. Pumping speeds of a typical two-stage Roots blower pump (nominal inlet diameter = 4 inches) when connected in series with three different compound fore-pumps of speeds 2.5, 5, and 10 liters per second. (From Dresser Vacuum Co.)

the vacuum system in such a way that its temperature can be controlled. When cooled, this material adsorbs gas; when heated, it desorbs the gas. Normally, the lower temperature is the boiling point of nitrogen ($-196°C$), and the higher temperature is room temperature or above.

Sorption pumps are frequently used as roughing pumps for sputter-ion pumps when a completely oil-free system is desired. In this application, the sorption pump is valved off from the system as soon as the sputter-ion pump starts pumping. Sometimes a system will have several of these sorption pumps attached, so that they can be chilled and valved off sequentially, in order to produce the lowest ultimate pressure.

It is very difficult to specify the pumping speeds of sorption pumps, since it depends so much on the history of the sorbent and the type of gas being pumped. The manufacturers usually present pumpdown curves instead of pumping speed curves. Figure 2-8 shows some actual pumpdown curves of small sorption pumps on chambers of 20, 100, and 200 liters. It is clear that, with only one such pump,

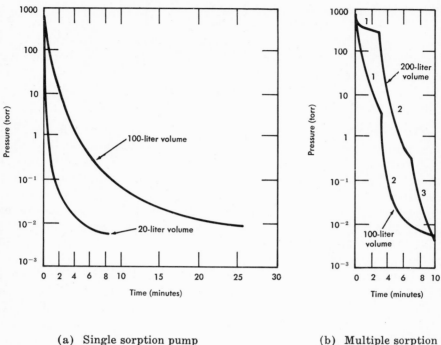

(a) Single sorption pump

(b) Multiple sorption
     pumps (in sequence)

FIG. 2-8. Pumpdown curves of sorption pumps after a 10-minute prechill. (After
Varian Associates.)

the volume that can be pumped out to 10^{-2} torr in a reasonable time is
rather small. (The pressure required to start a sputter-ion pump is
about 10^{-2} torr.)

Sorption pumps have no moving parts, no vibration or noise, and
no oil fumes, either in the room or inside the system. However, there
are some disadvantages. Any sorbent will eventually become saturated and quit pumping, so continuous pumping in dynamic systems
is not feasible (unless the throughput is very small). When the material becomes saturated, it must be warmed to be reactivated; if it
has pumped water vapor, it must be heated well above room temperature to remove the water. The pumping action also depends strongly
on the gas to be pumped. The ultimate pressures obtained with these
pumps are always limited by the presence of the inert gases, which
is most unfortunate because sputter-ion pumps also are least efficient
when pumping inert gases.

2.4   HIGH VACUUM PUMPS

Pumps that operate in the molecular flow region (below 10^{-2} torr) include diffusion pumps, getter and ion pumps, and turbomolecular pumps. The most common of these is the diffusion pump.

Diffusion Pumps

The operation of a typical diffusion pump is illustrated in Fig. 2-9. The pump fluid is boiled at the bottom of the pump, and the vapor is forced up through the center chimney and out through narrow apertures called nozzles (jets), directed generally downward. The heavy, high-speed molecules of pump fluid collide with any gas molecules in the stream and knock them downward, thus producing a net pumping action. The pump fluid molecules then continue until they hit the pump wall, where they condense and flow back to the boiler by gravity.

The temperature of the walls is usually maintained a few degrees below room temperature by water-cooled coils. The gas molecules which have been knocked downward are pumped away by the fore-

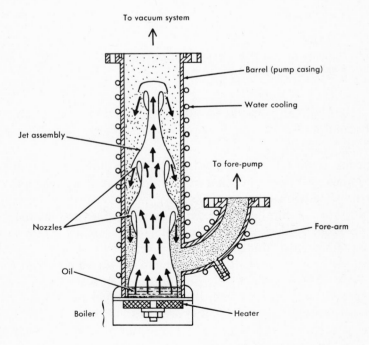

FIG. 2-9. Three-stage nonfractionating diffusion pump.

pump through the fore-arm, which is also watercooled. It may contain a baffle so arranged that any condensed pump fluid will flow back into the boiler.

The jet assembly may contain from one to five nozzles. Figure 2-10 shows a fractionating three-stage diffusion pump. This pump is called fractionating because the more volatile fractions of the fluid are boiled in the outer section of the boiler. The principle involved here is that during the operation of the pump, all oil returning to the boiler does so via the external walls of the pump. The lightest fractions can then boil off soonest, so that only the higher boiling fractions ever reach the center section of the boiler, which feeds the top stage of the pump. This action tends to lower the ultimate pressure obtainable, since any volatile products of thermal cracking are kept from the top stage. The pump shown in Fig. 2-9 has a common chimney which supplies vapors to all three nozzles, so there is no fractionation of the oil.

The heater must raise the temperature of the pumping fluid in the boiler so that its vapor pressure is about 0.5 torr. The heater should be designed to avoid local hot spots which might decompose the oil. Most commercial heaters are external to the pump for ease of re-

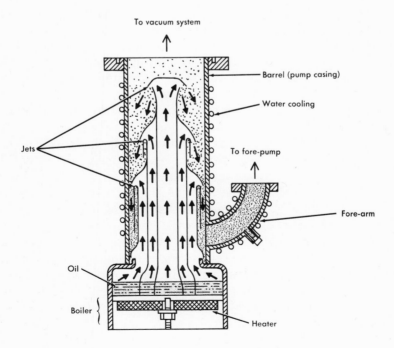

FIG. 2-10. Three-stage fractionating diffusion pump.

moval and replacement. Unfortunately, this arrangement decreases the efficiency of heat transfer to the fluid. Some pumps are designed with immersion heaters which are finned so as to improve the heat transfer to the oil, with the heater extending considerably above the liquid level in the boiler to superheat the vapor and achieve higher pumping speeds [5].

Diffusion Pump Fluids

For optimum pump performance, the diffusion pump fluid should have the following properties (as noted by Steinherz [3]):

1. The fluid should have a low vapor pressure to minimize backstreaming of oil into the vacuum system. Mercury has a vapor pressure at room temperature of a few microns, so it must be trapped with liquid nitrogen. Oils are available with vapor pressures of 10^{-9} torr at room temperature; therefore, for some applications, oil diffusion pumps can be used without extensive cold trapping, provided such oils are used.
2. The fluid should have low enough viscosity to flow back into the boiler at cooling water temperature. All commercial oils as well as mercury satisfy this requirement.
3. The fluid should have a high molecular weight, relative to the pumped gases, to increase the efficiency of removal by the vapor jets. The molecular weight of commonly used oils is in the 300 to 500 range, and the atomic weight of mercury is 201.
4. The fluid should not decompose. Decomposition depends primarily upon the operating temperature, which is determined by a requirement of 0.5-torr vapor pressure in the boiler. If very low-vapor-pressure oils are used, the required temperature could be high enough to decompose the oil. This problem makes impractical those oils which have extremely low vapor pressures. The decomposition of oils (especially hydrocarbons) is a definite problem, especially if the oil vapor can reach the vicinity of hot filaments or electrical discharges. Mercury, of course, being an element, does not decompose.
5. The fluid should be chemically inert and noncorrosive in the presence of common metals, glass, elastomers, and gases found in vacuum systems. Silicone oils are extremely good in this respect. Hydrocarbons, however, will rapidly oxidize or decompose in the presence of air at operating temperatures, and mercury will oxidize under the same conditions. Also, of course, mercury will amalgamate with many metals and alloys.

6. The fluid should be free from toxic vapors. The common pump oils are nontoxic, whereas mercury presents a definite hazard.

Some of the most common pump fluids are listed in Fig. 2-11. For normal vacuum use down to pressures of about 10^{-7} torr, the hydrocarbon oils are adequate. For best results at lower pressures and for resistance to atmospheric exposure, DC-705 is recommended even though it is more expensive. The polyphenyl ethers have low vapor pressures, but they require special procedures due to their high viscosity and tendency to adsorb gases.

Diffusion Pump Performance

Commercial diffusion pumps are designed to operate below a certain maximum forepressure. If the fore-pump is adequate for the diffusion pump, the forepressure will stay well below 100 microns, and small variations in the foreline will not appreciably affect the diffusion pump inlet pressure. If, however, a given throughput is established and the speed of the fore-pump is steadily reduced, the forepressure will eventually increase enough to affect the diffusion pump inlet pressure. The forepressure at which this occurs is listed by the various manufacturers for each diffusion pump, and is usually in the range of from 0.2 to 0.5 torr. It is desirable to maintain a generous safety factor in the operating forepressure of a system, however, as the backstreaming rate can increase sharply if the forepressure is allowed to become this high.

Probably the most significant parameter of a diffusion pump is how fast it can remove gas from the system; i.e., what is the pumping speed? Figure 2-12 shows pumping speed curves for a typical diffusion pump. In general, these pumps begin to pump effectively below 10^{-2} torr. As the pressure is lowered, the curve rises to a flat portion which exists from about 10^{-3} torr down to the vicinity of the "ultimate pressure." In an untrapped system which is leak-tight and clean, the ultimate pressure will theoretically be the vapor pressure of the diffusion pump fluid; therefore, the low-pressure ends of these pumping speed curves are determined primarily by the fluid being used. Some manufacturers show these curves as flat down to 10^{-9} torr because the pump still removes gas at the same rate in that region. There is also a backstreaming rate (due to oil vapor, decomposition products, and outgassing of pump walls), however, that eventually limits the *net* pumping speed to zero. Of course, when the pump is well baffled by a low-temperature trap, the ultimate pressure in the system can be well below the room-temperature vapor

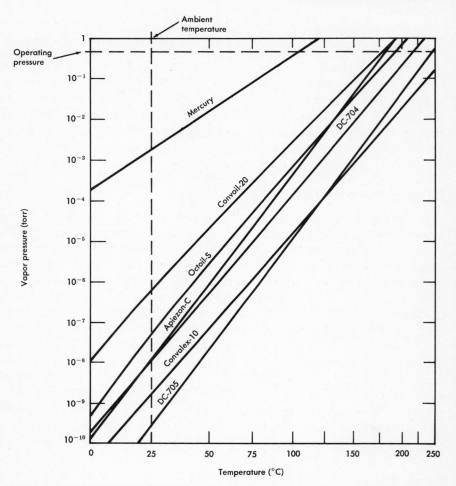

Fluid	Type	Molecular weight	Viscosity centipoise (80°F)
Mercury	Atomic species	201	1.5
Convoil-20	Hydrocarbon mixture	400*	69
Octoil-S	Ethyl hexyl sebacate	427	17
DC-704	Silicone (single species)	484	42
Apiezon-C	Hydrocarbon mixture	574*	150
Convalex-10	Polyphenyl ether	454*	1200
DC-705	Silicone (single species)	546	170

*Approximate average molecular weight.

FIG. 2-11. Properties of some common diffusion pump fluids.

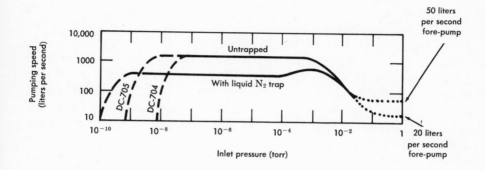

FIG. 2-12. Pumping speed curves for a diffusion pump. The cold trap reduces
the effective speed but permits lower absolute pressures. The speed is
independent of fore-pump size below 10^{-2} torr inlet pressure.

pressure of the pump fluid. Mercury diffusion pumps must always
be trapped since mercury has a vapor pressure of about 10^{-3} torr at
room temperature. The high-pressure performance (above 10^{-2} torr)
is independent of the trap but is a function of the fore-pump speed, as
indicated by the dotted lines in Fig. 2-12.

The maximum pumping speed is determined primarily by the pump
diameter. If a pump could be constructed so that it would permanently remove every molecule that entered its throat, its pumping
speed would be $59D^2$ liters per second,* where D is the diameter in
inches. In practice, the speed of modern diffusion pumps is about
$40D^2_{nom}$, where D_{nom} is the nominal diameter of the pump. The ratio
of these speeds is called the speed factor (Ho coefficient). In this
case it is almost 0.7. This coefficient is artificially inflated, however,
because the actual inner diameter of typical pumps is greater than
D_{nom}. For example, a "6-inch" pump may have an inner diameter of
7 inches or more. Actual speed factors are in the range 0.5 to 0.6.

In practice, diffusion pumps for thin film production are usually
trapped with an optical baffle (no line-of-sight paths available for
backstreaming molecules). The conductance of such a baffle is likely
to be much less than the pumping speed of the diffusion pump. In
such a system, the effective pumping speed is determined mostly by
the baffle, and increasing the efficiency of the pump will have only a
very small effect on the net pumping speed.

*Based on air at 25°C in the molecular flow region.

In any diffusion pump, there will be a certain amount of backward flow of gases. Some of this is due to outgassing of the walls of the pump, and some is due to the diffusion pump oil.

The walls of the diffusion pump in the section above the top jet can be thought of as an extension of the vacuum chamber. If the diffusion pump is blanked off while operating, it must still pump on a certain volume of gas. The walls of this residual volume are usually watercooled so that any oil that hits will condense and drip down to the boiler. This means, however, that this section cannot be conveniently baked out in order to reduce the outgassing rate. In fact, if the vacuum vessel above it is well baked and the rest of the system is clean and tight, virtually the entire gas load being handled by the diffusion pump may originate from the outgassing of the pump itself after the system has approached its ultimate vacuum.

The other source of backstreaming molecules is the diffusion pump oil (and decomposed products thereof). Some of these molecules are scattered upward out of the jet stream, near the nozzle. This is especially significant at the high-pressure end of the operating range (above 10^{-2} torr), where there are so many molecules of gas being pumped that they can disrupt the stream of oil molecules coming out of the top jet. The increase of backstreaming rate at higher pressures is shown graphically in Fig. 2-13 for a typical 32-inch diffusion pump with no trapping of any kind. In this case, if the pump is operated above 10^{-2} torr, it will lose half of its charge in just a few days, whereas if it is kept below 10^{-3} torr, it will operate for years without significant depletion of the charge of oil. Actually, the situation is even worse than shown in Fig. 2-13 because of "forestreaming." This is additional loss into the foreline due to oil swept along by the high throughput existing at high inlet pressures. Thus, it is wise to minimize the amount of time the pump is operated in the high end of its pressure range.

Another source of backstreaming, particularly important at low pressures, is the reevaporation of oil molecules from the walls of the pump. It is this effect which limits the ultimate vacuum to the vapor pressure of the oil at cooling water temperature.

Oil is objectionable in the vacuum chamber for several reasons: it can contaminate the film being formed, it can poison an ionization gauge and cause incorrect pressure readings, and it can limit the ultimate pressure of the vacuum system. An effective method of reducing backstreaming is the use of a very small cold cap placed just above the top jet of the pump, completely covering it, and extending a little over its edge. The cap may be directly water-cooled, cooled

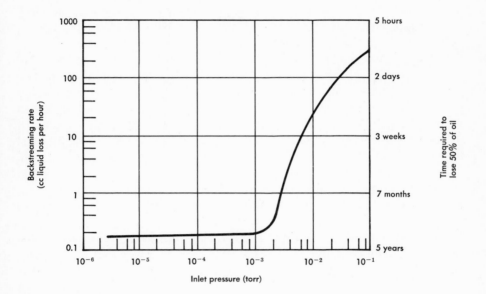

FIG. 2-13. Backstreaming rate versus inlet pressure for a 32-inch pump. (From Hablanian and Steinherz, "Testing Performance of Diffusion Pumps," in *Trans. 8th Nat. Vac. Symposium, 1961*, 1962; courtesy Pergamon Press.)

by conduction through copper supports to the walls of the pump, or cooled by radiation to a cold trap above it. In the latter case, the cold cap is silvered on the bottom and blackened on the top so that it will reflect radiation from the hot top jet and still efficiently emit radiation to the cold trap. Any oil molecules which may condense on the lip of the top jet will reevaporate at a high rate due to their high temperature. Those molecules which are reevaporated in an upward direction are intercepted by the cold cap. Evaporation from the cold cap is not nearly so fast, so the oil has a chance to accumulate there until it drips off onto the next stage jet. This type of cold cap can effectively reduce the backstreaming by one or two orders of magnitude.

The backstreaming rate can be reduced still further by a cold trap above the pump. This type of trap may be cooled by water, liquid nitrogen, or any other refrigerating liquid. Cold traps are usually optically dense so that no molecule can get through without hitting a cold surface, unless it hits another molecule on its way. If the pressure is so high that molecular flow does not exist, the molecules can diffuse through the trap without ever touching any of the cold surfaces. This is another good reason for minimizing the time the inlet pressure of the diffusion pump is above 10^{-3} torr.

In operation, the cold trap will collect oil preferentially on its lowest surfaces. Thus, when it is warmed up again, the oil molecules will have a greater chance of returning to the pump than of going up into the vacuum chamber. This is especially true if the diffusion pump is operating when the trap is allowed to warm up, because then molecular flow will exist and the molecules will have less chance of diffusing up into the chamber. Of course, the trap should be allowed to warm up with the diffusion pump hot only when the high vacuum valve is closed. If the trap were kept cold all of the time, it would eventually collect enough ice so that it would be extremely inefficient. This is especially true for systems which are frequently opened to atmosphere. The build-up of water during each operational cycle can be minimized by pumping with the roughing pump until the pressure in the chamber is at least 10^{-2} torr or lower, before opening the main vacuum valve above the diffusion pump. When the trap is allowed to warm up periodically, it will desorb the water that has formed on it, and there will be no long-term build-up of ice.

Diffusion Pump Operation

In operating a diffusion pump, the following should be avoided:
1. Too high an inlet pressure. The main vacuum valve should never be opened until the pressure is less than 0.1 torr in the system, and a hot diffusion pump should never be vented to air. A great many diffusion pumps have been badly contaminated because they were exposed to atmosphere when hot. It is possible to arrange an interlock using, for example, a thermocouple gauge as a sensor to minimize the problem. Even a momentary surge of gas through the pump, however, can sweep out a considerable amount of oil into the foreline, and should be avoided.
2. Insufficient cooling. All water-cooled pumps should have a flow switch (not a pressure switch) installed in the water line (outlet side) to turn off the heaters if the water flow drops below specifications. This protects against operator error as well as interruption of water service during unattended operation.
3. Too high a forepressure. A mechanical interlock can be placed on the fore-pump (not the motor) so that the diffusion pump

cannot be operated unless the fore-pump is rotating. This protects against operator error as well as breaks or excessive slipping in the fore-pump belts.

Getter and Ion Pumps

Getter and ion pumps have been developed for applications requiring oil-free high vacuum. Getters are materials which remove gas by sorption. Normally, some method is provided for ionization of the gas, since ionization is required for the sorption of inert gases and significantly increases sorption of other gases. Gases are pumped by direct chemical combination, ion burial, chemisorption, and adsorption. A fresh surface of sorbent (gettering agent) must be continuously deposited in order to maintain a constant pumping speed. The pumping speed depends upon the rate at which fresh getter is produced and also upon the area on which it is formed. One problem with this type of pump is that some of the gas which is originally in the getter will be liberated during operation and must be pumped along with the other gases. Titanium is commonly used as the gettering agent because of its high sorptive capacity for common active gases and its negligibly low vapor pressure. The inert gases are not readily captured, and they are usually pumped by ionizing them and driving them into a cathode with sufficient energy to penetrate and be captured. The atmosphere contains about 1 percent of argon; thus, if large amounts of air are to be pumped, reemission from the cathode can become very important, especially for the sputter-ion pumps. These pumps are not effective above 10^{-2} torr because ionization becomes more difficult as the pressure increases into the viscous range. Also, the high currents required may excessively heat the pump.

By far the most popular pump of this type is the sputter-ion pump. Figure 2-14 shows the principle of operation. The anode structure is an open lattice so that electrons which are accelerated toward it pass right through it. They are then repelled by the cathode on the other side, and are accelerated back toward the anode. Thus, they oscillate back and forth. Their trajectories are further lengthened by the application of a magnetic field (a few kilogauss) more or less parallel to their oscillations, making the trajectories spiral, thus increasing the effective distance traveled before finally reaching the anode. They therefore have a high probability of ionizing a gas molecule, and a high-voltage discharge can be maintained even at low pressures. The ions which are formed are accelerated to the cathode, where they may become imbedded. They also

54 THIN FILM TECHNOLOGY

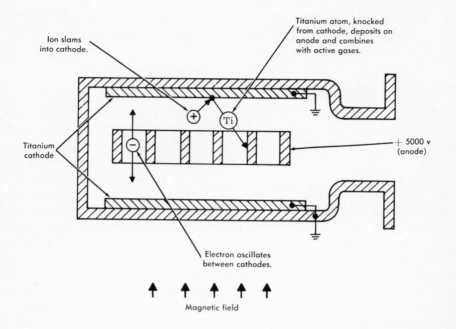

Fig. 2-14. Schematic diagram of a sputter-ion pump.

sputter atoms from the cathode (usually titanium) which are deposited on the anode (and elsewhere) and form a fresh gettering surface. This gettering surface does most of the pumping of the common active gases, while ion imbedment in the cathode is relied on for pumping inert gases. Inert gas atoms which have been thus imbedded, however, can be reemitted as the cathode is slowly eroded.

Advantages and Disadvantages of Sputter-Ion Pumps

This type of pump has many important advantages for use in evaporators, especially where clean oil-free operation is desired. For dynamic sputtering systems, however, these pumps are not recommended, except perhaps for initial pumpdown. The advantages of sputter-ion pumps include:

1. There are no moving parts to wear out or become misaligned.
2. No services are required except electricity. No liquid nitrogen is required because no baffle is necessary. In fact, these pumps can be sealed off under vacuum at the factory, then shipped, and plugged in as received.
3. No oil of any kind is used.

4. The current is proportional to the gas density, and this may be used to measure the pressure in the pump.

5. The pumping speed and the ion current are sensitive to the type of gas being pumped, so the pump can be used as a leak detector with an appropriate search gas.

6. The pump can be baked out for ultrahigh vacuum use. With magnets in place, it can be baked at about 300°C and without magnets at about 500°C. After bakeout, sputter-ion pumps are capable of pumping into the 10^{-10}-torr region.

7. The pumping speed is essentially constant over the range of 10^{-5} to 10^{-8} torr.

8. The internal pump elements are replaceable when they are consumed.

9. The pump can be exposed to atmosphere during operation without damage.

10. One power supply can be used for several pumps.

11. Power failure does not damage the pump or vacuum system.

The following are the principal disadvantages of sputter-ion pumps:

1. The system must be rough-pumped to 10^{-2} torr before the sputter-ion pump can be turned on. When first turned on, the power dissipated in the pump is high enough to raise its temperature appreciably. This can cause outgassing, which raises the pressure above the start-up pressure, and the pump should be turned off. It may take several such cycles before the pump can "take hold" and reduce the pressure on its own. Some of these pumps have water cooling available to minimize this problem. Another approach is to use an evaporated charge of titanium as a getter to bring the pressure down to about 10^{-4} torr, at which pressure the sputter-ion pump can be started easily.

2. The pump is relatively inefficient. For air, the speed factor (Ho coefficient) may range from 0.1 to 0.2, which is considerably less than that of diffusion pumps. A diffusion pump, however, usually must pump through a cold trap, in which case its net efficiency is comparable to that of the sputter-ion pump.

3. The sputter-ion pump does not pump the inert gases well. The pumping speed for helium may be only 10 percent of that for air. Also, inert gases are sometimes reemitted after having once been pumped. A modification of this pump, called a triode pump, has been designed to alleviate this problem. For

thin film applications, the problem of the inert gases is not very serious since these gases do not combine with the films, and in many cases may be ignored.

4. The internal pump structure has a finite lifetime. At a continuous pressure of 10^{-6} torr, a lifetime of about 5 years may be expected, while in a system which is continuously cycled up to atmosphere, a typical life might be 500 cycles. In continuous operation, the lifetime is essentially inversely proportional to the pressure.

Turbomolecular Pumps

The turbomolecular pump is used in applications requiring high throughput and minimal oil contamination. One example of such a pump is shown in Fig. 2-15. In the pumping chamber, rotating blades are alternately arranged with stationary slotted discs on a single axis like the stages of a turbine. The blades have a linear speed of about 14,000 cm per second, which is about 30 percent of the average speed of air molecules at 25°C [see Eq. (2-6)]. There are 19 stages in each of two parallel paths to the foreline. The large number of stages relaxes the clearance requirements on the moving parts (to about 0.040 inch) so that high speeds (about 16,000 rpm) can be used without danger of the rotating parts hitting the stationary parts. The gas is made to flow outward (axially) toward the ends of the axle from one stage to the next, until a compression ratio of as much as 10^6 is established. The seal at the driving end of the axle is effectively double-pumped, since the oil which lubricates it operates at forepressure. The bearings are purposely situated at the forepressure side of the pump so that if any oil should leak past the seal, it would have a negligible probability of getting past the turbine stages into the high vacuum region.

In all pumps of this type (i.e., those which depend on a compression ratio for pumping), there is a backleak which is proportional to the foreline pressure. If the conductance of this leak is C, the net pumping speed is given by

$$S = S_0 - \frac{p_f C}{p_{bj}}$$

where p_f is the forepressure, p_{bj} is the bell jar pressure, and S_0 is the pumping speed for $p_f = 0$. In the molecular flow range, the quantity S_0 is essentially independent of pressure and is almost independent of

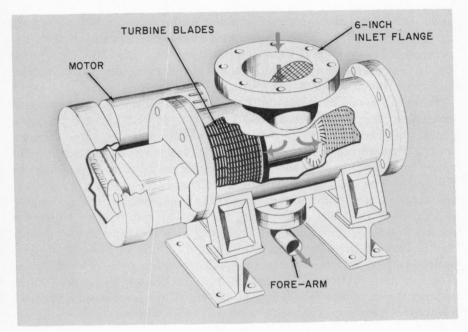

TURBINE BLADES

MOTOR

6-INCH
INLET FLANGE

FORE—ARM

(a)

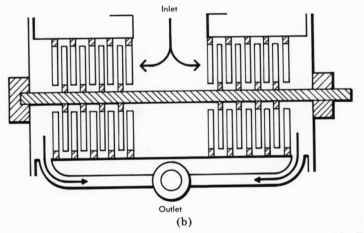

Inlet

Outlet
(b)

FIG. 2-15. Turbomolecular pump. (Courtesy The Welch Scientific Co.)

the type of gas being pumped (S_0 in liters per second is 260 for air, 300 for He, and 320 for H_2). The effective backleak conductance, C, is also essentially independent of pressure as long as p_f is less than about 10^{-2} torr. C, however, is dependent on the mass of the gas molecules being pumped. It is clear from the above equation that even if there were no other source of gas, the pump would have an ultimate pressure given by $p_f C / S_0$. Thus, the ultimate pressure depends on the fore-pump speed as well as the gas composition. The ratio S_0/C is 400 for H_2, 10^4 for He, greater than 10^6 for N_2, and even larger for heavier molecules (such as oil vapor).

Pumping speed curves for air are shown in Fig. 2-16 for a turbomolecular pump with two different fore-pumps. At higher inlet pressures, in the transition to viscous flow, the pump becomes less efficient, with a throughput which is essentially independent of pressure in the range from 10^{-2} to 10^{-1} torr. This is evidenced by the slope of -1 in Fig. 2-16. The "plateau" throughput in this region is roughly proportional to the fore-pump speed, such that if the speed of the fore-pump is expressed in liters per second, the throughput in torr liters per second is $0.3\ S_f$. For example, with a 20-liter-per-second fore-pump, the throughput plateau would be at about 6 torr liters per second. Also plotted in Fig. 2-16 is a pumping speed curve for a "6-inch" diffusion pump in combination with a typical liquid nitrogen cold trap (which reduces the exhaust speed from 1500 to 400 liters per second). With an adequate fore-pump, the useful range of the turbomolecular pump extends to higher pressures than does that of the diffusion pump. Thus, the turbomolecular pump is preferred for applications where the throughput must be high, especially since oil backstreams rapidly from a diffusion pump in this region. Aside from this, the two curves are markedly similar. In both cases, the pumping speed is essentially

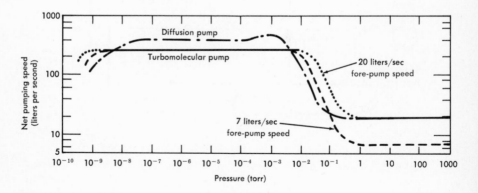

FIG. 2-16. Pumping speed curves for air for a turbomolecular pump in conjunction with fore-pumps of 7 and 20 liters per second. Also included is a curve for a typical 6-inch diffusion pump in combination with a cold trap and a 20-liter-per-second fore-pump. (Courtesy The Welch Scientific Co.)

constant from about 10^{-8} to 10^{-3} torr, and the throughput is essentially constant in the region of transition to viscous flow.

Following are some advantages of the turbomolecular pump:
1. No bypass is needed. The system may be roughed out directly through the turbomolecular pump. This eliminates the expense and maintenance of three vacuum valves.
2. No pumping fluid is required. Therefore, no oil gets into the system, even in the case of power or fore-pump failure.
3. No liquid nitrogen is required. This saves considerable trouble and expense, both initial and operating.
4. The speed is a true, usable pumping speed, since the pump may be connected directly to the system with neither a valve nor a baffle interposed.
5. Gases are not reemitted. All gases are pumped and removed from the system.
6. High throughputs can be achieved without danger of back-streaming oil.
7. There is no uncertainty. If the rotor is rotating, the pump is pumping.
8. The pump is not damaged by accidental exposure to atmosphere during operation.

The turbomolecular pump, like every other vacuum pump, has certain disadvantages:
1. The turbomolecular pump should not be turned on until the pressure is below 10 torr.
2. Full-speed operation (rpm) is limited to the pressure range below 0.2 torr. This is not usually a problem since the turbine takes about the same time to reach full speed as a normal vacuum system takes to pump down from 10 torr to 0.2 torr. If the system pumpdown takes too long due, for example, to a large leak, the motor may overheat. (A thermal cutoff should be provided so that overheating will not damage the motor.)
3. A coolant is required for the bearings. This may be either cold water from the mains or a permanent coolant circulated through a refrigeration system.
4. The initial cost of the pumping system is about 40 percent higher than that of a diffusion-pumped system of the same pumping speed. The expense of supplying liquid nitrogen to the baffles used on diffusion pumps is avoided, however, and these savings will offset the initial cost disadvantage.

2.5 VACUUM SYSTEMS

By far the most common type of vacuum system for thin film use
is the bell jar system, shown schematically in Fig. 2-17. A fore-pump,
foreline valve, diffusion pump, cold trap, high vacuum valve, and bell
jar are all connected in series (and in that order). A bypass (through
a roughing valve) which connects the bell jar directly to the fore-
pump is also common for the roughing part of the pumpdown. The

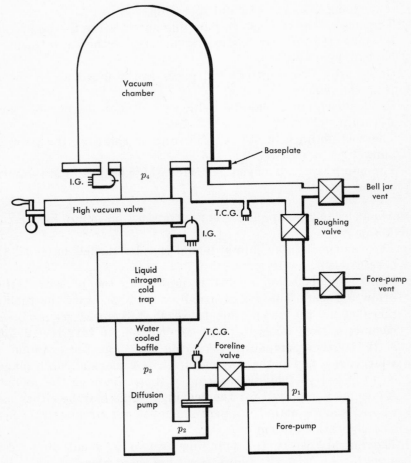

I.G. indicates positions of ionization gauges.

T.C.G. indicates positions of thermocouple or Pirani gauges.

Fig. 2-17. Typical bell jar system.

gas flow in such a system will now be analyzed. The analysis will be general, but will be illustrated by presenting calculations for a specific system with a bell jar volume of 220 liters, a "6-inch" diffusion pump, and a fore-pump with a speed of 21 liters per second.

Gas Flow Analysis

Assume that all the gas load originates in the bell jar, and that the system is sufficiently close to steady state that the throughput (gas flow) is the same at all points in the pumping line; both of these assumptions are normally justified. Thus, the number of molecules leaving the bell jar per unit time is the same as that passing through any other element in the pumping line.

Assume now that a certain throughput, Q, is established at the bell jar. This might be done, for example, by a controlled leak from the atmosphere. The throughput will be expressed as

$$Q = C \ (p_{atm} - p_{bj})$$

where p_{atm} is atmospheric pressure, p_{bj} is the pressure in the bell jar, and C is the conductance of the leak. In general, of course, p_{bj} is negligible compared to p_{atm}. As long as this is true, the throughput, Q, will be independent of the conductances and pumping speeds of the components in the line. If the conductance is altered, e.g., by throttling the high vacuum valve, this will (in general) affect pressures at all points upstream of the conductance, but will not affect pressures downstream because the throughput is unchanged. The same is true of turning on the diffusion pump, which will, of course, reduce pressures at points upstream but will not affect conditions downstream. Thus, the logical place to begin an analysis is at the downstream end, i.e., at the fore-pump. Given a throughput, Q, then, the pressure at all points in the pumping line will be calculated, starting at the fore-pump.

Fore-pump Pressure. The pressure at the fore-pump inlet is given by

$$p_1 = \frac{Q}{S_f}$$

where Q is the throughput, and S_f is the fore-pump speed. S_f is specified as a function of pressure by the manufacturers, so it is possible to construct a graph of pressure versus throughput for the fore-pump. Such a curve is shown as a dotted line in Fig. 2-18 for a specific

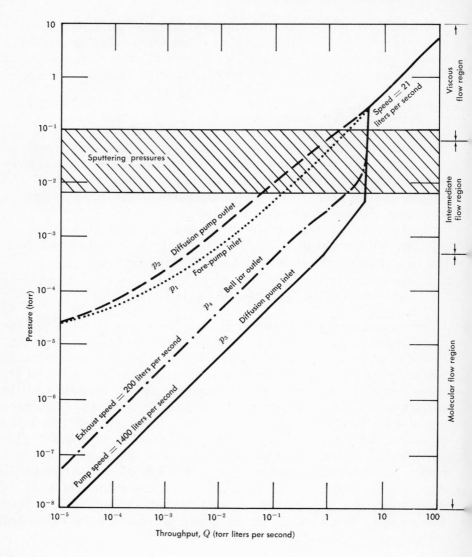

FIG. 2-18. Operating curves for a typical bell jar system.

fore-pump (Welch Scientific Co., Chicago, Ill., model 1398). At higher pressures (above 10^{-2} torr), the line has a slope of 1.0, corresponding to a constant pumping speed (21 liters per second).

Foreline Pressure. By the definition of conductance, the pressure at the diffusion pump outlet (i.e., at the other end of the foreline) is

$$p_2 = p_1 + \frac{Q}{C_1} \tag{2-21}$$

where C_1 is the foreline conductance. C_1, however, may be a function of pressure, so the calculation of C_1 is divided into three regions: molecular flow, intermediate, and viscous flow.

1. Molecular flow region. In the molecular flow region, the conductance can be calculated by Eq. (2-19), and p_2 can be calculated directly.
2. Intermediate region. In the intermediate region, the value of p_1 is used in Eq. (2-20) [or Fig. 2-3 along with Eq. (2-19)] to estimate C_1. This value of C_1 may be used to estimate p_2 using Eq. (2-21). A more accurate value for C_1 may then be obtained by using $\bar{p}$ [where $\bar{p} = (p_1 + p_2)/2$] in Eq. (2-20). Then p_2 can be reevaluated, and the iteration can be continued until the desired accuracy is achieved. Normally, one iteration should suffice.
3. Viscous flow region. In the viscous flow region, C_1 is proportional to the average pressure, such that

$$C_1 = B\left(\frac{p_1 + p_2}{2}\right) \tag{2-22}$$

where (for a long pipe)

$$B = \frac{180 \, d^4}{l} \qquad \text{liters/torr sec}$$

and where d is the diameter (cm) and l is the length (cm). Equation (2-22) can now be substituted into Eq. (2-21). The resulting equation can then be solved for p_2:

$$p_2 = p_1\left(1 + \frac{2Q}{Bp_1{}^2}\right)^{1/2} \tag{2-23}$$

If, however, C_1 is much greater than S_f, then $2Q/Bp_1{}^2$ is much less than 1, and p_2 can be approximated by

$$p_2 \cong p_1 + \frac{Q}{Bp_1}$$

This is equivalent to

$$p_2 \cong p_1 + \frac{S_f}{B}$$

This result could have been obtained by substituting p_1 for $(p_1 + p_2)/2$ in evaluating C_1 from Eq. (2-22). In a well-designed system these approximations will be valid, because C_1 will be much greater than S_f in the viscous flow region.

The pressure p_2 can now be plotted as a function of throughput, Q, over the entire pressure range from atmospheric to the ultimate pressure of the fore-pump. Such a plot is shown as a dashed line in Fig. 2-18. To generate this curve, it has been assumed that the fore-line consists of 15 feet of pipe that is 3 inches in diameter, such as might be installed if the fore-pump were in another room or on another floor. Figure 2-18 shows that, except for the intermediate range, p_2 is approximately equal to p_1. At high pressures, C_1 is proportional to p, and S_f is constant; thus, p_2 is about the same as p_1 because the conductance is so high. At low pressures, C_1 becomes constant, and S_f drops off to zero; thus, p_2 is about the same as p_1 because S_f is so low.

Diffusion Pump Pressure. As explained in Sec. 2.4, a diffusion pump should never be exposed to pressures above its tolerable forepressure. Thus, the diffusion pump is either turned off or valved off when the pressure is above about 0.2 torr, as during the roughdown part of the cycle. If p_2, as calculated above, is less than the tolerable forepressure, the diffusion pump may be permitted to pump on the bell jar, and the pumping speed curves for the diffusion pump may be used as given. The diffusion pump inlet pressure, p_3, can then be found as a function of throughput:

$$p_3 = \frac{Q}{S_d}$$

where S_d is the pumping speed of the diffusion pump. The solid curve in Fig. 2-18 shows the diffusion pump inlet pressure under these conditions for a typical "6-inch" diffusion pump, as given by the manufacturer (Consolidated Vacuum Corp., Rochester, N. Y.). Note that over most of the range, the slope of this curve is 1.0, corresponding to a constant pumping speed. At a throughput of more than 4 torr liters per second, pumping stops, and the bell jar pressure increases very rapidly and unstably until it finally crosses the fore-pump line.

The pressure in the bell jar will then be even greater than in the fore-line, just as it would be if the diffusion pump were turned off. Need-less to say, this is a condition to be studiously avoided.

Bell Jar Pressure. The pressure in the vacuum chamber is given by

$$p_4 = p_3 + \frac{Q}{C_2}$$

where C_2 is the conductance of the traps and high vacuum valve (the valve may be throttled, so this conductance is adjustable from very nearly zero up to full value). Since the pressure in this section is usually in the molecular flow range, C_2 is essentially independent of pressure. The conductance of a cold trap, however, does depend on how cold the trap is. The situation is also obscured by the fact that the cold trap can contribute to the pumping, especially if con-densable gases are in the system. In order to simplify the problem, assume that C_2 is equal to the room temperature conductance of the given cold trap (230 liters per second). The arrangement consists of a water-cooled baffle, a liquid nitrogen cold trap, a "6-inch" valve, and a short neck below the baseplate of the bell jar. The dot-dash curve in Fig. 2-18 shows the pressure at the throat of the baseplate (p_4) as a function of the throughput. This curve departs from a straight line in the region around 10^{-2} torr because the conductance of the trap begins to increase as the pressure approaches the viscous flow range. The same effect is evident in Fig. 2-12.

Curves of p_1, p_2, p_3, and p_4 have now been generated, all as a func-tion of throughput, Q. Theoretically, one pressure measurement at one point in the system is sufficient to determine the throughput and the pressures at all the other points in the system. For example, if the fore-pump pressure is 4×10^{-2} torr, the throughput must be 1.0 torr liter per second, and the pressure at the diffusion pump outlet is 7×10^{-2} torr, which is below the tolerable forepressure. The pres-sure at the diffusion pump inlet is 7×10^{-4} torr, and the pressure at the throat of the baseplate is 3×10^{-3} torr, provided the high vacuum valve is wide open.

Pumpdown

We now turn to an analysis of the dynamics of pumpdown. Assume that the bell jar is roughed out through the roughing valve, and that the diffusion pump is valved into the line when the pressure is below 10^{-2} torr. The gas load through the pumping line may be considered as being due to three terms:

$$Q = Q_L + Q_G - \frac{V dp}{dt} \tag{2-24}$$

where Q_L is the leak rate into the system (including permeation), Q_G is the outgassing rate of all the surfaces in the chamber, and the last term is the volume of the chamber times the rate of pressure reduction. During pumpdown in a clean, tight system, the first two terms are not significant until the pressure reaches about 10^{-6} torr or less. Therefore, for the roughdown and for the first part of the high vacuum operation, the following equation holds:

$$Q = - \frac{V dp}{dt}$$

If the effective speed of the pump (corrected for conductance losses) is S, then $Q = pS$. Inserting this into the above equation gives

$$\frac{1}{p} \frac{dp}{dt} = - \frac{S}{V}$$

For most of the roughing range of pressure, the speed of the fore-pump is a constant,* so the equation can be integrated to find the time-dependence of the pressure:

$$p = p_0 \exp\left(\frac{-St}{V}\right) \tag{2-25}$$

Note that the pressure will decrease one full decade every 2.3 V/S seconds (provided V is in liters and S is in liters per second). In the illustrative example, the volume is 220 liters, and the speed of the fore-pump is 21 liters per second. Thus, the fore-pump can reduce the pressure at the rate of one decade every 24 seconds. There are five decades from atmospheric pressure to 7.6×10^{-3} torr, so the roughing time would be about 2 minutes. In practice, the roughing time is about 3 minutes because the assumption of constant pumping speed is not really valid below 10^{-1} torr.

---

*The effective pumping speed falls off at lower pressures (below 10^{-1} torr) because the pump itself loses efficiency, and the conductance losses become significant.

The diffusion pump is then opened into the line, and again the speed is relatively constant from about 10^{-3} torr down. Therefore, Eq. (2-25) applies until the terms Q_L and Q_G become significant. The pumping speed in this case is 200 liters per second (corrected for conductance losses), so the time for a decade drop in the pressure is 2.3 V/S, which is only 2.5 seconds per decade. This is really too fast for the ionization gauge to follow. In an actual system similar to that described above, the measured rate was 6.5 seconds per decade (this was measured while pumping on argon). At this rate, the pressure would reach 7.6 $\times$ 10^{-9} torr in only 39 seconds. This does not happen in practice because outgassing, permeation, leaks, and backstreaming start to limit the pressure at about 10^{-6} torr. Thus, even though the pumping speed of the diffusion pump may remain constant, the pressure will level off and asymptotically approach a limiting value called the ultimate pressure. The point at which this occurs depends on the care with which the system has been designed and the care that has been taken to keep the system clean. The outgassing rate can be materially decreased by a few hours of bakeout at about 400°C. This will decrease the ultimate pressure by perhaps one more decade, after cooling back down to room temperature. The ultimate pressure for the system is given by

$$p = \frac{Q_L + Q_G}{S_d} \qquad (2\text{-}26)$$

where the backstreaming rate is included in Q_L.

In one actual system, the measured ultimate pressure was 10^{-8} torr. By extrapolating the dot-dash curve in Fig. 2-18 to this pressure, the gas load (throughput) is determined to be about 2 $\times$ 10^{-6} torr liters per second (0.12 micron liter per minute). This gas load was determined experimentally by closing the high vacuum valve and measuring the pressure rise. The pressure rise was observed to be linear, with a rate of about 1.5 $\times$ 10^{-8} torr per second, which corresponds to 3 $\times$ 10^{-6} torr liters per second for Q. This value is about the same as that determined above, indicating that the system was still pumping at about the same speed even at this low pressure (10^{-8} torr). The fact that this throughput was even larger than expected is probably because some pumping was being done by the cold trap.

Selection of Pump Sizes

The sizes of the pumps in the pumping line affect the pumpdown time, the steady-state throughput, and the ultimate pressure.

The roughing-pump speed required for roughdown is given by

$$S = 2.3 \frac{N_d V}{t}$$

where t is the required time for roughdown, V is the volume of the chamber, and N_d is the number of decades of pressure drop required, which is usually about four since there are four decades from atmosphere to 7.6×10^{-2} torr. In practice, roughdown will take a little (perhaps 30 percent) longer than estimated here because of conductance losses and pump speed fall-off at lower pressures, especially if a single-stage pump is being used.

The steady-state throughput is determined primarily by the diffusion pump speed. During sputtering, it may be desirable to maximize the throughput in order to "flush out" any gases that may be desorbed from the walls. With the system described by Fig. 2-18, the maximum safe throughput is about 3 torr liters per second. At this throughput (with no throttling), the bell jar pressure is 1.5×10^{-2} torr, and the diffusion pump inlet pressure is 2.5×10^{-3} torr. If the desired sputtering pressure is 1.5×10^{-2} torr or greater, this pump would be adequate. However, if a higher throughput or a lower sputtering pressure without sacrifice of throughput is desired, greater pumping speed must be provided (this can be done by increasing either the cold-trap conductance or the speed of the pump). Many manufacturers provide a curve of throughput versus inlet pressure in the high-pressure range to help the customer decide if a given pump will be adequate. These curves can depend on the fore-pumping speed, and the steady-state throughput specification can sometimes demand a larger fore-pump than the roughing time requirement. Again, it must be emphasized that diffusion pumps are not designed to operate continuously at these high pressures, and the backstreaming rate can become very high; thus, it is recommended that during sputtering, the diffusion pump be kept at inlet pressures of 10^{-3} torr or less.

In a system designed for evaporation, estimating in advance the throughput required during deposition is very difficult. It will depend strongly on what material is being evaporated, how pure it is, whether it can be outgassed before evaporation, the desired evaporation rate, the filament design, and the maximum permissible pressure rise during film formation. For evaporation of metals to be used as contacts and conductors, the requirements are not so great as when the films are used for circuit components. Generally, a nominal 6-inch diffusion pump has been found to provide satisfactory results on a 24-inch diameter bell jar.

The ultimate pressure also depends on the speed of the diffusion pump, as shown in Eq. (2-26). However, the outgassing and leak rates are extremely difficult to estimate in advance (especially for a system containing fixturing). As a rule of thumb, typical systems require at least a 6-inch diffusion pump in order to obtain an ultimate pressure of 10^{-7} torr or better in a 24-inch bell jar.

Operation of a Bell Jar System

There is not unanimous agreement on the best condition in which to leave a vacuum system overnight (or for longer periods). The main objective is generally agreed to be the minimization of the amount of backstreaming oil that is allowed to get into the vacuum chamber. The best way would probably be to keep the cold trap cold at all times. However, in a system which is frequently opened to the atmosphere, a coating of ice would build up on the trap and would eventually reduce its trapping efficiency.* Besides, liquid nitrogen is expensive, and there can be a safety hazard due to freeze-up in some automatic fill systems if they are left unattended. Because of this, the cold trap should periodically be allowed to warm up; the common practice is to do this every night (if the system is used only during the day). Naturally, the main vacuum valve will be closed to keep oil out of the bell jar as much as possible. The bell jar may be filled with 0.9 atmosphere of argon or dry nitrogen, or it may be evacuated and sealed off overnight. The latter method has the advantage that the overnight pressure rise can be measured, and this is a helpful indicator of system performance. In either case, the object is to minimize the adsorption of gases on the walls and fixturing, and thereby to minimize the pumpdown time the next day.

Assuming the cold trap is allowed to warm up, the question then arises whether to turn off the diffusion pump. This is not a simple decision, and arguments can be made to support either position. It is assumed here, however, that the diffusion pump remains on.

In the standby condition, then, the bell jar system shown in Fig. 2-17 has all valves closed except the foreline valve, and both pumps are operating. The water-cooled baffle will be cooled, but the liquid nitrogen trap will be at room temperature. The bell jar

---

*As previously mentioned, the ice buildup from atmospheric water vapor can be minimized by roughing to as low a pressure as possible before opening the main vacuum valve.

has been left sealed under high vacuum, but the overnight pressure rise has brought its pressure up to about 10^{-3} torr. To start the system, a little liquid nitrogen is put into the cold trap. When the violent boiling stops, indicating that the metal parts are chilled, the trap can be filled and an automatic system can be put into operation. Following a check to make sure the high vacuum valve is closed, the bell jar is vented to (preferably) either dry nitrogen or argon. If the bell jar can be kept warm (50° to 100°C) it will adsorb less water when opened to atmosphere, and the pumpdown and bakeout times will be substantially reduced. The substrates are then put into the bell jar and preparations are made for the desired processing (e.g., sputtering). At this point, it may be necessary to clean any particulate matter from the substrates and other places in the vacuum system with a vacuum cleaner. Also, the baseplate and its gasket should be wiped clean with a solvent. (Acetone should *not* be used on neoprene or viton. A preferable procedure would be wiping with trichloroethylene, and rinsing with isopropyl alcohol.) The bell jar can then be lowered, and the roughdown begun.

First, the foreline valve is closed. The diffusion pump outlet pressure should be observed for a few seconds to make sure no leak in the diffusion pump area can cause this pressure to exceed the tolerable forepressure while using the fore-pump for roughing. Sometimes a ballast volume (or even an auxiliary fore-pump) is installed at the diffusion pump outlet to minimize this problem. After making sure the bell jar vent is closed, the roughing valve is opened and the bell jar is roughed for a few minutes until the pressure is below 10^{-1} torr. It is preferable to rough the bell jar to 10^{-2} torr or lower if this can be done in a reasonable time.

The high vacuum part of the sequence is started by closing the roughing valve and opening the foreline valve. Finally, the high vacuum valve is opened,* and by the time the ionization gauge in the bell jar is turned on and outgassed, the pressure should be around 10^{-6} torr. At this point, the bakeout procedure (if any) can be started. The details of bakeout will vary considerably but, typically, the bell jar might be baked for a few hours with substrate heaters and filaments as hot as practical. Finally, the bell jar is cooled and the processing is performed. In production situations, bakeout of the system may not be practical because of the time involved.

---

*In some systems, it may be advisable to open the high vacuum valve slowly, while monitoring the foreline pressure to ensure that the transient pressure rise does not exceed the tolerable forepressure. If the bell jar is roughed out below 10^{-2} torr, however, this should not be necessary.

After everything has cooled down again, the high vacuum valve is closed and the rate of pressure rise in the bell jar is observed. This rate is a good indication of any deterioration of performance due to an increase of leaks, permeation, or gassy contaminants. At this point, the system is in the standby condition mentioned above, except that the cold trap is cold. The trap can be left cold for the next processing sequence, or it can be allowed to warm up if this sequence is scheduled for a later date.

Maintenance and Leak Detection

Cleanliness is one of the most important considerations in maintaining a vacuum system. If possible, every part should be rigorously cleaned (ultrasonic tanks are very helpful) before insertion into the system, especially if there has been any contact with solder flux or machining oils of any kind. It is good vacuum practice to avoid fingerprint residues by wearing gloves during handling, especially if operation in the 10^{-8}-torr range is required.

After a large number of thin film depositions in a vacuum system, there will normally be a buildup of deposited material on the fixturing and walls. These layers may "flake off" in small particles which can land on the substrates, causing holes in the film being deposited. Also, a loosely layered structure like this, with its high surface area, can constitute a "virtual leak" which increases the pumpdown time. Therefore, good maintenance procedures may require occasional scraping or etching of these deposits.

Diffusion pumps require very little maintenance if they are not operated at high pressures. If a pump shows poor performance, the jets can be cleaned and the oil changed, but this should be very infrequent.

The oil level in a rotary oil pump should be checked periodically, and the oil should be changed at least once every few months of operation. Also, the belts should be inspected, and changed before they are worn enough to break while in operation.

Another routine maintenance procedure that is beneficial is to change elastomer gaskets (e.g., o-rings) periodically, especially on dynamic (i.e., nonstatic) seals. Static seals that are kept sealed most of the time can be changed whenever they must be removed for other purposes.

Every vacuum system should have a log book in which is recorded all significant data about its operation and the processes that are performed. In this way, it is much easier to detect incipient problems before they get serious, and to determine if a change in procedure or the introduction of a new item has affected the performance of the system as a whole. A log book is especially helpful in calling attention to leaks which may occur. Leaks will show up as an increase in the ultimate pressure and an increase in the rate of pressure rise when the bell jar is valved off from the pumps.

An increase in the ultimate pressure can be caused by either an increase in the outgassing rate or a real leak from the atmosphere. An increase in the outgassing rate, however, is seldom the cause unless some new item has been introduced into the system. If so, it is generally easy enough to remove the suspected offender to see if that is the case. Usually, the cause of an increased gas load is a real leak from the atmosphere.

Many methods exist for finding leaks in vacuum systems. Perhaps the easiest method is to spray acetone on the outside of the system under vacuum. If the spray is at the point of the leak, the presence of acetone can be detected on the inside with either an ionization gauge or a Pirani gauge, by an increased pressure indication. If helium is used as a search gas, the pressure indication will increase on a Pirani gauge and decrease on an ionization gauge. The current through a sputter-ion pump will increase when helium is sensed, and will decrease when carbon dioxide is sensed.

The best device for finding very small leaks is generally agreed to be the helium mass spectrometer. In this instrument, the system is maintained at a vacuum while helium gas is sprayed on the outside of a suspected leak. The mass spectrometer is tuned to respond only to helium in the gas and can be made very sensitive. Leaks of 10^{-10} torr liters per second can be detected in this manner. A leak this size will be completely negligible in any practical thin film operation. The commercial helium mass spectrometer leak detectors are equipped with their own complete vacuum system so that they can be operated at a low pressure, even if the system under test is too leaky to be pumped out very much. In this case, only a small flow of gas is accepted from the system under test, but detecting leaks with the mass spectrometer is still possible.

Among the significant parameters affecting leak detection are the response time and the clean-up time. The response time is defined as the time required for the detector to reach 63 percent of its peak reading after application of the helium gas. The clean-up time is the time required for the detector to return to only 37 percent of its peak meter reading after the helium probe is removed from the leak (that is, the time required before it can sense the leak again). Normally, these times are determined by the system under test, and not by the leak detector. When a system of volume, V, is being continuously pumped with a (helium) pumping speed, S, the time constant associated with the partial pressure of helium in the chamber is given by V/S, where the delay in the detector has been neglected. The response time and the clean-up time will both be equal to V/S (independent of the size of the leak). The detector itself measures only the *amount* of helium in the gas being sampled, so anything that can be done to increase the pressure (while maintaining the same percent of helium) will increase the effective sensitivity of the detector. For example, the pump can be throttled down so as to reduce S, thereby increasing the pressure. This will, however, increase the response and clean-up time. The same principle holds for the "accumulator valve" on the top of the diffusion pump of the leak detector. When this valve is throttled, the sensitivity is increased at the expense of the time constants.

There are two time constants associated with a bell jar system. Let V_1 be the bell jar volume, and S_1 be the diffusion pump speed (after correcting for conductance losses). Let V_2 be the foreline volume, and S_2 be the speed of the fore-pump. Then, $t_1 = V_1/S_1$ and $t_2 = V_2/S_2$. If the detector is connected to the bell jar, the clean-up time and the response time will be t_1. If the detector is connected to the foreline, however, the time constants will still be approximately t_1, provided that $t_1 >> t_2$. This will normally be the case since the bell jar volume is so much larger than the foreline volume. When the detector is connected to the foreline, it samples gas at a much higher pressure, with a higher partial pressure of helium giving greater detector sensitivity. The conclusion, then, is that unless $t_2 > t_1$, the leak detector should be connected in to the foreline, with the diffusion pump operating, provided, of course, that the leak is not so great as to cause pressures exceeding the safe operating range of the diffusion pump.

When a small item is to be tested, it can be connected directly to the leak detector without any external pump. Then, the time constant is V/S, where V is the volume of the item plus the internal volume of the leak detector, and S is the speed of the diffusion pump in the leak detector. Besides having the advantage of a fast response time, this arrangement removes the possibility of confusion due to helium gas diffusing through air into a leak some distance from the point where the probe is being applied. It should also be noted that the clean-up time may be substantially increased if long connecting lines of rubber hose or thick deposits of sealing grease are used because the helium can permeate them, and it may take considerable time to remove the helium. This is primarily a problem for small leaks, however.

Continuous-Feed Vacuum System

Thus far, all the discussion in this chapter has concerned bell jar systems. These systems have the economic disadvantage that they are basically adaptable to "batch" processes only, and therefore are difficult to integrate into a production line. It is possible, however, to establish a continuous flow of materials into and out of the vacuum, thus creating an "in-line" machine. Because in-line machines are becoming important in large-scale production of thin films, the vacuum engineering requirements of such a system will now be discussed.

One such type of system, part of which is shown schematically in Fig. 2-19, has been designed by the Western Electric Company. It

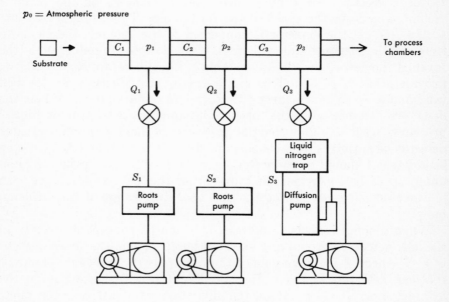

FIG. 2-19. Pumpdown part of the Western Electric in-line machine. The substrate rides the track into each of the chambers in turn, by passing through close-tolerance conductances C_1, C_2, and C_3.

consists of a track along which a continuous series of substrate holders is pushed. The holders carry the substrates into each of several separate vacuum chambers in succession. A large pressure differential between adjacent chambers must be maintained while still leaving enough clearance for the substrates and holders to enter without excessive friction. The passageway between adjacent chambers consists of a duct which closely matches the dimensions of the substrate holders, and which is long enough so that at least one of the substrate holders is always inside the duct. The clearance is about 0.005 inch. With reasonable values for the other dimensions, this permits a pressure differential of about three orders of magnitude from one chamber to the next. Thus, three chambers are required to go from atmosphere down to 10^{-6} torr, and three more to return to atmosphere.

The substrate starts at atmospheric pressure, passes through conductance C_1 into chamber one at pressure p_1, which is being pumped by a pump with a speed S_1. We assume at the start that $p_0 >> p_1 >> p_2 >> p_3$. Then, for reasonable pumping speeds, it follows that $Q_1 >> Q_2 >> Q_3$. Thus, at steady-state conditions,

$$Q_1 = p_1 S_1 = p_0\, C_1, \text{ where } C_1 \text{ is evaluated at } \frac{p_0}{2}$$

$$Q_2 = p_2 S_2 = p_1\, C_2, \text{ where } C_2 \text{ is evaluated at } \frac{p_1}{2}$$

$$Q_3 = p_3 S_3 = p_2\, C_3, \text{ where } C_3 \text{ is evaluated at } \frac{p_2}{2}$$

Rearranging, we see that the various pressures can be found from the conductances and pumping speeds:

$$p_1 = p_0 \left(\frac{C_1}{S_1} \right)$$

$$p_2 = p_1 \left(\frac{C_2}{S_2} \right)$$

$$p_3 = p_2 \left(\frac{C_3}{S_3} \right)$$

Thus, pumps are needed with a speed at least 10^3 times the corresponding conductances. The next step is to calculate the conductances, using Fig. 2-3. In this case, let d be the clearance; $d = 0.005$ inch, or 0.013 centimeter. Then,

$$\frac{p_0 d}{2} \text{ is of order 5 torr cm}$$

$$\frac{p_1 d}{2} \text{ is of order } 5 \times 10^{-3} \text{ torr cm}$$

$$\frac{p_2 d}{2} \text{ is of order } 5 \times 10^{-6} \text{ torr cm}$$

From Fig. 2-3, we see that $p_0 d/2$ is in the viscous range, and that $p_1 d/2$ and $p_2 d/2$ are in the molecular flow range. The conductance of a rectangular slit of cross section $a \times b$ (where $a >> b$) and length l is given by

$$C = \frac{600 a b^3 \, \overline{p}}{l} \qquad \text{liters/sec (viscous range)}$$

$$C = \frac{27 a b^2}{l} \log\left(\frac{l}{b}\right) \qquad \text{liters/sec (molecular flow range)}$$

where the second equation is valid for $a > l > b$, and where a, b, and l are in centimeters and $\overline{p}$ is in torr (the logarithm is to base 10). In this case, a is 10 cm and b is 0.013 cm; l_1 is 32 cm, l_2 is 7.6 cm, and l_3 is 5.1 cm.

Since p_0 is atmospheric pressure, $\overline{p}$ will be 380 torr, so C_1 can be calculated directly as $C_1 = 0.16$ liter per second.

For purposes of illustration, we have ignored a slit at the top with $a = 0.95$ cm and $b = 0.026$ cm. This would bring the conductance to

$$C_1 = 0.275 \text{ liter/sec}$$

A Roots blower was installed which had a speed of $S_1 = 150$ liters per second, after correcting for conductance losses. Then the pressure in chamber one can be calculated:

$$p_1 = 760 \text{ torr} \left(\frac{0.275}{150}\right) = 1.4 \text{ torr}$$

The actual measured pressure was 3.5 torr, which is in satisfactory agreement with the calculated pressure.

From the molecular flow formula, the conductance C_2 is calculated to be 0.017 liter per second. With the inclusion of the top slit, the conductance becomes

$$C_2 = 0.023 \text{ liter/sec}$$

Chamber two was pumped by a Roots blower which had a speed of 33 liters per second (again correcting for conductance losses). The pressure in this chamber is then given by

$$p_2 = 3.5 \text{ torr} \left(\frac{0.023}{33}\right) = 2.5 \times 10^{-3} \text{ torr}$$

The pressure observed in chamber two was 3×10^{-3} torr, which again shows satisfactory agreement with the calculated pressure.

The conductance C_3 is calculated to be 0.023 liter per second (also from the molecular flow formula). The top slit conductance brings this up to

$$C_3 = 0.031 \text{ liter/sec}$$

Chamber three is pumped by a diffusion pump of 47 liters per second (after conduction losses), so the pressure in chamber three is calculated to be

$$p_3 = 3 \times 10^{-3} \text{ torr} \left(\frac{0.031}{47}\right) = 2 \times 10^{-6} \text{ torr}$$

The observed pressure was 1×10^{-6} torr, which again is in reasonable agreement with the calculated pressure.

Once this system of calibrated leaks has been established, almost any type of vacuum processing can be performed in a continuous production line. In addition to the advantage of economy, such an arrangement can lead to increased reproducibility (as opposed to a batch process, in which variations from one batch to the next are almost certain).

When the in-line system must be shut down and opened to the atmosphere, the diffusion pumps are valved off and left pumping, so it only takes a few minutes to pump back down to operating pressures. In a process like sputtering, however, it may take a few hours of sputtering before the parameters of the sputtered films are stable.

2.6 VACUUM GAUGES

The vacuum engineer is continually faced with the problem of measuring the pressure in the vacuum system. As with vacuum pumps, no gauge is available that is suitable over the entire range of interest (this may include a span of 11 orders of magnitude). Thus, the engineer is faced with selecting from a large number of types of gauges. Among these are:

1. Mechanical gauges, such as Bourdon tube dial gauges.
2. Hydrostatic gauges, such as mercury manometers.
3. Thermal conductivity gauges, such as thermocouple gauges.
4. Ionization gauges, such as thermionic gauges.

Figure 2-20 shows the pressure ranges in which some of the most common gauges are useful.

Mechanical Gauges

Mechanical vacuum gauges rely on the actual physical pressure of the gas to produce a force which can be measured. The force is usually balanced against the restoring force of a bellows, a diaphragm, or a Bourdon tube. One important feature of these mechanical gauges is that they have the same calibration, regardless of the gas whose pressure is being measured. This is not true of most of the low-pressure gauges. A mechanical linkage can be made to translate the deflection into rotary motion of a dial on the face of the gauge. This type of gauge can measure pressure changes as low as 10^{-1} torr, but the range of greatest usefulness is above 1 torr. Recently, a mechanical gauge has been commercially introduced (Texas Instruments Co., Dallas Texas) which consists of a long quartz spiral Bourdon tube; at the end of this tube is a mirror for an optical level arm. It is claimed that this gauge can reproducibly measure pressures below 250 torr to within 5×10^{-3} torr. Another commercial gauge (MKS Instruments, Inc., Burlington, Mass.) uses a metal diaphragm which forms a common electrode for two back-to-back capacitors. The capacitors are in the arms of a sensitive bridge, so that any deflection of the diaphragm unbalances the bridge. In this manner, any pressure below 1 torr can be read to an accuracy of better than 5×10^{-4} torr and a sensitivity of 10^{-5} torr. With this arrangement, a deflection of only 10^{-8} centimeters of the diaphragm can be detected. This instrument could be very useful for calibrating thermocouple and Pirani gauges, and for measuring sputtering pressures very accurately.

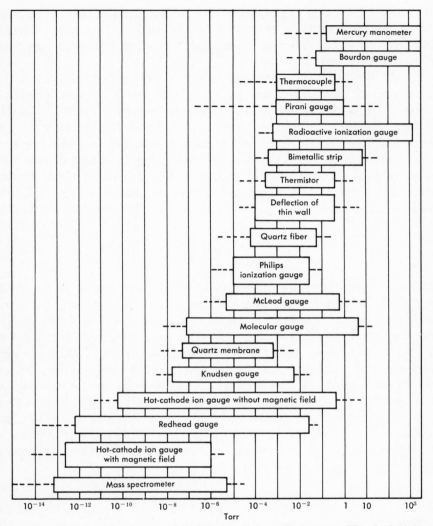

FIG. 2-20. Ranges of pressures for different types of gauges. The dashed lines
indicate the regions where usefulness can be extended under special
conditions or with special techniques. (After S. Dushman, *Scientific
Foundations of Vacuum Technique*, 2nd ed. 1962, J. M. Lafferty, Ed.;
courtesy John Wiley and Sons, Inc.)

Hydrostatic Gauges

This type of vacuum gauge measures pressure by determining the
hydrostatic head that the unknown pressure can set up in a liquid.
The classical example is the mercury u-tube manometer. Many elaborate devices have been designed to measure accurately the difference

in the levels of two liquid surfaces. Refinements of this type have increased the sensitivity down to a few times 10^{-3} torr, but a further extension of the sensitivity seems unlikely since the vapor pressure of mercury is about 10^{-3} torr.

Below about 1.0 torr, the generally accepted standard of pressure measurement is the McLeod gauge, which also measures pressures using the hydrostatic head of a liquid (generally mercury). In the McLeod gauge, a definite volume (V_1) of gas at the pressure (p_1) to be measured is compressed into a smaller volume, V_0. The pressure after compression (p_0) is read as the difference in height of two mercury columns. The original pressure is then calculated using Boyle's law (which states that for a given mass of gas at a constant temperature, the product of the pressure and the volume is a constant):

$$p_1 = p_0 \left(\frac{V_0}{V_1} \right)$$

The operating principle of one type of McLeod gauge is shown in Fig. 2-21. In standby position, the mercury level is maintained below the cutoff point, D. When a pressure measurement is required, the mercury level is raised until the level in the capillary C is even with the top of the closed capillary B. Since these two capillaries are very uniform, and also very much the same in bore size, surface tension effects are almost eliminated. Thus, the difference in the two heights is a reliable measure of the difference in the two pressures. Since the compression ratio is always many times greater than 100, the pressure on the surface of the mercury in capillary C will always be negligible compared to the pressure in capillary B (after compression). The pressure is found from

$$p_1 = p_0 \left(\frac{V_0}{V_1} \right) = h \left(\frac{ah}{V_1} \right) \tag{2-27}$$

where h is the difference in heights of the two columns of mercury (h must be measured in mm of Hg for it to be equal to p_0 in torr units). The volume of the gas after compression is h times a, where a is the cross-sectional area of the capillary tube.

A scale for reading the pressure is usually provided. As can be seen from Eq. (2-27), this scale will be quadratic; that is, it will depend on h^2. This type of scale has the advantage that the low-pressure end is expanded as compared to a linear scale. If, for some reason, a linear scale is preferred, this is easily accomplished by

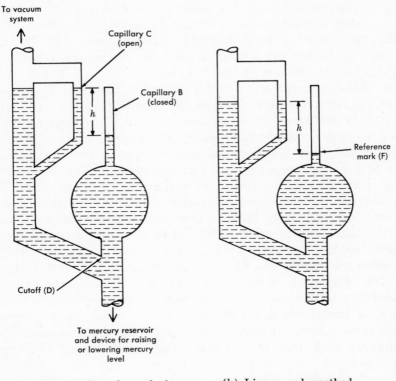

(a) Quadratic scale method   (b) Linear scale method

FIG. 2-21. McLeod gauge. If the mercury is raised until the level in capillary C
is even with the top of capillary B, the result is a quadratic scale.
If the mercury is raised to the reference mark (F) on capillary B,
the result is a linear scale.

always compressing the gas to the same volume; for example, one can
raise the mercury only to the mark F on the closed capillary. The
volume above this mark will be some constant V_2, and the pressure
before compression will be calculated by

$$p_1 = p_0 \left(\frac{V_2}{V_1}\right) = h \left(\frac{V_2}{V_1}\right)$$

where, again, h must be measured in millimeters in order for p_0 to
be expressed in torr. Some McLeod gauges have one capillary with

a linear scale and one with a quadratic scale. McLeod gauges also differ in the method used for raising the level of mercury in the gauge. Some control the level by having a reservoir attached to a flexible hose that can be raised or lowered by hand; others control the level by tilting the gauge on its side, or by air pressure (or vacuum) on the surface of the reservoir.

The McLeod gauge has several desirable features. It is simple physically, requiring only that the gas obey Boyle's law. It requires no auxiliary equipment or connections, and it has the same calibration for all common noncondensable gases. There are limitations, however. There should be a cold trap between the gauge and the system in order to keep mercury vapor out of the system. This trap will, of course, trap condensable vapors as well as the mercury. Even if the trap did not remove the condensable gases, the gauge might not measure them precisely because they would no longer obey Boyle's law. Another limitation of the McLeod gauge is that it cannot monitor a rapidly changing pressure; it may take as much as a minute between successive readings. Also, these gauges must be kept free from dust and dirt. Any particle that gets into the capillary tube can cause the mercury to separate in the capillary, causing erroneous readings until it is joined again, which sometimes is a laborious procedure. Lastly, these gauges are made of glass and may contain a sizable mass of mercury such that in operation, the possibility of breakage is always present. In summary, these gauges are very satisfactory for standards with which to calibrate other gauges, but are not recommended for routine use in production.

Thermal Conductivity Gauges

Thermal conductivity gauges are based on the fact that the thermal conductivity of a gas is pressure-dependent in the region below about 2 torr. Each gauge contains a heater in a tube and measures the temperature of the heater to indicate the amount of heat transfer between the heater and the walls of the tube. Naturally, some of the heat transferred will be by conduction through the leads and supports, and some will be by direct radiation. These terms will dominate the heat transfer at pressures below about 10^{-3} torr, so these gauges (in their commercial forms) are not useful below that pressure. One major problem with these gauges is that the heat radiation depends strongly on the surface of the heater as well as the surface of the walls. Thus, any contamination which changes the emissivities involved will affect the gauge calibration. Also, different gases are known to have different thermal conductivities, so the gauge must be calibrated for each type of gas used.

Figure 2-22(a) is a schematic diagram of a thermocouple gauge. The heater is normally operated at constant current, and its temperature is measured by a thermocouple welded to it. The meter reads the thermocouple voltage directly, and the current is adjusted so that the meter reads full-scale for very low pressures. The rest of the scale is then calibrated with a McLeod gauge. The thermocouple

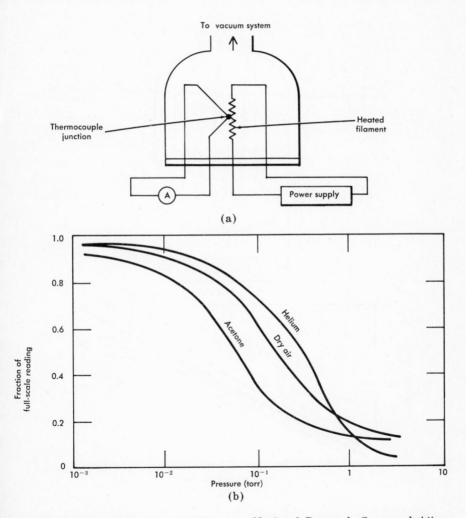

(a)

(b)

FIG. 2-22. Thermocouple gauge. (Courtesy National Research Corp., subsidiary of Norton Co.)

voltage will be very close to zero at all pressures above 2 torr. A typical response curve is shown in Fig. 2-22(b).

A Pirani gauge is shown schematically in Fig. 2-23(a). In this gauge, the heater is made of a metal with a high temperature coefficient of resistance, and the resistance of the heater itself is taken as a measure of its temperature. Many Pirani gauges have this heater element as one leg of a Wheatstone bridge; the other leg of the bridge consists of a similar element in a similar case placed nearby, except that it is sealed off at a very low pressure. The prin-

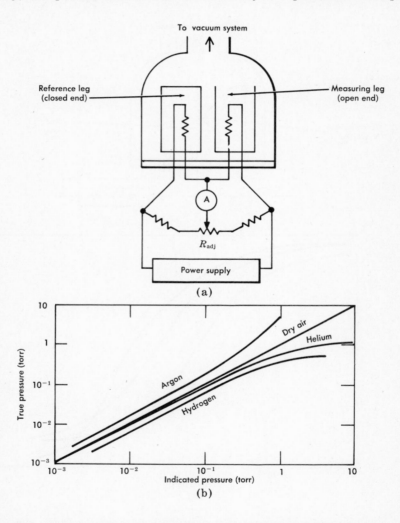

FIG. 2-23. Pirani gauge. (Courtesy Edwards High Vacuum, Inc.)

ciple is that ambient temperature changes will affect both elements equally and will not affect the meter reading. This type of detector allows an almost linear scale on the meter for pressures below about 10^{-1} torr (that is, where the bridge is almost balanced). The power supply may be operated at constant current or constant voltage but, in either case, the scale is found empirically and is simply marked on the meter face by the manufacturer. Many of these Pirani gauges have an expanded linear scale in the low-pressure region below 10^{-1} torr. These gauges are much more reliable when they can be zeroed each day by adjusting R_{adj} while the gauge is at a very low pressure. In a batch-type sputtering procedure, fortunately, this is easily done, so these gauges may be used to monitor pressure in the range of 10^{-2} to 10^{-1} torr during sputtering.

Because of their sensitivity to different gases and vapors [Figs. 2-22(b) and 2-23(b)], both thermocouple and Pirani gauges may be used for leak detection. The Pirani gauge is usually more sensitive for detecting small leaks than the thermocouple gauge. Both these gauges are easily adaptable for continuous recording and for use as sensors for interlocks to protect diffusion pumps and other equipment. Because they are small, rugged, and inexpensive, these gauges are standard equipment on many vacuum systems.

Thermionic Ionization Gauges

For pressure measurements below 10^{-3} torr, the most popular gauge is the ionization gauge. This gauge operates by ionizing a small fraction of the molecules in the system, and then collecting the ions at a cathode where they can be measured as a current. Over a wide range of pressures, this current is proportional to the density of molecules in the gauge, and therefore is a convenient measure of the pressure. The ratio between the current and the pressure depends on the physical configuration of the gauge and the ionization probability of the particular gas being measured. The ionization of the gas can be accomplished in a number of ways, but the most common is by the use of accelerated electrons thermionically emitted from a hot filament. In this case, the ionization rate is proportional to the rate of emission of electrons from the filament, so the ion current is proportional to the emission current.

One possible configuration for an ionization gauge is shown in Fig. 2-24. It consists of an emission filament, located in the middle of the gauge and surrounded by a grid. The grid is positively charged with respect to the filament, so the electrons are accelerated toward

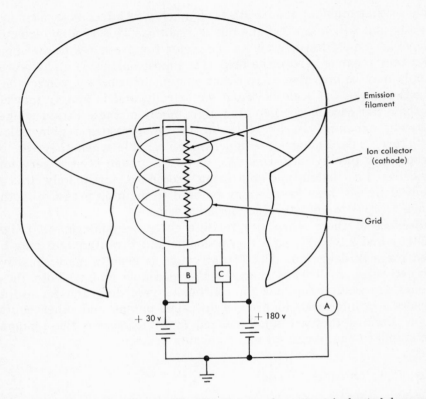

Fig. 2-24. Thermionic gauge. The filament in the center is heated by power
supply B to emit electrons which are accelerated toward the grid.
They pass through the grid and ionize molecules, which are repelled
to the collector and measured on the electrometer A. The power
supply C is used only for outgassing.

it. Most of them pass through the grid, however, and continue radially
outward. By this time, they have acquired sufficient energy to ionize
a molecule; thus, ions are formed in the region outside the grid. The
grid is more-or-less surrounded by a cylindrical cathode, and the ions
are accelerated toward the cathode and are collected on it. The
current to the cathode is read by a sensitive ammeter, and this current
is proportional to the density of molecules in the gauge.

In another configuration, the filament and plate are both inside the
grid. The plate is positive with respect to the filament, so the electrons are accelerated toward it without passing through the grid.
The grid is negative, and the ions produced by the collisions are
collected on the grid. In this configuration, the *grid* current is a
measure of the pressure.

A major advance in ionization gauge technology was the Bayard-Alpert arrangement, shown in Fig. 2-25. In this gauge, the filament is on the outside and the collector is on the inside of the grid. Again, the grid is positive to accelerate the electrons and the collector is negative in order to attract the ions; however, in the Bayard-Alpert gauge the collector is a very thin wire down the center, rather than a cylinder on the outside. This configuration dramatically extends the useful range of the gauge at very low pressures by lowering the X-ray limit (see below). Most ionization gauges sold today are of the Bayard-Alpert type.

Ionization Gauge Sensitivity. The sensitivity of the gauge to pressure is normally expressed as the ratio between the collector current and the pressure for a given emission current. An emission current of

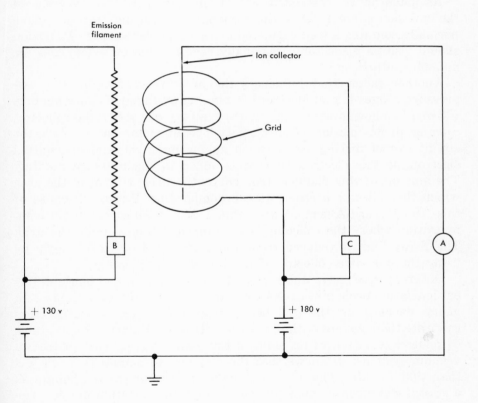

Fɪɢ. 2-25. Bayard-Alpert ionization gauge. Operation is inverse to the conventional gauge shown in Fig. 2-24. Power supplies B and C and electrometer A have the same functions as in Fig. 2-24.

10 milliamperes is typical. At this current, a representative ionization gauge has a sensitivity of 0.1 ampere per torr. Without special precautions, these gauges can be used in the range of 10^{-9} to 10^{-3} torr, which means that the ammeter on the collector must allow currents as low as 10^{-10} and as high as 10^{-4} amperes to be read. Needless to say, the collector lead wire and circuitry must be specially shielded to prevent leakage currents from adding to the meter reading at the very low end of the range. The gauge tubes are normally made of glass, and the collector is usually brought through the glass envelope at the opposite end of the tube from the other components in order to maximize the resistance between them. It is still helpful to wipe the glass tube on the outside with acetone occasionally, however, to remove any contamination which may cause leakage conductance that would disturb readings in the low-pressure ranges.

As noted above, the sensitivity of the gauge depends directly on the emission current; thus, the electronic control unit for the gauge normally contains a well regulated emission power supply. Provision should also be made for occasionally reading this emission, and for manually adjusting it.

Another factor that influences the pressure reading in the low-pressure ranges is that electrons impinging on the grid cause photons (X-rays) to be emitted. These X-rays in turn can strike the collector, causing photoemission of electrons. The ammeter in the collector circuit cannot distinguish between an incoming ion and an emitted electron, so this electron current is added to the pressure reading. The amount of this photoelectron current is proportional to the area which the collector presents to the bombarding X-rays. Because of this, the Bayard-Alpert gauge (Fig. 2-25) is superior to the configuration where the collector is a cylindrical tube outside the grid. The X-ray limit is reduced from about 10^{-8} to 10^{-11} torr simply by using the fine wire collector of the Bayard-Alpert gauge.

Different gases have different ionization energies, and hence different ionization probabilities. As a result, the sensitivity of an ionization gauge depends on the gas being measured. Most manufacturers calibrate their gauges with dry air or nitrogen. Figure 2-26 is a table of relative sensitivities for some of the common gases encountered in vacuum systems. It shows that the response to helium is much less than that for air. This property makes it possible to use helium as a search gas when looking for leaks using an ionization gauge. The sensitivities in Fig. 2-26 are for one specific gauge and may be slightly different for different gauges or different conditions.

Gas	Sensitivity at 15-ma emission (ma per torr)
Helium	14
Hydrogen	16*
Oxygen	85*
Air (dry)	100
Nitrogen	110
Carbon monoxide	112
Carbon dioxide	120*
Argon	162

*The figures for hydrogen, oxygen, and carbon dioxide are influenced by the gettering action of the gauge tube.

FIG. 2-26. Change of sensitivity with gas. (From A. Guthrie, *Vacuum Technology*, 1963; courtesy John Wiley and Sons, Inc.)

Limitations of Ionization Gauges. Naturally, an ionization gauge measures the pressure in the gauge itself. If this pressure is different from the system pressure, incorrect conclusions may be drawn. In general, wide, short tubulations are best for connecting the gauge tube to the system in order to minimize this problem. Ionization gauges are commercially available with a tubulation diameter of 1.5 inches, and some are even mounted on a flange such that the gauge elements actually protrude into the system. This latter type of gauge is called a "nude" ion gauge. If the gauge acts as neither a source of gas nor a pump, the pressure in the gauge will come to equilibrium with the rest of the system. If the gauge is at the same temperature as the system, a true pressure reading results. If not, gauge readings may differ from the true pressure due to thermal transpiration.*

The pressure in the gauge may also differ from that in the system for other reasons. The gauge may act as a source of gas or (if it is completely clean) as a pump. The evolution of gas is due to the hot filament. When the filament is turned on, temperatures inside the gauge will increase, causing increased evolution of gas from all surfaces. Most ionization gauge controls have arrangements for "degassing" the gauge while under vacuum so that most of the adsorbed gas will be desorbed. This can be accomplished by several means, the most common being simply to pass a current through the grid (which in this case would be in the form of a helix) until it glows red-hot from I^2R heating. The filament is heated by its own

*Dushman [1] discusses the correction required due to thermal transpiration.

current, and the collector and gauge envelope are heated by radiation from the grid and filament. After a few minutes of this heating, the gauge should be clean enough for practical purposes. It is also possible to outgas the gauge by torching it; this is a hazardous procedure, however, and is definitely not recommended for inexperienced operators. Other gauge control units have arrangements for electron bombardment of the grid and the collector, which may be necessary if the grid is a mesh structure instead of a helix.

Although the gauge is usually a source of gas, under some conditions it can actually be a pump. Several mechanisms have been proposed for this pumping action, and probably all contribute at times. The pumping action, however, is extremely dependent on the state of cleanliness of the gauge and its component parts, as well as the type of gas to which it is subjected. Bayard-Alpert gauges, when clean, may have a pumping speed for nitrogen as high as 1.0 liter per second. The pressure difference between the gauge and the system will be given by

$$p_s - p_g = \frac{S_g p_g}{C}$$

where p_s is the pressure in the system, p_g the pressure in the gauge, S_g the pumping speed of the gauge for the gas present, and C is the conductance of the tube connecting them. This is one more reason for having a large-diameter tube between the gauge and the system. If the diameter of the tube is 2 centimeters and the length is 5 centimeters, the conductance is 20 liters per second [see Eq. (2-19)]. If S is 1.0 liter per second, this effect will produce a 5-percent error in the pressure reading.

The upper pressure limit for most ionization gauges is about 10^{-3} torr, although the linearity may be questionable above 10^{-4} torr. Above these pressures, the ion density is so great around the collector that a space charge region is established and a glow discharge can be set up. The collector current is then no longer proportional to the pressure. Instead, the collector current levels off and becomes almost independent of the pressure. It can even reach a maximum and eventually decrease, causing the reading to be the same as for a lower pressure. This double-valued response can be very misleading unless there is some other reliable indication of the true pressure. Recently, a gauge has been introduced (by General Electric Company) which operates at a much lower emission current (10^{-5} amperes) in the high-pressure range. This decreases the ionization in the gauge and, with a suitable electronic control, this gauge is claimed to have a linear response up to 10^{-2} torr and to be usable up to 10^{-1} torr. This extended range makes this gauge very useful for measuring sputtering pressures. Another ionization gauge has been described which may be used up to 1.0 torr, and this is probably the greatest pressure that can be measured by an ionization gauge. This and many other special-purpose ionization gauges are described by Dushman [1].

As with any type of hot-filament triode, these gauges can fail due to filament burnout. Tungsten filaments are quite susceptible to burnout, especially if they are subjected to pressures above their operating range when hot, particularly in the presence of water vapor or oxygen. To avoid this problem, it is possible to specify "nonburnout" filaments. These are usually made of iridium metal coated with thoria, and can be safely exposed to air at operating temperatures for short periods of time. These filaments have the disadvantage that their emission characteristics can be readily altered ("poisoned") by cracked diffusion pump oil. It may thus become impossible to maintain the desired emission current.

Cold-Cathode Ionization Gauges

In addition to thermionic emission, electrons may also be produced by the ionic bombardment of a cathode. This principle is used in the cold-cathode gauge, also known as a Penning or Phillips gauge, shown schematically in Fig. 2-27. This gauge consists of two parallel plate cathodes with a ring-shaped anode between them. The principle is the same as in the sputter-ion pump. The electrons are generated at the cathode and are accelerated toward the anode. They pass through the anode ring and are then repelled by the other cathode, so they oscillate back and forth, eventually winding up at the anode ring. In the meantime, however, they have had a chance to collide with enough molecules to produce a certain number of ions. The ions impinge upon the cathode, releasing electrons which maintain the discharge. As in the sputter-ion pump, the effective path length of the electrons is drastically increased by a magnetic field more-or-less parallel to the electrons' oscillations. The resultant electron trajectories are helical and the probability of ionization is greatly increased, allowing the discharge to be maintained down to much lower pressures. The power supply for the cold-cathode gauge is a high-voltage (perhaps 2 kilovolts) d.c. supply with an ammeter and a limiting resistor in series. The current read on the ammeter is proportional to the pres-

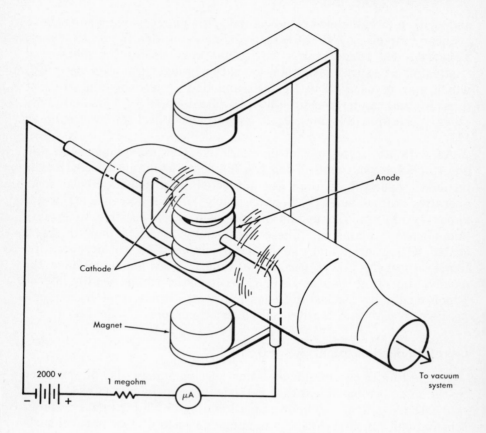

FIG. 2-27. Cold-cathode ionization gauge.

sure over the range from about 10^{-2} to 10^{-6} torr. In some cases, however, this has been extended to both higher and lower pressures. The calibration is dependent on the gas in the same way as the thermionic ionization gauges, and the sensitivity ratios are similar to those given in Fig. 2-26. The pumping speeds of these gauges are similar to those of the hot-cathode gauges (in the range of 0.1 to 1 liter per second), and the tubulation leading to the gauge should be of as great a conductance as practical.

Cold-cathode gauges are simple and rugged. They are often made with a metal enclosure, with the entire enclosure acting as the cathode. At low pressures, cold-cathode gauges can be difficult to start and erratic in operation. These gauges have found extensive use in applications where accuracy is not required, such as in protective circuits. They have advantages in that control circuits are simple, and the current is proportional to the pressure over several decades.

A modified form of cold-cathode gauge, known as a magnetron (or Redhead) gauge, has been constructed which is claimed to give accurate pressure readings down to the 10^{-11}-torr range and lower. This gauge operates on essentially the same principle, but has a guard ring around the edges of the cathode and a more completely enclosed active volume. It is finding application in many ultrahigh vacuum systems. All these cold-cathode gauges have sensitivities in the range of 1 to 10 amperes per torr. Since this is greater than the sensitivity of thermionic-emission ionization gauges, the pressure reading is not as dependent on special techniques for current measurement.

Gauge Locations

Figure 2-17 shows an ionization gauge connected to the baseplate. This is an accepted position for the measurement of the bell jar pressure. Another ionization gauge is shown just below the high vacuum valve. In general, this gauge will register a lower pressure than the one installed in the baseplate, and its reading should not be reported as the bell jar pressure. One reason for the lower pressure indication is the pressure drop across the high vacuum valve and the spoolpiece;* another reason is that this gauge may be "looking" directly at the cold surface of the cold trap. This causes an apparent decrease in pressure because condensables are removed. The main function of this gauge is for troubleshooting. For example, when the high vacuum valve is closed and the bell jar is being vented, this gauge will indicate the presence of a leak with a rate as small as 10^{-5} torr liters per second through the high vacuum valve. Also, if the ultimate pressure is higher than usual, closing the high vacuum valve and observing this gauge can help to determine whether the problem is above or below the high vacuum valve. At throughputs below 10^{-1} torr liters per second (in the illustrative case) this gauge can be used to measure the throughput or the conductance of the cold trap.

Every bell jar system should have at least two Pirani or thermocouple gauges, as shown in Fig. 2-17. One is necessary at the diffusion pump outlet to indicate whether the tolerable forepressure has been

---

*This is a term commonly used for the pipe connecting the main vacuum valve to the baseplate.

exceeded, and the other is used to indicate when to valve in the diffusion pump after roughing down the bell jar. The latter gauge can be connected on either side of the roughing valve but, if the gauge is on the high vacuum side, it can also be used to measure the pressure rise in the bell jar (when the bell jar is isolated for long periods) or to monitor pressure during sputtering. Probably the best arrangement is to have such a gauge on both sides of the roughing valve. A Bourdon gauge is also desirable to indicate when the bell jar has reached atmospheric pressure upon venting. Any of these gauges can be used to operate safety mechanisms for protection of personnel and equipment.

2.7   VACUUM VALVES

Of the various components that make up a vacuum system, vacuum valves constitute an important category. An ideal vacuum valve would have zero conductance when closed and very high conductance when open. It would have no outgassing materials, and no pressure bursts on opening or closing. Above all, it would be reliable for many closings without servicing or replacing parts. For particular applications, other properties such as bakability, fine control of conductance, adaptability to automatic operation, and ability to seal against atmospheric pressure differential in either direction may be desirable. Some of the most common types of vacuum valves, grouped according to their function in a vacuum system, are considered briefly.

High Vacuum Valves

In a diffusion pump system such as that shown in Fig. 2-17, there is usually a valve between the diffusion pump and the bell jar so that the bell jar can be vented to atmosphere without venting the diffusion pump. This valve, sometimes called the main vacuum valve, is probably the most important valve in such a system. Since the conductance of this valve is immediately reflected in the effective pumping speed at the baseplate (speed of exhaust), the conductance must be maximized; this means the valve opening should be at least as large as the diffusion pump throat. The valve should be leak-tight against an atmosphere of pressure differential in either direction, as there may be times when it is necessary to open the diffusion pump to air while the bell jar is under vacuum (this would probably occur only during initial checkout or leak detection). This valve may also be used to throttle the diffusion pump, in which case fine control of the conductance is necessary. This is especially true for dynamic sputtering systems. The valve should also have no additional outgassing or leaks during opening or closing.

The most popular and best all-around high vacuum valve is the gate valve, one style of which is shown in Fig. 2-28. In this valve, the seal is effected by means of a plate which has an o-ring in a groove near its perimeter. To close the valve, the plate is moved horizontally until the opening is almost closed. At this point, the track curves upward, directing the plate against the rim of the orifice to which it seals.

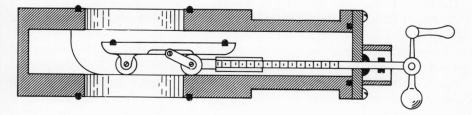

FIG. 2-28. High vacuum valve (partially open).

Because of the general design of the gate valve, it can be made quite thin; a typical case is a 7-inch diameter valve with a thickness of only 3.5 inches. This results in a conductance of 7500 liters per second, which is large enough that it does not significantly affect the exhaust speed of a 1500-liter-per-second (6-inch) diffusion pump.

The leak rate through these valves in the closed position can be made less than that detectable by a common helium mass spectrometer leak detector (that is, less than 10^{-10} torr liters per second with atmospheric pressure differential, corresponding to a conductance of the order of 10^{-13} liters per second). The force due to air pressure on a 6-inch diameter valve sealing against atmospheric pressure is over 400 pounds, so it makes quite a difference which way the valve is installed. Normally, the valve will be installed to seal on the bottom rim so that the pressure differential will assist in making the seal tight. If, however, the valve is able to seal reliably in the inverted position (as shown in Fig. 2-28), there is an advantage to installing it that way in a sputtering system. The advantage is that during sputtering, the interior parts of the valve (which may include screw threads, springs, and other moving parts) are not directly exposed to the glow discharge (except through the valve seat, which may be throttled until it is almost closed). Thus, any lubricant or diffusion pump oil which may be in the valve body is protected from decomposition by the plasma.

These valves are often designed to be fast-operating by using quarter-swing-manual, pneumatic, or hydraulic actuation. For some applications, however, the fine-control manual operation of a screw thread is preferable. One such case is a dynamic sputtering system, where a certain pressure and throughput may be required. A variable leak (of the sputtering gas) can be used to raise both the throughput and bell jar pressure proportionally, and the main vacuum valve conductance can be used to adjust the ratio between the throughput and the bell jar pressure. In this situation, fine control of the valve is often crucial in obtaining the desired operating parameters.

The outgassing properties of gate valves depend on the materials used for construction. Stainless steel is preferred for the metal parts, although aluminum is satisfactory for most applications. The body is sometimes cast as a single piece, except for a vertical plate at the end where the shaft goes in. Another body design (sometimes called the clam shell) consists of two halves joined at the edges with a horizontal o-ring seal. The former design has the advantage that the interior parts may be removed for cleaning or replacing without unbolting the valve itself. The other type (the clam shell) can be made thinner and both types can be formed from sheet metal, which is preferable to casting.

The seal which permits the motion to be transmitted into the vacuum system is a critical part of the high vacuum valve. One approach is to use two o-rings on a stem, with the area between them roughed out to pressures of the order of a few microns. Then, if the inner o-ring leaks, the pressure differential across it will be small so that only a small flow of gas will be introduced into the valve body. Another approach is to pack the area between the two o-rings with a high-grade vacuum grease. Unfortunately, the distances travelled make it impractical to use a bellows or a diaphragm to transmit the motion through the vacuum wall.

Foreline and Roughing Valves

Valves which are to be used in the foreline have approximately the same requirements as the high vacuum valve, although they are not so critical. For example, the opening need not be so large. Recall that conductances in the viscous flow range are much greater than in the molecular flow range (Fig. 2-3). Also, there is usually no advantage to having fine control of the conductance of a foreline valve. The most common valve for this purpose is a disk valve, of the type shown in Fig. 2-29. In this valve, the motion of the sealing disk is perpendicular to the disk. The seal is made by means of an o-ring or flat gasket. The amount of travel is small enough that a bellows can be used for a stem seal. From the point of view of vacuum reliability, it is almost always an advantage to have a permanent seal like a bellows, instead of an o-ring sliding or rotating seal. These valves are reliable; they seal tight (as checked by the helium leak detector), and their conductance is not much less than that of a pipe of equivalent diameter and path length (that is, the effective cross section of flow is comparable to that of the pipe to which it is attached). Disk valves are basically right-angle valves, but a slight modification allows them to be in-line with only a small offset. They are made with minor modifications by some thirty or forty different manufacturers, and are available in sizes from one-eighth of an inch to at least four inches. In addition to their use as foreline valves, they are frequently used for venting valves, or shutoff valves in gas-handling systems. Care must be taken, however, not to exceed the pressure rating of the bellows if they are used for pressures much greater than one atmosphere.

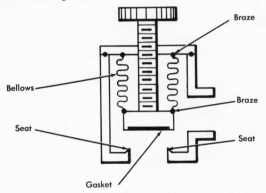

FIG. 2-29. Small disk valve with bellows.

Ultrahigh Vacuum Valves

Valves that are to be used at pressures below about 10^{-8} torr may be classified as ultrahigh vacuum valves. They have essentially the same requirements as other vacuum valves, but the need for reducing outgassing becomes critical. These valves should be bakeable to about

400°C, which implies that no elastomers such as neoprene or viton o-rings may be used. In fact, only metals (a very few metals) are acceptable in the construction of these valves. The motion must be transmitted to the inside by a bellows or metal diaphragm, permanent seals must be welded or brazed (using special techniques), and removable seals must be made with metal gaskets, usually of high-purity copper or gold. The valve seal itself is usually a knife edge of a hard metal, such as monel, which presses against a seat of a softer metal, such as silver. Many valves have been described for this pressure range, but few have become commercially available. One model, however, is available which is quite similar to the disk valve shown in Fig. 2-29. The main differences are that in the ultrahigh vacuum valve the bellows is permanently welded to the body, and the valve seal is made by compressing a very thin film of copper between a knife edge and a smooth, flat surface. Valves of this type are available (Granville-Phillips Co., Boulder, Colorado) in 4- and 6-inch sizes, with wide-open conductances of 350 and 650 liters per second. Valves of the same general type are also available with openings from 0.5 inch to 4 inches and wide-open conductances from 4 to 220 liters per second.

Another popular ultrahigh vacuum valve has been designed for small-diameter tubing, and a schematic diagram is shown in Fig. 2-30. The motion is transmitted by means of a metal diaphragm, and the seal is made by forcing a silver gasket into a groove machined in the monel body. This valve has a conductance of 1 liter per second when open, and about 10^{-14} liters per second when fully closed.

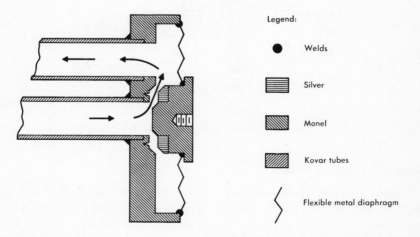

Legend:

● Welds

▦ Silver

▨ Monel

▨ Kovar tubes

〈 Flexible metal diaphragm

FIG. 2-30. Bakeable valve for ultrahigh vacuum. (Courtesy Granville-Phillips Company.)

Leak Valves

A leak valve is used to admit a controlled flow of gas into a line. The leak (throughput) through a valve is given by

$$Q = C(p_2 - p_1)$$

where C is the valve conductance, and p_2 and p_1 are the pressures on either side of the valve. Leak valves are usually rated in terms of atmospheric throughput (standard cc per sec at a pressure differential of one atmosphere). The conductance in liters per second is 10^{-3} times the leak rate in standard cc per sec at one atmosphere. It must be remembered, however, that for viscous flow the conductance can be a function of pressure; thus, if the input pressure is something other than one atmosphere, the conductance may be different from that specified at one atmosphere. If the leak acts more-or-less like an orifice, the conductance will be independent of pressure; if it acts like a tube, the conductance will be proportional to the pressure (see Sec. 2.2).

One type of leak valve is the needle valve (Fig. 2-31). The seal is made between a long needle and a conical surface that is machined to mate it. The needle is moved in and out of the seating surface by a fine-pitched screw, and the conductance can be precisely controlled. The needle is usually made of a harder material than the seat, so that it effectively makes its own seat upon closing. If too much force is applied, however, the seat can be damaged. Typically, fine-control needle valves can be adjusted to a desired conductance with a precision of about 10^{-4} liters per second, but there are some needle valves that are claimed to be adjustable "with certainty" to a

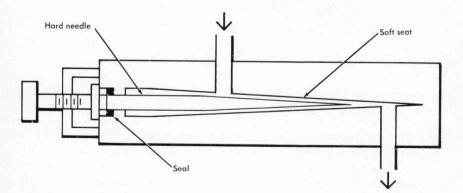

Hard needle   Soft seat

Seal

FIG. 2-31. Needle valve. (From A. Guthrie, *Vacuum Technology*, 1963; courtesy John Wiley and Sons, Inc.)

precision of 10^{-8} liters per second (Bendix-Balzers Vacuum, Inc., Rochester, New York).

Another approach to making a leak valve is to use a tapered groove in a rod which is passed through a sealing ring. At least one such model, constructed like a disk valve, is available (Vactronic Laboratory Equipment, Inc., East Northport, New York); it is shown schematically in Fig. 2-32. There is an o-ring in the disk so that it will have a positive close-off, just as in a disk valve. As the valve is opened, the conductance is limited by a second circular gasket of Kel-F, through which passes the rod containing the tapered groove. The seal length is 2.3 millimeters, and the effective cross section can be varied from zero to 1.5×10^{-4} square centimeters by withdrawing the stem. For one groove in the rod, the conductance is variable from zero to about 2×10^{-3} liters per second when one end is at atmosphere and the other is at vacuum. Thus, at atmospheric pressure differential, the throughput can be varied up to about 1.5 torr liters per second. Comparison with Fig. 2-18 reveals that this is about the maximum throughput for a 6-inch diffusion pump. This range is therefore convenient for such purposes as high throughput sputtering (higher conductance models can be made simply by adding more grooves). The curve of conductance versus number of turns open is roughly linear, and the maximum opening occurs at 25 turns; thus, the conductance can be controlled to a few times 10^{-6} liters per second, or 0.2 percent of wide-open conductance.

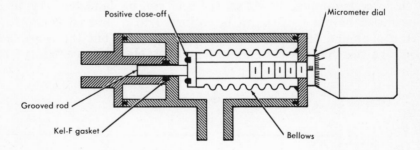

FIG. 2-32. Variable leak valve. Except at the groove, the rod seals to the gasket.

Another approach to the design of leak valves is to pass the gas through a flattened metal tube and then mechanically constrict the tube. In one case (Andonian Associates, Inc., Waltham, Mass.), the flattened tube is simply twisted by a mechanical device which twists the tube about one-quarter of a turn for a rotation of 20 turns on a micrometer dial. This valve can be made so that the elastic limits of the flattened tube are never exceeded, and the conductance is then repeatable for a given micrometer setting. The curve of conductance versus number of turns is roughly semilogarithmic, with the conductance changing by one decade for two revolutions of the dial. When fully "closed," the conductance is 10^{-10} liters per second, and the flow rate is 10^{-7} standard cc per sec with an atmosphere of pressure differential. The conductance when fully open may be about 10^{-3} liters per second, giving a flow rate of about 1.0 standard cc per sec. These valves are useful for the lower ranges of conductance values required, for example, in controlling dopant gases in a sputtering system. Because they do not have a positive shutoff, however, it is normally necessary to install another valve in series for that purpose. These valves have the distinct advantage that no moving seals or elastometer seals are involved, and there are no moving parts inside the vacuum. Only metal is exposed to the vacuum, and the valve may be baked out if desired.

Valves similar to the one shown in Fig. 2-30 are sometimes used as leak valves. A large mechanical advantage is gained either hydraulically or with a gear train so that it takes twenty or more revolutions of a knob to open the valve to its full conductance. The wide-open conductance may be from 10^{-2} to 1.0 liters per second, and control is possible over seven to ten decades of conductance below this. Accessory controls are available for these valves which permit automatic control of flow or pressure. These valves are bakeable and thus can be made very clean. In the low-conductance ranges, however, the conductance is not reproducible from the dial setting.

2.8 VACUUM MATERIALS AND TECHNIQUES

Besides structural integrity, certain material properties are of special importance in the design and construction of a vacuum system and associated internal hardware. Among these properties are vapor pressure, outgassing characteristics, and permeability to gases.

The vapor pressure is the pressure caused by the evaporating (or subliming) atoms of a material in equilibrium with its vapor. Thus, if a completely outgassed material is placed in a perfectly evacuated and sealed chamber and there is no pumping, the pressure will rise and approach a pressure equal to the vapor pressure of the material. Under normal conditions in a vacuum system, the materials used will have vapor pressures low enough that their contribution to the gas load will be negligible compared to the leak rate, outgassing rate, and permeation rate. In general, the vapor pressure of a material is very temperature dependent. An increase of 50 percent in the absolute temperature can produce an increase in the vapor pressure of from two to seven decades. The subject of vapor pressures will be discussed more fully in Chap. 3, which contains a figure showing the vapor pressures of some of the more common elements and materials as a function of temperature. In general, it is possible to find appropriate materials with sufficiently low vapor pressures for any thin film vacuum application, provided that the operating temperatures required are not excessive. Stainless steels (especially the 300 series, which are characterized by being nonferromagnetic), copper, and glass are satisfactory at room temperature and slightly elevated temperatures, while ceramics and refractory metals such as tantalum, tungsten, and molybdenum are recommended for incandescent temperatures. These refractory metals cannot be exposed to air at incandescent temperatures because they oxidize rapidly. In cases where high temperature exposure to the air may occur, platinum or platinum-rhodium alloys may be employed.

The outgassing of materials in a vacuum is a very complicated subject, and the data available are often inconsistent and difficult to apply. Several comments of practical interest, however, can be made. The outgassing of a surface depends on the surface finish, the history of the surface, and the temperature.

A fine machined finish is much to be preferred to a cast surface, since any porosity of the surface increases the effective surface area and tends to trap gases. Any metal casting to be used in a vacuum system should be cast in vacuum in order to reduce the porosity and the amount of entrapped gas. Porosity is a prime consideration in the selection of ceramic materials. Generally, a high-density, nonporous, and low-outgassing ceramic would be selected for vacuum use. For many special purposes involving such factors as high strength, high thermal shock resistance, or high electrical resistivity, a high-alumina-content ceramic with a high density is preferred.

Another important factor affecting outgassing is the presence of any coating of oil, grease, rust, or any other substance that can contribute to the gas load either by evaporation of the substance proper, or by desorption of gases that have been adsorbed upon exposure to air. This points up the importance of scrupulously cleaning all parts that are to be placed into the vacuum environment and of handling them with care to avoid recontamination. It also shows the advantages of using an absolute minimum of sealing greases and lubricants, and of selecting metals that do not rust or tarnish easily.

Increasing the temperature always increases the outgassing rate. If a material is baked out in vacuum, it will desorb gas at an accelerated rate, and its surface layers will become depleted in the gases being evolved. Following this, when the surface is cooled down, the outgassing rate will be considerably lower than it would have been without the bakeout.

Permeation is the process of gas atoms or molecules passing into, through, and out the other side of a solid wall. Permeation depends on the gas, the solid material, the pressure, and the temperature. In a glass vacuum system, or one with a glass bell jar, there may be a problem of helium diffusion through the walls, but this will normally become serious only in the very high vacuum range and only when the glass walls are heated. Metal walls can have a measurable rate of permeation of hydrogen, but in a vacuum chamber the walls are sufficiently thick that permeation through them is usually of no consequence, even at bakeout temperatures of 400°C. Most of the gas permeation into a typical vacuum system occurs through the elastomer seals; this permeation can be a definite limitation. Proper design of o-ring grooves is important, the object being to minimize the area of elastomer that is exposed to the vacuum. Some vacuum systems are built on the principle of double-pumping, in which the seals are made with two o-rings, with the space between being pumped out to fore-vacuum. Although this technique reduces the permeation and leak rates by several orders of magnitude, it introduces maintenance and troubleshooting difficulties. It is probably better to make one seal which may be tested and make it well, than to have two seals that may or may not be reliable but cannot readily be tested.

Elastomers

Demountable joints in thin film vacuum systems are usually made with o-ring seals made of elastomeric materials. An elastomer is a material, such as rubber, that can withstand large strains and still retain enough elasticity to recover, at least partially, when the stress is removed. Elastomers are used because of their ability to seal against the microroughness that is found in machined or drawn surfaces. Figure 2-33 contains a list of several of the elastomers most commonly used in vacuum systems. All of the properties listed are to be taken as indications of average or typical properties, which

Elastomer type	Natural rubber	Isobutylene isoprene rubber	Nitrile butadiene rubber	Chloroprene rubber	Silicone rubber	Fluorinated elastomers	Styrene butadiene rubber
Common names	Crude rubber	Butyl or GR-1	Buna N	Neoprene or GR-M	Silastic	Viton Kel-F Fluorel or Teflon	Buna S or GR-S
Durometer hardness	30-100	40-75	20-100	40-95	45-60	55-90	40-100
Low temperature range of rapid stiffening	$-45°C$ to $-30°C$	$-30°C$ to $-20°C$	$-30°C$ to $0°C$	$-30°C$ to $-10°C$	$-85°C$ to $-50°C$	$-35°C$ to $-5°C$	$-45°C$ to $-20°C$
Resistance to gas permeability	Good	Excellent	Good	Very good	Fair	Extremely good	Good
Maximum operating temperature	$75°C$	$110°C$	$140°C$	$125°C$	$250°C$	$220°C$	——

Fig. 2-33. Properties of elastomers. (After A. Guthrie, *Vacuum Technology*, 1963; courtesy John Wiley and Sons, Inc.)

may vary from batch to batch and from manufacturer to manufacturer.

The durometer range is a measure of the hardness or stiffness, with higher numbers indicating harder material. A durometer value of about 65 is ideal for vacuum seal applications. At low temperatures, all of these elastomers lose their elasticity and become brittle. Figure 2-33 lists the approximate temperature range at which the stiffness increases rapidly with decreasing temperature. All these materials will operate satisfactorily at room temperatures. For low temperature service, however, silicone rubber (sometimes called silastic) is exceptionally good.

All elastomers will eventually take a permanent "set" if they are stressed for long periods. This is especially pronounced at elevated temperatures; because of this, o-rings should be replaced regularly, as mentioned in Sec. 2.5. Because of this "cold-flow," o-rings for vacuum service are usually installed in grooves that contain their entire volume. This is especially important for teflon, silicone rubbers, and, to a lesser extent, viton.

Permeability to gases is one of the most important parameters of elastomers, because permeability may be a real limitation on the ultimate pressure in a vacuum system. For optimum results, the permeability factor limits the choice to butyl, neoprene, or fluorinated elastomers. Permeability can be greatly reduced by cooling the o-rings to about $-20°C$. In one case, a vessel containing six flanges with butyl o-rings was operated in the low 10^{-11}-torr range. This was made possible by cooling the o-rings.

Elastomers can be damaged by operation at high temperatures; therefore, the maximum operating temperature of the material should be considered if nearby elements are to be heated. Viton is recommended for use at temperatures up to about 200°C. If the bakeout temperature exceeds 200°C, however, even viton o-rings should be protected by water cooling. Viton has an additional advantage in that it can be preconditioned by baking for a few hours at about 100°C in air, which greatly reduces its initial outgassing.

A very light coat of a low-vapor-pressure grease on an o-ring can help to make a vacuum-tight seal. One such grease is apiezon L, which has a vapor pressure of less than 10^{-10} torr at 20°C. Other greases are available for higher temperatures. It is preferable, however, to make a seal without grease if possible, because each time the seal is exposed to atmosphere the grease will trap a certain amount of gas that will be evolved during the next pumpdown.

O-Ring Seals

Several typical o-ring seals are shown schematically in Fig. 2-34. The o-ring for a vacuum joint normally has a circular cross section before compression. The groove for a flange seal should be designed so that after compression, the cross section of the o-ring will be a rectangle with an aspect ratio (depth-to-width) of 0.65. Thus, the primary seals will be on the facing plate and on the bottom of the groove. The volume of the groove should be slightly greater than the volume of the uncompressed o-ring, but only by enough to ensure that the tolerance limits will not permit the o-ring volume to exceed the groove volume. Using this principle, the depth of the groove should be about 75 ± 5 percent of the actual diameter, w, of the o-ring cross section. The actual value of w is usually somewhat larger than the nominal value. The groove width should be about 115 ± 5 percent of w. The inner diameter of the groove should be 100.5 ± 0.5 percent of the nominal o-ring inner diameter, with the added condition that it

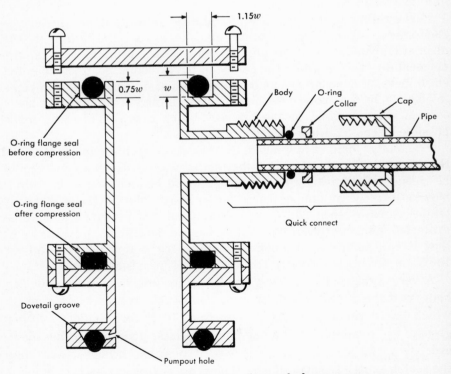

Fig. 2-34. Some examples of o-ring seals for vacuum use.

should never exceed the nominal o-ring inner diameter by more than 0.060 inch. These recommendations for dimensions are not the same as those recommended for pressure applications.

The o-ring at the bottom of the bell jar is a common source of difficulty and should receive special attention. If there is not enough squeeze allowed on the o-ring, there may be metal-to-metal contact between the bell jar flange and the baseplate. This could result in the baseplate being scratched, which could later cause a leak if the bell jar were lowered in a slightly different position. The bell jar o-ring groove may be slightly dove-tailed, instead of being a true rectangle, in order to keep the o-ring from falling out when the bell jar is raised; however, additional gas may be trapped in the corners and, if the seal against the inner diameter of the groove is not perfect, there may be a serious virtual leak problem. (This problem also exists to a lesser extent in a rectangular groove.) When this situation occurs, it is sometimes helpful to drill or saw holes into the corners from the inside of the vacuum system in order to facilitate pumping out this volume (see Fig. 2-34). Since the main seal is against the bottom face of the groove, this does not compromise the integrity of the seal, provided the bottom face of the groove is not marred or scratched. It is recommended that pumpout holes be specified when ordering or fabricating new vacuum equipment. They should be made on existing equipment only after it has been definitely established that this problem exists.

A wide variety of o-ring seals for various purposes is available commercially. One very useful seal is called a quick-connect, and is shown in Fig. 2-34. This is a method of sealing a pipe or gauge tube into the vacuum system. The o-ring is compressed by the collar and the cap, so that it seals against the pipe as well as the body (the male threaded piece). If the pipe surface is fairly free from defects, quick-connects can make reliable seals that are tight, as checked by a helium mass spectrometer leak detector.

O-ring seals are also used in the construction of feedthroughs, which permit entry for services into the vacuum system, usually through the baseplate. These feedthroughs permit entry of wiring for high current, for high voltage, and for thermocouple and other measurements and also permit entry of pipes for gases or cooling water. In ordering a baseplate, it should be decided in advance what services may be required, so that the necessary holes can be made for the appropriate feedthroughs since it is difficult to machine the holes later. Fortunately, "universal" feedthroughs are available.

These can be adapted to a wide variety of services for entry into the vacuum system, and it is strongly recommended that there be a few of these versatile feedthroughs planned for every vacuum system.

Metal Joining Techniques

Permanent joints between metals for vacuum use are usually made by soldering, brazing, or welding. Soft soldering should be avoided whenever possible. Soft solder can become brittle from work-hardening caused, for example, by vibrations. It will also cold-flow under continuous stress. The result may be a crack in the joint, which will produce a leak. In some cases, soldered joints may be necessary (for example, when the area cannot be heated to the temperatures required for brazing or welding). When soldering is necessary, the solder should contain no zinc, cadmium, or phosphorus, because these materials have high vapor pressures at relatively low temperatures (see Fig. 3-5).

Brazing is metal joining using a nonferrous filler material that melts at relatively high temperatures (above 400°C). In general, brazing materials have lower vapor pressures and greater strength than soft solders, and are much better for vacuum use. Alloys for brazing are made of materials such as silver, copper, indium, nickel, silicon, tin, and chromium, which have satisfactory vapor pressures at temperatures up to at least 500°C. One type of brazing that has been found especially suitable for vacuum use is called nicrobrazing. In this method, a powder of an alloy is positioned at the place where the fillet is desired and the entire assembly is heated in a hydrogen atmosphere until the alloy melts and fuses with the metal parts to be joined. A filler alloy with a melting temperature only slightly lower than that of the metals to be joined is selected. It is similar to nichrome but is rich in silicon (or boron) to depress the melting temperature. As the alloy melts, the silicon diffuses into the parts to be joined and the melting temperature of the alloy is raised above the oven temperature, so the alloy solidifies even before the oven is cooled down.

Stainless steel pieces for vacuum use are also readily joined by inert gas arc welding. The inert gas reduces the oxidation of the metal during welding and eliminates the need for a flux. When welding for vacuum use, smooth fillets are required to minimize the porosity and the surface area exposed to the vacuum. If the fillet is smooth and free of oxide and slag, it is preferable that the weld be from the vacuum side of the joint, so that no crevices are exposed to the vacuum. It is undesirable to make double welds with a trapped volume in between, since this makes it difficult to check the joint for leaks.

REFERENCES

1. Dushman, S. *Scientific Foundations of Vacuum Technique*, edited by J. M. Lafferty. New York: John Wiley and Sons, Inc., second ed., 1962.
2. Guthrie, A. *Vacuum Technology*. New York: John Wiley and Sons, Inc., 1963.
3. Steinherz, H. A. *Handbook of High Vacuum Engineering*. New York: Reinhold Publishing Corporation, 1963.
4. Knudsen, M. *The Kinetic Theory of Gases*. London: Methuen and Company, Ltd., 1934.
5. Stevenson, D. L. "A New Type of Boiler That Permits Improvements in the Performance of Oil Diffusion Pumps," *Trans. 6th Nat. Vac. Symposium, 1959*. New York: Pergamon Press, 1960. pp. 134-139.
6. Hablanian, M. H., and H. A. Steinherz. "Testing Performance of Diffusion Pumps," *Trans. 8th Nat. Vac. Symposium, 1961*. New York: Pergamon Press, 1962. pp. 333-339.

PRINCIPAL SYMBOLS FOR CHAPTER 2

Symbol	Definition
A	Area of an aperture
C	Conductance
d	Diameter of a pipe or aperture
d_0	Equivalent molecular diameter
E	Kinetic energy
h	Height of a mercury column
k	Boltzmann's constant
l	Length of a pipe or duct
m	Mass of a molecule
M	Molecular weight
n	Number of molecules/volume
n_M	Number of moles of gas
N	Total number of molecules
N_A	Avogadro's number
p	Pressure
q	Quantity of gas
Q	Gas flow or throughput
R_o	Gas constant
S	Pumping speed
S_E	Exhaust speed

Symbol	Definition
t	Time
T	Temperature
v	Speed
$\mathbf{v}$	Velocity
V	Volume
Z	Molecular arrival rate
α	Correction factor for the intermediate range of gas flow between molecular and viscous
β	Correction for conditions other than air at 25°C

Chapter 3
Evaporated Films

Evaporated films result from heating a material in a vacuum enclosure to such a temperature that large numbers of atoms or molecules leave the surface of the material and deposit on a substrate. Because of the low background pressure in the vacuum system, most evaporated molecules suffer no collisions with residual gas molecules and travel in straight lines to the substrate. The evaporation process, as discussed in this chapter, consists of the complete cycle of vaporization, transit, and condensation of the material on a substrate.

3.1 THE ENVIRONMENT FOR EVAPORATION OF FILMS

Because early workers in the field of vacuum evaporation did not appreciate the contaminating effects of residual gases, relatively crude vacua were considered sufficient. More recent research has revealed, however, that small amounts of active residual gases can cause contamination and can change significantly the properties of the evaporated deposit. A considerable amount of research effort has been devoted to reducing the amount of contaminants and controlling their effects on the evaporated film. Thus, as mentioned in Chap. 1, developments in evaporation technique were necessarily preceded by improvement in vacuum equipment.

An evaporation system is shown schematically in Fig. 3-1. The essential parts are: the vacuum enclosure, which provides a suitable environment for the evaporation process; a vapor source, where the material to be evaporated can be held and heated; the substrates, which are to be coated with evaporant; and a substrate holder and heater.

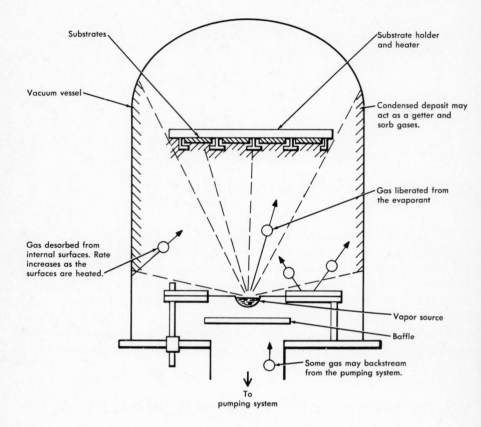

FIG. 3-1. Schematic of an evaporation system.

The number, n, of gas molecules per unit volume as a function of gas temperature, T (°K), and pressure, p, is

$$n = \frac{N}{V} = \frac{p}{kT}$$

where N is the total number of gas molecules in volume, V, and k is Boltzmann's constant. For air at room temperature and atmospheric pressure, $n = 2.5 \times 10^{19}$ molecules per cubic centimeter. Even at pressures corresponding to high vacuum, many gas molecules are present. For example, at 10^{-6} torr, $n = 3.2 \times 10^{10}$ molecules per cubic centimeter. Although this number is large, all but about $1/10^{12}$ of the volume is free space. These gas molecules move rapidly about the vacuum enclosure at an average speed of about 1040 miles per hour at room temperature.

As the molecules move, they collide with each other and with the walls and surfaces inside the container. The mean free path, L, of the gas molecules is inversely proportional to the pressure, p (torr),

$$L = \frac{5 \times 10^{-3}}{p} \qquad cm$$

At 10^{-3} torr, $L = 5$ centimeters, while at 10^{-6} torr, $L = 5000$ centimeters. Thus, at a pressure of 10^{-6} torr a molecule would travel about 165 feet between collisions with other molecules. At pressures common for evaporation, the mean free path is much greater than the source-to-substrate distance and the evaporant travels in line-of-sight paths.

Some fraction, N_1/N_0, of the initial number of emitted molecules, N_0, will collide at distances less than a given path length, l,

$$\frac{N_1}{N_0} = 1 - e^{-l/L} \qquad (3\text{-}1)$$

Figure 3-2 is a plot of the percentage of molecules colliding enroute versus the ratio of the actual path to the mean free path [1]. When the mean free path is equal to the source-to-substrate distance, 63 percent

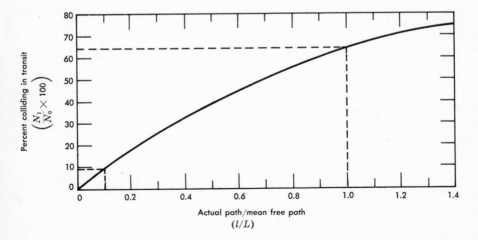

FIG. 3-2. Percent of molecules undergoing collision in transit versus ratio of actual path to mean free path. (After W. L. Bond, "Notes on Solution of Problems in Odd Job Vapor Coating;" courtesy *Journal of the Optical Society of America*.)

of the molecules undergo collisions; when the mean free path is 10 times the source-to-substrate distance, only 9 percent of the molecules undergo collisions. Thus, the mean free path must be considerably greater than the source-to-substrate distance in order to prevent collisions in transit. Common evaporation pressures are in the range of 10^{-4} to 10^{-7} torr, and source-to-substrate distances are of the order of 30 cm. Thus, except at the extreme upper end of this pressure range, only a very small fraction of the evaporant molecules enroute to the substrate undergo collisions with residual gas molecules.

Residual gas molecules are present in a leak-tight system primarily because of desorption from the surfaces inside the vacuum system, liberation from the evaporation source, and backstreaming from the pumps (negligible in a well designed system). Figure 3-3 gives approximate values for the ratio, N_s/N, of the number of molecules adsorbed in a monomolecular layer on the chamber walls to the number of molecules in the gas phase. Note that in a system of common size at a pressure of 10^{-6} torr, the number of molecules in a single layer on the inside surfaces greatly exceeds the number in the gas phase. Aside from gases released from the source during actual evaporation, adsorbed molecules which are desorbed are the main source of gas in a clean, tight, well designed vacuum system at pressures below about 10^{-6} torr. The rate of desorption of these molecules is reduced by the gettering action of the evaporated film, although this may be offset if the temperature of the surfaces inside the system rises sufficiently to increase the desorption rate.

Pressure (torr)	$N_s/N = n_s A/nV^*$
760	$2.0 \times 10^{-5} A/V$
1	$1.5 \times 10^{-2} A/V$
1×10^{-3}	$1.5 A/V$
1×10^{-6}	$1.5 \times 10^4 A/V$

*A is the surface area (cm^2) inside the vacuum system, V is the volume of the vacuum system (cm^3), and n_s is the number of molecules/cm^2 in a monomolecular layer.

FIG. 3-3. Ratio of the number of molecules adsorbed in a monomolecular layer to the number in the gas phase.

Residual gases bombard all surfaces inside the vacuum, including the growing film and the hot vapor source. The arrival rate, Z, of residual gas molecules (of molecular weight M) is given by

$$Z = 3.51 \times 10^{22} \, p \left(\frac{1}{TM}\right)^{1/2} \qquad \text{molecules/cm}^2 \text{ sec}$$

For air at room temperature and 10^{-6} torr, a sufficient number of molecules to form a monolayer will arrive at a surface in only 2.2 seconds. The fraction of incident molecules which stick (sticking coefficient) varies between 0 and 1, depending on the temperature and nature of the surface as well as the nature of the gas. In the production of films, then, the arrival rate of evaporant atoms at the substrate should be large with respect to the residual gas arrival rate if a high degree of purity is desired. This is especially true for active metals, since the sticking coefficient approaches one for fresh surfaces of these metals.

The deposition conditions can be further complicated by reactions between the residual gas and the evaporated film or the vapor source. For example, water vapor is a major constituent of the residual gases in most systems. This water vapor may react with a metal film to form an oxide and release hydrogen, or it may react with a hot source (e.g., tungsten) to form hydrogen and an oxide.

Even in a high vacuum evaporator, a dynamic situation exists where gas is constantly being released and either pumped away or readsorbed. High pumping speeds at low gas pressures permit fast operating cycles and insure rapid removal of any gases evolved during evaporation. Low residual pressures allow the evaporated molecules to travel to the substrate with no collisions and greatly reduce the molecular bombardment of the growing film.

3.2 THE VAPORIZATION PROCESS

As seen in Fig. 3-1, the evaporation system provides a vapor source, which holds the material to be evaporated. The evaporant is heated and the vaporization process is effected at this source.

The heat of vaporization is the energy which must be added to remove the atoms from the bulk substance. After atoms leave the bulk material they possess only kinetic energy, which on the average is $(3/2)kT$ per atom. For common source temperatures, $(3/2)kT$ is about 0.2 ev (electron volt) per atom or 4.6 kilocalories per mole (1 ev per atom $=$ 23.05 kcal per mole) and this amounts to only a small percentage of the total heat of vaporization. Thus, the heat of vaporization is mainly the energy required to overcome the attraction between atoms in the solid (or liquid).

Heats of vaporization decrease slowly with increasing temperature, and the decrease which occurs between room temperature and common evaporation temperatures normally does not exceed a few percent. Equations relating heats of vaporization to temperature are given in Fig. 3-4 for several common metals.

Material	Heat of vaporization* (cal/mole)
Al	$67,580 - 0.20T - 1.61 \times 10^{-3} T^2$
Cr	$89,440 + 0.20T - 1.48 \times 10^{-3} T^2$
Cu	$80,070 - 2.53T$
Au	$88,280 - 2.00T$
Ni	$95,820 - 2.84T$
W	$202,900 - 0.68T - 0.433 \times 10^{-3} T^2$

*All transformations are liquid-gas except Cr, which is solid-gas; T is in °K.

FIG. 3-4. Heats of vaporization for several common metals [2].

The pressure exerted by a vapor in equilibrium with its solid or liquid form is known as the vapor pressure, p_v, of the substance. The vapor pressure may be calculated thermodynamically from the Clapeyron-Clausius equation,

$$\frac{dp_v}{dT} = \frac{\Delta H_v}{T(v_g - v_l)} \tag{3-2}$$

where ΔH_v is the heat of vaporization, and v_g and v_l are the molar volumes of the gas and liquid phases, respectively. Since $v_g >> v_l$ and the vapor obeys the ideal gas law at low pressures, we may let $v_g - v_l \cong v_g = R_oT/p_v$. Then Eq. (3-2) becomes

$$\frac{dp_v}{dT} = \frac{\Delta H_v}{R_oT^2/p_v}$$

or

$$\frac{dp_v}{p_v} = \frac{\Delta H_v dT}{R_oT^2} \tag{3-3}$$

which can be rewritten as

$$\frac{d(\ln p_v)}{d(1/T)} = \frac{-\Delta H_v}{R_o}$$

If the natural logarithm of p_v is plotted against $1/T$, the heat of vaporization is given by the product of the slope and $-R_o$.

The heat of vaporization, ΔH_v, normally varies only slightly with temperature and may be assumed to be a constant. Thus, integration of Eq. (3-3) gives

$$\ln p_v = A' - \frac{\Delta H_v}{R_o T} \qquad (3\text{-}4)$$

where A' is an integration constant. Equation (3-4) is often written as

$$\log p_v = A_1 - \frac{B}{T} \qquad (3\text{-}5)$$

where $B = \Delta H_v/2.3\, R_o$. Equation (3-5) is a fairly accurate representation of the temperature dependence of the vapor pressure of most materials at temperatures for which the vapor pressure is less than about one torr.

Figure 3-5, which gives vapor pressure/temperature curves for several common metals, illustrates that vapor pressures increase rapidly with temperature. Vapor pressure data have been compiled for most common materials [3].

The equilibrium between a substance and its vapor is dynamic, with molecules continually evaporating from and returning to the surface of the substance. The rate at which molecules arrive at any surface was given in Chap. 2 as a function of gas pressure; thus, the arrival rate of molecules at the surface of a substance from its vapor is given by

$$Z = 3.51 \times 10^{22} \, p_v \left(\frac{1}{TM}\right)^{1/2} \qquad \text{molecules/cm}^2 \text{ sec} \qquad (3\text{-}6)$$

If all arriving molecules stick, then, at equilibrium, this is also the rate of evaporation. If the evaporating molecules are immediately removed such that none return to the substance, the net rate of evaporation, G, is a maximum and is given by

$$G = 0.058 \, p_v \left(\frac{M}{T}\right)^{1/2} \qquad \text{grams/cm}^2 \text{ sec} \qquad (3\text{-}7)$$

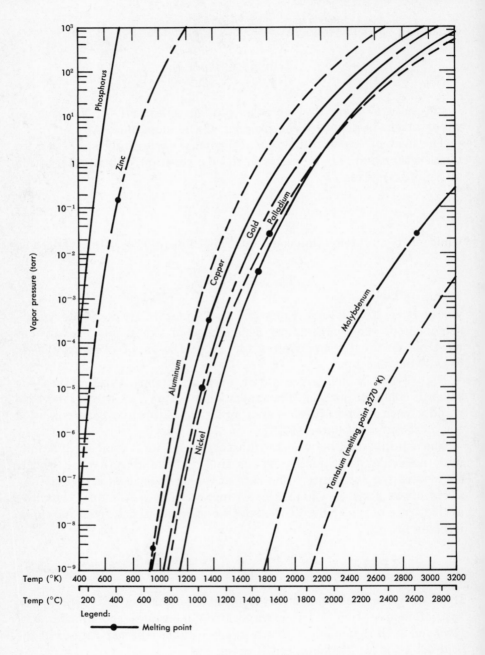

Fig. 3-5. Vapor pressures of various metals as a function of temperature.

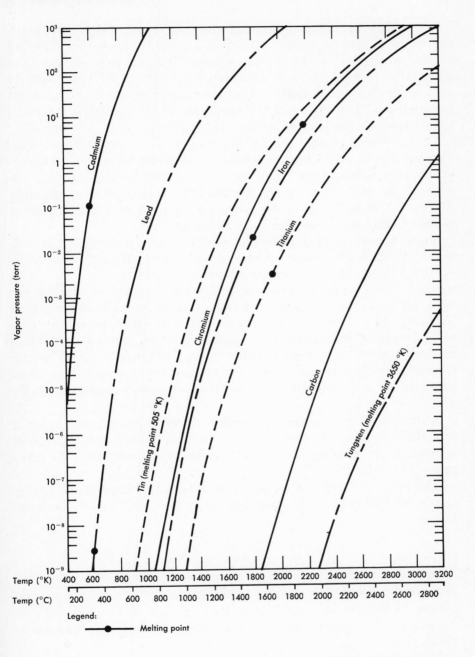

FIG. 3-5. Vapor pressures of various metals as a function of temperature (continued).(After R. E. Honig, "Vapor Pressure Data for the Solid
and Liquid Elements;" courtesy *RCA Review*.)

With a low background pressure (such as in vacuum evaporation), the vapor molecules escape without collisions if the source does not exceed an area of a few square centimeters and if p_v is less than about 10^{-2} torr. At high evaporation rates, the vapor beam may be sufficiently dense that collisions between vapor molecules occur, causing undesirable diffusion of the evaporated atoms and the return of some atoms to the surface of the evaporant.

Contaminants present on the evaporant lower the evaporation rate, particularly oxides, which can form a relatively impervious skin on evaporated metals. Oxide films are not a barrier to evaporation, however, if the oxide is more volatile than the evaporant (for example, SiO on Si), if the oxide decomposes when heated, or if the evaporant is able to diffuse readily through the oxide.

Vapor pressures of materials at high temperature are usually estimated from measurements of the evaporation rate of the material or the rate at which its vapor passes through a small orifice. One method consists of determining the surface area and weight loss of a piece of the material maintained at a constant temperature. Equation (3-7) is then used to calculate the vapor pressure. In a second method, the material is placed in a closed vapor source which is then heated to a known temperature; the vapor which passes through an orifice in the source is condensed and weighed. The vapor pressure of the material may then be calculated, again using Eq. (3-7).

Atoms leave a substance regardless of its temperature; in order to achieve reasonable rates, however, vacuum evaporation is typically carried out by heating the material to be evaporated until its vapor pressure is about 10^{-2} torr.

The rate of change of evaporation rate with temperature change can be estimated by combining Eqs. (3-5) and (3-7). Differentiation yields

$$\frac{dG}{G} = \left(2.3 \frac{B}{T} - \frac{1}{2} \right) \frac{dT}{T}$$

For metals, $2.3 \, B/T$ is usually in the range of 20 to 30, so a small change in temperature produces a large change in evaporation rate. Thus, accurate temperature control is necessary to control evaporation rates.

Heat should be supplied to the evaporant in such a way that large temperature gradients are avoided. The material to be evaporated must be in good thermal contact with the heater, or excessive heater temperatures may occur, causing erratic evaporation. Poor thermal contact is usually a problem only with materials which sublime, where the area of evaporant touching the heater is small. This problem may be avoided for metals which sublime by using filaments made of or plated with the material to be evaporated.

A splattering of evaporant sometimes results from rapid expansion of gas trapped in the evaporant. This trapped gas is released slowly from solid evaporants, even at temperatures near the evaporation temperature. When the evaporant melts, however, the gas may expand freely and cause splattering. To permit "smooth" evaporation, the evaporant must be extensively outgassed at temperatures as close as practical to the evaporation temperatures.

Local heater hot spots can cause vapor bubbles in liquid evaporants or violent evaporation in solid evaporants. Solid evaporants may shift about, bringing material in contact with a previously bare part of the heater which is very hot; rapid evaporation results at these points. This causes the evaporant to jump around and may even result in particles of evaporant being ejected. In practice, the maximum evaporation rate is limited by the rate that heat can be supplied to the charge, while still avoiding bubbles or rapid gas evolution which cause splattering of the evaporant.

3.3 GROWTH, STRUCTURE, AND ADHESION OF EVAPORATED FILMS

The vaporization and transit of evaporant atoms in the vacuum environment have been considered in the two previous sections. The evaporation process is completed by condensing evaporant atoms to form a film on the substrate. The initial condensation of the evaporated material involves adsorption of the evaporated molecules on the surface of a substrate, so the condition of that surface (including any adsorbed contaminants) is important in film growth. The nature of this initial deposit, in turn, has a strong influence on subsequent growth. The mechanism of growth and the structure of the deposited films have been investigated in recent years with the electron microscope. Since vacuum evaporation can be carried out inside this instrument, the film can be observed at any stage.

Since the substrate provides the structural support for the evaporated film, the film-to-substrate adherence must be sufficient to ensure that subsequent handling, processing, testing, or environmental conditions do not strip the film from the substrate. Adherence is closely related to the cleanliness of the substrate upon which the evaporated film is deposited. There are, however, no satisfactory nondestructive methods available to measure either substrate cleanliness or film adherence, so confidence must be placed in a cleaning procedure which has been verified to provide good adherence.

Adsorption Phenomena

At pressures common for evaporation, atoms travel from the source to the substrate without collisions or loss of energy in transit. When these incident atoms approach to within several atomic diameters of the substrate, they enter the force field of the substrate surface. At this point, several types of interaction are possible:

1. Reflection, in which the atom retains most of its kinetic energy and has only a short "dwell time" (on the order of a picosecond) on the substrate.
2. Physical adsorption, which involves only van der Waal's forces, which are present whenever any two molecules are brought close together. No potential energy barrier must be overcome in physical adsorption, so the heat of desorption is the same as the heat of adsorption (usually less than 10 kcal/mole).
3. Chemisorption, in which the binding forces involved are the same type as those involved in normal chemical bonds; hence, heats of chemisorption are greater than heats of physical adsorption. In extreme cases, the heat of chemisorption may be greater than 100 kcal/mole. A potential barrier must often be overcome before chemisorption can take place, and in these cases an activation energy is involved. Chemisorption is specific to certain material combinations, and ceases when the adsorbate can no longer make direct contact with the adsorbent surface.
4. Association with an atom or atoms of the same species already adsorbed on the surface.

After an atom has either been adsorbed or has joined another evaporant atom on the substrate surface, the atom may undergo additional processes which involve energy changes such as desorption, migration over the surface, or a change from physical adsorption to chemical adsorption.

Equations relating the rate of desorption to the heat of desorption usually contain an exponential term of the form $e^{-\Delta H_D/R_o T}$, where ΔH_D is the heat of desorption. Thus, desorption is more probable for weakly bound atoms (small ΔH_D). It should be noted that it may be impossible to distinguish between reflection and desorption when the adsorption-desorption process occurs in a time shorter than the measurement time.

The energy required to cause an adsorbed atom to migrate over the surface of the substrate is less than that required to remove the atom from the substrate. For physically adsorbed atoms, the activation energy of migration is typically of the order of one-fourth of the desorption energy. Adsorbed atoms often have sufficient energy to allow considerable surface migration, particularly at elevated substrate temperatures. This lateral migration is readily apparent when isolated islands of evaporant are observed during growth; when these islands touch, they exhibit a liquid-like mobility.

The term *sticking coefficient* describes the fraction of impinging atoms which "stick." For evaporated deposits, the sticking coefficient is normally defined as the fraction of incident atoms which enter strongly bound states, so that their dwell-time on the substrate is long. The "stuck" atoms are usually those atoms which have become associated with one or more like atoms, or those which have undergone chemisorption or chemical reaction with the substrate.

The energy change, ΔE_A, upon adsorption depends mainly on the strength of the adsorption bonds formed, and to a lesser extent on the temperature difference between the source and substrate. Thus, ΔE_A is equal to the evaporant-substrate binding energy plus $(3/2)(R_0)(T_{source} - T_{substrate})$. For atoms condensing on atoms of like material, $\Delta E_A \cong \Delta H_v$, since $(3/2)(R_0)(T_{source} - T_{substrate})$ is small in comparison with the binding energy.

The system of cadmium (Cd) on glass illustrates the phenomena discussed above. For a given incidence rate, a critical substrate temperature is observed, above which no deposit is formed. This is explained as follows: The heat of adsorption of Cd on glass is small ($\cong 4$ kcal/mole). Thus, at high substrate temperatures impinging atoms are either reflected or desorbed so that the number of atoms leaving the surface is equal to the number arriving, and no net buildup of deposit occurs. Condensation of a permanent deposit is not possible even though the evaporation rate of bulk Cd (at the substrate temperature) is much smaller than the incidence rate of Cd onto the substrate. This emphasizes that the initial binding is by adsorption of individual atoms, and that the film-to-substrate adherence depends on the strength of this adsorption bond.

If the incidence rate of Cd is increased so that appreciable numbers of evaporant atoms may join at the substrate before they can be desorbed, a permanent deposit builds up. The heat of desorption increases as the number of associated atoms increases; when the entire surface is covered, the heat of desorption reaches the value of bulk Cd (27 kcal/mole).

Nucleation and Initial Growth

Much research on the mechanism of thin film growth has been done with evaporated films, primarily because evaporation inside an electron microscope is easily accomplished. Because many aspects of growth observed in evaporated films are considered common to other thin film deposition techniques (e.g., sputtering, electrodeposition, and chemical deposition), the growth of evaporated films is discussed in detail. The growth and structure of single crystal films, however, is beyond the scope of this treatment.

Films with average thicknesses from a few angstroms to around 1000 A may be studied using transmission electron microscopy.* The upper limit is reached when the films become too thick to allow the necessary transmission of electrons.

Single crystal substrates have been used in studies of the initial stages of film growth because: (1) the surface of the substrate is well defined, (2) some of these substrates can be prepared thin enough to permit transmission microscopy with the film in place, and (3) a fresh, uncontaminated surface can often be prepared by cleaving in vacuum. Only a limited number of substrate materials can be prepared in single crystal form, however, and surfaces without cleavage steps (which promote nucleation on preferred sites) are difficult to obtain. The aspects of growth described in the following discussion appear similar for films deposited on single crystal, polycrystalline, or amorphous substrates.

Extensive information on initial growth has been published by D. W. Pashley [4, 5] and his co-workers and, because their work in this field is consistent with the work of others, their results will be considered as typical. Pashley's experimental conditions were:

1. Evaporation and examination of the films were conducted inside an electron microscope with a residual pressure of 10^{-5} to 10^{-6} torr.

*Transmission electron microscopy is required to obtain the resolution necessary for these studies.

2. All substrates were heated to at least 350°C because at lower substrate temperatures the electron beam caused crosslinked polymer films to form, indicating the presence of adsorbed hydrocarbons. While these polymer films were avoided in this manner, the hydrocarbons were present at the time of film deposition.

3. The substrates were very thin sheets of single crystal molybdenum disulfide (MoS_2), and transmission electron micrographs were taken without stripping the evaporated film from the substrates.

4. Deposition rates were below 10 A per second; higher rates were not used since this might have led to overheating the very thin substrates.

5. The evaporants were either gold or silver. Since only noble metals were investigated, the evaporant-substrate binding energies were small, allowing the forces between evaporated atoms to predominate.

6. A cine camera was used to record the rapid changes in structure during growth. (Some of these changes took place in less than 0.1 second.)

In the earliest observable stages of growth, a large number of discrete three-dimensional nuclei are seen. For gold evaporated on rock salt at 300°C, Pashley reports about 10^{11} nuclei per square centimeter, with a distance of about 20 A across a nucleus. Thus, the mean separation between nuclei is around 300 A. The nuclei are ordinarily distributed at random, but the nucleation is occasionally observed at preferred sites. The electron micrograph in Fig. 3-6 illustrates the preferential nucleation of gold on cleavage steps on single crystal rock salt substrates; Fig. 3-6 also includes a model of the substrate.

Continued deposition leads to an increase in the size of the nuclei without an appreciable increase in number. This suggests that the incident atoms move freely on the surface and join existing nuclei, or that atoms which strike other than a nucleation site are reflected or desorbed. When the nuclei have grown to the point that they touch, coalescence begins and rapid changes in geometry and orientation occur. Figure 3-7 illustrates the coalescing sequence of two nuclei. Shutting off the incident vapor stream stops the coalescing phenomena, but nuclei which have already started to coalesce continue their fusion.

Figure 3-8 illustrates reorientation (recrystallization) of an island (nucleus); this recrystallization is important in determining the

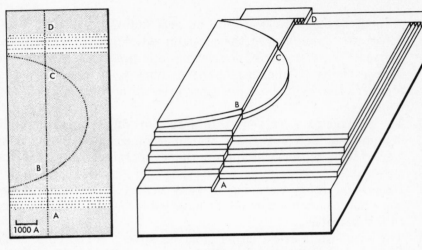

(a) Electron micro-   (b) Model of the substrate
graph of the substrate

FIG. 3-6. Preferential nucleation of gold evaporated (to an average thickness
of 5 A) onto a cleaved rock salt substrate. (After D. J. Alner, *Aspects
of Adhesion*, I, 1963; courtesy University of London Press, Ltd. and
The Chemical Rubber Co.)

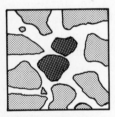  

(a) Arbitrary zero   (b) 0.06 second   (c) 0.19 second

  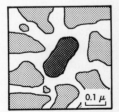

(d) 0.5 second   (e) 1.06 seconds   (f) 6.19 seconds

FIG. 3-7. Coalescing sequence of two gold islands on molybdenite (MoS_2) during
deposition at 400°C. The micrographs were exposed relative to (a),
the arbitrary zero, at the times indicated in (b) through (f) [5].
(After Pashley and Stowell, "Nucleation and Growth of Thin Films
as Observed in the Electron Microscope;" courtesy *Journal of Vacuum
Science and Technology*.)

final film structure. The amount of recrystallization and the crystallographic orientations which disappear depend partly on the relative size of the nuclei involved. Larger nuclei generally "consume" smaller nuclei.

As the nuclei continue to join, a network consisting of a film containing a maze of irregularly shaped openings is formed. A continuous film emerges as the average thickness further increases. The incident evaporant atoms now strike like atoms, the binding energy is high, and little reflection or desorption is observed.

The larger the number of initial nuclei, the sooner a continuous film is formed. Researchers have determined the average thickness required to form a continuous film by measuring the resistance between two terminals on the substrate during evaporation. According to Holland [6], aluminum films on glass are electrically conductive at an average thickness of only 9 A, while silver films are not conductive until the film thickness is 50 A. Bennett [7] found that gold on bismuth oxide (Bi_2O_3) became conductive at a thickness of 40 A, while gold on amorphous alumina (Al_2O_3 formed by anodization) did

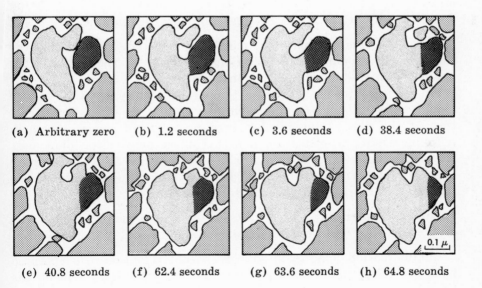

(a) Arbitrary zero   (b) 1.2 seconds   (c) 3.6 seconds   (d) 38.4 seconds

(e) 40.8 seconds   (f) 62.4 seconds   (g) 63.6 seconds   (h) 64.8 seconds

FIG. 3-8. Recrystallization sequence which occurs as a result of the growing together of gold islands on molybdenite at 400°C. The two orientations are revealed by the dark-light contrast. The micrographs were exposed relative to (a), the arbitrary zero, at the times indicated in (b) through (h) [5]. (After Pashley and Stowell, "Nucleation and Growth of Thin Films as Observed in the Electron Microscope;" courtesy *Journal of Vacuum Science and Technology*.)

not become conductive until the thickness was 70 A. The gold was harder to remove from the bismuth oxide substrate, giving qualitative confirmation to the higher binding energy for gold on bismuth oxide than gold on alumina.

Factors Affecting Film Growth and Film Properties

The growth and structure of evaporated films are affected by factors such as the nature of the substrate and any contamination present, as well as the parameters of the deposition process. The most important of these parameters and factors, excluding electrical and magnetic properties, will be discussed.

Substrate. An increase in the binding forces between the substrate and the evaporant atoms usually decreases the surface mobility of the evaporant atoms, increases the number of initial nuclei, and increases film adhesion. These factors affect coalescence and the size and structure of islands during initial growth. Mismatches in the coefficients of expansion between the substrate and the evaporated film produce stresses in the film when the temperature differs from that during deposition. Substrates are discussed in detail in Chap. 9.

Substrate Temperature. The most important effect of heating the substrate in vacuum is the additional cleaning of the surface that results from the desorption of contaminants.

An increase in the substrate temperature increases the surface mobility of the condensed atoms and may change the crystalline structure of the film. Interdiffusion between adjacent layers increases with increasing temperature, and this effect may cause alloy films that have been fractionated during evaporation to become more homogeneous. The probability of desorption of evaporant atoms also increases with increasing substrate temperature.

Source Temperature. The effects of the source temperature on the evaporated film will be considered in Secs. 3.4 and 3.5.

Contamination. The main sources of contamination in evaporated films are contaminants brought into the the vacuum system on the substrate, residual contaminant gases or vapors in the vacuum system, gases released from the evaporant, and the heater material itself (see Sec. 3.4). One catastrophic result of contamination is the loss of adequate adherence of the evaporated films; substrate contamination is particularly troublesome in this respect.

In spite of rigorous cleaning, the substrate brings some form of a surface layer of contaminants into the vacuum system. Since cleaning procedures performed outside the vacuum system are unable to eliminate the last monomolecular layer of adsorbed gases, the intent of these procedures is to remove everything but a controlled layer of adsorbed material, which may be removed in the vacuum system. Ideally, this layer would be completely removed by a short period of vacuum outgassing at room temperature. The rate of desorption of most contaminants, including adsorbed gases, is so low at room temperature, however, that heating of the substrates to about 100° to 400°C is necessary to effectively remove the contaminants. Some contaminants, particularly those found in fingerprints, are not completely removed by heating and may become "baked on." This "baking on" of the contaminant may be the result of chemical reactions with the substrate, or decomposition into products with low vapor pressures.

Methods for reducing the amount of contamination by residual gases have been discussed in Chap. 2. The gases released from the evaporant are usually permanent gases incorporated during refining and processing. These gases are normally released rapidly at or near evaporation temperatures, and in a typical case can cause a decade of pressure rise in the vacuum system. Metals may contain on the order of 5 percent gas by volume (at 1 atmosphere and 25°C), mainly CO, N_2, CO_2, H_2, and in some cases O_2. The principal method for outgassing these metals is the preliminary evaporation of a small portion of the charge onto a shutter prior to collecting the deposit on the substrates.

Matthews [8] observed that contamination of a clean substrate by exposure to air produced a larger number of initial nuclei. Minute amounts of any contamination may also influence nucleation, subsequent film growth, and structure. Specific interactions, however, are difficult to study because of the difficulty in detecting and identifying the small amounts of contaminants. The effort justified to reduce the amount of contaminants can be determined only after careful study of the effects of possible contaminants on the desired characteristics of the evaporated film. This study can only be made by first reducing contamination to a minimum and then purposely introducing specific contaminants into the system.

Evaporation Rate and Film Thickness. Since the rate of bombardment by residual gas is constant (for constant residual pressure), higher deposition rates produce films containing smaller amounts of residual gas. When an active metal is being deposited, the residual gas content of the film may be thickness dependent, since the film is gettering during deposition; the last layers deposited thus contain less residual gas than the first layers. Residual gas content can be minimized by depositing on a shutter for a relatively long time before starting to collect the deposit.

The intrinsic stress (stress arising from the growth process) is often a function of thickness and sometimes of deposition rate. A qualitative indication of the nonuniformity of the stress with thickness is obtained from the observation that films generally curl when detached from the substrate.

Angle of Incidence of the Evaporant. The angle of incidence, which is the angle between the normal to the surface and trajectory of the arriving atom, can affect the structure of the evaporated film. Shadow-casting techniques, employed in preparation of specimens for electron microscopy, use a large angle of incidence to produce shadows of projections.

A coarse texture develops more readily as the angle of incidence increases. As illustrated in Fig. 3-9, the incident vapor atoms are

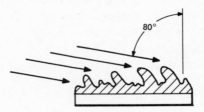

FIG. 3-9. Schematic diagram of an electron photomicrograph showing how the more elevated grains, which grow at the expense of their neighbors, tend to tilt toward the vapor beam. (Reproduced with permission from L. Holland, *Vacuum Deposition of Thin Films*, 1963, Chapman and Hall, London.)

shielded from depositing on the far side of peaks, which grow in the direction of the source. For some material, this change in texture affects the stress in the film. Figure 3-10 shows "stress lines" in films of ZnS [9]. The substrate of Fig. 3-10(a) was kept stationary during deposition. Extending the straight lines brings them to coincidence at a point just above the center of the source. The substrate of Fig. 3-10(b) was rotating about its normal during evaporation from the same source used to produce Fig. 3-10(a).

Electric Field. A d.c. electric field in the plane of the substrate

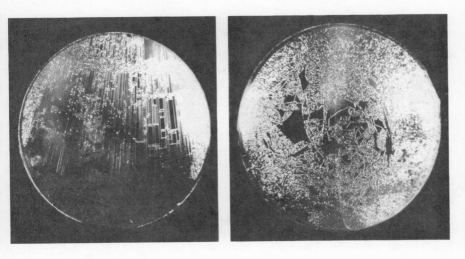

(a)   (b)

FIG. 3-10. Stress lines in films of ZnS.

surface can induce coalescence at an early stage of film growth, and the film thus becomes electrically conductive at a small average thickness [10]. Reorientation and recrystallization of the nuclei are also affected, which in turn affect certain film properties such as resistivity.

Adhesion of the Film

Little quantitative research on the adhesion of films to their substrates has been done because no satisfactory quantitative test method exists. Most research reported is concerned with the adhesion of evaporated metal films on glass substrates.

Adhesion of the evaporated film depends on the strength of the bond between the evaporant and the top layer of material on the substrate. The surface upon which condensation occurs may be the substrate proper, an adsorbed layer or layers of contamination, or another film previously deposited. Most bonds formed at the surface of a glass substrate are probably physical adsorption bonds, although the increase of bond strength with time [11] suggests that activated chemisorption, diffusion, chemical reaction, or some combination of these processes may be occurring.

Numerous observations have been made that the more chemically active metals adhere more tightly to glass or ceramic substrates than the less active ones. These active metals may react with adsorbed gases, such as water vapor or oxygen, to form an oxide which has good adherence to the substrate. This reaction is probably responsible for the good adherence of active metals sometimes obtained without substrate heating.

Regardless of the mechanisms of adhesion involved, there is general agreement that for maximum adhesion and good reproducibility, the condensed film should be immediately adjacent to the substrate, with no contaminating layer interposed. On rough substrates, such as unglazed ceramic, adhesion of the evaporated film is improved by mechanical interlocking in the pores and valleys. Binding energies also may be higher in pores than on a flat surface. The fundamental problem, however, remains one of proper surface cleaning to produce good adhesion.

Two qualitative methods which give an indication of film adherence are the Scotch® tape test, and the attachment of a lead (by soldering welding, etc.) to which a force is applied.

In the first method, a piece of pressure-sensitive adhesive (Scotch® tape) is firmly pressed onto the film and then peeled off rapidly. This test is more severe when the tape crosses a film-substrate step, where removal of the tape is more likely to initiate peeling of the film from the substrate. Rigorous cleaning of the film surface may be required after the tape is removed. Although test results depend on the "quality" of the adhesive, the pressure during application, and the speed and angle of pull, this is the only test which can be applied rapidly to a wide variety of films. In many cases, this test is not sufficient since it does not apply as much stress to the film-substrate bonds as functional leads do.

In the second method, the lead or leads used may either be special test leads or, in the case of devices, permanent functional leads. When a permanent lead is used, a predetermined force is applied to the lead as a "go, no-go" test. Though satisfactory test results indicate adequate adherence of the film (or films), care must be taken to prevent partial failure of any part of the structure during the test. Special test leads are usually tested to destruction, but this test measures film adhesion only at the point where the lead is attached. Results depend on joint geometry, joint-forming techniques, and the direction of the applied force. The lead bonding process, used to

® Registered trademark of Minnesota Mining and Manufacturing Company.

attach permanent or special test leads, may alter the adherence of the film. In many cases, film adhesion is such that failure occurs either in the lead itself or in the substrate.

Tests which tend to peel the film from the substrate usually produce failures at much lower forces than tests which pull the film normal to the substrate. This is particularly true when the stress is applied to the film edge.

A quantitative test of film adherence has been developed by Benjamin and Weaver [11]. A loaded point is drawn across the film on the substrate. The minimum load required to remove the film is determined and, from this, the force required to shear the film from the substrate is calculated. Benjamin and Weaver derived an expression relating the shear force to the normal force, assuming a spherical point. In practice, a point of 0.002-inch or less radius is required, and the point must be hard enough to plow a channel in the substrate itself. It is difficult, however, to obtain such a small spherical point which is sufficiently robust. Occasionally, the film is not cleanly removed from the channel and a judgment of what constitutes removal of the film must be made.

3.4   TYPES OF EVAPORATION SOURCES

Most materials to be evaporated require source temperatures in the range of 1000° to 2000°C, and some means of heating the materials to these temperatures must be provided. Perhaps the simplest technique utilizes resistance heating, where a simple filament or boat of tungsten often suffices. Many types and shapes of resistance heaters are available. The choice of a heater becomes more difficult as the evaporation temperature increases, since the heater must neither soften nor react appreciably with the molten charge. The development of electron bombardment heating has allowed evaporation of high-melting-point materials since the evaporant may be heated directly, and the holder may be water cooled. For specific applications, one of several special-purpose methods of heating might be used, including high-frequency induction, radiation, and levitation heating.

Resistance Heated Sources

Commercially available resistance heated sources (i.e., containers from which the evaporant is vaporized) range from the simple filaments described below to the complex sources used to sublime dielectrics. The requirements for resistance heated sources are:

1. They must supply the required heat at evaporation temperatures while still retaining their structural stability.
2. Their vapor pressure must be low enough that a negligible amount of source material is evaporated.
3. They must hold the charge to be evaporated.
4. They must not alloy excessively with or chemically combine with the charge, which is usually molten.

The common forms of resistance heated sources are helices and conical baskets made from wire, and boats of various shapes constructed from foil. The construction material is usually a refractory metal such as tungsten, tantalum, or molybdenum. The melting points and vapor pressure versus temperature curves of these metals are given in Fig. 3-5.

The usual final step in cleaning these sources is to heat them to high temperature in vacuum prior to use. Tungsten, when heated to evaporation temperatures, becomes brittle as a result of recrystallization. This recrystallization determines the maximum life of a tungsten heater, because it produces slip and preferential attack (and eventually fracture) along grain boundaries. After initial heating, tungsten must be handled carefully to avoid breakage. Tantalum usually does not become brittle; molybdenum may or may not, depending on its purity.

Excessive alloying between the source and the evaporant leads to redistribution of the dissolved heater material, which causes thinning and eventual failure. Because tungsten has some solubility in most common evaporants, a large ratio of tungsten to evaporant helps prolong heater life.

Tungsten reacts with water vapor to form a volatile oxide, WO_3. Thus, a continuous loss of heater material occurs when tungsten is heated in the presence of residual water vapor. Though the loss of heater material is small at low residual pressures, the contamination of the deposit may still be important.

The emission rate of radiant energy, E_s, from the surface of an evaporation source is given by

$$E_s = \epsilon \sigma T^4 \qquad \text{watts/m}^2$$

where σ is a constant equal to 5.67×10^{-8} watts/m² deg⁴, T is the temperature in °K, and ϵ is the emissivity of the surface. Emissivities range from 0 to 1, depending on the material and the nature of the surface, and vary with temperature; for example, for tungsten, $\epsilon(1000°C) \cong 0.15$, and $\epsilon(2000°C) \cong 0.3$. Other emissivities (approximate) are: Nichrome®, 0.9; aluminum, 0.1; copper, 0.2; gold, 0.05; titanium, 0.3; tantalum, 0.2.

Since the evaporation rate is a steeper function of temperature than T^4, the ratio of the amount of power consumed in vaporizing the evaporant to the amount of power radiated increases with increasing temperature. In other words, the evaporating efficiency increases with increasing temperature; however, the possibility of splattering of the evaporant and heater burnout must be kept in mind.

Filament Sources. Sources wound from wire are usually called filaments. Figures 3-11(a) and (b) depict typical filaments.

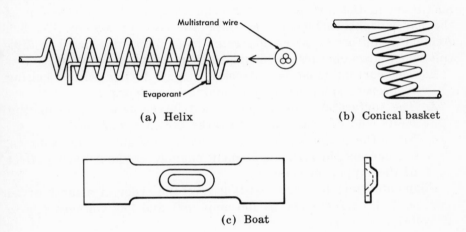

(a) Helix   (b) Conical basket

(c) Boat

FIG. 3-11. Common filament and foil evaporation sources.

The helix is used as a source for evaporants available in wire form. The charge should alloy slightly with the helix so that it "wets" the filament and does not fall off upon melting. Multistrand wire is commonly used because it acts like a wick and provides a larger surface area for evaporation. The evaporant should cover as much of the filament surface as possible to maximize the ratio of the amount of material evaporated to the amount of heat radiated (for a given filament temperature). If the filament is loaded too heavily, some of the evaporant may drop off when it melts. The maximum amount of molten evaporant which can be held on a helix is limited by the surface tension of the evaporant and the amount of alloying between the evaporant and the helix. The evaporant is loaded onto the helix by either hanging loops of wire onto the turns, inserting pieces of wire inside the turns, wrapping the turns with evaporant, or electroplating. Materials which sublime or are not available in wire form (e.g., chromium) can be evaporated from filaments using the latter technique.

The conical basket is the most common of the other varieties of wire sources and is often used for evaporants supplied in pellet or coarse granular form. It is not necessary that the evaporant wet the basket if the turns are spaced close enough that the molten material is held by surface tension.

Vapor emission from a helical source tends to be multidirectional. For a conical basket, the relative intensity of the emitted vapor is at maximum in the direction of the cone axis. Further discussion of the emissive characteristics of these sources appears in Sec. 3.6. Achieving uniform evaporation rates from filament sources is difficult when large charges are evaporated to completion because:

1. The current must be increased slowly during initial melting to prevent splattering or dropping of evaporant.
2. The surface area of the evaporant increases as the evaporant melts, and decreases as the charge becomes depleted.
3. Since the power supplies are usually constant-voltage sources, the current through the filament changes as the shunting effect of the evaporant changes.

Filaments typically have resistances on the order of a tenth of an ohm, so high currents (up to 400 amperes) and low voltages (up to 40 volts) are used.

Foil Sources. The refractory metals can be formed into many shapes of boats or troughs for use as sources. The main advantages of boats over filaments are that evaporants which do not wet the heater can be evaporated, and boats are more efficient since the vapor emission is limited to a hemisphere. The emissive characteristics of this type of source approach those of the plane source discussed in Sec. 3.6.

With the type of boat shown in Fig. 3-11(c), evaporation normally must be upward, as contrasted with other configurations which permit evaporation in various directions. If the evaporant appreciably wets the heater, it will creep around the heater and evaporate in all directions. Though they are more efficient in the use of charge, the power consumption of foil-type heaters is usually higher than that of filaments.

Refractory Metal-Oxide, Ceramic, and Carbon Crucibles. Nonconducting crucibles which react a minimum with the evaporant are also used as sources. Heat is generally supplied to the evaporant by a heater surrounding the crucible, or by direct heating of the evaporant by radiation or induction.

Common crucible materials are zirconia (ZrO_2), thoria (ThO_2), beryllia (BeO), magnesia (MgO), and alumina (Al_2O_3). Most of these materials can be used at temperatures up to about 2000°C. The limiting factor is their reactivity with the evaporant and their rate of vaporization.

A typical method of constructing a source of this type is to surround the crucible with a tungsten heater. The low thermal conductivity of the crucible and particularly the poor thermal contact generally present between the heater and the crucible may require excessively high heater temperatures in order to rapidly heat the evaporant. The crucible may also be fractured by thermal shock or it may react with the heater material.

Carbon may be used as a crucible material up to temperatures of about 2300°C and, since it will conduct electricity, may be heated directly by passing an electric current through it. Many metals react with carbon, however, and some compounds are reduced by carbon at evaporation temperatures.

Electron Bombarded Evaporation Sources

Evaporation sources heated by bombardment with high energy electrons have recently become commercially important. These sources usually incorporate a hot cathode to supply the electrons, a high voltage to accelerate the electrons, an anode (usually the evaporant), and perhaps a method of focusing the electrons. Potential advantages are:
1. The heat can be generated directly in the evaporant.
2. The holder for the evaporant may be cooled, eliminating interactions with the holder and unwanted evaporation of the holder material.
3. Refractory materials such as tantalum and tungsten can be evaporated.
4. The amount of heat radiated from the source is reduced since only the evaporant is heated.

Disadvantages are:
1. The equipment is more complicated, and modification by the user is more difficult than with resistance-type sources.

2. Most compounds are at least partially dissociated by the electron bombardment.

3. Residual gas molecules and part of the evaporant vapor may become ionized by the electrons.

Source Components. A hot tungsten cathode is commonly used as the source of electrons; an oxide-coated cathode is avoided since it is easily poisoned by residual gas ions.

The current drawn from the cathode is limited by the electron emissivity of the cathode, and the negative space charge produced by the stream of emitted electrons. Electron accelerating voltages usually range from about 5 to 10 kilovolts, with electron currents of the order of a few hundred milliamperes and power consumption of about 1 to 6 kilowatts.

The evaporant is often the only anode ("work-accelerated" sources). In other cases, electron acceleration is provided by a separate anode ("self-accelerated" sources). Self-accelerated sources minimize the problems, listed below, due to ionization of residual gas.

A crucible or boat may be used as the anode but, if heat is to be generated in the holder, resistance heating is usually simpler and less expensive. The advantages of using a crucible heated by electron bombardment are that the massive current leads necessary for resistance heating can be avoided, and that the electron beam can be directed onto the underside of the holder to avoid passing the evaporant through the electron stream.

The evaporant support is often water cooled to eliminate attack by the evaporant. Thermal contact between the evaporant and holder is usually sufficient to prevent melting at the interface, but poor enough to prevent a large loss of energy.

Some sources include electrodes or magnets to focus or bend the path of the electron beam. The amount of focusing (i.e., the maximum current density) that can be obtained in the electron beam is limited by mutual repulsion of the electrons. Also, if the focus is too sharp, holes may be drilled in the evaporant, altering the distribution of the evaporated atoms. The arrangement of the elements of a focused source should provide line-of-sight shielding between the cathode and the vapor source in order to prevent the formation of coatings on the cathode. Magnetic focusing is not affected by ionization of the residual gas or by the formation of insulating deposits (for example, from the evaporation of dielectrics), but both of these could affect electrostatic focusing.

Ionization of Residual Gas and Evaporant. An increase in pressure occurs at the beginning of evaporation because of the gas released from the evaporant. If the pressure is allowed to exceed about 10^{-4} torr, sufficient positive ions may be produced by the electron beam to:

1. Partially neutralize the negative space charge, causing uncontrollable fluctuations in the electron current.
2. Change the focus of the electron beam.
3. Change the distribution of the evaporant atoms leaving the source.
4. Produce erroneous readings on a deposition rate monitor.

When self-accelerated sources are used, the fraction of gas molecules released from the evaporant which is allowed to reach the cathode-anode region is small, and the number of ions produced in the accelerating region is small.

Glow Discharges. The possibility of a glow discharge exists if the background pressure rises because of desorption of gases, though the spacing between the cathode and anode is usually too small to allow a glow discharge in this region (see Chap. 4). If the cathode is grounded, however, the high potential also appears between the anode and any grounded surface (for example, the walls of a metal bell jar), which would allow a sustained glow discharge. For this reason, the anode is usually grounded, and the cathode and cathode heater supply are at high potential.

Examples of Sources. Figures 3-12(a) and (b) are examples of the melted-drop type of source. These sources are simple, but are limited to metals with fairly high surface tensions and relatively high vapor pressures at their melting points. In Fig. 3-12(b), a water-cooled anode permits the use of a larger molten droplet than is possible with a suspended droplet.

The source of Fig. 3-12(c) evaporates upward, so the molten drop does not have to support its own weight. Some focusing is obtained by the electron shields, and only a few evaporating atoms travel through the electron stream.

Figure 3-13 depicts three different focused-beam sources. Figure 3-13(a) shows a "self-accelerated" source. Spherical electrodes provide the proper field for focusing. Figures 3-13(b) and (c) show typical bent-beam sources. The source of Fig. 3-13(b) is electrostatically focused; the source of Fig. 3-13(c) is magnetically focused. The cathode is shielded from the evaporant in both cases, but many of the evaporated atoms must travel through the electron stream.

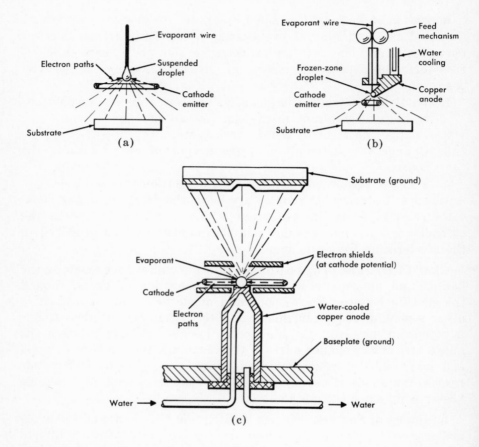

FIG. 3-12. Unfocused electron bombarded evaporation sources. ((a) After L. Holland, Ed., *Thin Film Microelectronics*, 1965; courtesy John Wiley & Sons, Inc. (b) After R. W. Berry, "Electron Bombardment Evaporation of Tantalum for Thin Film Components," in *Third Annual Electron Beam Symposium;* courtesy AGV Div., Nuclide Corp.)

Miscellaneous Sources

Several special-purpose methods of heating are available for specific applications. Three of these methods, namely, high-frequency induction, radiation, and levitation heating, are discussed in the following paragraphs.

High-Frequency Induction Heated Sources. Metals may be heated to evaporation temperatures by induction heating. A crucible containing the evaporant is held in the center of a helical coil (without touching it), and high-frequency current flowing in the coil induces current in the evaporant. The frequency required for efficient coupling

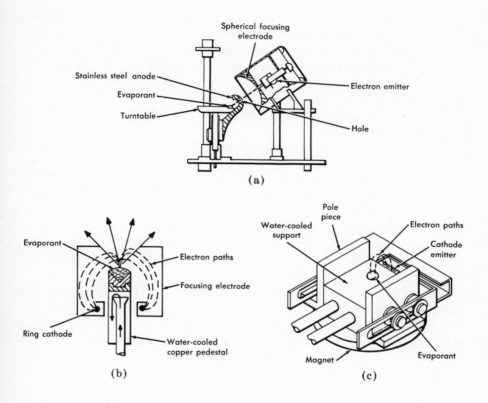

FIG. 3-13. Focused electron bombarded evaporation sources. ((a) After R. Thun
and J. B. Ramsey, "A New Electron Gun for the Vacuum Evaporation
of Metals and Dielectrics," in *Trans. 6th Nat. Vac. Symposium;*
courtesy Pergamon Press, Inc. (b) After Unvala and Booker, "Growth
of Epitaxial Silicon Layers by Vacuum Evaporation, I, Experimental
Procedure and Initial Assessment," in *Philosophical Magazine and
Journal of Science;* courtesy Taylor and Francis, Ltd. (c) After
L. Holland, Ed., *Thin Film Microelectronics,* 1965; courtesy John
Wiley and Sons, Inc.)

between the induction coil and the evaporant is higher for smaller
masses of evaporant. If the evaporant has a mass of several grams,
frequencies of about 10 to 500 kilocycles can be used; for masses in
the milligram range, however, frequencies of several megacycles must
be used.

The induction coil is usually constructed of copper tubing to allow
water cooling of the coil.

The advantages of induction heated sources are:

1. When the evaporant is a metal, heat is produced in the evaporant. Thus, the crucible does not have to conduct electricity or

heat to the evaporant, and a crucible which reacts minimally with the evaporant can be chosen.

2. Heat loss by conduction can be minimized since the power is delivered to the evaporant without electrical or mechanical connections.

Disadvantages are:

1. Relatively complicated and expensive high-frequency generators are required.
2. The induction coil occupies considerable space inside the vacuum system.
3. If the pressure near the coil rises above about 10^{-4} torr, the high-frequency field may ionize residual gas, resulting in a loss of power.
4. The high-frequency generating system must be shielded to prevent radio interference.
5. The crucible may be fractured by thermal shock, or it may react with the evaporant.

A typical induction heated source is shown in Fig. 3-14(a). The double-crucible arrangement prevents large temperature gradients across the inner crucible. This arrangement also allows the evaporant to be centered in the induction coil to provide uniform heating. Nonmetals may be evaporated in an induction heated source by generating the heat in the crucible.

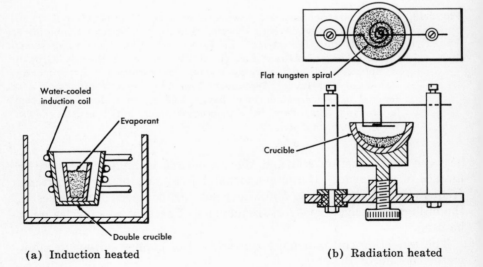

(a) Induction heated

(b) Radiation heated

FIG. 3-14. Miscellaneous vapor sources. ((b) Reproduced with permission from L. Holland, *Vacuum Deposition of Thin Films*, 1963, Chapman and Hall, London.)

Radiation Heated Sources. Materials with high absorptivity for infrared radiation may be evaporated by radiation heating, and many dielectrics have been evaporated using this technique. Metals, which have high reflectivity for infrared radiation, and quartz, which has a low absorptivity, are not easily evaporated with radiation heating.

The main advantage of this type of source is that heating of the evaporant occurs at the surface, so adsorbed gases are liberated at the surface and do not cause splattering. The temperature gradient established from the surface to the interior of the evaporant also permits generally smooth, progressive degassing.

A typical radiation heated source is shown in Fig. 3-14(b). About one-half of the heat from the bare tungsten filament is radiated away from the evaporant. Radiation heat shields can be mounted above the heater to improve the heating efficiency, but they must have a central hole for vapor emission. The vapor stream, however, must pass the heater, which is hotter than the evaporant surface, and this may be a disadvantage since some compounds would become dissociated.

Induction Heated Sources with Levitation. This technique uses an electromagnetic field produced by a coil or coils to keep a ball of material (metal or semiconductor) freely floating in order to eliminate the need for a holder. Eddy currents induced in the material by the levitating fields heat the material to evaporation temperatures.

Disadvantages of this technique are that many coil turns are required to produce the high fields necessary, and the coils often almost surround the charge. Ionization of evaporant vapors or gas between turns of the coils can also be a definite problem.

Van Audenhove [12] has used levitation heating to evaporate silver, titanium, aluminum, iron, cobalt, copper, and uranium. Heating the evaporant while levitating it finds greatest usefulness for deposition of semiconductors or other materials which may become contaminated by a holder. This technique remains primarily an experimental tool.

3.5 EVAPORATION OF SPECIFIC TYPES OF MATERIALS, SPECIAL TECHNIQUES

The preceding sections have provided information concerning the formation of evaporated films by the vacuum evaporation process. This section discusses the evaporation of certain specific types of materials (metals, metal alloys, and insulators and dielectrics). The technique of evaporating in the presence of a glow discharge is also discussed.

Metals

Some of the common, useful techniques developed for evaporating specific metals are described below.

Aluminum. (Melting point $= 660°C$; vapor pressure $= 10^{-2}$ torr at $1217°C$; $M = 27.0$). Films of this metal are used for conductors, capacitor electrodes, and reflectors.

The recommended resistance heated sources are tungsten and tantalum filaments.* Electron beam heating is not recommended since the surface tension of aluminum at vaporization temperatures is too small to prevent spreading of a drop. Aluminum alloys with tungsten and tantalum to form low-melting-point alloys; therefore, heavy filaments are recommended. When tungsten filaments are used, the dissolved tungsten apparently precipitates back onto the filament as the aluminum evaporates, and very little tungsten is evaporated. Aluminum reacts with carbon crucibles and most oxide crucibles, but thoria and zirconia crucibles may be used. For evaporation of large quantities of aluminum, sources have been constructed in which the aluminum is fed slowly onto a heated filament to evaporate on contact. Attack of the source is thus reduced, since only a small amount of molten aluminum comes in contact with the heater.

The adherence of aluminum to glass and ceramic substrates is usually adequate for handling and attachment of leads which are subjected to small forces, and the adherence tends to increase with aging. The initial adherence of aluminum on glass and ceramic substrates, however, is normally less than that of titanium or chromium.

Chromium. (Melting point $= 1900°C$; vapor pressure $= 10^{-2}$ torr at $1397°C$; $M = 52.0$.) Evaporation occurs without melting. This metal is used as an "adhesive" for less adherent evaporated metals. It, or one of its nickel alloys, is also used for resistors.

Since chromium is commercially available in chip or pellet form, a basket filament or boat is recommended. The source material should be tungsten because of the high evaporation temperature of chromium. Another useful method is to electroplate tungsten helices with chromium. This technique improves the thermal contact and increases the evaporation area, but the filaments must be well outgassed after plating.

---

*Boats may be used where a filament is recommended; however, creep of evaporant to the back side of the boat may occur. When a filament is recommended, wetting of the filament by the evaporant is implied.

The adherence of chromium to glass and ceramic substrates is better than any other common metal evaporant except titanium, which shows about the same adhesion as chromium. The adhesion of chromium is often sufficient to cause failure in the substrate when the film-glass interface is tested to destruction.

Copper. (Melting point = 1084°C; vapor pressure = 10^{-2} torr at 1257°C; $M = 63.6$.) Copper is often used for conductors because of its high electrical conductivity and excellent solderability. A practical lower limit for sheet resistance of evaporated copper films is about 0.01 ohm per square. Since copper oxidizes easily, it is often protected with an evaporated layer of a more noble metal.

The recommended resistance heated sources are a tungsten helix or tungsten, tantalum, or molybdenum boats. Wetting of a tungsten helix is not good, but is adequate to hold the evaporant onto the filament. Electron bombardment heating, although it may be used, is not recommended since the thermal conductivity of copper makes evaporation temperatures difficult to maintain.

The adherence of copper on glass or ceramic substrates is usually not adequate to permit the attachment of leads without failure at the copper-substrate interface during handling; an "adhesive" under-layer such as titanium or chromium is recommended.

Gold. (Melting point = 1063°C; vapor pressure = 10^{-2} torr at 1397°C; $M = 197$.) This metal is used as a top layer to protect underlying evaporated layers, as a conductor, and as a capacitor electrode.

Helices or baskets of tungsten or molybdenum are recommended as resistance heated sources. Gold wets tungsten, molybdenum, and tantalum, but tantalum is attacked more than the other two materials. Electron bombardment sources can be used, but the same problems exist as for copper.

Gold adheres poorly to glass and ceramic substrates, but may be made to adhere by using a titanium or chromium underlayer.

Nickel. (Melting point = 1452°C; vapor pressure = 10^{-2} torr at 1527°C; $M = 58.7$.) Nickel is often alloyed with chromium to form Nichrome, which is used as an "adhesive" for less adherent metals or as a thin film resistor material. The alloy has a lower melting temperature than chromium and, since it is available in wire form, is easier to handle. A general discussion of alloy evaporation is given below.

A heavy-gauge tungsten helix is recommended as a resistance heated source. Above 1500°C, nickel forms partial liquid phases with tungsten at all concentrations; as a result, it rapidly attacks tungsten filaments. To limit the amount of attack, the weight of nickel should not exceed 30 percent of the filament weight. Beryllia and alumina crucibles are satisfactory evaporation sources. Electron bombardment evaporation is also suitable for nickel.

The adherence of nickel to glass and alumina ceramic substrates is less than that of chromium or titanium.

Palladium. (Melting point $= 1550°C$; vapor pressure $= 10^{-2}$ torr at $1462°C$; $M = 107$.) This metal is usually used as a protective layer to prevent the oxidization of terminations and conductors.

The recommended resistance heated source is a heavy tungsten helix. Palladium alloys readily with tungsten and produces erosion of the filament. Difficulty is experienced in evaporating the charge to completion, so a charge larger than that calculated is usually used. Alumina and beryllia crucibles can be used. Electron bombardment evaporation is also suitable.

The adherence of palladium on glass or ceramic substrates is similar to that of gold; an underlying layer of titanium or chromium is usually required.

Titanium. (Melting point $= 1667°C$; vapor pressure $= 10^{-2}$ torr at $1737°C$; $M = 48.1$.) Titanium is normally used as an "adhesive" for less adherent evaporated materials and sometimes as resistor or capacitor films.

A tungsten helix or basket is recommended as a resistance heated source, but traces of tungsten may be found in the evaporated titanium. Tantalum is a satisfactory filament material if care is taken to prevent filament burnout because of overheating on the ends. Carbon may be used for a crucible source, and electron bombardment sources also are adequate.

The adherence of titanium to glass and ceramic substrates is about the same as that of chromium. Titanium is a good getter and can sorb large quantities of residual gas; thus, for pure film deposition a shutter is recommended.

Tungsten, Tantalum, and Molybdenum. (Tungsten: melting point $= 3377°C$; vapor pressure $= 10^{-2}$ torr at 3227 °C; $M = 184$. Tantalum: melting point $= 2997$ °C; vapor pressure $= 10^{-2}$ torr at 3057 °C; $M = 181$. Molybdenum: melting point $= 2617$ °C; vapor pressure $= 10^{-2}$ torr at 2307 °C; $M = 96$.) Electron bombardment sources are the only practical means for evaporating these metals. All form oxides which are more volatile than the metal proper, so oxidizing conditions in the vacuum system produce a continued evaporation of the oxides.

Tantalum, in particular, getters oxygen from the residual gas. The adherence of tantalum to glass or ceramic substrates is good, being about the same as that of chromium.

Zinc. (Melting point $= 420$ °C; vapor pressure $= 10^{-2}$ torr at 408 °C; $M = 65.4$.) The principal use of evaporated zinc films is for metallizing paper or similar dielectrics used in capacitors.

Satisfactory zinc films can be deposited with background pressures up to about 0.1 torr [13]. Thus, high vacuum equipment is not required and operating cycles are shortened for batch-type systems. Zinc films evaporated at background pressures approaching 0.1 torr have smaller grain sizes than those formed in higher vacua [6]. The exact effects of higher background pressure on grain growth are not well understood.

Evaporated silver is often used to presensitize the dielectric material surface, with an average film thickness less than 1 A being sufficient. The silver atoms serve as nucleating sites for the zinc atoms, and thus affect the growth, film structure, and electrical properties of the zinc film.

If the silver is selectively deposited, zinc films grow only on the areas containing silver nuclei. This phenomenon is used as a masking technique, but the arrival rate of zinc must not exceed a certain critical value; otherwise, a permanent deposit of zinc will be formed on the entire surface (see Sec. 3.3).

Most of the common boat or crucible sources can be used to produce high beam intensities. Zinc can be evaporated from the solid state at relatively low evaporation temperatures, with negligible attack or evaporation of the container. Steel crucibles may even be used if the zinc is not melted.

Cadmium. (Melting point $= 321$ °C; vapor pressure $= 10^{-2}$ torr at 217 °C; $M = 112$.) The uses and evaporation procedures for cadmium are similar to those for zinc.

Metal Alloys

When alloys are evaporated, fractionation (partial separation of the components) may occur. From Eq. (3-7), the evaporation rate of an individual component (A) of an alloy is

$$G_A = 0.058\, p'_A \sqrt{\frac{M_A}{T}} \qquad \text{grams/cm}^2 \text{ sec} \qquad (3\text{-}8)$$

where p'_A is that part of the vapor pressure of the alloy contributed by component A, and M_A is the molecular weight of component A. Since p'_A is not known, it is usually estimated. Dushman [14] and Holland [6] have used Raoult's law to estimate p'_A,

$$p'_A = X_A p_A \qquad (3\text{-}9)$$

where X_A is the mole fraction of component A, and p_A is the vapor pressure of the pure substance A. Equation (3-8) becomes

$$G_A = 0.058 \, X_A p_A \sqrt{\frac{M_A}{T}} \qquad \text{grams/cm}^2 \text{ sec} \qquad (3\text{-}10)$$

Often Raoult's law does not hold for metal alloys, and a factor S_A (an activity coefficient) is introduced to modify Eq. (3-10),

$$G_A = 0.058 \, S_A X_A p_A \sqrt{\frac{M_A}{T}} \qquad \text{grams/cm}^2 \text{ sec}$$

with S_A being determined experimentally.

Since activity coefficients are often not known, Eq. (3-10) is usually used to estimate the amount of fractionation of alloys. As an example, consider Nichrome® (80% Ni—20% Cr by weight) at a temperature of 1527°C ($p_{Cr} = 10^{-1}$ torr; $p_{Ni} = 10^{-2}$ torr). The mole fraction of chromium is

$$X_{Cr} = \frac{W_{Cr}/M_{Cr}}{(W_{Cr}/M_{Cr}) + (1 - W_{Cr})/M_{Ni}} = \frac{W_{Cr}}{W_{Cr} + (1 - W_{Cr})(M_{Cr}/M_{Ni})}$$

where W_{Cr} is the fraction, by weight, of chromium.

$$X_{Cr} = \frac{0.2}{0.2 + (1 - 0.2)(52/58.7)} = 0.22$$

and

$$X_{Ni} = 1 - X_{Cr} = 0.78$$

Thus,

$$\frac{G_{Cr}}{G_{Ni}} = \left(\frac{0.22}{0.78}\right)\left(\frac{10^{-1}}{10^{-2}}\right)\sqrt{\frac{52}{58.7}} = 2.8$$

The initial evaporation rate of chromium from the alloy would be about 2.8 times that of nickel. As the chromium becomes depleted, the ratio G_{Cr}/G_{Ni} eventually becomes less than unity since four times as much nickel must be vaporized when the charge is evaporated to completion. This fractionation of Nichrome® produces better adherence to the substrate since the film adjacent to the substrate is chromium-rich.

The composition of evaporated alloy films may be controlled by several techniques:

1. Flash evaporation: Finely divided particles of the alloy are dropped onto a surface so hot that the particles evaporate almost instantaneously. The finely divided metal has a large gas content, and the rapid release of this gas produces higher background pressures and may cause unmelted particles to leave the source and strike the substrate.

2. Evaporation from separate sources: Alloys may be prepared using several sources, each with a component of the alloy to be formed. In order to obtain uniform thickness distributions, the substrates are normally rotated.

3. Alloy sublimation: In a solid, the more volatile component is restored to the surface of the alloy only by diffusion. In many cases, the evaporation rate is much higher than the diffusion rate, and the evaporation eventually reaches a steady state. This technique has been used to form evaporated films of Nichrome® which exhibit bulk composition [15].

Insulators and Dielectrics

Most insulators and dielectrics are compounds rather than elements, and their evaporation presents problems such as dissociation (caused by heating in vacuum or interaction with an electron beam) and chemical reaction with the heater material. Since both depend on the source temperature, the composition of the condensed film may vary with source temperature.

Most insulators and dielectrics used in thin film form are oxides. At any given temperature a dissociation pressure exists, representing an equilibrium between the oxide, the gas (oxygen), and the element or a lower oxide. If the partial pressure of oxygen is reduced below the dissociation pressure, the compound loses oxygen and thus dissociates.

The temperatures at which several oxides show a vapor pressure of 10^{-2} torr, and their dissociation pressures at that temperature, are listed in Fig. 3-15. Note that Al_2O_3 does not dissociate at evaporation temperatures, while nickel oxide dissociates appreciably at 1596°C if in a vacuum environment. The oxygen pressure may be deliberately increased to prevent dissociation, but an increase in oxygen pressure to the levels necessary may adversely affect the film structure.

Lower metal oxides (less oxygen) are often more volatile than higher oxides (AlO and SiO are more volatile than Al_2O_3 and SiO_2, respectively); thus, the decomposition products evaporate more readily when the higher oxide is heated. Restoring the evaporated film to the original oxide requires oxidation at the substrate. The amount of oxidation in flight is normally small and may be neglected.

Metal oxide	Temp. at which p_v of the oxide is 10^{-2} torr (°C)	Dissociation pressure, p of O_2 (torr)
$Al_2O_3 \rightarrow 2\ Al + 1.5\ O_2$	1781	1.5×10^{-18}
$CuO \rightarrow Cu + 0.5\ O_2$	—	1.3×10^{-1} at 1027°C
$MoO_3 \rightarrow Mo + 1.5\ O_2$	730	1.1×10^{-14}
$NiO \rightarrow Ni + 0.5\ O_2$	1596	3.9×10^{-1}
$WO_3 \rightarrow W + 1.5\ O_2$	1122	3.3×10^{-10}
$Fe_3O_4 \rightarrow 3\ FeO + 0.5\ O_2$	$\cong$1300	8.7×10^{-9} at 1127°C
$3\ Fe_2O_3 \rightarrow 2\ Fe_3O_4 + 0.5\ O_2$	—	7×10^{-8} at 1127°C

FIG. 3-15. Dissociation pressures of metal oxides. (Reproduced with permission from L. Holland, *Vacuum Deposition of Thin Films*, 1963, Chapman and Hall, London.)

A metal oxide may also be reduced by chemical reaction with the heater. Metal oxides in contact with a carbon crucible at evaporation temperatures may either be reduced by the carbon to a lower oxide or free metal (also producing CO gas), or they may chemically react to form carbides. Tungsten, tantalum, and molybdenum may also reduce metal oxides and form a volatile oxide of the heater material, such as WO_3. Al_2O_3 is reduced by tungsten above 2000°C. When more than one oxide of a metal exists, any free metal present may react with a higher oxide to form a lower oxide. For example, TiO_2 is partially reduced by metallic titanium at high temperatures.

In Situ Reactions. Evaporated dielectrics with reproducible properties are usually prepared by compensating for the dissociation which occurs at the source by a chemical reaction at the substrate. Evaporation of a metal oxide usually produces a film consisting of mixed oxides and free metal. The film may be oxidized as it grows by maintaining a partial pressure of an oxidizing gas in the system during evaporation. Even at pressures as low as 10^{-6} torr, enough molecules bombard a surface to form a monolayer in approximately 2.2 seconds. Thus, the film may be reoxidized or even become more oxidized than the evaporant material. The amount of oxidation depends on the ratio of the arrival rate of oxidizing gas molecules to that of evaporant atoms and the affinity of the film for the oxidizing gas. Stoichiometric oxide films are often difficult to obtain.

Evaporation of Silicon Monoxide. Silicon monoxide (SiO) was one of the first evaporated dielectrics studied because of its ease of evaporation. The vapor pressure of SiO is high enough to produce practical evaporation rates at relatively low temperatures (1050° to 1400°C), as can be seen from its vapor pressure versus temperature curve in Fig. 3-16 [16, 17]. Since SiO sublimes, contact with the heater is reduced and little contamination of the deposit with heater material results. SiO forms a uniform, amorphous layer with good adherence to most metals and dielectrics, and is chemically inert to many environmental conditions.

Control of the source temperature, the background pressure of oxidizing gases, and the angle of incidence is necessary to deposit films with reproducible properties because these factors affect the composition, structure, and stresses of the condensed film. Stresses sufficient to cause crinkling (compressive stresses) and cracking (tensile stresses) are not uncommon [18]. Heat treatment, partic-

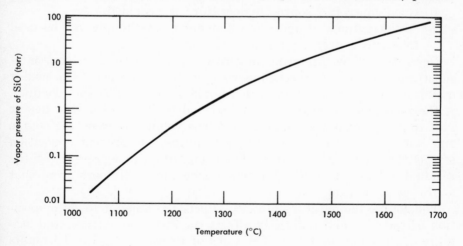

FIG. 3-16. Vapor pressure versus temperature curve for SiO [16, 17].

ularly in atmospheres containing oxygen or water vapor, can also severely alter the properties of the film.

SiO may be produced by several reactions at temperatures around 1100°C, such as

$$SiO_2 + Si \rightarrow 2\ SiO$$

$$SiO_2 + H_2 \rightarrow SiO + H_2O$$

$$SiO_2 + C \rightarrow SiO + CO$$

The difficulties involved in bringing the correct proportions of the reactants together in an evaporation source and preventing any of the reactants from evaporating render the simultaneous production and evaporation of SiO impractical. Therefore, the usual procedure is to load a source with SiO.

Drumheller [19] has designed an adequate source for evaporating SiO. This source, shown in Fig. 3-17, has the following desirable features:

1. Spitting of particles of SiO is almost eliminated. Any particles of SiO ejected from the charge through small perforations in the heater tube collide with the heater tube and are vaporized.
2. Large charges (several dekagrams) can be used. The source is easily disassembled for loading.
3. Multiple radiation shields are provided to avoid large temperature gradients in the charge and to limit the amount of heat lost by radiation.
4. Good control of source temperature is provided.
5. A conventional evaporation filament power supply can be used to supply power for the source.

Deposition rates up to 60 angstroms per second at a source-to-substrate distance of 80 centimeters have been obtained. The heater material is usually tantalum, which can be fabricated by spot welding. A thermocouple is sometimes spot-welded to the wall of the heater tube to measure source temperature. The volume between the source and substrate is often enclosed by a "chimney" to prevent deposition of SiO throughout the system. The composition and pressure of the residual gas may be quite different inside these chimneys from that at the vacuum gauge.

The source temperature directly affects the amount of decomposition of the SiO, and indirectly affects the structure, stress, and stability of the condensed film. The work of Priest et al. [18, 20] points to the following conclusions:

1. Source temperatures less than about 1250°C favor the reaction

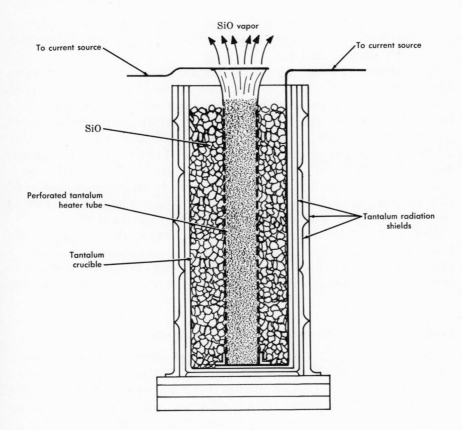

FIG. 3-17. Drumheller evaporation source for SiO. (After C. E. Drumheller, "Silicon Monoxide Evaporation Techniques," in *Trans. 7th Nat. Vac. Symposium, 1960*, 1961; courtesy Pergamon Press.)

$SiO \rightarrow Si + SiO_2$, producing a condensed film which is a mixture of Si, SiO, and SiO_2. These films usually are under tensile stress and appear to be more porous than films produced at higher source temperatures. The Si in the film oxidizes easily upon exposure to O_2 or H_2O. This oxidation reaction produces a volume increase in the film, resulting in compressive stresses which cause the film to buckle.

2. Source temperatures above about 1400°C favor the evaporation of SiO. The condensed film is mainly SiO, is relatively insensitive to exposure to H_2O or O_2, and is normally under compressive stress. These films apparently contain a large number of defects and often crack upon annealing at room temperature.

3. Source temperatures in the range of 1250° to 1400°C produce film with good mechanical properties. The films are mainly SiO, with some Si and SiO_2.

The amount of oxidizing gases, particularly water vapor, present during evaporation affects the film properties. As one might expect, the effect of these residual gases increases with decreasing source temperature (since the lower source temperatures produce a more porous film rich in Si). Partial pressures of H_2O as low as 10^{-7} torr affect film stress at source temperatures of 1183°C [18]. Deliberate, controlled addition of water vapor or oxygen has been used to produce films which oxidize less upon exposure to atmosphere; however, this addition of an oxidizing gas also changes the film stress and may produce films which are not satisfactory mechanically.

The angle of incidence of the SiO molecules also influences the stress in the deposited film. Anisotropy in the plane of the film is produced by other than normal incidence. There is general agreement that normal incidence of the vapor stream produces the most durable film.

Evaporation in the Presence of a Glow Discharge

Mattox [21, 22] has developed a variant of the evaporation technique which he has termed "ion plating." After being evaporated from the filament, the charge passes through a glow discharge region set up between the filament and the substrate, with the substrate being the cathode. As the vapor atoms pass through the glow discharge, some are ionized and accelerated to the substrate. Mattox has reported improved film-substrate adhesion of many materials, such as gold, which are normally nonadherent.

The improved adhesion may result either from cleaning (perhaps due to sputtering) of the substrate produced by ion bombardment, from penetration of the substrate by the energetic evaporant ions, or from both of these causes.

A schematic of the apparatus used for ion plating is shown in Fig. 3-18. The procedure is:

1. Load the evaporant into the filament and the substrates into the holder.
2. Evacuate the bell jar to remove residual gases.
3. Introduce an inert gas and establish a glow discharge. Continue the glow for as long as necessary to clean the substrate.
4. Pass current through the filament to vaporize the evaporant while continuing the glow discharge.

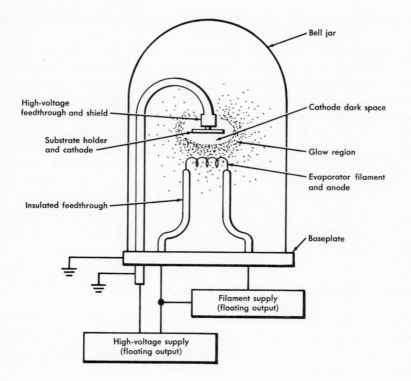

Bell jar

High-voltage
feedthrough and shield

Cathode dark space

Substrate holder
and cathode

Glow region

Evaporator filament
and anode

Insulated feedthrough

Baseplate

Filament supply
(floating output)

High-voltage supply
(floating output)

Fɪɢ. 3-18. Simple ion-plating apparatus using a conventional glass bell jar
vacuum system.   (After D. M. Mattox, "Film Deposition Using
Accelerated Ions;" courtesy *Electrochemical Technology*.)

The ions produced by the glow discharge bombard the substrates
while the electrons in transit to the filament ionize some of the inert
gas and evaporant atoms.

The ion bombardment may sputter material from both the substrate
and the substrate holder. In order for the film thickness to increase,
the rate of condensation on the substrate must be greater than the
sputtering rate from the substrate.

Collisions between the evaporant atoms and the inert gas molecules
cause the evaporant to diffuse to points not in line-of-sight of the
filament. This diffusion may produce a better thickness uniformity in
the deposited film; however, the final thickness distribution can be
modified by sputtering. This diffusion has disadvantages in that high-
voltage insulators may become coated, and material may be deposited
under masks which do not fit tightly against the substrate.

3.6   EVAPORATED FILM THICKNESS AND THICKNESS DISTRIBUTION

Evaporated films range in thickness from about 20,000 angstroms down to an apparent thickness of only several angstroms. While films thicker than 20,000 angstroms have been prepared, problems of handling large charges, long evaporation times, and excessive heating of surfaces inside the vacuum system must be overcome.

Theoretical calculations of thickness distribution are often used to determine how the source (or sources) should be placed relative to the substrates and to estimate the amount of evaporant required. These calculations involve approximations, so in-process monitors or post-deposition thickness measurement techniques are often used as a supplement to or in place of the calculations. Several common in-process monitors and thickness measurement techniques are discussed in this section.

Theoretical Film Thickness Distribution

The thickness of an evaporated film at any point on the substrate depends on the emissive characteristics of the source, the geometry and arrangement of the substrate relative to the source, and the amount of material vaporized. Thickness distributions produced by two simple ideal sources are considered in this section; these results allow a first approximation of charge sizes and film thicknesses. In-process monitors or post-deposition measurements may be used to determine the actual film thickness or the electrical, magnetic, optical, or mechanical property controlled by the film thickness.

Point Source. Consider a small sphere evaporating an equal amount of material in all directions. Such a source is called a point source. Let the total mass of material evaporated be m_1. The amount of material, dm_2, incident upon a small receiving surface, dS_2, is equal to the amount of material passing through the solid angle, $d\Omega$, subtended at the source by the surface (see Fig. 3-19). Since $d\Omega = dS_2 \cos \beta / r^2$,

$$dm_2 = Cm_1 d\Omega = \frac{Cm_1 \cos \beta dS_2}{r^2} \qquad (3\text{-}11)$$

where C is a proportionality constant which can be evaluated by integrating over the entire receiving surface. Let the receiving surface

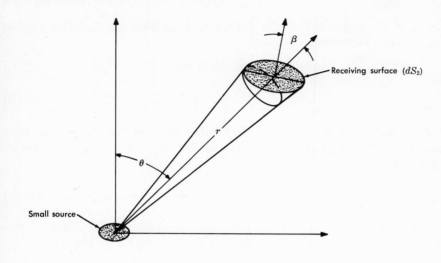

Note:
The small source may be either a point or plane source.
The angle θ is the angle between the normal to the plane source and
the line between the source and the receiving surface.
The angle β is the angle between the normal to the surface and the
line between the source and the receiving surface.

FIG. 3-19. Effect of receiving angles on the deposit thickness.

be a sphere with its center at the source. Then $\beta = 0$, $dS_2 = 2\pi r^2$
$\sin \theta \, d\theta$, and integration of Eq. (3-11) over the sphere gives

$$\int dm_2 = Cm_1 \int_{\theta=0}^{\pi} (2\pi \sin \theta \, d\theta)$$

or

$$m_2 = Cm_1 4\pi$$

Since $m_2 = m_1$, then $C = 1/4\pi$ and Eq. (3-11) becomes

$$dm_2 = \frac{m_1 \cos \beta \, dS_2}{4\pi r^2} \qquad \text{(point source)} \qquad (3\text{-}12)$$

Small Plane Source. Now replace the small sphere with a small
plane source. Knudsen has shown that this source has directional
emissive properties such that the amount of material evaporated at
angle θ is proportional to $\cos \theta$. As shown in Fig. 3-19, θ is the
angle between the normal to the plane source and a line between
the source and the receiving surface. Equation (3-11) must be

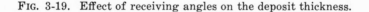

160   THIN FILM TECHNOLOGY

changed to account for the directional emissive properties of the source, and becomes

$$dm_2 = Cm_1 \cos \theta \, d\Omega$$

$$= \frac{Cm_1 \cos \theta \cos \beta \, dS_2}{r^2} \qquad (3\text{-}13)$$

For a plane source, emission is limited to a hemisphere, and integration of Eq. (3-13) over a hemisphere gives $C = 1/\pi$. Equation (3-13) becomes

$$dm_2 = \frac{m_1 \cos \theta \cos \beta \, dS_2}{\pi r^2} \qquad \text{(small plane source)} \qquad (3\text{-}14)$$

Equations (3-12) and (3-14) can be converted to an equivalent film thickness, t, by assuming the deposit has a density ρ; therefore,

$$dm_2 = \rho t \, dS_2$$

Thus,

$$t = \frac{m_1 \cos \beta}{4\pi \rho r^2} \qquad \text{(point source)} \qquad (3\text{-}15)$$

and

$$t = \frac{m_1 \cos \theta \cos \beta}{\pi \rho r^2} \qquad \text{(small plane source)} \qquad (3\text{-}16)$$

for the point and small plane sources, respectively.

A comparison of the relative thickness distributions for a point source and a small plane source is given in Fig. 3-20. The receiving surface is a plane surface, parallel to the small plane source ($\theta = \beta$). The equations in Fig. 3-20 are obtained by substituting $\cos \beta = \cos \theta = h/(x^2 + h^2)^{1/2}$ and $r^2 = x^2 + h^2$ in Eqs. (3-15) and (3-16). Although the distributions are similar, Eqs. (3-15) and (3-16) indicate that the maximum thickness for the plane source is four times that of the point source for a given amount of evaporant and a given source-to-substrate distance.

Two nomographs which can be used to estimate the amount of evaporant required for a specific application are given in Figs. 3-21 and 3-22. Note that both nomographs are for point sources and assume that the deposit has bulk density.

Expressions for the theoretical thickness distributions obtained with an extended strip source, a cylindrical source, and a ring source are given by Holland [6].

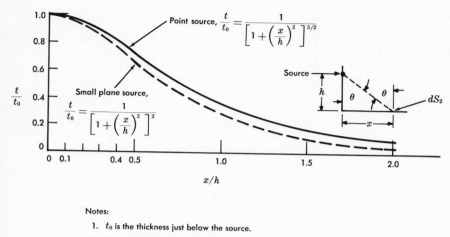

FIG. 3-20. Distribution of the deposit on a plane surface.

Emissive Characteristics of Practical Sources

A hairpin-shaped source or an electron bombarded source where the molten material assumes a spherical shape approximates a point source. A small amount of evaporant which does not wet the boat from which it is evaporating assumes a spherical shape, but it emits like a small plane source because the heater surface adjacent to the evaporant reevaporates all atoms which initially evaporate downward.

Emission from a helix approaches the ideal cylindrical source [6], provided the evaporant wets the entire helix. Since the expressions for a cylindrical source are rather complicated, the point source approximation is often used as an expedient. Evaporation of a charge to completion alters the emission since the center of a helix evaporates more material than the ends, and vapor-beaming from the open ends of closely wound helices has been observed.

Equation (3-16) can be used with fair accuracy to calculate the emissive properties of boats [Fig. 3-11(c)] and conical baskets [Fig. 3-11(b)], provided the turns are closely spaced, but some vapor-beaming does occur from the bottom of the basket.

An open crucible acts as a surface source. When the crucible is covered with an aperture in the form of a tube, however, marked beaming in the direction of the axis of the tube is observed. Thus, a crucible source may be varied from a surface source to a highly directed one, depending on the aperture used.

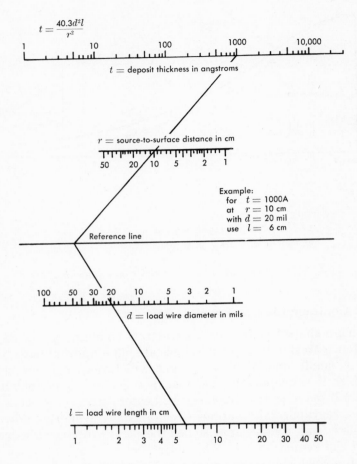

$$t = \frac{40.3 d^2 l}{r^2}$$

FIG. 3-21. Deposit thickness versus load wire size for a point source.

Methods of Producing Uniform Deposits

When the source is a single-point source, the substrates should be arranged in the shape of a sphere concentric with the source. Since $\cos \beta = 1$ and r is constant in Eq. (3-15), the thickness of the deposit is the same everywhere. The radius of the sphere must be large with respect to the greatest linear dimension of the substrate.

With all the substrates placed in a plane, the ratio x/h should be minimized (see Fig. 3-20). This is usually done by increasing h, resulting in a decrease in the deposit thickness for a given amount of evaporant.

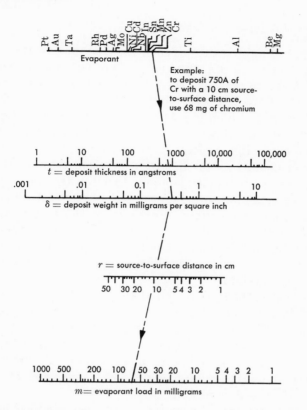

FIG. 3-22. Deposit thickness versus load weight for a point source.

When the source is a single small plane source, the substrates should be arranged on the surface of a sphere, with the source at some point on the same sphere. The result is a deposit with a thickness that is the same everywhere [6].

When multiple sources are used, the distributions are superimposed from the individual sources to estimate the combined thickness distribution. One particular arrangement for which an expression for the distribution is available consists of a number of small sources placed in a circle on a plane parallel to the receiving surface [6]. Good uniformity can be obtained with ratios of source-to-substrate distance to ring diameter as small as 1.5.

Other methods which have been used to produce more uniform deposits include rotating the substrates during deposition and interrupting the vapor stream with a rotating shutter whose cross-section is designed to cancel the directional properties of the source.

Film Thickness Measurement Techniques

Resistivity and light transmittance are examples of thin film properties which depend on film thickness. Thus, to characterize films it is often necessary to measure their thickness. Although direct measurement of thickness is preferred, it is sometimes easier to measure some film parameter which is related to the thickness. The thickness is then calculated from this parameter, or the parameter itself is used to control the deposition process.

Thickness is defined as the distance, perpendicular to the surface, from a point on one boundary surface, through the film, to the other boundary surface. If the boundary surfaces are rough or nonparallel, the thickness is not well defined. Fortunately, most vacuum deposited films have smooth and parallel, or nearly parallel, boundary surfaces.

The measurement techniques described in this section are limited to those which are commonly applied after the substrate is removed from the vacuum system.

The following methods have found the widest application and will be described in some detail:

1. Weighing. The mass of the film is determined by weighing; the thickness is then calculated, usually assuming bulk density.
2. Stylus. A stylus attached to a lever and fulcrum is made to traverse a substrate-film step. The movement of the stylus is amplified and recorded.
3. Multiple Beam Interferometry. Light rays, which have undergone multiple reflections between the substrate and a reference flat, interfere and form fringes. These interference fringes, which are sharp because of multiple reflections, are used to determine the film thickness.
4. Beta Particle Back-Scattering. Beta particles (high-energy electrons) are directed onto the film. The film thickness can be estimated from the number of beta particles which are back-scattered.
5. X-ray Fluorescence. Secondary X-rays characteristic of the film are excited by primary X-rays directed onto the film. The intensity of the secondary X-rays is used to estimate the film thickness.

Weighing. If A is the area, t the thickness, ρ the density, and m_2 the mass of the deposited film, $tA\rho = m_2$, or $t = m_2/A\rho$; then, m_2 can be determined by weighing the substrate before and after deposition, and A can be measured. Since both ρ and t are unknown, only the mass per unit area, ρt, can be calculated from the measured quantities.

If the deposit has bulk density, ρ_B, the equivalent thickness, t_e, is

$$t_e = \frac{m_2}{A\rho_B}$$

For reasonable-size areas, the error in determining the area, A, can be made small and considered to be negligible. Thus, the fractional error in thickness, dt_e/t_e, is

$$\frac{dt_e}{t_e} = \frac{dm_2}{m_2}$$

With care, masses can be determined to about $\pm$ 2 micrograms.

As an example, consider a substrate upon which aluminum has been deposited to produce a weight gain of 27 ± 2 μg on 10.000 cm². Since $\rho_B = 2.699$ g/cm³ at 20°C, then $t_e = 100 \pm 7.4$ angstroms.

The substrates are commonly removed from the vacuum system after deposition and are weighed. Because the freshly deposited film on the substrate may sorb gases (such as water vapor) upon exposure to atmospheric conditions, the change in weight due to sorption can easily be one or two orders of magnitude greater than the basic accuracy of the microbalance. This factor usually limits the accuracy of this method.

Another limitation of the weighing method is that determination of thickness distribution on a single substrate is not possible, since an average thickness is obtained for the entire area weighed. In addition, the equivalent thickness is not the actual thickness when the density of the film differs from bulk density.

Stylus. In principle, this method consists only of measuring the mechanical movement of a stylus as it traverses a film-substrate step. Practical aspects of the method will be illustrated by describing a commercially available stylus instrument manufactured by Taylor-Hobson, Ltd., and marketed under the tradename Talysurf®.

The diamond stylus has a tip radius of 0.0001 inch and bears on the specimen being measured with a force of about 0.1 gram. The stylus traverses a substrate-film step, and the vertical motion of the stylus relative to a reference plane is converted to an electrical signal which is amplified and recorded on rectilinear paper. Thus, a profile graph is produced which represents a cross-section of the film step as well as substrate surface irregularities.

The available magnification that is normal to the surface ranges in eight steps from $500\times$ to $100,000\times$ (see Fig. 3-23). Standard magnifications parallel to stylus travel are $20\times$ and $100\times$.

Magnification	Step height (A) for full-scale chart deflection of 5 cm
500	10^6
1,000	5×10^5
2,000	2.5×10^5
5,000	1×10^5
10,000	5×10^4
20,000	2.5×10^4
50,000	1×10^4
100,000	5×10^3

FIG. 3-23. Magnification and scale ranges.

Traces of the two standard steps (3950 A and 24,000 A) provided to check instrument calibration are shown in Fig. 3-24. When repeated traces of a single, well defined step on a flat, smooth surface are taken, a one-sigma limit of about ±40 A is obtained [23]. Errors are a combination of machine errors, errors due to the width of the writing line, and reader error in determining the step height. Good shock mounting must be provided to minimize background vibrations. The film thicknesses measured with the stylus method compare well with those obtained with multiple beam interferometry.

Several factors which limit the usefulness and accuracy of this instrument are:

1. The diamond stylus may penetrate and scratch a groove in soft films, such as tin. This will destroy the films or introduce a large error.
2. Substrate roughness may introduce "noise" which causes errors because of the uncertainty in the position of the substrate and film surfaces, as shown in Fig. 3-25.
3. "Wavy" substrates introduce slopes and curvature in the traces on each side of the step; this causes error in the determination of the step height.
4. A film-substrate step is required.

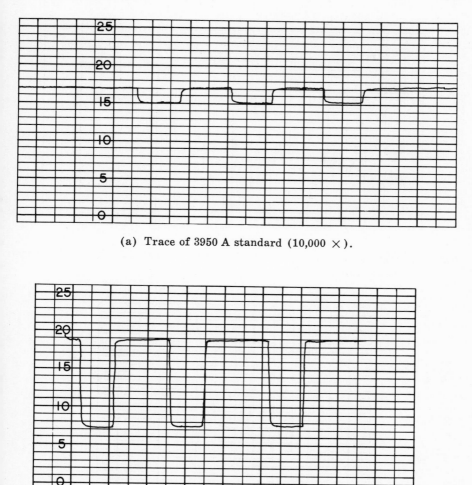

(a) Trace of 3950 A standard (10,000 ×).

(b) Trace of 24,000 A standard (10,000 ×).

Fig. 3-24. Talysurf® traces of standard steps.

A Talysurf® trace of tantalum deposited on a "wavy" glazed ceramic substrate is shown in Fig. 3-26. The proper method of reading the step height is explained with the aid of Fig. 3-27, which shows schematic representations of the film-substrate step and the trace obtained from that step. The film thickness is the perpendicular distance, BC, between the film surface and the substrate surface. The corresponding distance on the trace [Fig. 3-27(b)] is $B'C'$; however, this dis-

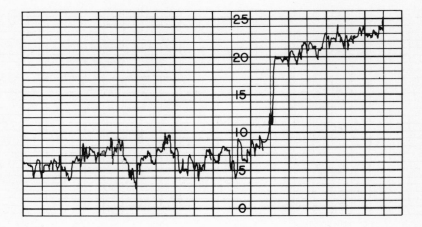

FIG. 3-25. Talysurf® trace of a film deposited on a rough substrate (vertical magnification 10,000 ✕).

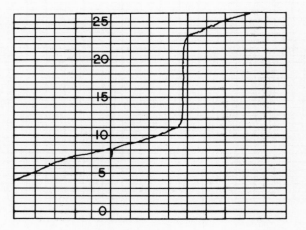

FIG. 3-26. Talysurf® trace of a film deposited on a wavy substrate (vertical magnification 20,000 ✕).

tance cannot be scaled directly from the chart because the horizontal and vertical scales are different. Only lengths parallel to one of the axes can be read directly.

The film thickness, BC, is $BC = BD \cos \gamma$, where γ is the angle between the surface of the film and the surface of the reference plane; however, $BD = B'D'/M_v$ and $\tan \gamma = (M_h/M_v) \tan \gamma'$, where M_v and M_h are the vertical and horizontal magnifications on the trace,

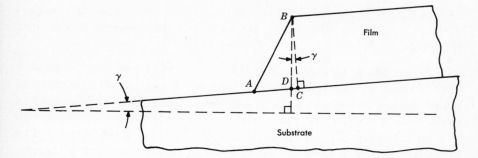

(a) Schematic of substrate-film step on a wavy substrate. (γ is the angle between the surface of the film and the reference plane).

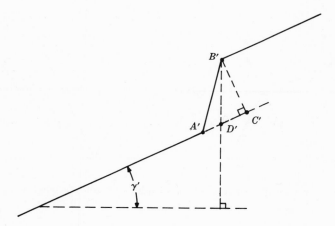

(b) Schematic of Talysurf® trace of the film-substrate step above.

FIG. 3-27. Talysurf® trace on a wavy substrate.

respectively. Thus,

$$BC = \frac{B'D'}{M_v} \cos\left[\tan^{-1}\left(\frac{M_h}{M_v}\tan\gamma'\right)\right]$$

$$= \frac{B'D'}{M_v\sqrt{1 + (M_h \tan\gamma'/M_v)^2}}$$

If $\gamma' \leqq 89°$, then $\tan\gamma' \leqq 75$, and if $M_h/M_v \geqq 10^{-3}$, then $1/\sqrt{1 + (M_h \tan\gamma'/M_v)^2} \geqq 0.997$. Thus, $BC = B'D'/M_v$ to an error of less than 1 percent in the worst case $(\gamma' = 89°)$, and the vertical distance, $B'D'$, on the trace is measured to determine the film thickness.

Multiple Beam Interferometry. Tolansky [24] and his co-workers have developed the technique of multiple beam interferometry, in which highly reflective surfaces are used. As shown in Fig. 3-28, a beam which is reflected many times between the two surfaces still has an intensity large enough to affect the total interference. The intensity pattern is then the summation of a large number of beams, with each one only slightly less intense than the previous one. With highly reflective silver surfaces, as many as 60 reflections can be obtained. Multiple beam interferometry also makes sharp fringes possible, such as those shown in Fig. 3-29.

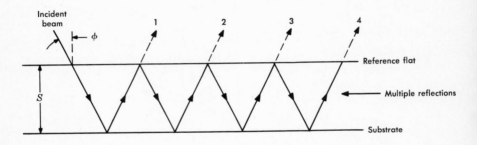

FIG. 3-28. Schematic of multiple beam interferometry.

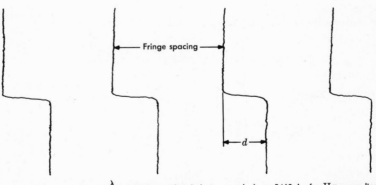

Fringe spacing $= \dfrac{\lambda}{2}$ when the incident light is normal. $\lambda = 5460$ A for Hg green line.

FIG. 3-29. Fringes produced by multiple beam interferometry across a film-substrate step.

Under the best conditions, this is the most accurate method available to determine film thickness, with a resolution of about ± 5 A in a direction normal to the substrate surface.

Two general methods of multiple beam interferometry are used to measure film thickness. The first produces Fizeau fringes of equal thickness, using a monochromatic light source. The second uses a white light source and produces fringes of equal chromatic order. The second method is preferred for films thinner than a few hundred angstroms since it is capable of higher resolving power.

Fizeau Fringes of Equal Thickness. The procedure used to produce these fringes is:

1. A sharp step is produced between the surface of the substrate and the surface of the film.
2. A high-reflectivity, low-absorptivity, opaque film which accurately contours the step is deposited by vacuum evaporation.
3. A multiple beam interferometer is formed by placing a half-silvered reference plate on top of the step at a small angle to the substrate surface. This is illuminated with collimated monochromatic light, and the fringe system produced is viewed with a microscope. A typical interferometer arrangement is shown in Fig. 3-30.
4. The fringe spacing and fringe displacement across the step are measured and used to calculate the film thickness.

The fringes occur where the beams interfere constructively, i.e., when the phase difference between successive beams is an integral number times 2π. When measuring film thickness, the angle ϕ of Fig. 3-28 is very small and, due to path difference, the phase difference between successive beams is approximately $(2S/\lambda) 2\pi$, where λ is the wavelength of the incident light and S is the distance between the reference flat and the substrate. The condition for constructive interference is

$$\frac{2S}{\lambda} 2\pi + 2\delta = a_1 2\pi \tag{3-17}$$

where a_1 is an integer and δ is the phase change accompanying reflection (assumed to be the same at both surfaces). For high reflectivity films, $\delta \cong \pi$, and Eq. (3-17) becomes

$$S = \frac{a_1 - 1}{2} \lambda = \frac{a_2 \lambda}{2} \tag{3-17a}$$

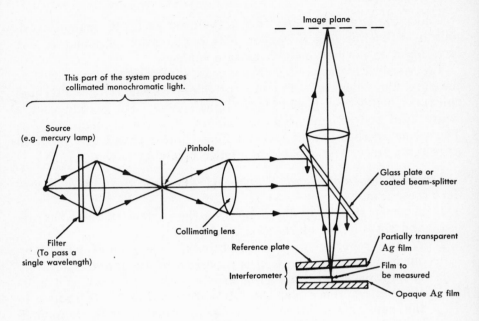

FIG. 3-30. Interferometer arrangement for producing reflection Fizeau fringes
of equal thickness. (Reproduced with permission from L. Holland,
Vacuum Deposition of Thin Films, 1963, Chapman and Hall, London.)

Thus, the distance between the maxima of successive fringes corresponds to a distance S equal to $\lambda/2$. Destructive interference also
occurs to produce a black background for the series of sharp bright
fringes.

When a sharp step exists on the substrate, the distance S is abruptly
changed and the fringe system is abruptly displaced by an amount d,
as shown in Fig. 3-29. The amount of displacement is proportional to
the film thickness, t. Whether the fringe system has been displaced
by an amount d or by an amount d plus an integral number of fringe
spaces must usually be determined by an auxiliary method. The film
thickness is given by

$$t = \frac{d}{\text{fringe spacing}} \frac{\lambda}{2} \qquad (3\text{-}18)$$

For highly reflective surfaces, the width of a fringe is about 1/40th
of the fringe separation. Displacements of about 1/5 of a fringe
width can be measured under the best conditions. If the source is the

Hg green line ($\lambda = 5640$ A), the resolution is

$$\left(\frac{1}{40}\right)\left(\frac{1}{5}\right)\left(\frac{5640}{2}\right) \cong 10 \text{ A}$$

Factors which influence the resolution are:

1. The overlying film on the step must be highly reflective. The number of beams determines the fringe width, and the number of beams is determined by the fraction, F_1, of the incident light which is reflected. An increase in F_1 from 0.9 to 0.95 reduces the fringe width to less than one-half of its former value.

2. The film on both the step and the reference flat must have low absorptivity. The fraction, F_2, of incident light absorbed determines the intensity of the fringes. When $F_2 = 0.04$, the intensity at a fringe maximum is 1/25 of the intensity of the incident light.

3. The thickness, S, of the air gap between the reference plate and the substrate surface must be small, preferably less than a few light waves. Decreasing S reduces the area from which the multiple beams originate and thus reduces the area over which the reference plate must be flat.

4. The angular spread in the incident beam should be less than 1° to 3°. The larger the angular spread, the greater the fringe width.

5. The incident light should be normal to the reference plate. Normal incidence produces minimum divergence of the multiple beams, allowing a maximum number of beams to be used.

6. The substrate should be flat, at least within the order of 0.001 centimeter on either side of the step. Lack of flatness causes the fringes to be curved and reduces the accuracy of the step-height measurement.

7. The step should be sharp so that the discontinuity in the fringe system covers a small area. An abrupt shift in the fringe system increases the accuracy of the fringe displacement measurement.

Silver is the only common metal which can be evaporated as an overlay on the step to produce an opaque surface with $F_1 \cong 0.95$ and $F_2 \cong 0.05$ (~ 1000 A of silver is usually used). The reference flat is commonly coated with silver so that $F_1 \cong 0.94$ and $F_3 \cong 0.01$, where F_3 is the fraction of light transmitted.

Fringes of Equal Chromatic Order. The procedure is the same as for Fizeau fringes, except that the parallel incident light is white light. Equation (3-17) applies, but λ is no longer single valued. For a given value of S, Eq. (3-17a) gives the condition for those wavelengths which will be present in the reflected light,

$$2S = a_2\,\lambda_1 = (a_2 + 1)\,\lambda_2 = \ldots = (a_2 + i)\,\lambda_{i+1}$$

These wavelengths are measured with a spectroscope. Thus, adjacent lines in the spectrograph correspond to different λ's and different orders, where $(a_2 + i)$ is the chromatic order of any given line.

When a step is present on the substrate, S changes to $S + t$ and, using Eq. (3-17a), $a_2\,\lambda = 2S$ and $a_2\,\lambda' = 2(S + t)$ where λ and λ' are the wavelengths which give constructive interference at the substrate and film surfaces, respectively. Or,

$$2(S + t) - 2S = 2t = a_2\,(\lambda - \lambda') = a_2\,(\Delta\lambda) \qquad (3\text{-}19)$$

When a_2 is known, t can be calculated since $\Delta\lambda$ can be measured with the spectroscope. The order, a_2, may be determined by measuring the wavelength of two adjacent fringes. Since $a_2\lambda_1 = (a_2 + 1)\,\lambda_2$, then $a_2 = \lambda_2/(\lambda_1 - \lambda_2)$ where λ_2 is the wavelength of the fringe on the shorter wavelength side. The wavelengths λ_1 and λ_2 are also measured with a spectroscope. For the fringes corresponding to λ_1 and λ_2, Eq. (3-19) becomes

$$t = \left(\frac{\lambda_2}{2}\right)\left(\frac{\Delta\lambda_1}{\lambda_1 - \lambda_2}\right) \qquad (3\text{-}20)$$

Fringes of equal chromatic order appear as in Fig. 3-31, which also contains a schematic representation of typical experimental apparatus.

Previous comments on factors which affect the resolution of Fizeau fringes also apply to this method, but are subject to the following:

1. In this method, the beams can be made to originate from a smaller area, so the requirements on the reference flat are less stringent. Selected fire-polished glass is often used for the reference flat.

2. The phase change, δ, accompanying reflection may vary with wavelength. Since the phase change for aluminum is fairly insensitive to wavelength, it is often used even though its reflectivity is lower than that of silver. The error introduced by δ being unequal to π decreases as the order, a, increases.

As previously stated, the maximum resolution of the multiple beam

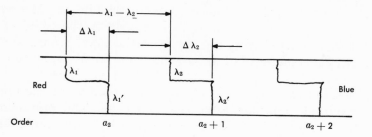

(a) Appearance of fringes of equal chromatic order

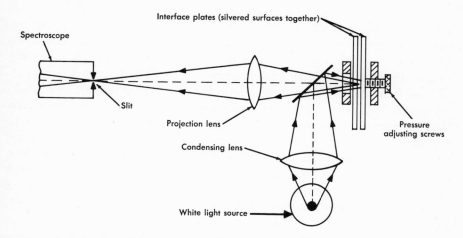

(b) Apparatus for producing fringes of equal chromatic order

FIG. 3-31. Fringes of equal chromatic order. (Reproduced with permission from L. Holland, *Vacuum Deposition of Thin Films*, 1963, Chapman and Hall, London.)

interferometer is about ± 5 A. The factors listed above may degrade the resolution, and a systematic error is introduced when $\delta \neq \pi$, so the overall error is commonly several times the maximum resolution. Reproducible results are not obtained for films which have an average thickness less than that necessary to produce a continuous film. Thus, the thinnest silver film which could be measured with multiple beam interferometry has a thickness of about 50 A. The thickest films which can be measured are on the order of 20,000 A.

Beta Particle Back-Scattering. When beta particles (high-energy electrons ejected from the nucleus of an atom during radioactive decay) are directed onto a film, some are scattered back in the general direction of the source. The thickness of the film (up to a certain maximum value) can be determined by comparing the count of beta particles back-scattered from the film with the count from a film (or films) of the same material of known thickness on the same substrate material.

Advantages of this method are that no preparation of the film (or step) is required, it is nondestructive and rapid, and minimum operator skill is required. A disadvantage, however, is that the thickness is not obtained directly; calibration against standards is required and, when the substrate has a rough surface, determination of the thickness of the standard is difficult. Other disadvantages are that certain film-substrate combinations produce unacceptably low accuracy, and the minimum area which can be measured is about 0.1 inch in diameter. The thickness of multilayer or alloy films usually cannot be measured with this method.

The fraction, F, of the incident beta particles which are back-scattered decreases with increasing energy of the incident beta particles, increases with atomic number of the target, and increases with target thickness up to about 0.2D. D is the minimum surface density, in grams per square centimeter, which stops all beta particles. The limiting value of F is designated F_∞ and occurs when the target surface density is somewhat greater than 0.2D. The general form of the dependence of F on these factors is illustrated in Fig. 3-32 [25]. For beta particles of energy 0.223 mev penetrating aluminum, D is about 0.05 g/cm^2. Assuming bulk density of 2.7 g/cm^3, the thickness corresponding to $D = 0.05$ g/cm^2 is 1.85×10^{-2} cm. The minimum surface density for most common substrates is the same order as for aluminum.

When the target consists of a film on a substrate, the general form of the dependence of the fraction of beta particles back-scattered on film thickness (for a given beta particle energy) is shown in Fig. 3-33. Note that when F_∞ (film) approaches F_∞ (substrate), the change in the fraction back-scattered when the film is added approaches zero. Thus, for maximum sensitivity, the difference between F_∞(film) and F_∞(substrate) should be as large as possible.

Muller [26] gives data on the dependence of F_∞ on the atomic number, Z', for certain elements. F_∞ is linear with Z' within each period of the periodic system. The back-scattering from a compound, G_xH_y, can be determined by using an effective atomic number, Z'_{eff}, which is given by Muller as

$$Z'_{eff} = \frac{x(A_GZ'_G) + y(A_HZ'_H)}{\text{mol. wt. of } G_xH_y}$$

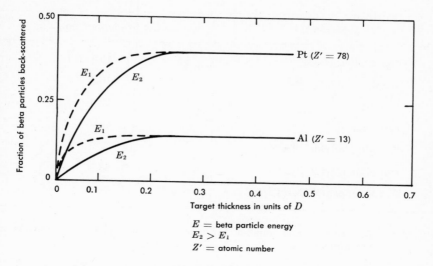

FIG. 3-32. General form of the dependence of the fraction of beta particles back-scattered on beta particle energy, target thickness, and atomic number of the target.

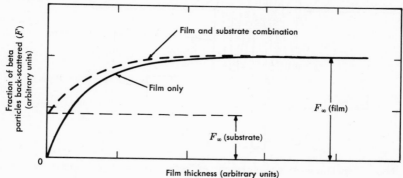

FIG. 3-33. Back-scattering of beta particles from a film-substrate combination.

where A_G and A_H are the atomic weights of G and H, respectively. Calculated values of Z'_{eff} for some common substrate materials and tantalum films are given in Fig. 3-34.

A typical arrangement of the apparatus used in this method is shown in Fig. 3-35. The radioactive source emits beta particles, mainly upward. These particles interact with the medium through which they travel (air, film, substrate, etc.), and three interactions are possible: transmission, absorption, or reflection (back-scattering).

Material	Effective atomic no., Z'_{eff}
Fused silica (SiO_2)	10.8
Sapphire (Al_2O_3)	10.8
7059 Glass	23.0
Tantalum	73.0
Tantalum nitride (Ta_2N)	70.5
Tantalum nitride (TaN)	68.2
Tantalum pentoxide (Ta_2O_5)	61.2

FIG. 3-34. Effective atomic number for some common substrate materials and tantalum films.

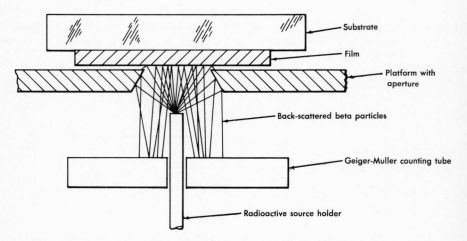

FIG. 3-35. Experimental arrangement of beta particle back-scattering thickness measuring apparatus.

Most particles which are back-scattered enter the Geiger-Muller tube, where they produce voltage pulses. These pulses are counted and summed over a selected period of time. Comparison of the counts thus obtained to the counts from a standard of the same material of known thickness, measured with the same experimental apparatus, allows the film thickness to be determined.

All beta particle sources release beta particles which range in energy from zero up to some maximum energy E_o, the end-point energy. Three common sources are Carbon 14, Promethium 147, and Thallium 204. The type of decay, the end-point energy, and the half-life for these materials are given in Fig. 3-36.

Material	Type of decay*	E_o (mev)	Half-life (years)
C^{14}	β^-	0.155	5700
Pr^{147}	β^-	0.223	2.26
Tl^{204}	β^- plus K_c	0.765	2.7

*β^- is the emission of an electron. K_c is the capture by the nucleus of an electron from the K shell. This process is followed by emission of X-rays as the K shell is filled.

FIG. 3-36. Common sources for beta particles.

Figure 3-32 indicates that the amount of back-scattering for thin targets is higher for lower-energy beta particles. Thus, C^{14} should give higher sensitivity, but practical difficulties are encountered in the use of this source. The specific activity (disintegrations per second per unit mass) is inversely proportional to the half-life, so, for a given size source, the C^{14} has a much lower disintegration rate.

The lower end-point energy and the absence of X-ray emission which accompanies K shell electron capture make Pr^{147} the most widely used source. Sources of about 10^{-5} curies are common (1 curie $= 3.7 \times 10^{10}$ disintegrations/sec). A source of this activity can be handled without special precautions, but must be used in accordance with Atomic Energy Commission regulations. The source should be close to the film to minimize the number of beta particles back-scattered by the air.

Radioactive decay is a random process, so the number of beta particles emitted per unit time fluctuates. Relatively long measurement times (1 to 10 minutes) average these fluctuations and produce larger total counts which can be measured more accurately.

The platform, which contains the aperture, should back-scatter a minimum number of beta particles and allow no beta particles to pass outside of the aperture. Thus, the platform material should have a low atomic number and be thick enough so that no beta particles are transmitted (aluminum is commonly used). The platform must be in intimate contact with the film so that the area bombarded by the beta particles is reproducible.

The Geiger-Muller tube, which collects most of the back-scattered beta particles, is placed below the source so that it receives only back-scattered particles. Geiger-Muller tubes have a dead time on the order of 100 microseconds after each count, and particles entering the tube during this time are not counted. Maximum sensitivity is achieved when the average time interval between the arrival of beta particles is longer than the dead time. In practice, this condition can be achieved by placing a filter (such as a 0.0005-inch aluminum sheet) over the face of the Geiger-Muller tube to filter out a fraction of the beta particles, but much longer counting times are required.

Beta particles enter the Geiger-Muller tube from several sources such as the film, the substrate, the air, and the platform. Experimentally, two quantities can usually be measured: the number, N_s, of particles back-scattered from the bare substrate (plus background), and the number, N_t, of particles back-scattered from the substrate and the film (plus background).

If the substrate is homogeneous, thicker than the penetration depth of the beta particles, and one side contains no film, this bare side can be placed over the aperture to obtain N_s. For most substrates, a value of N_s obtained at any point on the substrate is typical of the substrate. For nonhomogeneous substrates (e.g., a ceramic substrate coated with a thin glaze), N_s may vary from point to point on the substrate. The quantity N_t is determined by placing the film side on the aperture.

Two parameters approximately proportional to the number, N_f, of particles back-scattered from the film only are often calculated. The parameter $\alpha_1 = N_t - N_s$ is approximately equal to N_f, but the relation is not exact since N_s with the film in place is less than N_s without the film. This is because the number of beta particles incident on the substrate with the film in place is smaller because of the back-scattering and adsorption in the film; therefore, the number back-scattered from the substrate is smaller. The magnitude of α_1 depends on the source activity. A parameter α_2 which does not depend on the source activity is obtained by defining $\alpha_2 = (N_t - N_s)/N_s \cong N_f/N_s$. Whether α_1 or α_2 is used is a matter of choice since either parameter must be compared to a like parameter obtained from standards.

The quantity N_f is approximately linear with film thickness for film mass per unit area much less than $0.2D$ (see Fig. 3-32), and the linearity of N_f with film thickness is valid up to thicknesses of several thousand angstroms for most thin film materials. Thus, α_1 and α_2 are approximately linear with film thickness, and only one or two standards are usually required for a given film composition.

By repeatedly measuring tantalum film on Corning Code 7059 glass, a one-sigma deviation of about $\pm$ 30 A can be obtained. To determine the absolute accuracy, the error in determination of the thickness of the standards must be added.

When the film to be measured is a compound or mixture, the amount of back-scattering depends on the relative amount of the constituents. With a given source and apparatus geometry, Figs. 3-32 and 3-33 indicate that the amount of back-scattering depends on the effective atomic number and the surface density of the film. Variations in either of these quantities change the amount of back-scattering for a given film thickness, so one must be fixed before the other can be determined.

X-Ray Fluorescence. When a thin film (or any other target) is bombarded with X-rays of the proper energy, it absorbs some of these primary X-rays and emits secondary X-rays of wavelengths characteristic of the target material. Thus, the film becomes a source of secondary X-rays. The intensity of the emerging secondary X-ray depends on the type and amount of film material, because these factors affect the amount of primary X-rays absorbed and the attenuation of the secondary X-rays as they emerge from the film. Up to a certain thickness, the intensity of the secondary X-rays can be used to determine the mass per unit area of film.

Advantages of X-ray fluorescence include: (1) no film preparation, (2) it is nondestructive (discoloration, usually temporary, of glass substrates is sometimes observed), (3) for most substrates, the quantities measured are independent of the substrate materials, and (4) absolute values for mass per unit area can be obtained without comparison to standards of known thickness.

Disadvantages of this method are that the thickness is not obtained directly (calibration against standards or an assumption of the density of the film is necessary), and the equipment is the most expensive of any discussed. In addition, the thickness obtained is an average thickness over a small area, and the thickness of multilayer or alloy films usually cannot be measured unless the composition is known.

The instrument used in this method is an X-ray spectrograph, shown schematically in Fig. 3-37.

The wavelength, λ_p, of the primary X-rays must be less than the wavelength, λ_s, of the secondary X-rays in order for any secondary X-rays to be produced. Typically, the wavelength of the primary X-rays is less than 1 A.

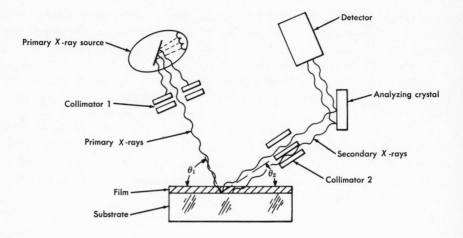

FIG. 3-37. Schematic representation of an X-ray spectrograph.

The primary X-rays pass through a collimator (collimator 1, Fig. 3-37) which limits the breadth of the beam and allows the angle, θ_1, to be well defined. The primary X-rays are partially absorbed in the film and excite secondary X-rays, which issue in all directions.

Collimator 2 allows secondary X-rays emerging at an angle θ_2 to impinge onto the analyzing crystal, which is oriented to reflect only secondary X-rays of the desired wavelength into the detector. A scintillation counter is used to detect the secondary X-rays.

The loss of power in this process is tremendous; in a typical arrangement, the number of X-ray quanta reaching the detector is only about 10^{-14} of that emitted by the tube.

The intensity of the secondary X-rays decreases with film thickness below a thickness of about 10^{-3} cm (100,000 A). The intensity measured depends on the geometry of the apparatus and certain properties of the film. Consider the ideal case where the primary X-rays are monochromatic, the film is a pure element, and excitation of secondary X-rays in the film by secondary X-rays from the substrate does not occur. The intensity, I_s, of the secondary X-ray beam can be expressed in terms of the intensity, I_o, of the primary X-ray beam and a factor K_1 (which depends on the conversion efficiency of primary to secondary X-rays) as $I_s = I_o K_1 \rho t$ where ρ is the density and t is the film thickness [27]. For thin films, K_1 is a constant. A standard of the same material and a known thickness is used to determine the product $K_1\rho$. Either I_s or I_s/I_o for the film of unknown thickness is then compared to the corresponding value for the film of known thickness.

Alternately, I_s can be expressed in terms of $I_{s\infty}$, the saturation value of I_s when the thickness is great enough that $dI_s/dt = 0$, as

$$I_s = I_{s\infty}(\mu_1 \csc \theta_1 + \mu_2 \csc \theta_2)\rho t$$

where μ_1 and μ_2 are the mass absorption coefficients for the primary and secondary X-rays, respectively, and θ_1 and θ_2 are the same as in Fig. 3-37. A piece of bulk material may be used to obtain $I_{s\infty}$; both μ_1 and μ_2 can be obtained from tables such as those in Ref. [27]. The product ρt for the film of unknown thickness can then be determined. To obtain the thickness t, a value of ρ must be assumed or comparison must be made with a standard.

The reproducibility obtainable with this method appears about the same as that obtained with multiple beam interferometry, and the absolute accuracy is comparable to that of the beta particle backscattering method.

The characteristic wavelengths and mass absorption coefficients of the constituents are affected very little by the chemical form of the material because the generation of characteristic X-rays involves electrons near the nucleus rather than the valence electrons. If the characteristic X-rays of one constituent of the mixture or compound do not interact with the other constituent, each constituent may be treated as if the other were not present. Interactions occur when both elements significantly absorb the primary X-rays (but in different proportions) or when one element absorbs the characteristic X-rays of the other. In this case, calibrations must be made against standards of known composition. The substrate may also contribute to the count if the characteristic X-rays of the substrate are of the proper wavelength to be reflected into the detector by the analyzing crystal or to excite characteristic X-rays in the film.

In-Process Deposition Monitors

The two general-purpose monitors commonly used to monitor the deposition of many vacuum-deposited films are quartz crystal oscillators and ionization gauges. These monitors are robust and require a minimum of preparation, maintenance, and operator skill. Neither measures film thickness; the quartz crystal monitor determines mass, and the ionization monitor determines the flow of vapor through an electrode system.

Most deposition monitors require deposition on or through an area other than the substrates, and allowances should be made for geometrical effects. Mechanical microbalances used inside the vacuum system, which are delicate and require considerable operator skill, are not discussed.

Quartz Crystal Monitors. Quartz crystal oscillators may be used to monitor the deposition of dielectric, resistive, and conductive films. They are simple to use, the sensing elements are small, and they have good sensitivity.

A thin quartz crystal, with previously deposited electrodes, receives a deposit at the same time as the substrates. The rate of change in resonant frequency is proportional to the deposition rate, and the total change in frequency during deposition is proportional to the mass of the deposit. When the crystal is made part of an oscillator, measurement of the change in oscillator frequency permits a determination of the deposition rate or the mass of the deposit. Indications of both quantities are usually provided in commercial monitors of this type.

The temperature coefficient of frequency for the crystal varies with the crystal cut, commonly being from a fraction of a part to a few parts per million per °C. Unless care is taken to prevent heating of the crystal, these frequency changes may introduce significant errors.

The crystal in Fig. 3-38 [28] has electrodes on the top and bottom. Contact is by a spring clip which allows for easy replacement of the crystal. The contact to the crystal must be of repeatable low resistance to obtain a high crystal Q in order to allow loading of the

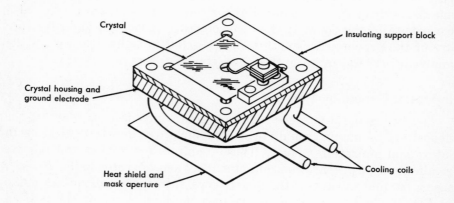

FIG. 3-38. Quartz crystal monitor.

crystal to large frequency shifts. Crystals can be cleaned and reused, provided the electrodes are undamaged or are redeposited. Frequency shifts of up to 5 percent of the fundamental crystal frequency (corresponding to many tens of thousands of angstroms) between cleanings are common.

Sometimes an aperture is used to limit the deposition area on the crystal to less than the crystal electrode area. The aperture should be positioned over the center of the crystal (for high sensitivity) and should always cover the same part of the crystal to insure repeatable calibrations.

Ionization Monitors. This method can be used to monitor most vacuum evaporation processes. It cannot be used to monitor sputtering rates, however, because of the large number of ions and electrons present.

Thermionic ionization gauges are the sensing elements. A shield is usually placed around the gauge electrodes to limit the access of residual gas molecules. The gauge is positioned so that evaporant streams through a hole in the shield into the gauge. The fraction of evaporant atoms ionized depends on the number of evaporant atoms, their ionization probability, and the electron current, as discussed in Chap. 2. Calibration for each type of evaporant is necessary.

Residual gas molecules are also ionized and contribute a background in current which may be of the same order as the ion current from the evaporant. Since the amount of residual gas changes during evaporation, its contribution to the ion current cannot be determined by a measurement taken before evaporation is started. Two techniques, which utilize the directional property of the evaporant beam, are used to minimize the background effects. One of these techniques uses two ionization gauges, one of which is exposed to the evaporant beam, with the other isolated from it. The difference between the two ion currents is assumed to be due to the evaporant ion current. The other technique uses a rotating mechanical shutter to modulate ion current by interrupting the evaporant beam just before it passes into the ionization gauge, and only the modulated ion current is amplified.

Ionization monitors are particularly suited for use with resistance heated sources, but, even with these sources, suppression of electrons from the sources by biasing the sources is often necessary. When electron bombarded sources are used, many evaporant ions are produced at the source, and ionization gauges must be specially shielded to prevent erroneous readings.

Another difficulty encountered with ionization monitors is the condensation of evaporant on the electrodes, which is particularly bothersome when the evaporant is a dielectric. This may be overcome by using electrodes which can be heated to reevaporate the deposit.

Resistance Monitors. Perhaps the simplest monitor is the use of a special substrate with conductors on each end to act as electrical terminals, with the area between the conductors receiving the film. A film thickness can be calculated by assuming a value for the resistivity; however, the sheet resistance measured on the special substrate is usually used for controlling the deposition process since the sheet resistance on the substrates is a quantity which must be directly controlled. Both wheatstone bridges and constant-voltage, current-measuring circuits have been used. In any case, it is desirable to know the temperature coefficient of resistance of the depositing film and the temperature of the special substrate if precise film resistance is required.

REFERENCES

1. Bond, W. L. "Notes on Solution of Problems in Odd Job Vapor Coating," *J. Opt. Soc. Am.*, 44, 429-438 (1954).
2. Kelley, K. K. "Contributions to the Data on Theoretical Metallurgy." Washington, D. C.: U. S. Government Printing Office, 1935.
3. Honig, R. E. "Vapor Pressure Data for the Solid and Liquid Elements," *RCA Review*, 23, 567-586 (1962).
4. *Thin Films*. American Society for Metals. Metals Park, Ohio, 1964.
5. Pashley. D. W., and M. J. Stowell. "Nucleation and Growth of Thin Films as Observed in the Electron Microscope," *J. of Vac. Sci. and Tech.*, 3, 156-167 (1966)
6. Holland, L. *Vacuum Deposition of Thin Films.* London: Chapman and Hall, Ltd., 1963.
7. Neugebauer, C. A., J. B. Newkirk, and D. A. Vermilyea, editors. *Structure and Properties of Thin Films.* New York: John Wiley and Sons, Inc., 1959.
8. Matthews, J. W. "Growth of Face-Centered-Cubic Metals on Sodium Chloride Substrates," *J. of Vac. Sci. and Tech.*, 3, 133-146 (1966).
9. Behrndt, K. H. "Angle of Incidence and Stress Effects on Rotating and Stationary Substrates," *J. Vac. Sci. Tech.*, 2, 63-70 (1965).
10. Chopra, K. L. "Influence of Electric Field on the Growth of Thin Metal Films," *J. Appl. Phys.*, 37, 2249-2254 (1966).
11. Benjamin, P., and C. Weaver. "The Adhesion of Evaporated Metal Films on Glass," *Proc. Roy. Soc.*, CCLXI, 516-531 (1961).
12. Van Audenhove, J. "Vacuum Evaporation of Metals by High-Frequency Levitation Heating," *Rev. Sci. Instr.*, 36, 383-385 (1965).
13. McLean, D. A. "Metallized Paper for Capacitors," *Proc. of the IRE*, 38, 1010-1014 (1950).

14. Dushman, S. *Scientific Foundations of Vacuum Technique*, edited by J. M. Lafferty. New York: John Wiley and Sons, Inc., second ed., 1962.
15. Holland, L., editor. *Thin Film Microelectronics*. New York: John Wiley and Sons, Inc., 1965.
16. Schäfer, V. H., and R. Hörnle. "Die Silicummonoxyd Drucke uber den festen Bodenkörpern Silicum und Silicumdioxyd," *Z. Anorg. Allgem. Chem.*, **203**, 261-279 (1950).
17. Tombs, N. C., and A. J. E. Welch. "Thermodynamic Properties of Silicon Monoxide," *J. Iron and Steel Institute*, 172, 69-78 (1952).
18. Priest, J., H. L. Caswell, and Y. Budo. "Mechanical Stresses in Silicon Oxide Films," *Vacuum*, 12, 301-306 (1962).
19. Drumheller, C. E. "Silicon Monoxide Evaporation Techniques," *Trans. 7th Nat. Vac. Symposium, 1960*. New York: Pergamon Press, 1961. pp. 306-312.
20. Priest, J. R., H. L. Caswell, and Y. Budo. "Mechanical Properties of Silicon Oxide Films," *Trans. 9th Nat. Vac. Symposium, 1962*. New York: The Macmillan Company, 1962, pp. 121-124.
21. Mattox, D. M. "Film Deposition Using Accelerated Ions," *Electrochem. Tech.*, 2, 295-298 (1964).
22. Mattox, D. M. "Design Considerations for Ion Plating," Sandia Corporation Reprint SC-R-65-997. Clearinghouse for Federal Scientific and Technical Information, Springfield, Va. (Jan., 1966)
23. Schwartz, N., and R. Brown. "A Stylus Method for Evaluating the Thickness of Thin Films and Substrate Surface Roughness," *Proceedings of the Second International Vacuum Congress*. New York: Pergamon Press, 1962. pp. 836-845.
24. Tolansky, S. *Surface Microtopography*. New York: Interscience Publishers, Inc., 1960.
25. Bleuler, E., and G. J. Goldsmith. *Experimental Nucleonics*. New York: Holt, Rinehart and Winston, 1952.
26. Muller, R. H. "Interaction of Beta Particles with Matter," *Analytical Chemistry.* 29, 969-975 (1957).
27. Liebhafsky, H. A., H. G. Pfeiffer, E. H. Winslow, and P. D. Zemany. *X-Ray Absorption and Emission in Analytical Chemistry*. New York: John Wiley and Sons, Inc., 1960.
28. Behrndt, K. H., and R. W. Love. "Automatic Control of Film-deposition Rates with the Crystal Oscillator for Preparation of Alloy Films," *Vacuum*, 12, 1-9 (1962).

PRINCIPAL SYMBOLS FOR CHAPTER 3

SYMBOL	DEFINITION
a_2	Integer
A	Area
C	Proportionality constant in film thickness equation
D	Minimum surface density which stops all beta particles
G	Rate of evaporation
G_A	Rate of evaporation of component A of an alloy
ΔH_v	Heat of vaporization
L	Mean free path of gas molecules
m_1	Mass of evaporant
m_2	Mass of deposit
M	Molecular weight
M_A	Molecular weight of component A of an alloy
N_s	Number of beta particles back-scattered from bare substrate (plus background)
N_t	Number of beta particles back-scattered from substrate and film (plus background)
p_A	Vapor pressure of element A
p_v	Vapor pressure
r	Distance from source to receiving surface
R_o	Gas constant

Symbol	Definition
S	Separation between substrate and reference flat
S_2	Area of receiving surface
t	Film thickness
t_e	Equivalent film thickness
T	Temperature in °K
Z	Arrival rate of residual gas molecules
ϵ	Emissivity
λ	Wavelength

Chapter 4
Sputtered Films

Sputtering entails the bombardment of a target with energetic particles (usually positive gas ions) which cause some surface atoms to be ejected from the target. These ejected atoms deposit onto any solid which may be close to the target. Much of the early work on sputtering was prompted by the difficulties encountered in cathode erosion in electron tubes, and was oriented toward learning the mechanism of the interaction between the bombarding ions and the target, rather than studying the properties of the deposited material. Recently, however, increasing effort has been placed on the study of the sputtering process as a means of depositing films, and on the interrelation between the film properties and the deposition parameters.

The sputtering phenomenon has been known for well over a century[1,2], and sputtering was used as a method for depositing a film as early as 1877. The vacuum pumps available at that time were mechanical pumps of the crudest sort (probably operated by the laboratory assistant in a manner similar to that of the pumps then used for pipe organs). Because of this, vacuum evaporation was not available to the experimenter, and the only sputtering gas available was the residual air in the vacuum chamber. Figure 4-1 shows the apparatus used in 1842 by W. R. Grove in the discovery of cathodic sputtering. It is not surprising, then, that the films produced had properties which could not be reproduced in the same laboratory, much less by other experimenters.

With the advent of better pumps, vacuum evaporation became available as a method of producing films, and the sputtering process fell not only into disuse, but into disrepute. For the next several decades, sputtering studies were confined to investigations related to

191

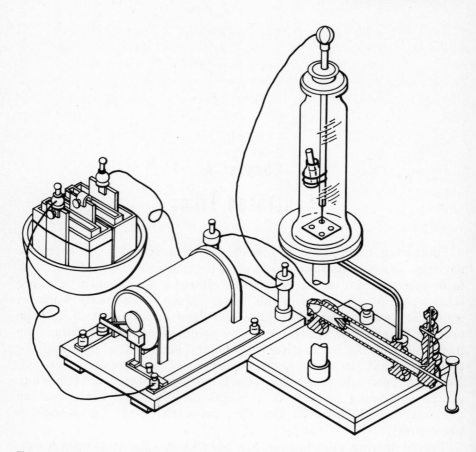

FIG. 4-1. The apparatus used by W. R. Grove in the discovery of cathodic sputtering.

the interaction between the target and the ions bombarding it, and to the related general phenomenon of gaseous discharge. That research which was done on film deposition by sputtering during these early years was primarily related to the noble metals, which could be deposited with an air plasma. The results of these experiments were such that much conflicting data about the properties of sputtered films entered the literature. Thus, most of the literature concerning sputtered films prior to 1950 is unreliable.

In recent years, the use of sputtering as a method of depositing films of controlled properties has gained considerable momentum. The increasing interest in the refractory metal films for electronic applications has resulted in a "rebirth" of sputtering as a convenient method of preparing films. Films may now be deposited with electrical properties comparable to those prepared by evaporation in an ultrahigh vacuum. More effort is being placed on the effects of various sputtering atmospheres on the properties of the films produced, and effort is also being expended on attempts to ascertain the relationship between the electrical parameters of the plasma and the structure of the film. Several review articles on the sputtering process have been published recently by Wehner[3], Kay[4], Moore[5], and Wolsky[6], and a recent book by Kaminsky[7] contains an extensive chapter on sputtering. Maissel[8] has most recently reviewed the use of sputtering for film deposition.

4.1 THEORIES OF SPUTTERING

Two theories have been proposed to explain the ejection of material from a target being bombarded by energetic ions. The first, proposed by Crookes[9] in 1891 and von Hippel[10] in 1926, postulates that sputtering is due to the creation of very high local temperatures by the bombarding ions, leading to the evaporation of the target material from those local areas. This theory predicts that the sputtering rate is a function of the heat of sublimation of the target, and of the energy of the bombarding ions. The evaporation theory also predicts that the ejected atoms of metal show a cosine distribution. This theory gained wide acceptance when the predictions appeared to be in agreement with the experimental data available. Most of this data, however, had been obtained from experiments performed in a glow discharge, and many of the fine points were obscure because the bombarding ions have a wide range of energies and directions in a glow discharge.

A more tenable theory, proposed by Stark[11] in 1909, suggests a mechanism involving the direct transfer of momentum from the bombarding ion to the atoms of the target. Considerable data have been collected which support this theory, including data on sputtering yield (the ratio of the number of atoms sputtered to the number of bombarding ions), and data on the angular distribution of ejected atoms which were sputtered from single crystal targets. The main observations in support of a momentum transfer mechanism are summarized below:

1. The distribution of atoms ejected from single crystal targets does not obey a cosine law, but tends to concentrate along the directions of closest packing in the crystal.

2. Sputtering yields do not depend solely on the energy of the bombarding ion, but also on its mass.
3. There is some threshold energy, below which sputtering does not occur.
4. Sputtering yields decrease at very high ion energies, probably because of the deep penetration of the ion beneath the surface of the target.
5. The energies of the sputtered atoms are many times higher than could be exhibited by thermally evaporated atoms.
6. Electron bombardment has not been observed to produce sputtering, even at very high energies.

The experiments which produced the above observations were done in a relatively high vacuum using a low-pressure plasma or ion beam bombardment, and the results were therefore not confused by the presence of a glow discharge.

Some of the experimental data which illustrate the above observations are outlined below.

Distribution of Sputtered Atoms

The noncosine behavior of the ejected material is illustrated in Fig. 4-2, which shows the angular distribution of molybdenum atoms ejected by obliquely incident Hg^+ ions at 250 electron volts (Wehner[12]). In this case, the target was polycrystalline. By bombarding single crystal targets with low-energy ions (300 electron volts), Wehner found that the atoms are preferentially ejected in the directions of close packing. The current view is that the energetic ions cause chains of focused collisions similar to those which occur when one billiard ball strikes a tightly packed array of similar balls. The following brief description shows how this focusing can occur. In order for an impulse to become focused, the ratio of the interatomic distances in the lattice to the atomic radius must be favorable. Figure 4-3 illustrates this concept. The upper part of this figure shows the location of the atoms before collision as solid circles, and their location at the time of impact as dotted circles. The angle chosen for the initial collision is represented by an arrow in the figure. The angle of collision becomes less and less as the distance proceeds down the line of atoms; i.e., focusing does occur. In the lower half of the figure, the same process is shown, except the collision radius (twice the atomic radius) is used. It can be shown, by using simple geometry, that the collision angle may grow smaller only if the interatomic distance is smaller than twice the collision radius, and will always grow smaller if the interatomic distance is smaller than $\sqrt{2}$ times the collision radius. In a crystal,

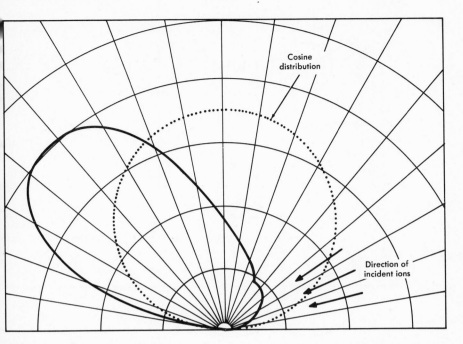

Fig. 4-2. Angular distribution of material sputtered from molybdenum target
under oblique incidence of the Hg$^+$ ions [12]. (After G. K. Wehner,
"Angular Distribution of Sputtered Material;" courtesy *Journal of
Applied Physics*.)

the interatomic distances are different in different directions, being
smallest in the closest packed directions, so that the chance is high
for focusing in these directions, and the sputtering yields are also
expected to be high.

Sputtering Yield

Experiments have been done during the last decade toward the
determination of the sputtering yield of many different target metals
bombarded with ions of many different elements, and over broad
ranges of ion energy. Wehner has done a large portion of the low-
energy work, and Almen and Bruce[13], as well as Kistemaker[14],
have provided information at the higher energies. Figure 4-4 shows
the sputtering yield of several metals as a function of the atomic
number of the bombarding ion. There is a general increase in yield
as the mass of the ions is increased, but there is a periodic undulation
superposed on this trend which coincides with the grouping of the
periodic table. It is interesting to note that the maxima occur for
the points representing the rare gases. As yet, there is no valid ex-

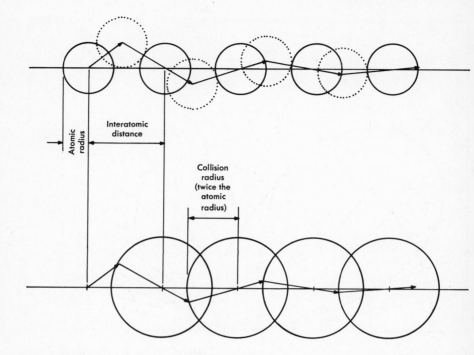

FIG. 4-3. A focusing collision chain in a row of atoms. (After Rol, Onderlinden, and Kistemaker, "Some Physical Aspects of Sputtering," in *Advances in Vacuum Science and Technology*, Vol. I, 1966; courtesy Pergamon Press.)

planation for this periodicity.

Figure 4-5 shows the typical relationship between sputtering yield and the energy of the bombarding ion. This curve may be divided into three general areas: first, the portion at low voltages, where little or no sputtering occurs; second, the area between about 70 ev and 10,000 ev, where the yield increases with increasing voltages, and which includes the area where most film deposition work is done; and third, the area above about 30,000 ev, where a decrease in yield is observed with increasing voltage. This decrease is assumed to be caused by penetration of the bombarding ion deep into the lattice, where a large portion of its energy is lost to the bulk of the target rather than to the surface. With heavier bombarding ions, this decrease occurs at higher energies.

Many measurements of threshold energies have been made by extrapolation of yield curves to zero yield. This is a difficult task, since the measurements must be very sensitive to detect very small amounts of sputtering. Older methods, which involved the detection

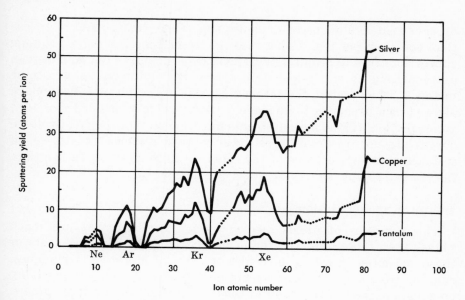

FIG. 4-4. Sputtering yield for silver, copper, and tantalum as a function of
bombarding ion at 45 kev. Experimental errors are ±10 percent. (After
Almen and Bruce, "High Energy Sputtering," in *Trans. 8th. Nat. Vac.
Symposium, 1961*, 1962; courtesy Pergamon Press.)

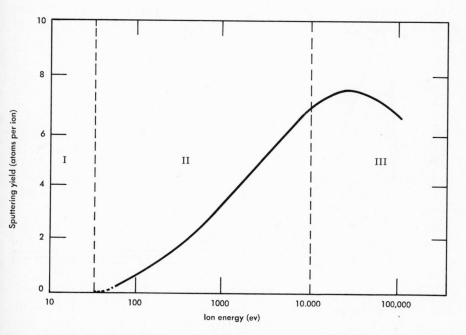

FIG. 4-5. Copper bombarded by argon ions. (After L. I. Maissel, "Deposition
of Thin Films by Cathodic Sputtering," in *Physics of Thin Films*, Vol.
III, 1966; courtesy Academic Press.)

of a deposited film, become inadequate when yields are much less than one atom per ion. Using these methods, the yield-versus-voltage curves appeared to be quite linear, and experimenters felt confident in extrapolating to zero yield. More recently, however, spectroscopic methods, which detect the target material in the plasma itself, have allowed measurements to be made down to yields of the order of 10^{-4} atoms per ion. These measurements have shown that the yield-voltage curves are not linear at these lower voltage regions, and threshold voltages now appear to be of the order of 20 to 25 volts, rather than the 30- to 150-volt figures obtained previously.

Energy of Sputtered Atoms

The velocities of atoms ejected by ion bombardment are much higher than those observed for evaporated atoms. Measurements of these velocities have been made by several different techniques, ranging from a direct measurement of the force of the atoms arriving on a plate to the measurement of the time of flight of sputtered atoms. Figure 4-6 gives some typical energy distributions for copper atoms bombarded with ions with energies below 1500 electron volts. It should be recalled, however, that these measurements were all performed

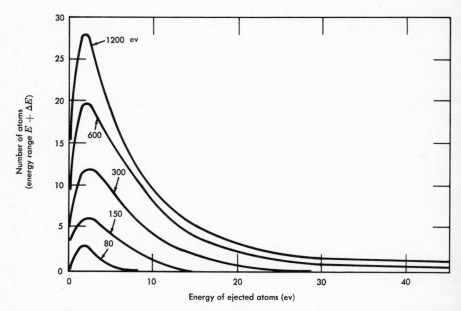

FIG. 4-6. Distribution of energy for copper atoms ejected in the [110] direction for krypton ion bombardment with energies from 80 to 1200 ev. (After Stuart and Wehner; courtesy *Journal of Applied Physics*, **35**, 1819-1826 (1964).)

at relatively low pressures, and that for deposition performed in a glow discharge, the energy of an atom arriving at a substrate, while still higher than that of evaporated atoms, will be decreased in relation to the number of collisions suffered with the molecules of gas enroute.

4.2 SPUTTERING IN A GLOW DISCHARGE

Although there are several techniques available for generating the positive ions necessary for sputtering, by far the simplest method is by establishing a glow discharge. This is usually done by setting up a high potential between two flat parallel electrodes in a low-pressure (1 to 100 millitorr) gas.

Glow Discharge Characteristics

Several appropriate reviews are available [15] which cover the complete subject of discharges in gases. Only those areas pertinent to sputtering phenomena will be discussed here.

There are several types of gas discharge, and the type which is observed between two electrodes depends upon several factors: the pressure of the gas; the applied voltage; and the electrode configuration, which influences the path length of the discharge and its current density. Figure 4-7 shows the current-voltage characteristics of a discharge between two flat plates in a gas with a pressure in the range of from several millitorr to several hundred millitorr. There is no appreciable current below some minimum voltage at which the gas "breaks down" abruptly. This region is known as the Townsend discharge, and the current may be increased within this region without a change in voltage. As the current is increased further, however, additional carriers are created and the discharge exhibits negative resistance. As current is further increased, a second constant voltage region, known as the "normal" glow, is reached. The operation of voltage regulator electron tubes is based on this type of discharge. If the current is increased beyond a certain level, the voltage rises with increasing current, and this region is known as the "abnormal" glow. It is in this region that most sputtering work is performed. If the current is further increased, the voltage drops abruptly and the discharge becomes an arc.

Figure 4-8 is a diagrammatic illustration of the appearance of a glow discharge for a gas in the pressure region between 10 and a few hundred millitorr. Shown below the diagram are plots of various parameters of the discharge along its length. The transport of cur-

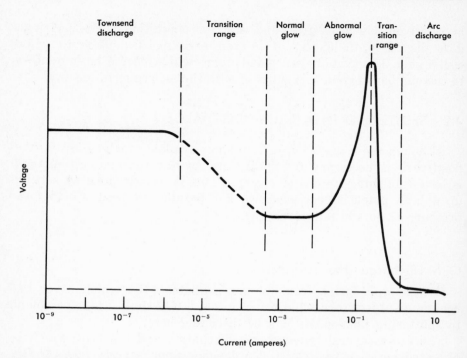

FIG. 4-7. Schematic of a gas discharge. The voltage is represented by a linear scale and the current by a logarithmic scale.

rent through a glow discharge occurs by the motion of electrons and positive ions parallel to the electric field. In order for the gas to be a conductor, however, some source of energy is necessary to continuously produce ions and electrons in the gas. Follow an electron emitted from the cathode, and consider its activities as it moves through space. This electron is first accelerated by the strong field adjacent to the cathode, but it initially makes few, if any, ionizing collisions because its energy is not sufficiently above the ionization potential of the gas. Further from the cathode, however, the electron has gained sufficient energy to cause ionization when it collides with a gas molecule, and, in effect, gives rise to electron multiplication. In order to have a steady state, each electron emitted from the cathode must produce sufficient ionization to effect the release of one further electron from the cathode. Most of the ionization which is necessary to sustain the glow occurs within the Crookes' dark space region. If the anode is moved toward the cathode into the region of the Crookes' dark space, the discharge is extinguished since there is now insufficient ionization to sustain the glow.

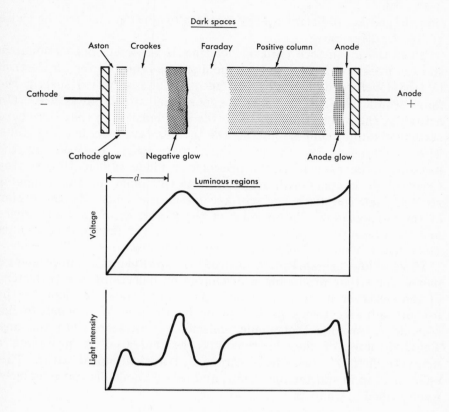

FIG. 4-8. Appearance of glow discharge at low pressure.

The various luminous and dark regions of a glow discharge arise in the following manner: An electron usually leaves the cathode with very small initial velocity, such that its energy is of the order of one electron volt. It is not able to excite gas molecules until its energy is as great as the excitation potential of the gas, and this results in a region called "Aston's dark space." The cathode layer is the region in which the electrons reach an energy corresponding to the ionization potential, and this is the luminous region closest to the cathode. At distances beyond the cathode layer, i.e., in the "Crookes' dark space," the electron energies are mostly far above the maximum excitation potential, so that little visible light is emitted. By the time the negative glow is reached, the number of slow electrons (i.e., those produced by an ionizing collision) has become very large and, while these electrons do not have sufficient energy to produce ionization, they do possess sufficient energy to cause excitation and are the cause of the negative glow.

Since the mean free path of electrons is inversely proportional to the gas pressure, it follows that the distance required for an electron to travel before it has produced adequate ionization to sustain the glow would also be inversely proportional to the pressure. (In actuality, the proportionality is to the gas density, so that the above statements apply only so long as the temperature remains constant.) The thickness of the Crookes' dark space, then, increases as the pressure is decreased. If the pressure is made sufficiently low, the Crookes' dark space will expand until the plane of the anode is reached, and the discharge will become extinguished. In the region of the normal glow, the product of the thickness of the dark space and the pressure is independent of current. For argon, this product (pd) has a value of about 0.3 torr centimeter.

If very low pressures are desired for sputtering, a supplemental means for either producing a discharge or increasing the trajectory of the electrons must be provided. The former may be done by the use of radio-frequency excitation or by the use of a separate hot cathode to provide thermionic emission of electrons into the gas. The trajectory of the electrons may be increased by providing a magnetic field to cause the electrons to travel in spiral paths. This effect is used in sputter-ion pumps and cold cathode discharge gauges, as described in Chap. 2.

Normal and Abnormal Glow Discharge

When a glow discharge is maintained between two plane electrodes, the potential drop across the discharge is a function of current, as was shown in Fig. 4-7. Over a certain range of current (which may be from two to three orders of magnitude), there is a constant voltage drop across the discharge. As was mentioned previously, this region is called the normal glow. In this region, the area of the cathode that is active in the discharge increases with increasing current such that the current density over the active portion of the cathode remains constant. When the entire cathode becomes active (completely covered by the glow), the current can be increased only by increasing the current density at the cathode. This increase in current density (j) increases the voltage drop across the discharge, according to

$$V = E + \frac{F\sqrt{j}}{p} \qquad (4\text{-}1)$$

and the region of the "abnormal" glow is reached. The increase in voltage produces a higher field at the cathode, and the dark space shrinks as shown in Eq. (4-2):

$$pd = A + \frac{BF}{V - E} \qquad\qquad (4\text{-}2)$$

A, B, E, and F are constants with values that depend on electrode material and geometry as well as on the gas composition.

The dark space continues to decrease in size with increasing voltage, until the region of an arc is reached. For deposition work, it is essential that sputtering be done in this region of the abnormal glow in order to assure the complete coverage of the cathode, and therefore a more uniform coverage of a substrate. The fact that the distance the dark space extends from the cathode is a function of the voltage in the region of the abnormal glow is often overlooked, but should be taken into consideration when designing a sputtering facility. If the cathode shield were spaced according to the position of the dark space for a normal glow discharge, it could become quite ineffective when operating well into the abnormal region. For example, 0.3 torr centimeter was quoted as the pd constant for argon in the normal glow region. This implies that at a pressure of 15 millitorr, the dark space would extend 20 centimeters from the cathode. At 5 kilovolts (well into the abnormal glow) and at the same pressure, however, the dark space is found to extend only about 3 centimeters.

In the abnormal glow region, a large number of the ions is produced in the negative glow. Any object, then, which interferes with the negative glow will of necessity affect the ion bombardment of that portion of the cathode being blocked. Ions and electrons do not recombine in a plasma at an appreciable rate since the difference in their masses is so great that it is difficult to conserve momentum. Because of this, ions and electrons outside the dark space can diffuse for appreciable distances. At the walls of the chamber (or any available surface), however, recombination may readily occur since their kinetic energy may be given up as heat. Walls or other structures can cause a serious distortion in the uniformity of the ion density, and of the sputtering rate if they happen to be in the vicinity of the cathode. Another effect which can arise from this action comes from the fact that ion bombardment is an effective method of removing adsorbed contaminants from surfaces. Any such contaminants, once released, become a constituent of the discharge and thus may be incorporated in any film which is being deposited. For this reason, extraneous hardware should be kept remote from both the cathode and the deposition area.

4.3   FILM DEPOSITION IN A GLOW DISCHARGE

The general properties of a glow discharge have been discussed, but as yet nothing has been said regarding the film deposition techniques utilizing these phenomena. A diagrammatic representation of a simple sputtering apparatus is shown in Fig. 4-9. The apparatus consists of a large area cathode and an anode holding the substrates in a plane parallel configuration, and it is contained in a bell jar vacuum system. The gas to be used is admitted to the system to provide a pressure of from 10 to 100 millitorr. The spacing of the electrodes is of the order of from 1 to 12 centimeters, and the

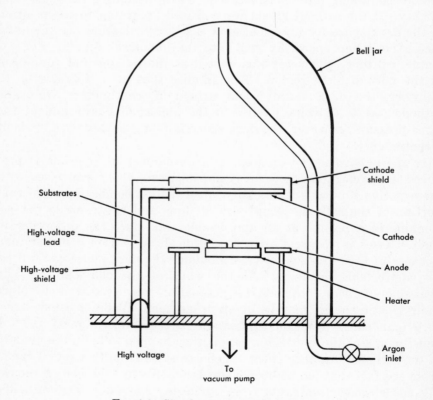

FIG. 4-9. Simple sputtering apparatus.

electrodes are typically from 5 to 50 centimeters in diameter. This type of system is usually operated with voltages of from 1 to 10 kilovolts.

Deposition Rate

The rate at which material is deposited at the substrate site is proportional to the rate at which it is removed from the cathode, and may be represented by

$$Q = CI\gamma$$

where Q is the deposition rate, C is a constant which characterizes the sputtering apparatus, I is the ion current, and γ is the sputtering yield. It must be recalled that the yield is itself a function of the sputtering voltage and the particular ion being used. In most work, it is desirable to operate at the highest sputtering rate commensurate with the particular type of film to be deposited. The gas chosen as the medium for establishing the discharge should be chosen with the sputtering yield in mind. Recalling Fig. 4-4, it is noted that, of all the readily available gases, argon is an excellent choice from this point of view as well as for its inertness. It would appear from the above equation that the maximum possible current should be used if the highest deposition rate is to be achieved. While this is generally true, for most applications the situation is considerably more complex. The power available is not without limit, and the only way to increase current without increasing power is to increase the pressure of the plasma. If the pressure is increased, however, there is a higher probability that sputtered atoms will return to the cathode by diffusion. As a matter of fact, at pressures in the region of 100 millitorr, only about 10 percent of the atoms sputtered from the cathode travel beyond the Crookes' dark space. This apparent drop in yield (as measured from the cathode) as pressure is increased is shown in Fig. 4-10. The net result of these observations is that the pressure chosen from sputtering rate considerations alone would be the highest pressure for which the yield was still close to the maximum.

The position of the substrate relative to the cathode is also an important consideration with respect to the deposition rate. The substrate should be as close as possible to the cathode without disturbing the glow discharge to collect a maximum of sputtered material. As the substrate approaches the cathode, however, the current will fall drastically, even before the edge of the dark space is reached,

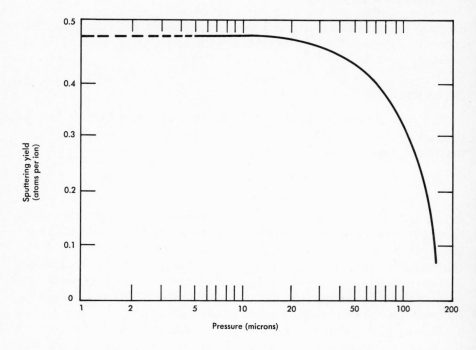

FIG. 4-10. Sputtering yield of nickel bombarded by 150-ev argon ions as a function of total gas pressure. (After Laegreid and Wehner; courtesy
Journal of Applied Physics, **32**, 365-369 (1961).)

and the sputtering rate will be reduced. In cases where this has been
done, the thickness distribution over the substrate shows a minimum
of film at the center, where the shielding of the cathode was greatest.
It should be noted that the substrate does not need to be at the anode
potential to produce this effect, and that it can act as a shield even
though it assumes the potential of the plasma in which it is located.
Maissel[8] has suggested that a convenient rule-of-thumb is to keep
the cathode-substrate distance about twice the length of the Crookes'
dark space. Any structure which may be required in the sputtering
system will also affect the characteristics of the glow if placed too
close to the cathode, and thus seriously disturb the distribution of
the depositing film.

The effect of such a structure is illustrated in Fig. 4-11, which
shows the distribution of ion current on a cathode surface when an
auxiliary tube is placed at varying distances from the surface. Even at
the relatively high pressure at which these data were taken, the effects
of the auxiliary wall do not become negligible until it is well outside
the dark space area. Another point which should not be neglected is

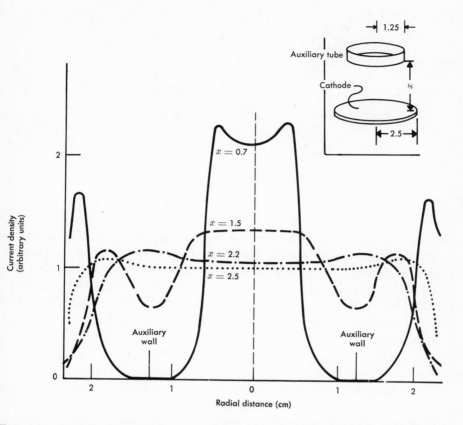

Fig. 4-11. Radial distribution of the positive ionic current density on the cathode surface. Variation with position of the auxiliary tube. Discharge in hydrogen. All dimensions in centimeters. (From Chiplonkar and Joshi, "Radial Distribution of the Positive Ionic Density on the Cathode Surface of a D. C. Glow Discharge;" courtesy *Physica*, **30**, 1746-1756 (1964).)

the current concentration at or near the edge of a cathode. This is brought about by the distortion of the electric field as the edge is approached, and results in much more cathode erosion near its borders. This can be circumvented by attaching thick strips of cathode material around the border, as shown in Fig. 4-12. Since deposition of film is enhanced near the edge of the cathode, substrates should be placed with a considerable cathode overhang.

With the use of a plane parallel configuration with large area electrodes, thickness uniformity over a large area is relatively easy to attain. For example, with a circular cathode of 35-centimeter diameter, and a cathode-substrate distance of 7.5 centimeters, a thickness uniformity of better than one percent was achieved over an

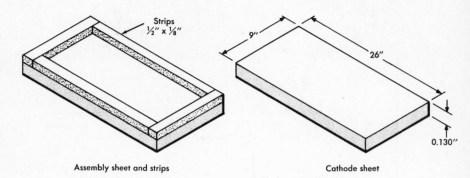

Assembly sheet and strips   Cathode sheet

FIG. 4-12. Cathode used in continuous in-line machine.

array of substrates that was 17.5 centimeters on a side. The equipment in which these results were obtained was designed so that no extraneous hardware approached the cathode, not even the bell jar walls. Depending upon sputtering conditions, deposition rates of from 100 to 500 A per minute are achieved even with materials such as tantalum, which has a low sputtering yield.

Contamination During Deposition

The need to reduce the pressure into the high vacuum region ($<10^{-6}$ torr) before admitting the sputtering gas has been widely recognized, and the partial pressure of active gases initially present should thus be quite low. Even so, there are many possible sources for contaminants, with some of the major ones being:

1. The walls of the vacuum chamber and other structures in the chamber may have adsorbed gases, water vapor, and hydrocarbons which can be desorbed by the action of the ion and electron bombardment present in the glow. Because of this, all surfaces which may be exposed to the glow should either be adequately cooled during deposition, of such low mass that they will reach thermal equilibrium within the first few minutes of deposition (while a shutter is presumably in place), or baked out at high temperature during pumpdown.

2. At the pressures normally used in sputtering, diffusion pumps are very inefficient and backstreaming of pump oil can become quite severe. Optically dense baffles alone will not suffice to prevent these vapors from entering the chamber, since the plates of such baffles are separated by distances equivalent to

several mean-free-path lengths at these pressures. It is usually necessary, therefore, to introduce some form of throttling between the discharge area and the baffle. This is most easily done by using the high vacuum valve in the system as a throttling valve. The throttling should be such that the pressure in the baffle does not exceed a few millitorr. Replacing the diffusion-pump, baffle combination with a "turbomolecular"-type pump would eliminate this source.

3. With the diffusion pump throttled as above, the pumping speed of the system is considerably reduced and the effect of "small" real leaks is thereby increased. It is perhaps more important, then, to have a good leak-free system for sputtering than for evaporation. Any measure of the "background" pressure before admitting the sputtering gas should be made with the high vacuum valve in its throttled position.

4. Perhaps the most obvious source for contamination is the substrate onto which the film is being deposited. The effects of particulate matter on the film are such that pinholes are produced, as well as a general contamination of the deposit. If patterns are to be produced in the film, these pinholes can cause serious difficulty at that stage. Because of this, as well as the general problem of contamination of the deposit, substrates should be rigorously cleaned before deposition, and every precaution taken to assure that the substrate is as free as possible from a contaminating layer or particulate matter.

Mass spectrometers have been used to analyze the composition of the sputtering atmosphere both before, during, and after deposition[16]. The important observations were that the reactive gases such as nitrogen, oxygen, and water vapor showed an immediate and dramatic *decrease* in concentration as soon as the glow was initiated. The only constituent to exhibit an increase in concentration was hydrogen. This was possibly due to the reduction of water vapor or to the cracking of hydrocarbons by the glow discharge. The decrease in the concentration of the reactive gases is undoubtedly due to the fast pumping action of the freshly deposited metal. (In the case referred to above, the depositant was tantalum, an especially good "gettering" agent.) This rapid pumping which occurs during deposition has made it difficult to assess quantitatively the actual amount of gas desorbed by the glow. Theuerer and Hauser[17] have taken advantage of this pumping action and have developed a technique called "getter sputtering," and tantalum films of near bulk properties have been obtained using this method.

Reactive Sputtering

The intentional inclusion of reactive gas in the sputtering atmosphere in order to alter or control the properties of the deposit has been termed "reactive sputtering." Insulating and semiconducting compounds of various metals have been deposited using this technique. Most of the work done in this area has been aimed at the production of oxides by incorporating oxygen in the sputtering gas, or by using pure oxygen as the glow discharge medium. Nitrides, carbides, and sulphides have also been prepared [18], however, by using nitrogen, methane or carbon monoxide, and hydrogen sulphide. Although the pure reactive gas may be used without dilution, it is more common to use a mixture of an inert gas with a relatively small amount of the reactant. For the deposition of insulators, this can prove to be quite an advantage, for with a high concentration of reactant the surface of the cathode becomes covered with the insulating compound, resulting in a considerable reduction in the sputtering rate.

The exact mechanism by which compounds are formed during reactive sputtering is not presently known. Depending on the pressure of the reactive gas present, the reaction may occur either at the cathode (with the compound being transported to the substrate) or by a reaction at the surface as the film deposits. At low pressures, the latter mechanism is more likely; conversely, at high pressures, the former is more likely. It has also been suggested that the reaction may occur in the vapor phase between atoms of the sputtered material and the gas atoms; however, it would appear that this possibility is quite remote since some means for the release of the energy of the reaction, as well as the kinetic energy of the atoms, would be necessary to prevent spontaneous decomposition.

In many cases, the composition of the film may be varied all the way from the metal through semimetal and semiconductor to an insulator simply by varying the proportion of reactive to inert gases in the discharge. As a matter of fact, in many cases the reactant is incorporated interstitially into the structure of the metal films, rather than forming compounds, until quite a high concentration of gas in the metal is reached. For example, it has been found in the case of tantalum films sputtered with either oxygen or nitrogen that the bulk solubility limit may be exceeded by almost a factor of two before there is any evidence of compound formation. This technique is one of the most important processes involved in the manufacture of thin films for use in electronic circuits, since it allows for the necessary control and flexibility of film properties. The actual proportions of reactant to inert gas in the discharge required to produce films of any particular property are dependent on many factors, among which some of the most important are:

1. *Deposition rate.* The rate at which the film is being deposited must be matched with the availability of the reactant gas. That is, other things being equal, the faster the film is being deposited, the more reactant gas is needed to maintain any desired film property.

2. *Total throughput of the system.* Since the reactive gases are gettered very rapidly under the conditions of sputtering, the percent of reactant gas must be decreased as the system throughput is increased, while the total pressure in the system is maintained the same.

3. *Total pressure during sputtering.* The amount of reactant which is available to the film being deposited must be controlled. If the total pressure used for sputtering is changed, the proportion of reactant must also be changed in order to maintain the same film properties. In general, as the total pressure is increased, the concentration of reactant would need to be decreased.

It is apparent from these factors that the important element to be maintained constant is the ratio of the arrival rate of the depositing film to the arrival rate of the reactant gas at the substrate.

Deposition Parameters and Their Influence

Film properties can be seriously affected by the choice of the particular conditions to be used for sputtering, even though all the factors mentioned above are optimized. The film properties may be influenced by the voltage used for deposition and the electrical status of the substrate (i.e., grounded, floating, or biased) as much as by impurities. The sputtering voltage, in addition to its effect on the deposition rate, can dramatically affect the structure of the deposited film. The crystallite size of sputtered tantalum films has been found to be quite dependent on the sputtering voltage, particularly below about 3000 volts. Also, tantalum films deposited at low voltages (<3000 volts) can show a considerable porosity. These films will be discussed in more detail in a later section.

The potential which appears on the substrates also affects the film being deposited because it controls the flux of incidental ions or electrons bombarding the substrate while the deposition is taking place. Again in the case of tantalum, the crystalline habit of the crystallites depends on the potential appearing on the substrates. If the substrates are electrically isolated from the other structures in the sputtering chamber, they will assume the potential of the plasma in which they are immersed. In this case, there should be no appreciable attraction for either electrons or ions, and the incidental bombardment by charged particles should be minimized. If the substrates are placed at the anode potential (usually ground), they should receive electron bombardment to the same extent as the anode. An intentional bias may be placed on the substrate and cause it to be the recipient of either electrons or ions, depending on the polarity of the bias. If the substrate is biased so that it is bombarded with a significant flux of either electrons or ions, the crystallites of the film show the normal crystal structure of bulk tantalum (body-centered cubic). If the bombardment is minimized, however, the crystallites exhibit a different structure (probably tetragonal, and called beta tantalum) with quite different electrical properties. Although these complications might appear to cause difficult control problems, properties of films produced are quite reproducible as long as care is taken to prevent these parameters from being unintentionally varied. As a matter of fact, the control of these factors, as well as those previously mentioned, allows for the tailoring of film properties to a particular application.

The mechanisms involved in these structural changes are far from being fully understood, but the interaction of foreign particles with those of the depositant would undoubtedly have a significant effect on the nucleation and growth of the film, and thereby on the structure of the deposit.

Alloy Deposition

The use of sputtering for the preparation of alloy films has several advantages over evaporation. The chemical composition of a sputtered film will usually be the same as that of the cathode from which it was sputtered, at least after some short equilibration period. This is the case even though the various components of the alloy exhibit quite different sputtering yields. The reason for this is that after a period of time during which the component with the highest sputtering rate is preferentially sputtered, the surface of the cathode becomes enriched in the other component until a "steady state" surface composition is reached. When this condition is reached, the composition of the sputtered film is the same as that of the cathode. In the case of evaporation, this effect is not observed because the high temperature of the source allows for rapid diffusion from the bulk to the surface of the source. If an extremely high cathode temperature were to be employed in sputtering such that bulk diffusion were rapid, the composition of sputtered films would no longer be identical to that of the cathode.

Another method for preparing alloy films is to synthesize them directly from the individual components by using multiple cathodes at the same potential in the discharge. The relative areas of the cathode can be arranged such that almost any composition can be achieved. One simple technique which has been utilized to incorporate small percentages of one metal in another is the use of wires of one metal strung across a cathode of the other metal. Strips or discs of the second metal may be attached to or plated on the cathode for including larger amounts. For this method of depositing alloys, the composition of the alloy film may be estimated from the ratio of the areas of the component cathodes and their sputtering yields (as long as those yields are known for the gas and sputtering voltage used). The ratio of the two film components may be expressed as

$$\frac{N_1}{N_2} = \frac{A_1 \gamma_1}{A_2 \gamma_2}$$

where N represents the number of atoms, A is the area of the cathode, and γ is the sputtering yield, with the subscripts denoting the particular component of the alloy.

A second method employing separately powered cathodes has also been used. Various compositions are achieved by varying the voltages applied to each cathode during deposition.

These techniques have been combined with reactive sputtering in oxygen by Sinclair and Peters[19] for the deposition of multicomponent glasses utilizing cathodes of aluminum and silicon.

Film Adherence

Almost no quantitative work has been done on the adhesion of a thin film to its substrate. In general, those data which have been reported are related to the adherence of evaporated films to glass substrates. Many have attempted to relate film adhesion to the free energy of oxide formation for the film material, and at least a qualitative trend is observed. This has led to the belief that most bonding occurs through the formation of some form of oxide-metal bond at the substrate-film interface. It has also been observed that the refractory metals (Ta, W, Mo, etc.) seem to form a better bond than metals with low melting points, even though the free energy of oxide formation may be similar. One of the problems associated with studies of film adhesion is that there is no satisfactory method available for a quantitative measurement of the adhesive forces.

Regardless of the mechanisms involved in film adhesion, it should be obvious that it is necessary to have the film immediately adjacent to the substrate, with no contaminant layer interposed, if maximum adherence is to be achieved. There is essentially universal agreement that considerable care must be taken to assure as clean a surface as possible in order to achieve good adhesion. Unfortunately, there is not such agreement on a method for obtaining such a surface. As a matter of fact, there must be almost as many cleaning methods as there are researchers in the field of thin films, and not all of these methods can be equally effective. A further discussion of cleaning will be provided in Chap. 9. Sputtering does provide some advantages in this respect, however, in that the depositing atoms are more energetic than those deposited during evaporation, and thus can penetrate or dislodge small amounts of contaminants on the surface. It has been stated many times that sputtered films exhibit better adherence than evaporated films under similar circumstances, but no controlled experiments have been performed to substantiate this statement.

4.4   SPUTTERING WITH AN ARTIFICIALLY SUPPORTED DISCHARGE

As was mentioned in Sec. 4.2, when the pressure is lowered, the dark space expands until it reaches the anode plane, at which point the glow extinguishes and sputtering ceases. There are several advantages which might be anticipated, however, if sputtering could be done at lower pressures; two advantages would be the possibility of lowering the amount of inert gas trapped in the film, and the possibility of achieving better adhesion because of the higher energy of the sputtered atoms when they reach the substrate. Some of the techniques which have been developed for sputtering at low pressures are described briefly below.

Thermionic Electron Emission

A discharge is extinguished at low pressures because of inadequate ionization by the secondary electrons emitted at the cathode. In order to sustain the discharge, the ionization must be increased, and this may be done by injecting electrons from a source other than the target cathode. One of the simplest methods for injecting electrons into the low-pressure system is by the use of a hot filament. A system based on this technique is illustrated in Fig. 4-13. The operation of this system may be explained in the following manner: A flow of electrons is established between the filament cathode and an anode placed at the top of the bell jar. The filament is protected from any sputtered material by being placed in an elbow shown at the bottom of the bell jar. The voltage used to accelerate the electrons to the anode is usually of the order of from 150 to 200 volts. The plasma is usually confined to a cylindrical region between the cathode and anode either by the use of a physical "chimney" or a magnetic field provided by an external coil. The target and the substrate are usually placed on opposite sides of the active plasma region. No sputtering occurs until a negative potential (usually from 200 to 1000 volts with respect to the anode) is applied to the target. This potential, then, repels electrons and attracts ions from the plasma region, and sputtering occurs. The ions bombard the target in essen-

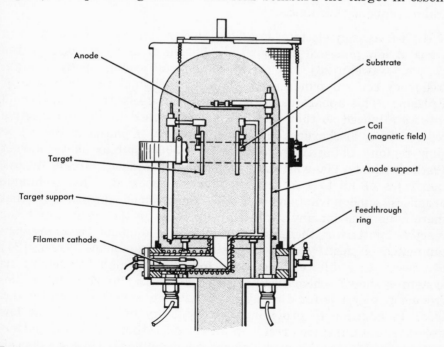

FIG. 4-13. Commercial version of low-energy sputtering unit. (From Consolidated Vacuum Corp.)

tially the same way as in standard (sometimes referred to as "diode") glow discharge sputtering. The sputtered atoms deposit on the substrate in the same fashion as before. The density of ions in the plasma may be controlled by varying either the electron emission current or the voltage used to accelerate the electrons, while the energy of the bombarding ions is controlled by controlling the target voltage. This arrangement has allowed sputtering experiments to be performed at pressures from about 0.1 to 1 millitorr.

A system using a hot-filament source of electrons to sustain a glow at low pressures has one main disadvantage. This type of system does not produce uniform sputtering from flat targets of large size, since the ion density (and thus the sputtering rate) is greatest along the axis of the electron beam and at the end nearest the hot filament. Another disadvantage is that filament life can be short if reactive sputtering is to be performed. For some laboratory experiments, this technique appears to provide some advantages. Additional information on this type of sputtering may be obtained in Refs. [20], [21], and [22].

Radio-Frequency Excitation

High-frequency electromagnetic radiation can excite a discharge in a gas at low pressure without the addition of any electrodes in the system. Gawehn[23] has described a small system using a high-frequency coil surrounding the chamber to induce the formation of a plasma. This apparatus is shown in Fig. 4-14, with the target and substrate placed so that they are not immersed in the most active region of the discharge. The anode, however, is placed in the active glow region. Using this system with a 500-volt bias on the target (cathode) and a 200-watt input to the high-frequency circuit, deposition rates of up to 1 A per second were reported. This technique promises some advantage over the hot-filament approach in that there is no need for any additional electrodes in the system, and, for reactive sputtering, there is no problem of a filament becoming contaminated or destroyed by the action of the ambient. Vratny[24] has described a different technique for utilizing high frequency; his system is shown schematically in Fig. 4-15. In this case, the radio-frequency power is fed directly to the cathode, superposed on the d.c. bias. In addition to providing the capability of sputtering at low pressure, these methods result in higher deposition rates for sputtering at normal pressures, particularly at low voltages. Uniform deposition from large-area cathodes is possible using this technique.

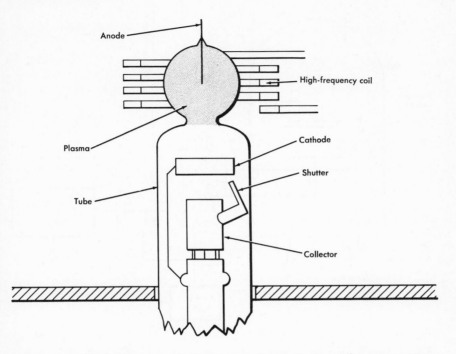

FIG. 4-14. High-frequency plasma system for sputtering at low pressures.

Direct Sputtering of Dielectrics

Conventional sputtering techniques have not proved to be useful when the target material to be sputtered is an insulator, as was implied in the discussion of reactive sputtering. The reason for this is that since the charge on an ion bombarding the surface of the target cannot be neutralized, the majority of the field will be concentrated across the insulator, with little left in the gas. Under these conditions, neither the rate of ion arrival nor the ion energy would be sufficient to cause appreciable sputtering. Anderson et al[25] developed a technique utilizing radio-frequency sputtering, by which insulators are sputtered by the alternate action of ion and electron bombardment of the surface. The arrangement used is shown in Fig. 4-16. The dielectric target is mounted onto a conducting plate to which the high-frequency power is applied. This arrangement allows the positive charge on the target to be neutralized by the electrons of the plasma during that half of the cycle when the target is positively charged. The operation of this plasma in sputtering may be described as follows: Because of the difference in mobility be-

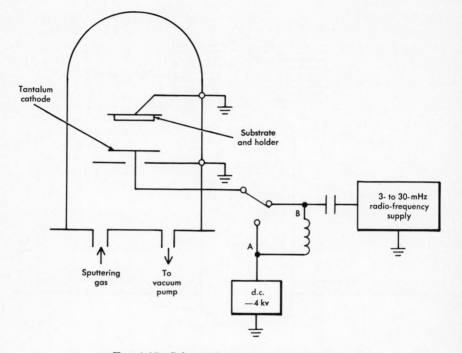

FIG. 4-15. Schematic of deposition system.

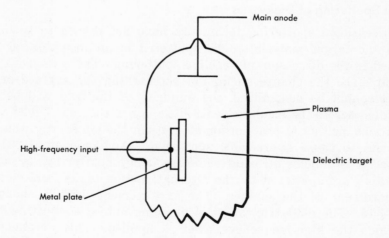

FIG. 4-16. Typical arrangement used for radio-frequency sputtering of a dielectric target [25]. (After Anderson, Mayer, and Wehner, "Sputtering of Dielectrics by High Frequency Fields," courtesy *Journal of Applied Physics*.)

tween ions and electrons, a "sheath" enriched in ions is formed around the electrodes. These ions bombard the target and produce the sputtering. Frequencies employed for this type of sputtering are usually of the order of 10 megahertz. As far as the electrical characteristics are concerned, the behavior is quite similar to that of a capacitor, and the current, I, may be calculated from standard formulae such that

$$I = \omega C V$$

where C is the capacitance of the target, ω is the angular frequency, and V is the root-mean-square voltage of the radio-frequency source. The effective ion current will be only half this figure, however, since the ions are being accelerated toward the target only every half-cycle.

Davidse and Maissel[26] have applied this technique to the deposition of thin insulating films and have reported excellent properties for the resulting films. Deposition rates of up to 30 A per second were obtained for various glasses and fused silica. These authors also reported the use of a magnetic field, and claim that this aided in the support of radio-frequency discharge as well as in stabilizing the glow. The radio-frequency technique can be sensitive to the design of feedthroughs for the radio-frequency power input, and reflective losses between the power supply and the sputtering system should be kept to a minimum.

The system illustrated in Fig. 4-15 may also be utilized for the direct sputtering of insulators, and promises to provide higher deposition rates than the system employing radio frequency alone. This technique has proved to be of particular value in the preparation of insulating films via reactive sputtering, because the insulating layer which forms on the cathode no longer limits ion bombardment.

4.5 EFFECTS OF SPUTTERING CONDITIONS ON FILM PROPERTIES

In the previous sections of this chapter, it has been stated that various film properties may be related to the conditions during deposition. In this section, the effects of the various parameters on the properties of tantalum films will be described. Tantalum has been chosen as the example to relate these dependences because of its importance in the field of electronic circuits and because there is more information relating the properties of tantalum films to the deposition parameters than for perhaps any other metal. The properties which are of the greatest interest for device application are the resistivity, the thickness, the temperature coefficient of resistance, and the sheet resistance, as well as the aging behavior of the films. The relationships of these properties with the deposition conditions are of primary interest.

Sputtering Atmosphere

The particular gas chosen for the atmosphere used during sputtering is usually selected with three points in mind: First, the gas should be inert to the material being deposited; second, it should provide a high sputtering yield; and third, it should be relatively inexpensive and readily available in high purity. Argon is usually chosen as the gas that most nearly achieves all these requirements. The effects of various gaseous reactive additives on the properties of sputtered films can be quite extensive. Sputtering in the presence of reactive additives is termed "reactive sputtering" and is discussed below. The discussion will show (a) that the properties of sputtered films may be controlled by the intentional addition of "doping" agents, and (b) that it is necessary to eliminate all unintentional contaminants from the area of deposition.

The first system to be discussed is that of tantalum films containing nitrogen. These films are made by the addition of small amounts of nitrogen into the argon sputtering atmosphere [27]. A complete range of compositions may be produced, depending on the amount of nitrogen introduced into the system. Figure 4-17 illustrates the relationship of various properties of tantalum films to the amount of nitrogen present in the system during deposition. The amount of nitrogen is plotted as the partial pressure of nitrogen in arbitrary units, since the scale factor which applies is dependent on the characteristics of the particular system being used. The lowest partial pressure indicated on the plot represents the background pressure in the bell jar just prior to the admission of the argon sputtering gas and, as such, represents the ultimate pressure of the system with the diffusion pump throttled as described in Sec. 4.3. As is indicated in the figure, either of two types of tantalum film may be deposited at the point of no intentional doping. The resistivities of these two types of tantalum are quite different because they have different crystal structures. The films with the lower resistivity show a structure corresponding to the normal body-centered cubic (bcc) tantalum,

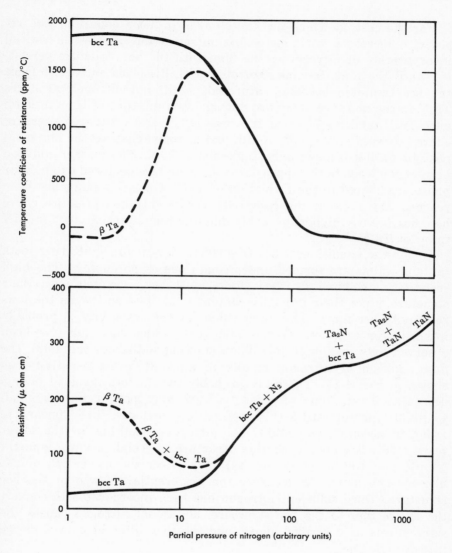

FIG. 4-17. Resistivity and temperature coefficient of resistance as a function of nitrogen pressure during sputtering.

while those with the higher resistivity show a tetragonal structure and have been labeled beta tantalum. As shown later in this chapter, the particular operating parameters chosen for sputtering determine which of these two types of "undoped" tantalum films is deposited.

For the case in which bcc tantalum is initially deposited, the resistivity increases with increasing nitrogen due to the interstitial incorporation of nitrogen in the film, with the bcc structure persisting until the films become saturated with dissolved nitrogen. Once the bcc tantalum becomes saturated, additional nitrogen produces Ta_2N in the film; the structure now shows a mixture of bcc tantalum with Ta_2N, and the slope of the resistivity versus nitrogen pressure curve becomes quite small. Additional nitrogen changes the film composition until it is essentially single-phase Ta_2N. The further addition of nitrogen leads to the appearance of a new face-centered cubic (fcc) phase, attributed to the formation of TaN. Once this point has been reached, the slope of the resistivity versus nitrogen pressure curve becomes slightly higher, probably due to a higher resistivity for TaN than for Ta_2N.

Under the conditions most frequently chosen for sputtering, beta tantalum films are formed under conditions of no doping. The high resistivity exhibited by these films must be associated with their structure, since their impurity content is at least as low as the low-resistivity bcc films. This high value for the resistivity is probably intrinsic for tantalum of this structure, which has thus far been observed only in film form. When starting with beta tantalum, the first addition of nitrogen results in films of lower resistivity, as shown in Fig. 4-17. There are probably two factors involved in this decrease: First, films of this structure may permit only a small solubility for interstitials; therefore, the bcc structure appears in order to accommodate additional nitrogen, resulting in the lower resistivity associated with this structure. Second, a small amount of nitrogen (or other active gas) may affect the nucleation of the film in such a way as to cause the preferential growth of the bcc structure. Once sufficient nitrogen has been incorporated to convert the entire film to the bcc structure, additional nitrogen causes the same effects as described above, and the two different curves merge into a single line.

In addition to the high resistivity exhibited by the beta tantalum, the temperature coefficient of resistance is near zero, while that of the bcc tantalum is highly positive. Thus, when starting from beta tantalum and adding nitrogen, the temperature coefficient first goes through a large maximum before it returns to a near-zero value for the Ta_2N films.

The highest resistance stability is observed at nitrogen levels in the range of Ta_2N. Figure 4-18 presents a plot of the change in resistance after accelerated load-life testing of resistors made from

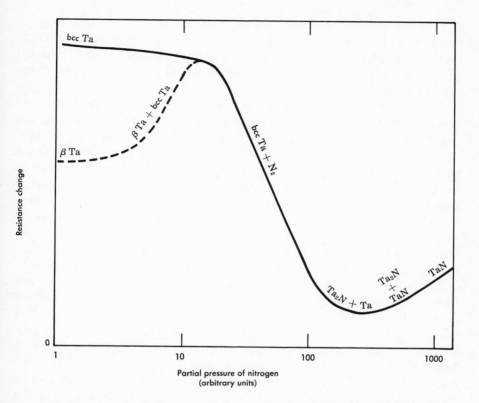

FIG. 4-18. Resistance change due to aging as a function of the partial pressure
    of nitrogen during deposition.

tantalum sputtered at various nitrogen levels. Fairly large changes
in resistance are noted at the point which corresponds to the lowest
resistivity in the film. In this region, the nitrogen content is well
below the solubility limit. Presumably, upon exposure to atmosphere
during life, additional nitrogen and oxygen can be dissolved into the
film, causing large changes in the resistivity. When beta tantalum
is present in the films, negative changes may even be observed since
added oxygen or nitrogen may cause a conversion to bcc, as mentioned previously. If the initial doping is sufficient to saturate the
film with interstitial nitrogen (indicated by the presence of Ta_2N),
further interstitial dissolution is inhibited. The resistance is then
relatively stable. An additional reason for this stability is that even
if gases from the air are incorporated into the film, their effect on
the resistivity is very small, as evidenced by the small slope of the
curve of resistivity versus doping level in this region (see Fig. 4-17).

Once sufficient nitrogen has been introduced into the film to produce the first TaN, however, additional nitrogen incorporated during deposition produces a more significant change in resistivity, and for probably similar reasons the films show larger increases in resistance during aging.

It is apparent from the above discussions that the properties of undoped tantalum films are somewhat indeterminate, since they are dependent on the operating parameters of the deposition system. One of the major advantages of nitrogen-doped films in the Ta_2N region is that some of the control problems in manufacture tend to be minimized. This comes about primarily because the properties are controlled by the amount of nitrogen incorporated, rather than by the character of a particular deposition system. Although the precise values of the resistivity may vary from one system to another, there is always some "operating point" which yields the maximum stability, and which correlates with the structures reported above.

Similar studies have been made with oxygen, methane, and carbon monoxide [18] (these gases being chosen as representing the other types of background gases likely to be in a vacuum system). The results of these studies are quite similar to those for nitrogen, particularly in the region of low partial pressures. When starting with beta tantalum, the same decrease in resistivity is observed for any of these gases. It would appear that small amounts of reactive species have similar effects on the stability of the beta tantalum structure and on the nucleation of a film. Those effects which are peculiar to the particular reactant do not begin to dominate the properties of the film until a relatively high concentration is reached, such that compounds begin to be formed. The relationship of the resistivity with the concentration of oxygen in the sputtering atmosphere is shown in Fig. 4-19, and the similarity with Fig. 4-17 at low pressures is apparent. As the partial pressure of oxygen is increased to relatively high levels, the composition of the deposited film approaches that of tantalum oxide and the resistivity increases rapidly. This reactive sputtering technique has been used to prepare the dielectric for thin film capacitors, but the tantalum oxide films produced in this way have much lower breakdown strengths and higher leakage currents than do films prepared by anodization. Other insulating films such as silicon dioxide, aluminum oxide, and titanium oxide have been prepared by reactive sputtering but, again, their breakdown strengths leave something to be desired. The direct sputtering of insulators by using one of the radio-frequency techniques appears to provide some advantages over the reactive techniques.

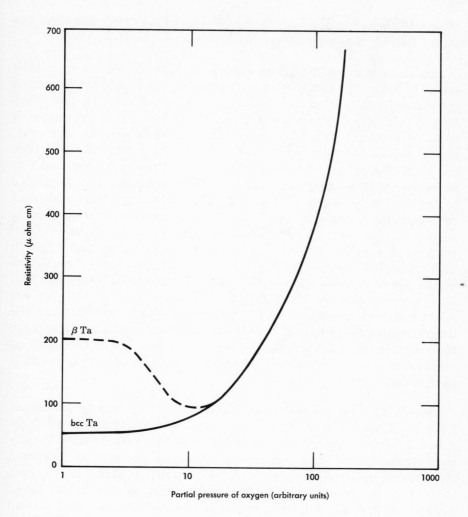

FIG. 4-19. Resistivity of tantalum films as a function of partial pressure of oxygen.

Deposition Parameters

The particular conditions which are chosen for performing the deposition can influence the properties of the deposited film, as was mentioned previously. Assuming that a sputtering station exists, the pressure, current, and voltage at which to work must still be chosen, and the substrate may either be grounded, floated, or biased. The effects of these variables on the properties of tantalum films are the topics for the following discussion.

Sputtering Current and Deposition Rate. The deposition rate itself seems to have little if any effect on tantalum film properties except as it relates to the amount of contamination which may be picked up during deposition. For pure films, then, a high rate of deposition is to be preferred, with other things being equal. When reactive sputtering is involved, however, a change in the deposition rate will alter the ratio of reactant to depositant arriving at the substrate, and film properties will be changed accordingly. The deposition rate is related to the sputtering current, and Fig. 4-20 shows the relationship of resistivity and temperature coefficient of resistance with sputtering current for tantalum films sputtered at a constant voltage of 5000 volts and a constant influx of nitrogen (the pressure of argon had to be increased in order to increase the sputtering current). Also shown in the figure are the approximate areas of transition from one film structure to another. The resistivity changed from 600 to 400 micro-ohm centimeters over the range of currents studied, and the film structure ranged from TaN to bcc tantalum. In order to achieve reproducible results, then, the deposition rate as well as the input of reactant gas must be maintained under tight control during reactive sputtering.

Sputtering Voltage and Pressure. All the films whose properties have been discussed so far were deposited using voltages of from 4 to 6 kilovolts. Voltages much lower than this may be used for depositing tantalum films, but unless the argon pressure is increased the current falls off and, with it, the deposition rate. It is usual, then, for the pressure to be increased when sputtering at lower voltages so that the deposition rate may be maintained at a reasonable level. Schuetze[28] performed a series of investigations on the effects of voltage on film properties. The resistivity of the films becomes quite high at low voltages as observed in Fig. 4-21, which shows the resistivity and temperature coefficient of resistance of the films as functions of the sputtering voltage. An interesting observation is that films of quite high resistivity can be obtained which have near-zero temperature coefficients. These films have a porous structure with a pore size of the order of 40 to 50 A. An electron micrograph of a film sputtered at 1500 volts is reproduced in Fig. 4-22, which shows the porous nature of the structure. The islands of metal between the pores consist of "grains" containing a large number of

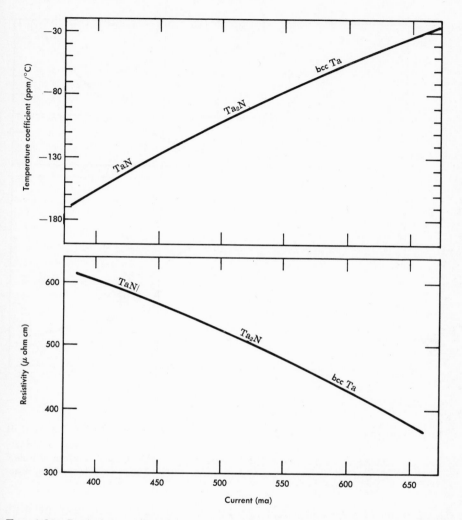

FIG. 4-20. Resistivity and temperature coefficient of resistance as a function of
sputtering current for nitrogen-doped tantalum films. The influx of
nitrogen is constant and the sputtering voltage is constant.

exceptionally small crystallites of beta tantalum. This combination
of small crystallites and a porous structure provides only narrow
tortuous paths for current flow, and this is the main reason for the
high apparent resistivity exhibited by these films.

The porous nature of these films is corroborated by density measurements, with the density decreasing as the sputtering voltage
decreases, as is illustrated in Fig. 4-23. Because of the high porosity,

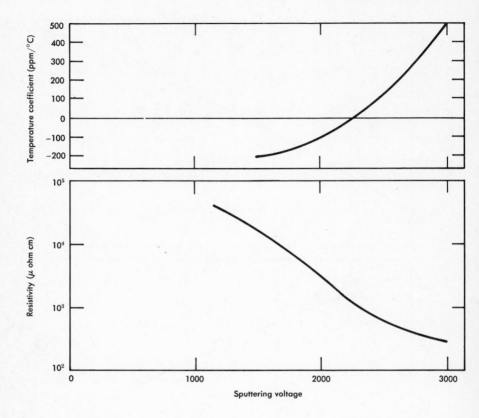

FIG. 4-21. Resistivity and temperature coefficient of resistance as a function of
sputtering voltage at constant current [28]. (After Schuetze,
Ehlbeck, and Doerboeck, "Investigation of Thin Tantalum Films,"
Trans. 10th Nat. Vac. Symposium, 1963, 1964; courtesy The Macmillan Co.)

these films absorb and react rapidly with oxygen, especially at high
temperatures. This oxidation causes a rapid increase in resistivity
and also shifts the temperature coefficient from zero down to values
around −300 parts per million per degree C.

Sputtering with Asymmetric Alternating Current. Another method
for sputtering is the use of an asymmetric a.c. source rather than
the normally used d.c. [29]. A schematic diagram of the electrical configuration used for this type of deposition is shown in Fig. 4-24.
With this arrangement, both the target and the substrate are alternately bombarded with electrons and ions. The circuit is designed
such that the current flowing during the half cycle that the target
is negative is appreciably greater than that flowing when the sub-

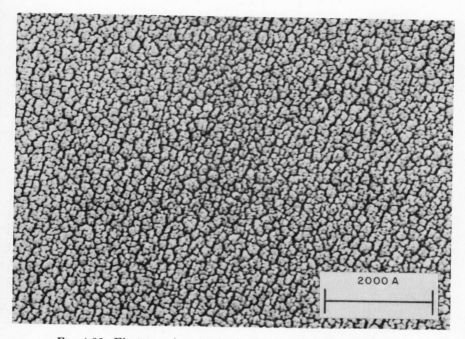

FIG. 4-22. Electron micrograph of a film sputtered at 1500 volts.

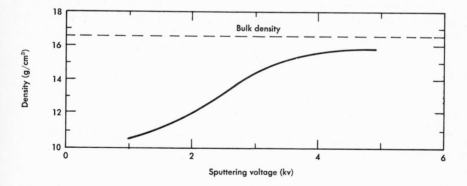

FIG. 4-23. Density of beta-tantalum films sputtered at various voltages.

strate is negative. A net transfer of material occurs, and a film is grown on the substrate. It is claimed that during the half cycle that the substrate is negative, ion bombardment removes adsorbed gases, and films of higher purity may be produced. The deposition rate suffers, however, and there is still opportunity for adsorbed gases to be incorporated in the film so that, generally speaking, purer films

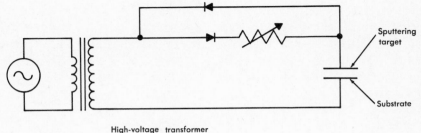

High-voltage transformer

FIG. 4-24. Schematic of circuit for asymmetric a.c. sputtering.

may be prepared by the technique of biased sputtering (described below) than by asymmetric a.c. sputtering. The power supply chosen for d.c. sputtering can also affect the film properties since the amount of ripple can be such that it behaves like an asymmetric a.c. source, and subtle differences in film properties have been observed when using different power supplies.

Biased Substrate. Another method of obtaining ion bombardment of the substrate is to provide an electrical bias to the substrate. Figure 4-25 illustrates the circuit used for bias sputtering. If the substrate is made negative with respect to the anode potential, it will be subjected to a steady bombardment by ions from the gas during the entire time that the film is being deposited. This ion bombardment

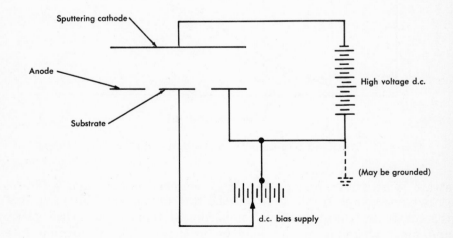

FIG. 4-25. Schematic of circuit for bias sputtering.

can act to clean the film surface of adsorbed gases which might otherwise have been incorporated in the film. Because of this effect, more reactant gas is required for sputtering on biased substrates than for unbiased substrates to achieve the same degree of gas inclusion. The use of a negative bias during reactive sputtering, then, provides a means for obtaining an easier control over the film composition since it is easier to control a relatively large amount of reactant than a small amount. Difficulties may arise, however, since it is difficult to assure a uniform ion current density across the substrate, especially at the start of the deposition, when the conductivity of the substrate is very low. The technique of bias sputtering has been put to good advantage by IBM for the preparation of films of molybdenum [37] and nickel-iron alloys [38] with sufficient purity to exhibit properties near those of the bulk materials.

In the case of tantalum films, the structure of the deposit can also be changed by the use of biasing. Figure 4-26 shows the relationship of the resistivity of tantalum films with the bias voltage on the substrates during deposition [39]. The data from which this curve was plotted were taken in a system which provided a minimum of contamination, and the decrease in resistivity observed with a biased substrate was due to a change in crystal structure; that is, from beta tantalum in the region of zero bias to body-centered cubic tantalum for the biased films. One point to be noted from this curve is that the effects of bias become noticeable at lower voltages when the substrate is made positive with respect to the anode than when a negative bias is used. This implies that electron bombardment is more effective than ion bombardment in controlling or determining the structure of the film.

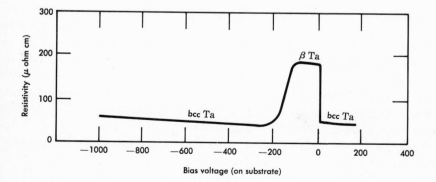

FIG. 4-26. Resistivity of tantalum films as a function of substrate bias.

Floating Substrate. When the substrate potential is floating with respect to the anode, the substrate will assume a charge and reach a potential equal to that of the plasma surrounding it. This potential may vary broadly, depending on the substrate-cathode distance, the sputtering voltage, and the pressure in the system, but will commonly be of the order of from 5 to 20 volts negative with respect to the anode. Under these conditions, the bombardment of the substrate by either ions or electrons is minimized since there is no tendency for the substrate to extract charged particles from the plasma, and the only bombardment received will be that of incidental ions and electrons. Under these conditions of deposition and with a large-area cathode, uniform film properties have been achieved over a 17.5-centimeter square, such that the total spread in sheet resistance over that area was only 1.7 percent. Since there is a minimum bombardment under these conditions and since there can be no net current flow, the small amount of bombardment which is received would tend to be distributed uniformly over the substrate area; therefore, any effects on properties would be uniform. If, however, the substrates are placed too close to the dark space, the plasma itself may not be at a uniform potential over the entire area, and this would allow for current to flow from the plasma into the film and back out into the plasma at different points on the substrate. The film properties become quite varied over the substrate area under these conditions. Figure 4-27 shows a profile of the sheet resistance for two runs. The first, with the substrates floating and well outside the dark space region, produced beta tantalum with quite good uniformity. The second deposition was performed under the same conditions except the pressure was reduced such that the dark space became distorted; this run produced bcc tantalum with quite poor film uniformity.

Substrate Temperature. Although the substrate temperature does not seem to have a large influence on the properties of tantalum films (except as it may relate to the cleanliness of the substrate) in the normal operating region (200° to 400° C), it can play a very important role in determining the structure at higher temperatures. Above about 700° C, for example, bcc tantalum is deposited where beta tantalum was deposited at the lower temperatures. Another example is that of sputtered germanium films deposited on single crystal germanium substrates, where both the substrate temperature and the deposition rate are involved in determining whether the films are "single crystal," polycrystalline, or amorphous [30]. This is illustrated in Fig. 4-28, which is a pseudo phase diagram relating

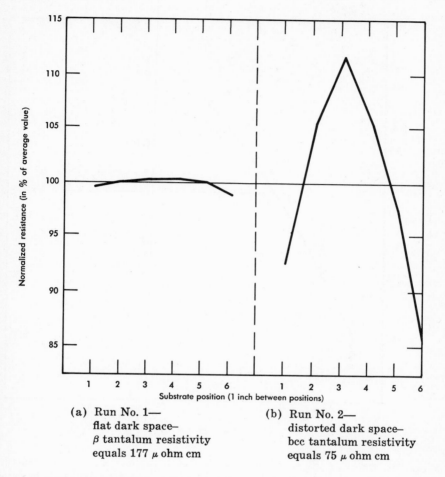

(a) Run No. 1—
flat dark space—
β tantalum resistivity
equals 177 μ ohm cm

(b) Run No. 2—
distorted dark space—
bcc tantalum resistivity
equals 75 μ ohm cm

FIG. 4-27. Influence of glow distortion on resistance uniformity.

the structure of the deposit to both substrate temperature and the deposition rate.

Geometric Influences. Sputtered atoms suffer frequent collisions with gas molecules in transit, and the deposition rate is at least partially controlled by diffusion processes. Because of this, the distribution of the deposit on the substrate may be influenced by nearby fixturing, even though the precautions mentioned in Sec. 4.3 are observed and the glow itself is not disturbed. This effect comes about because essentially all surfaces are traps for the sputtered atoms and, as such, reduce the local concentration of these atoms within

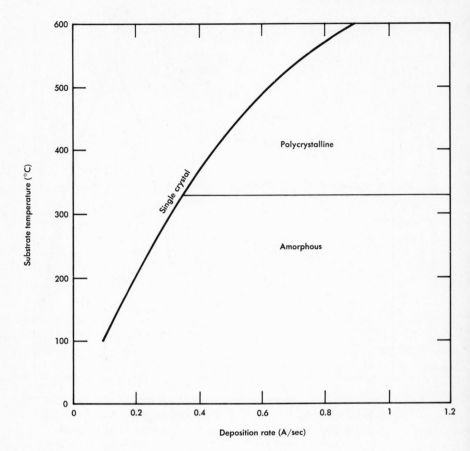

FIG. 4-28. Effect of substrate temperature and deposition rate on film structure for germanium on the 111 face of single crystal germanium substrates. (After Krikovian and Sneed, "Deposition Parameters Affecting Epitaxial Growth of Single Crystal Films," *Trans. 10th Nat. Vac. Symposium, 1963,* 1964; courtesy The Macmillan Co.)

the nearby areas. The particular method chosen for holding the substrate, then, may influence the distribution of the sputtered deposit. This is illustrated in Fig. 4-29, which shows a number of possible substrate placements and the relative distribution of the deposit for each placement. The substrate which is supported by a holder that is smaller in area than the substrate will collect considerable deposit on its edges and even on the under side, while the substrate supported on a large-area holder collects much less deposit on its edge and, of course, none on its under side. This effect may extend to the surface of the substrate as well, with the deposit being

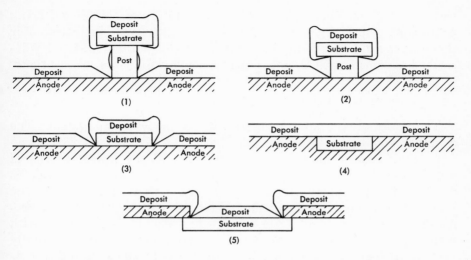

FIG. 4-29. Influence of substrate position on deposition uniformity.

somewhat thicker around the periphery of an isolated substrate as shown in the figure. Substrate holders are either made approximately the same diameter as the cathode or are mounted such that the anode plane is flush with the surface of the substrates, with the anode being of the same diameter as the cathode. Other nearby structures, such as an open shutter, can act in a similar fashion and can locally starve the neighboring substrate area of deposit. For this reason, the shutter should be as close as possible to the substrate plane when it is in the open position.

Substrate Influences

Substrates, their properties, and their influence on film properties will be discussed in detail in a later chapter. A point which may be made here is that crystalline substrates may influence the nucleation of the film by epitaxy. "Single crystal" films of several materials have been prepared using single crystal substrate materials such as sapphire, silicon, germanium, etc., and the mechanisms involved are very similar to those discussed for evaporated films.

4.6 SPUTTERING TECHNOLOGY

The principles involved in sputtering and the relationship of some of the sputtering variables to the properties of the films produced have been discussed in the previous sections. This section relates these basic principles to the design and operation of actual sputtering equipment, especially for batch- or laboratory-type facilities. For the purpose of illustration, one laboratory-type facility will be described in considerable detail. Other systems will be discussed briefly.

Laboratory or Bell Jar Sputtering Facility

The heart of any sputtering facility is the vacuum system. The chamber should be large enough that its walls will be well away from the Crookes' dark space region, even at the lowest pressures which may be used. Generally speaking, a minimum diameter is 60 cm (24 in.), although for small cathodes a 45-cm (18-in.) jar may be used. The jar should be constructed of a high-quality nonferromagnetic stainless steel (300 series) and should be supplied with water cooling. It is also advantageous to provide means for rapidly heating the bell jar to temperatures in excess of 200°C to aid in outgassing. The entire system should be constructed using the vacuum construction techniques outlined in Chap. 2.

The pump, baffles, and valves should be chosen to maintain the effective pumping speed to as high a pressure at the input as possible. This is required to minimize backstreaming or instability of the pressure. It must be recalled that a sputtering system requires pump qualities which are not normally considered in the design of diffusion pumps, and some compromise will probably have to be made. The baffle should be capable of being cooled with liquid nitrogen, and the high vacuum valve must have a reproducibly adjustable opening to provide controlled throttling. Some of the difficulties involved with the use of a diffusion pump may be avoided altogether by replacing it with a mechanical high vacuum pump such as a "turbomolecular" pump. In this case, the liquid nitrogen baffle may be eliminated, and there is no need for a high vacuum valve since rough pumping is done through the high vacuum pump itself. If this type of pump is chosen, however, it is still desirable to provide some form of throttling since the pumping speed begins to fall off rapidly above about 5 millitorr, and the operation of the pump would tend to be unstable above this pressure. This may be done by installing a butterfly-type throttle in the line to the pump. Whatever pump type is chosen, the vacuum attainable in the system should be of the order of 10^{-7} torr or better. It is not unusual to obtain pressures of 10^{-8} torr in 60-cm bell jars. In order to achieve pressures in this range, the total leak rate (both virtual and real) into the system should be less than 2×10^{-5} torr liters per second.

Structure. One design of a sputtering apparatus which embodies the principles outlined in Chap. 2 and the previous sections is illustrated in Fig. 4-30. It consists of a large-diameter cathode with a grounded back shield to prevent sputtering from the back of the cathode, a shutter, a substrate heater, and a large-diameter anode

FIG. 4-30. A laboratory sputtering facility.

plane. In order to minimize the effects of ion bombardment of the walls and their subsequent outgassing, the diameter of the entire structure is limited to keep the active area at least 12 cm from the walls of the chamber. A brief description of each of the substructures is provided in order to illustrate some of the design principles.

Cathode. The cathode for this system is a flat (±2 mm) sheet that is 35 cm (14 in.) in diameter and 2.50 mm (0.100 in.) in thickness. It is held within a ring (40 cm in diameter and made of 3/4- by 3/4- by 1/8-in. stainless steel angle stock) by means of tantalum hooks which are electrically isolated from the ring itself by high alumina standoffs. The cathode is maintained in tension by interposing "Belleville" springs (made of Inconel) between the ceramic part and the ring support. These springs supply a 60- to 100-pound force on each of the hooks when fully compressed, and tend to maintain the cathode in a flat configuration during sputtering, in spite of thermal expansion of the cathode. Electrical connection to the cathode is made via one of the tantalum support hooks. The back side of the cathode is shielded from sputtering by placing a piece of sheet stainless steel on the support ring. All of the hooks must be shielded on the outside of the ring to protect them from sputtering. This is accomplished by means of stainless steel cups covered on the open end with either stainless steel or Nichrome mesh. The insulating standoffs were designed with reentrant cavities to prevent an electrical short if they are coated with sputtered material. The distances employed for all shields are of the order of 1.25 cm, which is sufficiently close to remain effective at high voltages, even for relatively high sputtering pressures (i.e., up to 70 millitorr). This cathode assembly is shown in Fig. 4-31. All materials used in the assembly were selected to have low outgassing characteristics. The assembly was designed to minimize the total heat capacity by using the minimum mass consistent with mechanical and strength requirements. The low heat capacity is desired so that the system will reach thermal equilibrium rapidly, thus minimizing temperature variations during sputtering. This cathode assembly is supported in the system by means of small posts placed on a bolt circle with a diameter of about 45 cm (at least 5 cm from the cathode).

This cathode design will meet the needs of most laboratory studies, including the coating of large substrates. For experiments in which tantalum deposition rates greater than about 300 A per minute are desired, however, the amount of power dissipated at the cathode may become such that the temperature in the chamber will become ex-

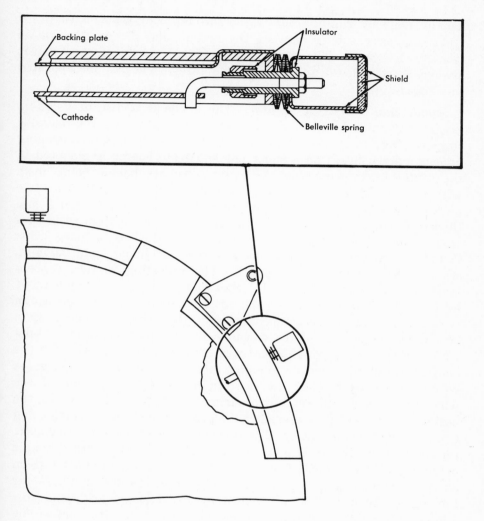

FIG. 4-31. Cathode assembly.

cessive. The maximum power which can be safely employed with this type of cathode (and of this size) is of the order of 2 kilowatts, assuming that the bell jar is well cooled. The addition of water cooling to the cathode would allow for considerably more power to be used during deposition, but the added complexity to the system design generally dictates against it for laboratory systems. When designing for water cooling the cathode, considerable care must be taken to provide adequate electrical isolation from the water mains. This is usually done by running the water through long lengths (6 to 10 feet) of plastic tubing between the system and the supply line as well as the drain. The dangers of a lethal electrical shock must always be considered and appropriate safety features incorporated into the design.

Anode and Substrate Heater. The anode is a disc of 0.062-in. grade "A" nickel sheet of the same diameter as the cathode support ring; located in the center of the disc is a square opening approximately 20 cm on a side. The substrate heater is mounted in the center of this opening, with the substrate holder being flush with the surface of the anode. The heater itself is made from two pieces of plate-fused silica (approximately 15 cm square and 3 mm thick), one of which is slotted to accept platinum-rhodium resistance wire, with the second plate serving to hold the wires in place. This heater is supported in a shallow box reflector made from 0.018-in. stainless steel with a high polish. The reflecting surface concentrates the heat toward the substrates and minimizes the amount of heat which reaches the other parts of the system. This heater design is capable of attaining temperatures of the order of 1000°C in a few minutes, with a power consumption of approximately 2 kilowatts. Platinum—20% rhodium was chosen as the material for the heater wire to minimize any possible interaction of the reactant gases with the heating element and to minimize the vaporization of the wire during operation, thereby assuring long element life. Again, the design is such that all materials have good outgassing properties and a minimum of heat capacity. The substrate holder is made of 0.050-in. tantalum sheet that is 17.5 cm square and is provided with a handle extending off the center of one side. The holder will support sixteen 25- by 75-mm substrates. Nickel was chosen as the material for the anode because it is not etched by hydrofluoric acid, and this allows accumulated tantalum to be stripped from its surface by etching. Tantalum was selected for the substrate holder in order to minimize warping during substrate heating, which occurs in the case of many materials that might anneal at the temperatures used for substrate heating. The size of the cathode and anode were chosen to be as large as possible, leaving at least as much clearance to the bell jar walls as the cathode-anode separation, with the area covered by the substrates selected so as to leave a border of free anode also approximately equivalent to the cathode-anode separation.

Shutter. A shutter is desirable in a sputtering system in order to allow time for equilibrium conditions to be reached before collecting the depositing film. When reactive sputtering is involved, it is also desirable to be able to remove the shutter while the deposition is in progress, and not to have to quench the glow at this time, destroying equilibrium. The shutter shown in Fig. 4-30 is made of relatively thin (0.032-in.) annealed nickel sheet with some light bracing to provide rigidity, and meets these requirements as well as being simple in operation. The mechanism is operated by means of stranded molybdenum cable wound around a mandrel on a rotary feedthrough in the baseplate of the vacuum system.

The cathode assembly, anode, and substrate heater are all supported from a single support plate which also supports the necessary pulleys for actuating the shutter. All of the support structure is arranged such that the active area between the cathode and the anode is free from any disturbances, with the distance between the cathode and anode being approximately 9 cm. In this particular design, in order to maintain flexibility in the choice of the mode of operation (i.e., whether to use bias sputtering, floating substrate, or any other electrical configuration), all of the support rods are provided with insulating members in order to provide mutual isolation. These insulators are equipped with deep drawn nickel cups to protect them from being coated with metal and losing their insulation. The heater wires are isolated from the remaining structure by means of an isolation transformer with insulation rated for better than 10 kilovolts. If this precaution is not taken, spurious glows, which can have an unwanted influence on the deposit, are set up around the heater. The high-voltage lead-in wire is surrounded with a stainless steel tube at ground potential in order to prevent sputtering from this source. All screws used in the construction have slotted shafts, and all blind holes are provided with a means of being pumped out so as to minimize the time for pumping on the system to reach the low pressure desirable before admitting the sputtering gas. The success which has been achieved with this design is indicated by the fact that in at least six different systems which have been built with this design, the properties of the films were reproducible. This type of hardware has been installed in both diffusion-pumped systems with "6-inch" pumps of high speed and "turbomolecular"-pumped systems with similar results. In all cases, the ultimate pressure in the vacuum system was lower than 10^{-7} torr, and the background pressure with the pumping port throttled was lower than 2×10^{-6} torr.

Gas Handling. Attention must be given to the "plumbing" used to transport and control the gases used in sputtering. The same cautions required in the construction of the sputtering system must be taken in the design and construction of the gas manifolding to prevent contamination of the sputtering gas in transit to the system itself. All valves and tubing should be constructed using the materials and techniques recommended for the vacuum system in Chap. 2. In order to purge the manifold of all air, water, and other volatiles after its installation, the entire gas-handling system must be capable of being pumped to pressures of the order of from 1 to 10 millitorr. The requirements for the component parts of this system are somewhat different from those of the usual vacuum components, however, in that they must be capable of operating at pressures as high as 200 pounds per square inch, as well as at vacuum conditions.

A schematic of a manifold system for handling two gases is shown in Fig. 4-32. In this manifold design, any restriction in the line which would increase the time required to pump the line down is bypassed

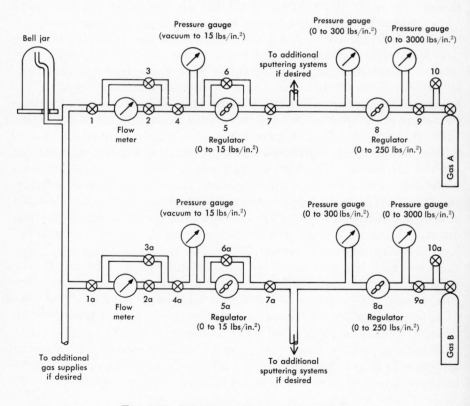

FIG. 4-32. Schematic of gas supply system.

by a valve with an opening at least as large as the diameter of the piping itself. The function of and requirements for each component will be described, starting at the vacuum station and proceeding toward the gas supply tank. The first valve (1) is the main shutoff valve, followed by the gas flow control system, which consists of a gas flow meter and a flow control valve (2). Many systems are operated without the use of a flow meter and depend on the flow characteristics remaining constant during a particular run. This is usually adequate, provided the flow control valve chosen is one of the types recommended in Chap. 2. The flow control system is bypassed with a valve (3) for the reasons mentioned above. Another shutoff valve (4) is provided to allow the line to be pumped back to the gas control valve when checking the system leak rate. This is required since many of the control valves are not capable of being completely closed.

A pressure regulator (5) is required in order to provide a constant pressure drop across the control valve, since its throughput will depend on the pressure difference as well as on its "setting." A gauge should be provided to measure this pressure, and this gauge should allow a pressure change of only one percent of the set value (usually about 10 pounds per square inch) to be detected. All-metal construction is preferred for the gauge and regulator, both of which should be well cleaned before installation. In many cases, this will have to be done by the manufacturer. A "single-stage" regulator is preferable here since one stage of reduction will already have been incorporated upstream. In specifying the regulator, care must be taken to be certain that it may be evacuated without damage to its diaphragm. The regulator is also provided with a bypass valve (6). Another valve (7) is shown in the illustration, but may be omitted if the gas supply is to serve only one sputtering station.

Another pressure regulator (8) and gauge are required to reduce the high pressure in the gas supply tank to a more convenient level. The construction of this regulator should be similar to that of the low-pressure regulator, except that it should be capable of outlet pressures in excess of 200 pounds per square inch. A pressure this high is required to allow the low-pressure regulator to provide adequate regulation; lower inlet pressures are insufficient to guarantee seating of the seal, and gas will continue to flow through the low-pressure regulator, causing the pressure to rise above the set point. A high-pressure bellows-sealed valve (9), capable of handling pressures up to 3000 pounds per square inch, is placed between this regulator and the gas supply tank. This valve, used when new gas tanks are to be installed, prevents the atmosphere from entering the gas lines. A tee and a small bleed valve (10) are placed in the line here to allow the volume of piping which is exposed to air when the gas tank is changed to be purged by bleeding off gas from this volume when it is pressurized from the new tank. It is also helpful to provide this section with a thermocouple-type vacuum gauge to allow the pressure to be read when the line is being pumped out. The outlet of the bleed valve should be provided with a length of small bore tubing in order to prevent a rapid release of gas, which might cause injury to personnel.

The piping itself should be of sufficiently large bore that the time required to pump it out is not inordinately long, but the size should be held to within reasonable limits. One-half inch stainless steel or copper tubing has proven to be a good compromise. With stainless steel construction, connections are preferably welded but an acceptable alternate, particularly applicable for copper, is the use of swage-type connectors. For relatively long runs of piping, many connections may be avoided by using a single length of "refrigerator"-type copper tubing. Whatever type of tubing is used, the interior must be thoroughly degreased and cleaned before installation. Prior to the first use of the manifold, it should be evacuated and leak-checked with a mass spectrometer-type leak detector, and preferably should be heated while being pumped in order to remove as much adsorbed water and gases as possible.

For increased flexibility it is advisable to provide manifolding to handle two different dopant gases, as well as the manifolding for the argon.

In order to provide for a smooth gas flow pattern in the sputtering chamber, the gas entry should be at the center and the top of the bell jar, rather than directly at the baseplate.

Power Supply. The power supply used for sputtering must meet special requirements. A main consideration is that the supply must be designed to prevent the possibility of a transition of the discharge from that of the abnormal glow into that of an arc. Several methods are available for achieving this. Among them are the use of a large ballast resistor in series with the discharge, the use of a power supply whose input transformer is designed to limit the output current when it is completely short-circuited, and the use of a saturable reactor in series with the primary of the high-voltage transformer. Although the efficiency of a high-reactance transformer is low in comparison with normal transformers, it is high when compared to that of a supply using a limiting resistor since the resistor itself may consume as much as 30 percent of the output power. All of these techniques allow the power supply to recover immediately after the cause of high current has ceased to exist. This is a distinct advantage over supplies which have an overcurrent tripping device which needs manual resetting, since spurious arcs of short duration, caused by particulate matter, often occur when first starting a system. Typical voltage-current characteristics of a high reactance transformer are shown in Fig. 4-33 for several different input voltages. Although this type of supply provides no filtering, it has proven quite adequate for the production of high-quality films.

The ideal solution is probably one which involves not only the power supply, but also the argon leak valve. A constant voltage power supply, equipped with some form of electronic current-limiting device

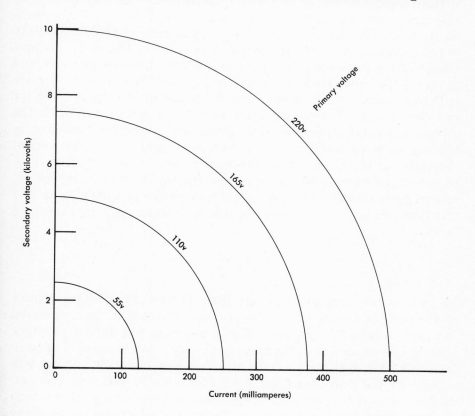

Fɪɢ. 4-33. Current-voltage relationship for a high-reactance transformer.

to prevent arcing without shutting off the supply, would assure that film properties would not be affected by varying voltage. The current should also be kept constant in order to maintain a constant deposition rate, and is normally done by adjusting the sputtering pressure during the run with the argon leak valve. If one sensed the current electronically and used this signal to drive some form of servomechanism coupled to the argon leak valve, slow drifts in current could automatically be compensated. This would require a rapid response time between a change in leak rate and the corresponding change in sputtering pressure. Since the amount of doping achieved in films is a function of the throughput of reactive gas and the deposition rate, the flow of active gas must also be maintained constant. For this reason, separate gas controls are used for the two gases rather than using a premixed gas with a single control.

Operation. Before a sputtering system may be operated with assurance, the pumping characteristics should be verified experimentally in order to insure that the gas load will not cause instability in the pressure. A simple procedure is as follows: The system is first pumped for a sufficiently long time to check the ultimate pressure and to insure that no extraneous leaks are present. The argon leak valve is then opened a small amount and the pressure in both the bell jar and the foreline is recorded after allowing a few minutes for the system to reach equilibrium. This procedure is continued by opening the leak valve in small increments and plotting the pressures against the valve setting. The actual throughput for each valve setting may be determined by closing the high vacuum valve and measuring the rate of pressure rise in the bell jar. Knowing the bell jar volume (V) and this rate of rise in pressure, the throughput may be calculated from

$$Q = \frac{pV}{t}$$

where Q is the throughput in torr liters per second, V is the volume of the bell jar in liters, p is the pressure increase in torr, observed in time t in seconds. An instability in pumping will appear as either a rapid rise or an instability in pressure with only a small change in leak rate. The bell jar pressure at which this instability occurs will depend on the characteristics of the pumping system and the conductance of the line between the bell jar and the pump. If the high vacuum pump and the fore-pump have matched characteristics, the bell jar pressure will be between 5 and 50 millitorr, with a foreline pressure of about 150 millitorr. The maximum throughput used when sputtering should be kept below this critical throughput by about a factor of two in order to assure stable operation.

Once the throughput to be used for sputtering has been determined, the system should be operated with this throughput, and the pressure in the bell jar should be brought to the approximate sputtering pressure by means of the throttling valve. The power supply is then turned on and the voltage slowly increased until the desired sputtering voltage (e.g., 5000 volts) is reached. The pressure should now be readjusted (again by means of the throttling valve) to provide a dark space configuration as close to the substrate holder as possible while still flat and parallel to the substrate plane. To assure stability in the system, the pressure should next be increased to move the dark space boundary approximately 0.5 cm farther away from the substrates than the above procedure provided. This will usually result in the dark space being approximately 2 cm from the substrates. Once this condition has been achieved, the position of the throttling valve should be marked so that its setting can be accurately reproduced in all subsequent sputtering runs. A consistent setting of the throttling valve is particularly important during reactive sputtering in order to attain reproducible throughput.

The operating pressure chosen by this method, then, is a function of the particular separation between the cathode and substrate holder. For the laboratory sputtering system described above, the operating argon pressure is approximately 27 millitorr, which gives an indicated pressure reading of 18 millitorr on a Pirani gauge calibrated for air. If it is desired to operate the system at some other pressure while still maintaining the "optimal" substrate-dark space distance, the substrate-cathode separation must be changed. The proper separation may be estimated since, for a constant voltage, the product of cathode dark space distance (d) and the pressure (p) is a constant. For the above example, this product is about 0.19 torr cm. Thus, if it is desired to operate at 50 millitorr instead of 27, the cathode-to-substrate distance should be changed from 9 cm to 5.8 cm, keeping the same 2-cm distance between the edge of the dark space and the substrates. In many cases, however, systems are operated at pressures slightly higher than the optimum pressure without seriously affecting the film properties. Most experiments involving changes in pressure during sputtering have been performed with a constant cathode-to-substrate distance, because of the difficulty of changing the spacing for each pressure being investigated. If a broad pressure range is involved in the investigation, however, erroneous conclusions may be reached since the sputtered atoms suffer many more collisions in transit to the substrate at the higher pressure.

Other Designs

In order to deposit several layers of different films, several systems involving multiple cathodes have been designed. These are based on moving either the substrate between fixed cathodes or the cathodes over a fixed substrate. If it is desired to monitor the film during deposition, the latter method offers an advantage since slip rings or commutators are not required for a multitude of monitoring leads. Typical designs of these two types are described by Maissel et al [31].

Engineers at the Western Electric Company have designed an "open-ended" system for the continuous deposition of tantalum films [32] and, although this is covered in detail in a later chapter, a brief description of this equipment will be provided here. The arrangement of the vacuum system and the chambers of this equipment is shown in Fig. 4-34. With any "batch"-type system, the enclosure must be opened to atmosphere between each deposition run. This is not necessary with the "in-line" system since the substrates move continuously through the chamber. This type of operation is made possible through the use of a series of continually pumped chambers interconnected

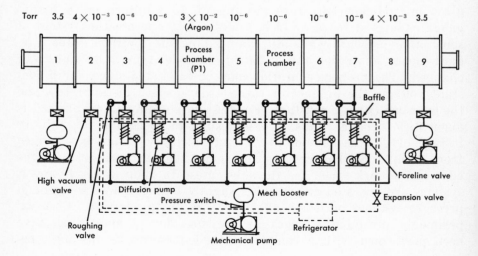

FIG. 4-34. The vacuum pumping system of a continuous, open-ended, in-line sputtering machine.

only via restricted openings through which the substrates and their carriers pass. The substrate carriers or "boats" along with the track on which they move provide a partial closure of these openings, as was illustrated in Chap. 2. No shutter is required for such a system since empty boats are passed through the machine until an equilibrium condition is established, at which time substrates are started through. This equilibrium condition is maintained until the operating conditions are changed or until the system is shut down.

A disadvantage of this continuous system is the use of substrate carriers which provide for the restricted flow of gases between chambers. These boats, which are relatively massive, are exposed to air between each use, allowing for the sorption of contaminants. Although some of the contaminants are desorbed in the "preheat" chambers, complete outgassing is unlikely and some unintentional doping occurs. A simple closed machine for the continuous deposition of tantalum has been designed by Shockley et al. [33], and is shown schematically in Fig. 4-35. Substrates (0.7 in. square) are loaded

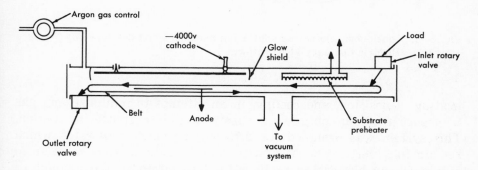

FIG. 4-35. Basic diagram of a closed-end continuous sputtering machine. (After Shockley, Geissinger, and Svach, "Automatic Sputtering of Tantalum Films for Resistor and Capacitor Fabrication," in *IEEE Transactions*, 1964; courtesy The Institute of Electrical and Electronics Engineers, Inc.)

and unloaded through rotary valves which serve as locks and prevent the admission of air into the system. The substrates are carried through a preheat section and then through the sputtering region on a continuous belt. A system of this type is being used commercially by the Collins Radio Company in the production of tantalum thin film circuits. More elaborate closed-end machines have been designed both by Stetka et al [34] and Viola et al [35]. The Viola-designed system, shown in Fig. 4-36, is divided into three chambers, one each for

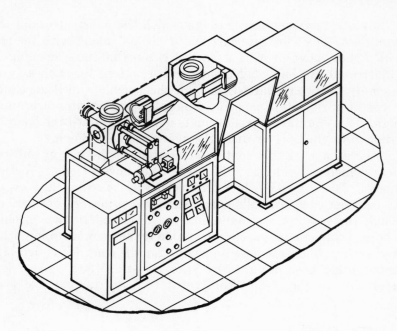

Fig. 4-36. Continuous, closed-end sputtering machine being developed by Western Electric Company at Allentown.

heating, depositing, and cooling, in an attempt to minimize any gas transport from an outgassing substrate to the sputtering region. This system was designed for 3.75- by 4.50-inch substrates which remain in a vertical position with no substrate carrier in order to minimize any outgassing problem, and to minimize the amount of particulate matter which might collect on the substrate prior to deposition.

Bell jar facilities have been designed to accommodate large "batches" by the use of either a magazine feed for substrates or by using a conveyor system with a high capacity. One such system designed by Needham [36] is shown in Fig. 4-37. This system was designed to handle fifteen 6-inch by 6-inch substrates in one cycle by means of a conveyor. The system permitted bakeout at temperatures as high as 300°C, and pressures of the order of 10^{-8} torr were achieved. While this system appears to be capable of producing films of high purity, there were several disadvantages reported. The conveyor mechanism caused considerable difficulty, particularly during system checkout, and stringent bakeout was required in order to reduce the outgassing of the massive interior hardware.

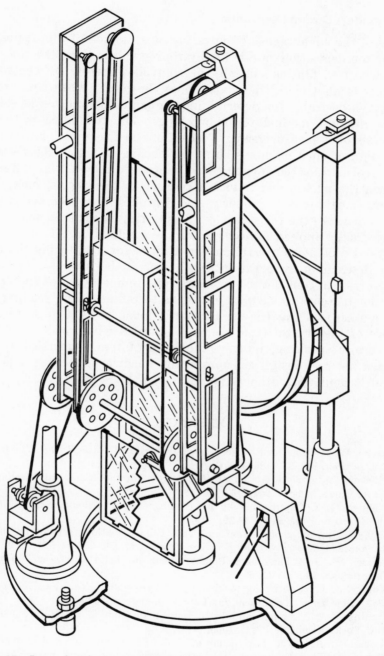

FIG. 4-37. Large-batch arrangement with conveyor mounted within a bell jar. (After J. G. Needham, "A High Capacity Vacuum Sputtering Machine," *Trans. 10th Nat. Vac. Symposium, 1963,* 1964; courtesy The Macmillan Co.)

Film Monitors During Deposition

One of the advantages of sputtering as a method of film deposition is that the deposition rate is highly reproducible; thus, the thickness of a deposited film can usually be controlled merely by controlling the deposition time. There are times, however, when an internal monitor is desirable for determining the point to stop the deposition.

The simplest method for monitoring conductive films is the use of a resistance measurement. Precautions are required, however, in the design of such monitors since the plasma in which the measurement probes must be immersed may produce a potential difference between the probes and induce a current in the circuit, making all readings meaningless. Properly placed probes have been successfully used to measure the resistance of films during deposition, with a correction factor applied to account for the conductivity of the plasma.

Crystal oscillators are frequently utilized for monitoring the deposition in vacuum evaporation, but they can be used for sputtered films only with considerable caution. The temperature of the crystal must not increase to such an extent as to influence the reading, and some method, such as the use of a grid, must be used to prevent the surface of the crystal from being bombarded with ions.

Generally speaking, internal monitors of film properties are not included in the design of sputtering hardware because adequate control is achieved without them.

REFERENCES

1. Grove, W. R. "On the Electrochemical Polarity of Gases," *Phil. Trans. Roy. Soc. London,* 142, 87 (1842).
2. Plücker, J. "On the Action of the Magnet on the Electrical Discharge in Dilute Gases," *Pogg. Ann.,* 103, 90 (1858).
3. Wehner, G. K. "Sputtering by Ion Bombardment," *Advances in Electronics and Electron Physics,* 7, 239-298 (1955).
4. Kay, E. "Impact Evaporation and Thin Film Growth in a Glow Discharge," *Advances in Electronics and Electron Physics,* 17, 245-322 (1962).
5. Moore, W. J. "The Ionic Bombardment of Solid Surfaces," *American Scientist,* 48, 109-133 (1960).
6. Wolsky, S. P. "Sputtering Mechanisms," *Trans. 10th Nat. Vac. Symposium, 1963.* New York: The Macmillan Company, 1964. pp. 309-315.
7. Kaminsky, M. "Atomic and Ionic Impact Phenomena on Metal Surfaces." New York: Academic Press, 1965.
8. Maissel, L. I. "The Deposition of Thin Films by Cathodic Sputtering," *Physics of Thin Films,* vol. III, edited by R. Thun. New York: Academic Press, 1966.

9. Crookes, W. "On Electrical Evaporation," *Proc. Roy. Soc. London*, **50**, 88 (1891).
10. von Hippel, A. "On the Theory of Cathode Sputtering," *Ann. Physik*, **81**, 1043-1075 (1926).
11. Stark, F. "On the Theory of Sputtering by Atomic Rays," *Z. Electrochem.*, **15**, 509-512 (1909).
12. Wehner, G. K. "Angular Distribution of Sputtered Material," *J. Appl. Phys.*, **31**, 177-179 (1960).
13. Almen, O., and G. Bruce. "High Energy Sputtering," *Trans. 8th Nat. Vac. Symposium, 1961*. New York: Pergamon Press, 1962. pp. 245-251.
14. Rol, P. K., D. Onderlinden, and J. Kistemaker. "Some Physical Aspects of Sputtering," *Advances in Vacuum Science and Technology*, vol. 1. London: Pergamon Press, 1966. pp. 75-82.
15. Cobine, J. D. "Gaseous Conductors." New York: McGraw-Hill Book Company, Inc., 1941.
   Penning, F. M. "Electrical Discharge in Gases." New York: The Macmillan Company, 1957.
16. Schwartz, N. "Reactive Sputtering," *Trans. 10th Nat. Vac. Symposium, 1963*. New York: The Macmillan Company, 1964. pp. 325-334.
17. Theuerer, H. C., and J. J. Hauser. "Getter Sputtering for the Preparation of Thin Films of Superconducting Elements and Compounds," *J. Appl. Phys.*, **35**, 554-555 (1964).
18. Gerstenberg, D., and C. J. Calbick. "Effects of Nitrogen, Methane, and Oxygen on the Structure and Electrical Properties of Thin Tantalum Films," *J. Appl. Phys.*, **35**, 402-407 (1964).
19. Sinclair, W. R., and F. G. Peters. "Method for Controlled Multicomponent Sputtering," *Rev. Sci. Inst.*, **33**, 744-746 (1962).
20. Blevis, E. H. "Hot Cathode and Radio Frequency Sputtering." Monograph distributed by the R. D. Mathis Company, Long Beach, California, 1965.
21. Edgecumbe, J., L. G. Rosner, and D. E. Anderson. "Preparation and Properties of Thin Film Hard Superconductors," *J. Appl. Phys.*, **35**, 2198-2202 (1964).
22. Nickerson, J. W., and R. Moseson. "Low Energy Sputtering," *Research/Development*, **16** (3), 52-56 (March, 1965).
23. Gawehn, H. "On a New Cathode Sputtering Process," *Z. Angew. Physik*, **14**, 458 (1962).
24. Vratny, F. "Deposition of Tantalum and Tantalum Oxide by Superimposed RF and DC Sputtering," *J. Electrochem. Soc.*, **114**, 505-508 (1967).
25. Anderson, G. S., W. N. Mayer, and G. K. Wehner. "Sputtering of Dielectrics by High Frequency Fields," *J. Appl. Phys.*, **33**, 2991-2992 (1962).
26. Davidse, P. D., and L. I. Maissel. "Dielectric Thin Films through rf Sputtering," *J. Appl. Phys.*, **37**, 574-579 (1966).
27. Gerstenberg, D., and E. H. Mayer. "Properties of Tantalum Sputtered Films," *Proceedings 1962 Electronics Components Conference*, 57-61 (1962).
28. Schuetze, H. J., H. W. Ehlbeck, and G. G. Doerbeck. "Investigation of Thin Tantalum Films," *Trans. 10th Nat. Vac. Symposium, 1963*. New York: The Macmillan Company, 1964. pp. 434-439.
29. Frerichs, R. "Superconductive Films made by Protected Sputtering of Tantalum or Niobium," *J. Appl. Phys.*, **33**, 1898-1899 (1962).

30. Krikorian, E., and R. J. Sneed. "Deposition Parameters Affecting Epitaxial Growth of Single Crystal Films," *Trans. 10th Nat. Vac. Symposium, 1963.* New York: The Macmillan Company, 1964. pp. 368-373.
31. Simmons, J. G., and L. I. Maissel. "Multiple Cathode Sputtering System," *Rev. Sci. Inst.*, 32, 642-645 (1961).
    Maissel, L. I., and J. H. Vaughn. "Techniques for Sputtering Single and Multilayer Films of Uniform Resistivity," *Vacuum*, 13, 421-423 (1963).
32. Charschan, S. S., R. W. Glenn, and H. Westgaard. "A Continuous Vacuum Processing Machine," *Western Electric Engineer*, 7(2), 9-17 (April, 1963).
33. Shockley, W. L., E. L. Geissinger, and L. A. Svach. "Automatic Sputtering of Tantalum Films for Resistor and Capacitor Fabrication," *IEEE Trans. Comp. Parts*, CP11, 34-37 (1964).
34. Stetka, D. G., and H. Westgaard. Personal communication, 1964.
35. Viola, F. J., and R. D. Kaufmann. Personal communication, 1966.
36. Needham, J. G. "A High Capacity Vacuum Sputtering Machine," *Trans. 10th Nat. Vac. Symposium, 1963.* New York: The Macmillan Company, 1964. pp. 402-406.
37. Glang, R., R. A. Holmwood, and P. C. Furois. "Bias Sputtering of Molybdenum Films," *Advances in Vacuum Science and Technology*, Vol. II, London: Pergamon Press, 1966. pp. 643-650.
38. Flur, B. L. "On the Magnetic Properties of Sputtered NiFe Films," *IBM J. Res. Develop.*, 11, 563-569 (1967.)
39. Vratny, F. "Anodic Tantalum Oxide Dielectrics Prepared From Body Centered Cubic Tantalum and Beta Tantalum Films," *Electrochem. Technology*, 5, 283-287 (1967).

Chapter 5

Chemical Methods of Film Formation

Fabrication methods for thin film components and circuitry are not limited to vacuum evaporation and sputtering for film deposition. The conductance of interconnections is increased by electroplating, thin film nickel resistors are made by electroless plating, silicon and germanium transistors are deposited by chemical vapor plating, and in tantalum thin film circuitry both resistor adjustment and capacitor dielectric formation utilize the process known as anodization.

Chemical methods are generally faster, use simpler equipment, and thus may be more economical. These methods, however, are often complex and difficult to control. Side reactions occur, along with nonequilibrium conditions, and it is difficult to reproduce film properties. Some of these techniques demand that the supporting substrate withstand high temperature; others require that substrates be exposed to various solvents. These complications have tended to limit the application of chemical deposition.

5.1 ELECTROPLATING

Some chemical reactions are spontaneous and, under proper conditions, may be used to produce electrical energy; others require the addition of electrical energy to proceed. The use of an electric current to produce a chemical change is called electrolysis, with electroplating a particular example. Electroplating may be defined as the production of metal coatings through the action of an electric current. These metal coatings are used to improve appearance, corrosion resistance, hardness, bearing quality, or some similar property of a basis metal.

An electroplating system, Fig. 5-1, consists of a plating bath, a battery, and two electrodes. The electrode connected to the positive battery terminal is called an anode and is often made of the metal to be deposited. Electrons flow from the anode metal to the battery and leave positively charged metal cations which dissolve in the plating bath. The electrons arriving at the negative electrode (cathode) from the battery neutralize the positive charges on the metal cations in the surrounding plating bath and convert them to metal atoms. These atoms become attached to the cathode and are thus removed from the solution. The removal of metal cations from the plating bath creates a concentration gradient adjacent to the cathode. This provides a force causing neighboring ions to diffuse toward the cathode. The electric field between the electrodes provides another force which causes the cations to migrate toward the cathode and the anions to move toward the anode. These moving ions conduct a current through the solution and complete the electrical circuit. Thus, electrical energy is supplied by the battery to cause chemical change (electrolysis) in the plating bath. The net result is that a metal is plated onto the cathode from a solution of metal ions, and metal dissolves from the anode.

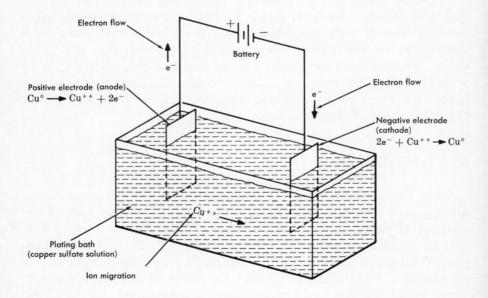

FIG. 5-1. Electroplating system.

Deposition Parameters

Properties of electrodeposited films are affected by the chemistry and physics of the electrolyte and the electrodes and by the current density. The bath characteristics include the concentration and nature of the positive metal ions (cations), the negatively charged ions (anions), and the organic and inorganic additives or impurities. Temperature, agitation, viscosity, and surface tension of the solution are also important. The shape and surface condition of the cathode affect the uniformity of growth. Both the anode and the cathode may become coated with an extraneous film, resulting in unwanted electrochemical reactions. The current density controls the deposition rate and thus has a pronounced effect on the character of the deposited film.

The electrolyte must provide metal ions to the region surrounding the cathode as the concentration of these ions is depleted by the plating process. The forces which act on the metal ions to transport them to the depleted region are those of diffusion, migration in the electric field, and convection currents in the solution. Diffusion is caused by the concentration gradient across the depleted electrolyte layer and is generally more important than the migration due to the applied electric field. Convection is caused by density and temperature differences in the bath as well as by agitation.

The nature of the electrolyte anion influences the film structure. For example, significant variations result, depending on whether the metal salt used is a nitrate (NO_3^-), sulfate ($SO_4^=$), or acetate (CH_3COO^-). In some plating baths, growth perpendicular to the cathode surface is more rapid than lateral growth, and needle-like crystals are produced. Silver nitrate and stannous chloride baths behave in this manner. In silver cyanide, on the other hand, the outward growth is inhibited and a more useful coating results—hence the use of cyanide baths in plating. This particular effect has been attributed to preferential adsorption of anions on the cathode. The anion present in the bath can also control the free metal ion concentration and consequently side reactions and deposition rate.

Impurities in the electrolyte may be adsorbed at the cathode surface, and one added molecule or ion may affect many thousands of depositing metal ions. Sometimes this effect is beneficial, resulting in a smoother, brighter film. This smoothing is caused by preferential plating in crevices and valleys, since the impurities tend to be adsorbed at peaks and inhibit plating there. A similar phenomenon retards the growth of trees and nodules, and chemicals which perform this function are often purposely added to the plating bath. Ionic impurities may precipitate metal salts which, if attached to the cathode, may cause either nodular growth or pit formation.

The surface tension of the bath is another important parameter. For example, if hydrogen is evolved at the cathode, pits will result if the bubbles remain attached to the cathode. Wetting agents added to the bath promote the release of these bubbles and thus prevent pit formation.

Plating Technology

The practice of electroplating begins with the preparation of the material to be plated, and it is generally agreed that poor preparation is the source of most plating trouble. Control of the deposited film properties also depends on the plating bath composition, the electrodes, and the current density. The discussion here is purposely brief, and Ref. [1] as well as the general references listed at the end of this chapter should be consulted for greater details and other examples.

Cleaning. Film adherence, smoothness, brightness, and porosity are all affected by the cleanness of the basis metal. Grease, dirt, oxide, and other nonmetallic materials must be removed before plating. Cleaning methods may be classified as (1) solvent cleaning, (2) emulsion cleaning, (3) alkaline cleaning, and (4) acid cleaning.

Solvent cleaning makes use of organic solvents such as hydrocarbons (or chlorinated hydrocarbons) to dissolve grease and oils. Stoddard solvent and mineral spirits are widely used because of their low cost and high flash points. Trichloroethylene is an excellent solvent and is nonflammable, but it is more toxic and expensive than the hydrocarbons. Articles may be cleaned with these solvents by soaking, spraying, or vapor degreasing. Ultrasonic agitation is also used to improve the effectiveness of the soaking technique. The use of ultrasonics involves the formation and collapse of microscopic bubbles in the solvent, generated by transmitting ultrasonic vibrations through the liquid. This bubble formation results in a powerful scrubbing action which separates contaminants from the surface.

Emulsion cleaning uses an organic solvent dispersed in water with the aid of an emulsifying agent, and functions by removing dirt and grease and holding them in suspension at the oil-water interface. The organic solvents generally used are petroleum-based hydrocarbons such as kerosene, with the emulsifier being a soap soluble in the oil phase. Emulsion cleaning may be brought about by spraying or immersion and agitation. This cleaning technique leaves a film of oil on the article being cleaned, which must then be removed by one of the other methods.

Alkaline cleaners contain materials such as sodium carbonate, sodium hydroxide, trisodium phosphate, and sodium silicate in combination with various wetting agents. Again, cleaning is performed by either soaking or spraying, generally at temperatures from 70° to 95°C. The cleaning action is a combination of preferential wetting, emulsification and dispersion of dirt, and the conversion of oils to soap-like compounds. The cleaning ability of alkaline cleaners may be improved by making the object to be cleaned either anodic or cathodic while immersed in the cleaning solution. This produces rapid evolution of oxygen or hydrogen, providing an efficient scrubbing action.

Acid cleaning is used to remove metal oxides from the surface of the object to be plated. Mineral acids, organic acids, and acid salts may be used separately or in combination. Wetting agents and inhibitors are often included to improve surface coverage and retard the attack on the metal itself.

The various cleaning techniques are used in different combinations, depending on the metal to be cleaned and the expected contaminants. It is difficult to determine the degree of cleanness of a sample, but it should at least pass the water break test. This test is performed by withdrawing the surface to be tested from a container of clean water. The water draining from the surface should form a continuous sheet.

The Plating Bath. In a plating bath, the metal may be present as free ions from a simple salt such as copper, tin, or nickel sulfate or may be present in combination with other atoms as part of a complex ion as are gold or silver in cyanide plating baths. Other chemicals are frequently added to plating baths to control the acidity (pH) of the bath, to implement dissolution of the anode, and to control various properties of the deposited coating. The bath pH affects the evolution of hydrogen, the precipitation of some salts, and the composition of the complex ion from which the metal is deposited. These factors are not all exactly predictable, and the choice of proper pH is often empirical. Viscosity, agitation, and temperature of the bath all affect the rate of diffusion to the cathode and therefore the properties of the deposit.

As an example of the contents and properties of plating baths, some of the principal systems for plating copper will be discussed.

Copper may be plated from either alkaline or acid baths, and one bath of each type is discussed.

Alkaline Copper Plating Baths. Cyanide baths are the most widely used of the alkaline types since they can be controlled to produce deposits of uniform thickness. Also, it is often necessary to deposit a "strike" coating of about 0.05 mil of copper from a "low concentration" cyanide bath in order to prevent the spontaneous displacement of copper ions by a more active metal. A "high concentration" bath is used for the rapid plating of thicker coatings (up to 2 mils).

Copper cyanide itself is insoluble in water except in the presence of excess cyanide ions, usually furnished by either sodium or potassium cyanide. In the presence of this excess cyanide, a soluble copper-cyanide complex ion is formed. This complex ion has a composition corresponding to $Cu(CN)_3^=$ at room temperature and $Cu(CN)_2^-$ at high temperatures, and both are present at normal plating temperatures. Cyanide not bound in this copper complex is called free cyanide. A typical dilute cyanide bath contains 15 grams of copper and 7-1/2 grams of free cyanide per liter of solution, and a high concentration bath typically contains 85 grams of copper and 4 grams of free cyanide per liter. Copper anodes are used in both baths, and the potential applied to each is about 6 volts. The pH of the dilute solution is brought to a value of 12 and that of the concentrated solution to 13 by the addition of sodium hydroxide.

Dilute cyanide baths may be operated at room temperature but 30° to 40°C is more common. The free cyanide in this bath must be controlled, since a low concentration causes rough deposits and a high concentration lowers the cathode efficiency and rate of deposition. Agitation and continuous filtration of the bath are recommended. The filtration removes any suspended matter which would cause roughness. The only additive generally used in this bath is sodium carbonate (to buffer the pH).

The "high concentration" cyanide bath operates at nearly 100 percent current efficiency, and its composition may vary up to 10 percent without adverse effects on its deposition behavior. A typical bath temperature is 60° to 65°C and the deposition rate is some 3 to 5 times greater than that of the dilute cyanide bath. Mechanical agitation is necessary at high current density to avoid a dull, "burned" appearance of the coating. Filtration is also used in this bath. A diaphragm is often placed between the anode and cathode; for example, a nylon bag may be placed around the anode. Proprietary wetting agents are common in high concentration cyanide baths to reduce pitting and the effects of organic contamination. The bath may be continuously filtered through activated carbon to reduce the concentration of organic contaminants, and in this case a wetting agent is not employed.

Acid Copper Plating Baths. Acid baths are more tolerant of ionic impurities than alkaline baths. They also have better micro "throwing power" (ability to deposit inside holes or depressions) which may be effective in sealing small pores, but they have poorer macro "throwing power" and thus less uniform metal distribution. The acid bath will spontaneously deposit copper on more active metals such as iron or zinc when these metals are immersed in the bath. This action yields a nonadherent film and must be prevented by first applying a copper "strike" in a dilute cyanide bath.

Copper sulfate baths are used to deposit thick copper films, primarily in electroforming. A typical sulfate bath contains 225 grams of copper sulfate pentahydrate per liter and 50 grams of sulfuric acid per liter. Plating temperatures range from 20° to 30°C. Higher temperatures reduce tensile strength and increase grain size. Current density also affects grain size, and agitation is necessary at high current densities if a fine grained deposit is desired. The addition of gelatin, glue, phenolsulfonic acid, or glycine to copper sulfate baths also results in smoother, fine grain deposits. Most brighteners added to the bath are proprietary materials which may contain sulfonated aromatic compounds, thiourea, or a sodium salt of naphthalene disulfonic acid. During plating, the copper concentration increases and the free acid content decreases so that some of the bath is periodically discarded and replaced with more acid and water.

Silver, gold, arsenic, and antimony impurities in the bath will codeposit with copper. Small amounts of gelatin or tannin added to the bath inhibit the deposition of the latter two impurities. Organic contaminants from decomposition of the additives may cause the deposit to be brittle and discolored, and activated charcoal is often used to remove these contaminants.

Cathodes. Electrodeposition is influenced by the uniformity of current density over the cathode area which is predetermined, at least in part, by the shape and location of the electrodes. The uniformity of current density affects the uniformity of the plated film. Edges and projections have a higher current density than do crevices and hollows, and these areas of high current density receive thicker deposits. If a hollow (five sided) box is plated in a copper cyanide bath, the resulting cross sectional appearance is similar to that shown in Fig. 5-2. Some problems in cathode geometry have been solved by placing a "thief" or "robber" cathode near the cathode to be plated. In addition to irregular current densities due to geometry effects, the variation in cathode current efficiency with current density can also affect the film uniformity.

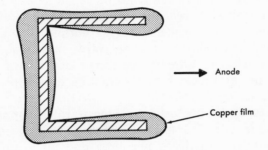

FIG. 5-2. Cross section of box plated in copper cyanide bath [1].

The power supply required depends on the area to be plated (i.e., on the total current requirements) and may vary all the way from a dry cell to a motor generator.

Plating on Thin Film Circuits

An example of the use of the electrodeposition technique in a tantalum thin film circuit was the use of a 100,000A thick copper film on conductor areas to minimize transmission losses at microwave frequencies. The procedure followed for this application involved several steps. The first layer of the thin film circuit, above the substrate, was sputtered tantalum. Since the oxide on the tantalum film base layer was found to interfere with the adherence of the electrodeposit, Nichrome and gold were evaporated onto the tantalum to produce a layer to which the plated copper adhered. Because of intervening processing, a stringent cleaning of this gold film was necessary to assure adherence. The plating bath was an acidic solution of copper sulfate formulated to provide films with a minimum of internal stress. A current density of 15 ma/cm^2 was applied for about 20 minutes. Following copper plating, the circuits were rinsed with distilled water and then plated with 10,000A of gold to prevent oxidation of the copper. Bath agitation was achieved by continuous circulation through a filter, and bath density and pH were maintained by periodic measurement and adjustment.

5.2 ELECTROLESS PLATING

Continuous deposition of a metal film from a plating solution without the use of an external source of electric current is called electroless plating. The depositing metal atoms must be formed from the metal ions in the plating bath only at the surface of the object being coated. This is accomplished by the use of a chemical reaction which will occur only when catalyzed by a freshly formed metal film. Thus, each layer of depositing atoms becomes the catalyst for deposition of the next layer. Since only certain surfaces act as catalysts, this method of film deposition may prove to be advantageous in applications where only selected areas are to be coated. An advantage over electroplating exists where coatings of uniform thickness and quality are desired on irregular surfaces because there is no problem of maintaining uniform current density. Nonconductors may be coated using this method, and thin film resistors have been prepared on both glass and ceramic substrates. Electroless plating has been utilized for several metals, but a majority of the development has been applied to nickel films; for this reason, the discussion here is limited to the electroless plating of nickel.

Deposition Process

Nickel ions (Ni^{++}) may be reduced to metallic nickel (Ni^0) by several chemical techniques, but the reduction by hypophosphite ion ($H_2PO_2^-$) is the only one which has yielded an adherent coating of nickel. Other methods generally precipitate the metal throughout the solution instead of continuously depositing a film onto a single object. The positively charged nickel ions (Ni^{++}) in the plating bath must receive two electrons to convert them to the neutral metal atoms (Ni^0), and these electrons are supplied as a result of the hypophosphite anion ($H_2PO_2^-$) reacting at the catalyst surface. The catalysts which promote this reaction are freshly precipitated nickel, palladium, rhodium, and ruthenium in acid solutions and cobalt in alkaline solutions. Electroless plating may be used to coat nonconductors if they are first coated with evenly dispersed nuclei of one of the catalysts. This is most often done by dipping the substrate in a solution of palladium chloride and then in an aqueous solution of a reducing salt such as stannous chloride to reduce the palladium salt to metallic palladium.

When a metal more reactive than nickel is immersed in the plating bath, a nickel coating is spontaneously formed on it, and, in the presence of hypophosphite ion, this initial nickel film acts as the catalyst to continue the coating process. If a metal less active than nickel is placed in the plating bath, no reaction will occur, but if it is momentarily made cathodic, nickel will electroplate out and initiate the electroless deposition.

A typical electroless plating bath contains the following in 1 liter of solution: 21 grams of nickel sulfate hexahydrate ($NiSO_4 \cdot 6H_2O$); 26.5 grams of sodium hypophosphite (NaH_2PO_2); 33.8 grams of an 80-percent solution of lactic acid ($CH_3CHOHCOOH$); 0.002 gram of lead nitrate [$Pb(NO_3)_2$]; and sufficient sodium hydroxide (NaOH) to adjust the pH to 4.6. If the bath temperature is held constant at 99°C, nickel will deposit on a catalytic surface immersed in the bath at a rate of nearly 4000A per minute.

The mechanism of electroless nickel plating has been described by Gutzeit [2]. He suggests that the first reaction to occur is the conversion (oxidation) of hypophosphite to metaphosphite:

$$H_2PO_2^- \xrightarrow{\text{catalyst}} PO_2^- + 2H \text{ (on catalyst)} \qquad (5\text{-}1)$$

It is the active hydrogen adsorbed on the catalyst that reduces the nickel ions in solution to metallic nickel:

$$Ni^{++} + 2H \text{(on catalyst)} \rightarrow Ni^0 + 2H^+ \qquad (5\text{-}2)$$

At the same time, the active hydrogen reduces some hypophosphite ion to phosphorus:

$$H_2PO_2^- + H \text{(on catalyst)} \rightarrow P^0 + H_2O + OH^- \qquad (5\text{-}3)$$

This results in the codeposition of some phosphorus to yield a nickel-phosphorus alloy. Gaseous hydrogen is also formed during the reaction by combination of active hydrogen atoms on the catalyst:

$$2H \text{(on catalyst)} \rightarrow H_2 \qquad (5\text{-}4)$$

The useful life of the plating bath may be extended for continuous operation by replenishing the hypophosphite and nickel ions as they are used up. The metaphosphite anion formed in Eq. (5-1) is converted to orthophosphite ($HPO_3^=$) by water:

$$PO_2^- + H_2O \rightarrow HPO_3^= + H^+ \tag{5-5}$$

and the presence of this orthophosphite ion limits the quantity of nickel ions in solution by precipitating nickel phosphite. Thus, an additional material must be added to prevent the nickel from precipitating. This material (lactic acid in our example) forms a compound in which the nickel ions are loosely held and easily released as the nickel ion concentration in the plating bath is depleted by the deposition process. It is also necessary to remove the hydrogen ions produced by Eqs. (5-2) and (5-5) by either periodic or continuous neutralization. The optimum deposition rate has been found to be in the pH range of 4.5 to 5.5.

Deposition Parameters

The important bath parameters are temperature, nickel and hypophosphite ion concentration, and pH. The rate of deposition increases exponentially with bath temperature, and the preferred operating temperature is above 95°C (provided that local heating is avoided). The optimum concentration of hypophosphite is 0.25 mole per liter with a usable range being from 0.15 to 0.35 mole per liter. The optimum concentration of nickel is 0.30 to 0.45 mole per liter or from 0.25 to 0.60 times the concentration of hypophosphite ion. The pH is maintained in the range 4.5 to 5.5 by the addition of a buffer such as sodium acetate.

Sufficient lactic acid or malic acid is added to complex the free nickel ions present and prevent precipitation of nickel orthophosphite. Various organic acids are used to increase the deposition rate (it is thought that they react with hypophosphite to replace oxygen and facilitate the removal of hydrogen). Electroless nickel plating baths may spontaneously decompose unless traces of stabilizing agents, which act as catalyst poisons, are added.

Properties of Electroless Plated Films

Films produced by the chemical reduction of nickel by hypophosphite ion are alloys of nickel and phosphorus. The phosphorus content of an electroless nickel film is a function of the bath composition and can range from 3 to 15 percent (by weight). Photomicrographs

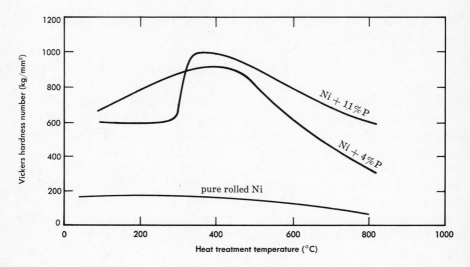

FIG. 5-3. Effect of heat treatment on Vickers hardness number of electroless
nickel that contains various amounts of phosphorus, expressed in weight
percent [3]. (Courtesy *Plating*.)

reveal laminar film structures with layers of varying composition,
and this has been interpreted as an indication that the phosphorus is
incorporated in the film in a cyclic fashion. These layers disappear
on heat treatment, and the structure reverts to a more homogeneous
mixture of nickel and the intermetallic compound Ni_3P.

Only a diffuse band is observed on X-ray diffraction indicating an
amorphous, liquid-like structure. As might be expected, the mechanical properties of the film vary with the phosphorus content. The
effect of annealing on the hardness of films of various phosphorus
content is shown in Fig. 5-3 [3]. The melting point and corrosion resistance are also affected by the amount of phosphorus in the films.
Parts plated with electroless nickel may be soldered or brazed, but
welding is not recommended since welds on plated areas may be
embrittled by the phosphorus present in the film. Some other properties of these nickel films are listed in Fig. 5-4 [4].

5.3   CHEMICAL VAPOR PLATING

In vapor plating, a volatile compound of the substance to be deposited is vaporized, then thermally decomposed at the substrate to
yield the desired deposit. In the case of metals, the volatile compounds are usually halides (chlorides or iodides), metal carbonyls,

Resistivity	60 μ ohm cm
Thermal conductivity	0.0105 to 0.0135 cal/cm sec °C
Reflectivity	45 to 50%
Density	7.85 gm/cm^3
Coefficient of thermal expansion	13 $\times$ 10^{-6} per °C
Melting point	890 °C
Hardness	500 Vickers C49 Rockwell
Adhesion to steel	30,000 to 60,000 psi

FIG. 5-4. Properties of a typical film of electrolessly deposited nickel.

or metal hydrides. Some of these compounds may be decomposed by high temperature alone; others require a combination of high temperature and a reducing agent such as hydrogen. Sometimes the deposition is brought about by reacting the incoming gas with the substrate itself or with some reactive coating on the substrate. Vapor plating is carried out at either atmospheric or reduced pressures. Coatings of materials with very high melting points may be deposited using this technique and it is also used in growing epitaxial single crystal films, particularly of silicon.

Film Deposition Mechanism

The vapor of the volatile plating compound is generated by evaporation. This evaporation may be accomplished either by heating a volatile compound or by passing a reactive gas over or through the material to be plated to generate a volatile compound. Gas flow provides a continuous supply of reactive gas to the area of the substrate, with the actual transfer to the substrate surface taking place by diffusion. This process is thus different from vacuum evaporation where the depositing molecules reach the substrate with few or no collisions. The reaction occurs on the hot substrate surface and results in the formation of the deposit and volatile by-products (which must then be transported away).

Deposition Parameters

Factors which affect the diffusion layer adjacent to the substrate also affect film growth. These include gas flow rates, pressure, and temperature. If there is a depletion of reactive gas next to the substrate, growth may proceed columnarly into the diffusion layer rather than as a smooth film. An increase in concentration of reactive gas or substrate temperature increases deposition rate. Increasing substrate temperatures or decreasing reactive gas concentration increases the crystallinity of the film.

Impurities affect film growth by modifying the nucleation rate and by causing structural defects. For example, carbon contamination during the deposition of silicon may cause stacking faults and whiskers, and impede step motion.

Film Properties

Chemical vapor deposition results in films with greater crystallite orientation than usual for other deposition techniques and thus accentuates stresses caused by temperature changes after deposition. Crystallites or grains are often columnar and perpendicular to the substrate, and if the film is strained laterally, these grains separate and the film ruptures. Difficulty in matching thermal expansion coefficients of coating and substrate has been a major obstacle to the wider use of chemical vapor plating. Of course, this problem disappears when the film and substrate are the same material, as in epitaxy.

Film Deposition Technique

A currently important application of chemical vapor deposition is the epitaxial growth of single crystal germanium and silicon films on single crystal substrates. The process is shown schematically in Fig. 5-5. The following two reactions occur:

$$I_2 + Ge \xrightarrow{\substack{450°- \\ 700°C}} GeI_2 \qquad (5\text{-}6)$$

$$2\,GeI_2 \xrightarrow{\substack{325°- \\ 400°C}} Ge + GeI_4 \qquad (5\text{-}7)$$

Films made using this thermal decomposition or disproportionation of germanium iodide have the disadvantage of having rough surfaces. Epitaxial germanium films have also been deposited by the hydrogen reduction of germanium tetrachloride. These films may be produced in layers of either p or n conductivity type by using doped source material, and p-n junctions formed in this way are comparable with

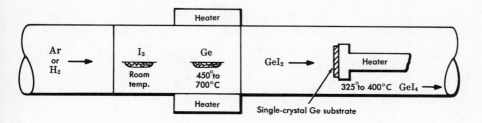

FIG. 5-5. Deposition of Ge by disproportionation of GeI_2.

similar structures formed more conventionally. Silicon single crystal films may be deposited by either of the methods described.

The reaction which describes the production of silicon films by hydrogen reduction is

$$SiCl_4 + 2\,H_2 \rightarrow Si + 4\,HCl \qquad (5\text{-}8)$$

An apparatus commonly used for this deposition is shown in Fig. 5-6. The incoming hydrogen gas is either passed over the surface of liquid $SiCl_4$ or bubbled through it. Some of the halide vapor is thus incorporated in the hydrogen, and this mixture is passed into the vertical reaction chamber. The heated substrate then causes reduction to take place on its surface. Substrate heating may be RF, as shown, or resistive. The substrate material is often highly doped single crystal silicon (for increased conductivity), but single crystal sapphire

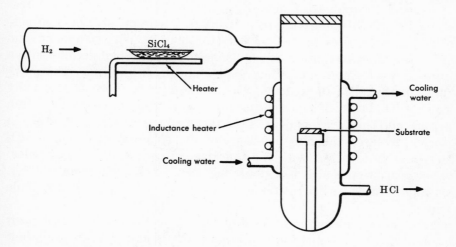

FIG. 5-6. Chemical vapor deposition of Si.

(Al_2O_3) and beryllia (BeO) are also used. For some time, there was a belief that epitaxy (oriented growth of one crystal upon another) was possible only where the misfit between the lattice of the deposited material and that of the substrate was very small (less than 15 percent), but epitaxial growth has been observed on substrates with a mismatch much larger than this. The growth of single crystal semiconductor films of controlled doping on heavily doped (high conductivity) substrates is used in production of transistors with improved frequency response, and deposits on insulating substrates provide devices which are well isolated from one another.

Refractory dielectrics have also been deposited as amorphous films by chemical vapor deposition. Silicon dioxide dielectrics, for example, may be formed by the hydrolysis of silicon tetrachloride:

$$SiCl_4 + 2\,H_2O \rightarrow SiO_2 + 4\,HCl \qquad (5\text{-}9)$$

The water vapor required may be formed in the gas phase by the thermal reduction of carbon dioxide with hydrogen:

$$H_2 + CO_2 \xrightarrow{400\,°C} H_2O + CO \qquad (5\text{-}10)$$

The silica films formed by the reaction of Eq. (5-9) are similar to thermally grown silica films and have been used as dielectric films for capacitors.

Another technique for depositing silica films is to pyrolyze tetraethoxysilane [$Si(OC_2H_5)_4$] in a hydrogen, helium, or argon atmosphere on substrates heated to 800° to 1000°C. This compound and other similar ones contain Si—O bonds so that thermal decomposition leads to the formation of silica (SiO_2) and hydrocarbons. The properties of the silica formed this way are similar to those of thermally grown SiO_2, but they sometimes contain some hydrocarbons, are often porous, and thus are less dense than thermally grown SiO_2. Heating in steam increases the film density and makes the vapor-plated SiO_2 more nearly identical with the thermally grown material. Chemical vapor deposition is also used to produce films of silicon nitride (Si_3N_4), aluminum oxide (Al_2O_3), titanium dioxide (TiO_2), and silicon carbide (SiC).

Some of the advantages of vapor plating are the ability to deposit pure films of high-melting-point materials and to obtain adherent films with a wide range of thickness. The main disadvantage is the high deposition temperature required; it places stringent requirements on substrate materials. Figure 5-7 is a list of some materials which have been vapor plated, together with the reaction conditions [5].

Coating	Reactant	Reaction temperature (°C)	Pressure (torr)
Ti	$TiBr_2 + H_2$	900-1400	760
Ti	TiI_4	1200-1400	50
Zr	ZrI_4	1300-1800	50
Zr	$ZrBr_4 + H_2$	900-1400	760
Ta	$TaCl_5 + H_2$	600-1400	760
Mo	$MoCl_5 + H_2$	800-1100	20-760
W	$WCl_6 + H_2$	500-1100	760
Si	$SiCl_4 + H_2$	1000-1200	760
Ge	GeI_2	300-400	100-3000
SiO_2	$SiCl_4 + CO_2 + H_2$	900-1200	760
Al_2O_3	$AlCl_3 + H_2 + CO_2$	900-1200	760
Si_3N_4	$SiH_4 + NH_3$	700-1200	760

FIG. 5-7. Chemical vapor deposition.

5.4 ANODIZATION

The production of a coating of metal oxide or metal hydroxide by the electrochemical oxidation of a metal anode in an electrolyte is called anodization. This type of oxidation is limited to a few metals, and those which most readily form a coherent film of metal oxide are listed in Fig. 5-8. These metals are often referred to as "valve metals" because of the rectifying characteristic of their anodic oxides. The electrolytic cell used for producing an anodic oxide film consists of an electrolyte in which a valve metal anode and an inert cathode are immersed. A battery with its positive terminal connected to the anode and its negative terminal connected to the cathode will cause electrolysis with the result that an oxide film grows on the anode. As the thickness of this film increases, so does its resistance, and the rate of growth eventually drops to a negligible value. The coating formed may be used to protect the underlying metal from atmospheric oxidation or chemical corrosion and is also useful for its dielectric properties as an insulator or capacitor dielectric.

Metal	Anodic oxide	Dielectric constant of anodic oxide
Aluminum	Al_2O_3	10
Antimony	Sb_2O_3 or Sb_2O_4	~20
Bismuth	Bi_2O_3	18
Hafnium	HfO_2	45
Niobium	Nb_2O_5	41
Tantalum	Ta_2O_5	22
Titanium	TiO_2	40
Tungsten	WO_3	42
Zirconium	ZrO_2	12

FIG. 5-8. Valve metals.

Tantalum is a good example of a valve metal since it readily forms an adherent high resistance oxide film when anodically polarized in a suitable electrolyte.

Anodization Mechanism

During anodization, the metal anode is oxidized and the metal ions formed may either dissolve in the electrolyte solution or form a stable oxide which remains attached to the anode. If the ions dissolve in the electrolyte, the surrounding solution may become saturated, causing precipitation onto the anode. Cadmium, zinc, and magnesium exhibit this dissolution and precipitation behavior, and the resulting coatings are not continuous. These porous films exhibit a low resistance to current flow and are crystalline in structure. If the metal cations formed at the anode react directly with oxygen or hydroxyl ions (as do the valve metals) and do not dissolve in the electrolyte, continuous amorphous films with high electrical resistance are formed. The resistance increases with film thickness and may become as high as thousands of megohms for 1 square centimeter.

Tantalum oxide is insoluble in the common electrolytes at all pH values, but the anodization of aluminum results in soluble products, except in a limited pH range. Data relating the stability of the products of the electrolysis of metals with electrolyte pH and electrode potential are available for some 40 metals. Graphical representations of these data are known as Pourbaix diagrams [6] and two are shown

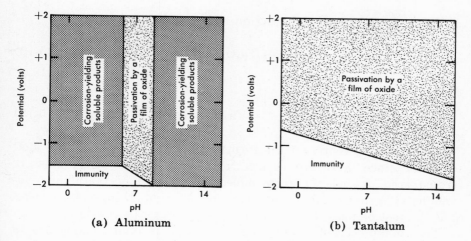

FIG. 5-9. Domains of corrosion, immunity, and passivation of aluminum and tantalum [6]. (Courtesy M. Pourbaix and G. Govaerts.)

in Fig. 5-9. These diagrams were constructed in conjunction with corrosion studies and are strictly applicable only at equilibrium conditions. Under the nonequilibrium conditions normally used for anodization, the rate of formation of a given phase may be well in excess of its rate of dissolution, and thus the phase to be expected at equilibrium may not be observed.

For a continuous oxide film to form, the volume occupied by the oxide must be at least as great as the volume of metal consumed in forming it. The metals in the first two columns of the periodic table form oxides with less volume and thus produce discontinuous oxide films. Aluminum, anodized in a sulfuric acid electrolyte, forms a composite film. A continuous, nonporous layer forms next to the aluminum anode, and then a noncontinuous, porous layer forms between the continuous film and the electrolyte. This composite film is excellent for many decorative and protective uses, but for capacitor dielectrics a more dense, amorphous film is preferred. Anodization in buffered solutions of tartrates, borates, phosphates, etc., results in the nonporous aluminum oxide used in capacitors.

The high resistance, continuous anodic oxide that forms on tantalum results from the combination of tantalum ions with oxygen ions:

$$2\,Ta^{+5} + 5\,O^= \rightarrow Ta_2O_5 \qquad (5\text{-}11)$$

Since the oxygen is present in the electrolyte in the form of water or hydroxyl ion, the half cell reaction is generally written as

$$2\,Ta + 5\,H_2O \rightarrow Ta_2O_5 + 10\,H^+ + 10\,e^- \tag{5-12}$$

or

$$2\,Ta + 10\,OH^- \rightarrow Ta_2O_5 + 5\,H_2O + 10e^- \tag{5-13}$$

These equations represent the formation of 10 equivalents or 1 mole of oxide since there are 10 electrons per molecule which must be transferred. A mole (1 gram molecular weight) of tantalum oxide weighs 442 grams, and 10 faradays (965,000 ampere-seconds) of electricity must be passed through the cell to effect this conversion.

The combination of metal ions with oxygen anions may occur at the anode, inside the oxide, or at the oxide-electrolyte interface, depending on whether cation, anion, or both migrate through the oxide. It has been experimentally demonstrated that both cations and anions migrate during the anodization of tantalum, niobium, aluminum, and tungsten [7].

Anodization Kinetics

There is strong evidence that the rate-controlling process for tantalum anodization is the migration of ions within the oxide film. It has been shown that the relation between the field intensity and the ionic current is not influenced by the crystallographic orientation. It is also not influenced by the film left on the metal after electrolytic polishing nor by different films produced by subsequent anodization in an ethylene glycol solution. Thus, neither the interface between the metal and the oxide nor that between the oxide and the electrolyte controls the rate of anodization.

Güntherschulze and Betz [8] used an empirical equation to relate anodization current to field strength:

$$j = \alpha e^{\beta E} \tag{5-14}$$

where α and β are constants, E is the field strength, and j is the current density. At high field strengths, the equation for ionic diffusion has the same form as Eq. (5-14). Thus, it is assumed that ionic conductivity in the metal oxide is the rate-controlling step in anodization. Values of α and β for tantalum, aluminum, niobium, and zirconium are in the range of 10^{-20} to 10^{-30} amp/cm^2 for α, and 10^{-6} to 10^{-7}

centimeters per volt for β. From Eq. (5-14), a plot of $\ln j$ versus E should be linear, and the slope of such a plot is called the "Tafel slope."

If the kinetics of anodization are examined from a mechanistic viewpoint, an equation similar to Eq. (5-14) may be derived. Assume, for simplicity, that only one type of ion is mobile, and that the field strength is high. Consider n mobile ions per unit volume, with each ion having a charge ze and vibrating in simple harmonic motion with frequency ν. These ions are impeded in their motion through the oxide, and this impedance may be represented by an energy barrier of height Q. The application of an electric field E to the oxide reduces the barrier height by an amount $ze\lambda E$. This factor represents the product of the charge on an ion (ze), the distance (λ) from the equilibrium position of an ion to the position of the maximum of the energy barrier, and the field strength (E). Since the ions are oscillating with a frequency ν, there are ν chances per second for an ion to surmount the energy barrier (Q) and thus travel the distance from one equilibrium site to the next (2λ). Thus, an electric current density of $2n\lambda\nu ze$ would flow if every ion crossed its energy barrier. The probability that any ion has sufficient energy to cross the barrier is not unity, but is given by the factor $\exp - [(Q - ze\lambda E)/kT]$ and the resulting current density, j, is

$$j = 2n\lambda\nu ze \quad \exp \quad -\left(\frac{Q - ze\lambda E}{kT}\right) \qquad (5\text{-}15)$$

Equation (5-15) is identical in form to the empirical equation [Eq. (5-14)] but with physical meaning attached to the constants α and β so that they may be estimated independently.

Equation (5-15) gives inadequate agreement in several respects, and various refinements have been made to obtain more reasonable agreement between the values of the parameters needed to fit the observed currents and fields and the values of the same parameters obtained from independent experiments. An example of this is the fact that values of about twice the expected distance must be substituted for λ in Eq. (5-15) to get agreement with experiment. This discrepancy may be overcome by permitting the number of charge carriers to change with field strength. Other refinements include the effect of interfaces and the amorphous rather than crystalline nature of the oxide films. Another complication not included in this derivation is the fact that both metal and oxygen ion migration are involved in film growth [7].

Film Growth Rate. The rate of increase of film thickness with time is proportional to the current density for constant current anodization. This is only true, of course, if the metal being anodized forms an oxide film to the exclusion of all other reactions. Most metals do not behave in this ideal manner, and a knowledge of the current efficiency is necessary before the film growth rate can be related to the current density. Tantalum, however, is an example of a metal for which the anodization current efficiency is 100 percent. Thus, Faraday's law provides the relationship between the quantity of electricity passing through the cell and the amount of tantalum oxide formed. From a knowledge of the equivalent weight of the oxide (G), the density of the oxide (D), and the current density (j), the rate of increase of oxide film thickness (dd/dt) may be found from

$$\frac{dd}{dt} = \frac{jG}{96,500D} \qquad (5\text{-}16)$$

where 96,500 is Faraday's constant in units of ampere-seconds per equivalent. The equivalent weight of tantalum oxide is one tenth its molecular weight or 44.176 gm/eq, and its density has been measured as 8.00 gm/cm^3. Thus, at a current density of 1.0×10^{-3} amp/cm^2, the rate of oxide growth calculated from Eq. (5-16) is 5.7×10^{-8} cm/sec.

The maintenance of constant current during anodization requires a nearly constant rate of increase of voltage. The differential field strength (E_d) may be obtained from a knowledge of this rate of voltage rise and the rate of oxide growth, i.e.,

$$E_d = \frac{d\phi/dt}{dd/dt} \qquad (5\text{-}17)$$

For tantalum anodization, the measured rate of voltage rise $(d\phi/dt)$ is 0.33 volt/sec at a constant current density of 1.0×10^{-3} amp/cm^2. The rate of oxide film growth (dd/dt) at this same current density is 5.7×10^{-8} cm/sec [as calculated from Eq. (5-16)]. Thus, from Eq. (5-17) the differential field strength is 5.8×10^6 volts/cm.

If anodization is carried out at constant voltage rather than at constant current, the field, current density, and growth rate decrease with time. Equation (5-16) and either Eq. (5-14) or (5-15) may be combined to relate growth rate with field strength. An approximate integration of the resulting equation provides a relationship between film thickness and anodization time at constant voltage:

$$\frac{1}{d} \cong a - b \log t \tag{5-18}$$

where d is in centimeters, t is in seconds, and a and b are experimentally determined constants. This equation is not valid for times so great that the value of $b \log t$ approaches a. The values of these constants have been determined by Vermilyea [9] to be

$$a = 6.165 \times 10^4 \, cm^{-1} \quad \text{and} \quad b = 0.372 \times 10^4 \, cm^{-1}$$

at 120 volts and 20°C. Differentiation of Eq. (5-18) yields the growth rate as

$$\frac{dd}{dt} = \frac{bd^2}{t} \tag{5-19}$$

The growth rate is also proportional to current density [Eq. (5-16)]; thus, a plot of d^2/t versus current density should be linear. Such a plot of experimental data was linear from 0.05×10^{-3} to 8.0×10^{-3} amp/cm², indicating the validity of Eqs. (5-18) and (5-19).

Anodic oxide growth is limited by film breakdown evidenced by scintillation or sparking and a change in appearance from the normal uniform interference color to a grey coloration or mottled effect. The very high local temperature at the points where arcing occurs leads to the formation of small crystallites of $\beta \, Ta_2O_5$ at the metal-oxide interface, mechanically displacing the previously grown amorphous oxide. This same crystallization has been observed on tantalum foil at potentials below the oxide breakdown potential, and its occurrence at these lower voltages is attributed to impurities or surface inhomogeneities.

The growth of this crystalline oxide occurs at different potentials, depending primarily on the structure and purity of the tantalum, and on the nature of the electrolyte system. As the crystalline oxide forms and lifts off the amorphous oxide, the film resistance decreases markedly. Thus, during constant current anodization, the potential would decrease sharply, but at constant voltage the current would increase.

Anodization Technique. Anodization, as applied to tantalum thin film circuits, is typically performed at constant current, with the specific procedure depending on the application. If the intent is to reduce the thickness of resistor films and adjust their electrical resistance, anodization is terminated when the proper resistance is reached. If the intent is to create tantalum oxide as a capacitor dielectric, however, the film is anodized at constant current density until some preselected voltage is reached, and then the anodization is continued at this constant voltage until the current has decreased to a small fraction (say one hundredth) of its initial value. It has been found advantageous, in practice, to carry out an additional step while forming capacitor dielectrics. This step consists of a constant voltage anodization at two-thirds the initial anodization voltage for 5 seconds, using an electrolyte in which the reaction products are soluble. Unanodized tantalum metal is attacked behind conducting spots in the oxide (points at which shorts are likely to develop) and this metal dissolves in the electrolyte (e.g., $AlCl_3$ in methanol) and remains in solution as the halide ($TaCl_5$). This operation may expose the underlying tantalum, and a subsequent reanodization is recommended.

Anodization Parameters

The variables to be controlled include current density, cell voltage, the anodization electrolyte, and the anodization electrodes.

Current Density. Anodic films are often formed at constant current density. Typically, the increase in potential necessary to maintain a constant current density is proportional to the increase in oxide thickness, and thus the field strength remains constant. Maintenance of constant current is easily achieved with commercially available constant current power supplies.

Cell Voltage. Anodization to form a capacitor dielectric, although initially performed at constant current, is completed at constant voltage. This final constant voltage anodization results in better quality dielectrics and in more precise control of dielectric thickness. Thickness control at constant voltage depends on anodization time as shown in Fig. 5-10 where thickness is plotted versus time at 120 volts [from Eq. (5-18)]. Thus, after 30 minutes at 120 volts, a 4-minute error in control of anodization time results in only 1/2 percent variation in thickness (and capacity). Control of voltage may be obtained either by limiting the voltage of the constant current power supply or by using a constant voltage supply.

Convenient factors for approximating the potential required to produce a given thickness of anodic film and for estimating the

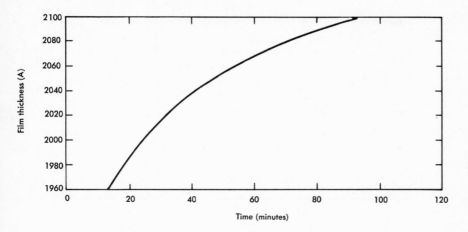

FIG. 5-10. Film thickness versus time for constant voltage anodization of tantalum at 120 volts and 20°C.

thickness of the metal converted to oxide are the voltage rate of growth of oxide and the voltage rate of decrease of metal thickness. These factors are commonly expressed in "angstroms per volt." The idea that the anodic oxide film approaches a limiting thickness when an anodization cell is held at constant voltage is erroneous. The rate of growth does, however, decrease rapidly so that, if the voltage is held constant for a long time, reproducible thicknesses are obtained. Under these conditions, the voltage rate of growth of tantalum oxide has been found to be 16A/volt, with the rate of decrease of tantalum thickness 6.3A/volt [10]. Tantalum nitride resistive films are consumed at different voltage rates depending on the nitrogen content of the film (see Chaps. 4 and 7). A value quoted for the voltage rate of decrease of metal thickness for tantalum resistor films of about 20 atom percent nitrogen is 4.5A/volt [11], and the oxide forms on this material at a rate of 16.7A/volt. This indicates a very low density for the anodic film formed on tantalum nitride. Figure 5-11 depicts the anodization voltage and current changes with time for consecutive constant current and constant voltage anodization used to form capacitor dielectrics.

The Anodization Electrolyte. The electrolyte solution may affect the formation of anodic films by dissolving the products or by initiat-

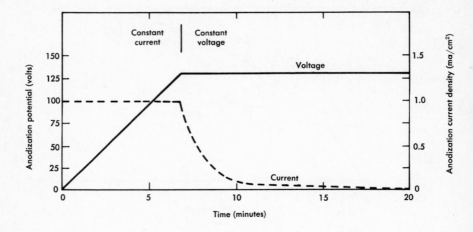

FIG. 5-11. Constant current—constant voltage consecutive anodization.

ing film crystallization and film breakdown at low voltages. Film breakdown is dependent on electrolyte concentration, and Young [23] suggests that it may be caused by heat produced by high current densities in the electrolyte at film imperfections. The relationship of the field strength, the anodization current, and temperature are reputed to be independent of the nature of dilute electrolytes. The first report of the fabrication of tantalum thin film capacitors[12] lists a hot oxalic acid-water-ethylene glycol electrolyte. This and many other electrolyte solutions have been used for forming satisfactory capacitor dielectrics. Commonly used electrolytes for the anodization of tantalum are aqueous solutions containing anions such as acetate, citrate, borate, phosphate, oxalate, and sulfate. The solution cation is usually either hydrogen or ammonium ion. A popular solution for anodizing tantalum is 0.01 percent citric acid (pH $\cong$ 3.5, resistivity $\cong$ 8000 ohm cm). This solution is not always satisfactory when lead (from a glazed ceramic substrate or lead glass) is present. In this case phosphoric acid is preferable. The anodization of aluminum cannot be performed in such an acidic electrolyte as 0.01-percent citric acid if a high resistance oxide is to be obtained, since aluminum oxide is soluble except in a pH range of 6 to 9 (see Fig. 5-8). Aqueous electrolytes for forming "high resistance" aluminum oxide are buffered to maintain this pH range, an example being a boric-acid sodium-borate combination. Some nearly nonaqueous electrolytes are also used to form thin high resistance aluminum oxide films, and a popular electrolyte of this type is ammonium penta-borate dissolved in ethylene glycol. The dissolution of aluminum in the electrolyte is also dependent on the current density, with high current density retarding dissolution.

Agitation of the electrolyte solution is unnecessary at the current densities normally used but is useful in some experiments, especially where temperature control is required.

Anodization Electrodes. For very low conductivity electrolytes, the electrode geometry may affect the uniformity of the oxide thickness, but for most thin film anodizations, it is relatively unimportant. The surface character of the metal to be anodized is important and should be neither rough nor contaminated. The surface roughness of thin films is essentially dependent on the character of the supporting substrate. It has been reported that, for capacitor dielectric formation, a surface roughness of 3000A peak-to-valley height can be tolerated in the substrate, but a 17,000A roughness appreciably reduces the yield of capacitors [13]. Surface contamination has been avoided more than studied. In general, however, clean anodes are required for successful anodization. It has been observed that incomplete removal of the grease used for area definition leads to an increase in the moisture sensitivity of completed capacitors. Another observation on the effect of contaminants is the lack of adherence of anodic films to tantalum which was electropolished in a fluoride bath without complete removal of adsorbed fluorides from the surface. The presence of a small amount of a fluoride in an anodization electrolyte is actually used to prepare anodic films which may be stripped off the underlying tantalum.

The cathode must not dissolve or contaminate the electrolyte solution, and its shape should allow easy cleaning. Capacitor grade tantalum sheet is commonly used. The character and purity of the metal to be anodized (particularly the surface) are the most important parameters affecting the quality of the anodic film. Beta-tantalum films are most often used for capacitor formation, but films doped with methane, oxygen, nitrogen, and hydrogen have also been found suitable for capacitor preparation. On the other hand, similar constituents present in bulk metals have been responsible for the formation of oxide films of poor quality, probably because they are segregated in the bulk metal in the form of compounds.

Properties of Anodic Oxide Films

Anodic oxide films may be characterized by electrical properties such as conductivity and dielectric constant; optical properties such as refractive index and reflectivity; and chemical properties such as their rate of dissolution in various solvents. The electrical characteristics of anodic films depend on film structure, and there is a large difference in behavior between continuous films and discontinuous ones. There also exist structural defects including voids, impurities, and misplaced ions which influence their properties, independent of the degree of crystallinity.

Conduction Properties. The conductivity of an ideal dielectric is zero, but actual dielectrics exhibit a finite current flow when an external field is applied. The ratio of the capacitance of a dielectric to its conductance is a normalized measure of this conductivity which depends only on the nature of the dielectric. This ratio is called "insulation resistance" and has dimensions of ohm-farads. Tantalum oxide thin film capacitors exhibit average insulation resistances of the order 1×10^4 ohm-farads at field strengths of 5×10^6 volts/cm [12]. The corresponding apparent resistivity is of the order of 10^{16} ohm-cm. This represents a lower limit for Ta_2O_5; at half this stress these figures may be as much as an order of magnitude higher. If the logarithm of the leakage current is plotted as a function of the reciprocal temperature, the linear relation shown in Fig. 5-12 is obtained. From this, the activation energy of conduction may be calculated as 17.1 kcal/mole. The conduction is apparently electronic in nature at these fields since the activation energy is independent of potential.

As the name implies, valve metal anodic oxides have rectifying properties in an electrolyte with the easy direction of current flow

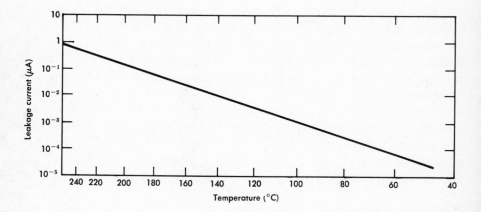

FIG. 5-12. Logarithm of leakage current (at 10 volts) versus reciprocal temperature (3000-pf capacitor anodized to 150 volts) [12].

occurring when the base metal is negative. A number of theories have been proposed to explain this behavior. The mechanisms have been investigated using aqueous solutions and metal films (or probes) as the counterelectrodes and with alternating and direct currents. The asymmetric conduction has been attributed to many things, especially to flaws in the oxide film. Young [23] proposed that these flaws are opened by hydrogen pressure to form holes through which the electrolyte may contact the base metal, and Vermilyea [14] suggested a second conduction mechanism that involves the formation of a semiconductor contact. He also proposed that, in addition to the flaw mechanisms for Nb_2O_5, Ta_2O_5, and WO_3, there is a bulk conduction mechanism in which electrons are injected from the base metal into the oxide. Schwartz [15] postulates that the asymmetric conduction results from an increase in anodic breakdown due to the presence of water at flaws in the film.

Dielectric Properties. In a capacitor made of an ideal dielectric with ideal electrodes, the voltage and current are precisely 90° out of phase and there is no dissipation of power. In a real dielectric, a deviation from this 90° phase difference is observed and this deviation is called the "loss angle" (δ). The tangent of this loss angle ($\tan \delta$) is termed the dissipation factor of the dielectric. A second parameter used to characterize a dielectric material is permittivity (ϵ). This constant is defined, for an isotropic medium, as the ratio of the electric displacement (D') to the electric field strength (E), i.e.,

$$\epsilon = \frac{D'}{E}$$

For a parallel plate capacitor, the electric displacement is equal to the electric field strength when no dielectric is present so that the permittivity is measured by the ratio of the capacitances with and without the dielectric.

Gevers [16] has derived, assuming a distribution of relaxation energies, a relation among the dissipation factor, the permittivity, and the frequency to agree with the experimental fact that the dissipation factor has only a very slight dependence on frequency. This expression was found applicable to both tantalum solid capacitors and tantalum thin film capacitors by McLean [17].

Temperature changes affect both the permittivity and the dissipation factor of dielectrics. An increase in temperature may either increase or decrease the dielectric constant (or permittivity) depending on whether the application of an external field to the dielectric increases or decreases its entropy (disorder).

Valve metal oxides exhibit interference colors similar to those shown by an oil film floating on water. Light incident on the air-oxide interface is partially reflected and partially transmitted, as shown in Fig. 5-13. The portion of the light that is transmitted through the oxide strikes the oxide-metal interface, is partly reflected, and is partly absorbed. The total light reflected from the oxide-metal combination may be considered as the sum of two contributions. The first of these is the light which is reflected at the oxide-air interface

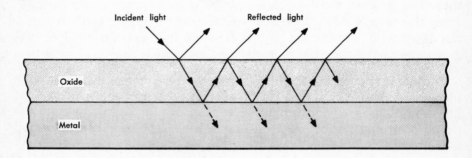

FIG. 5-13. Reflection from anodized valve metal.

and never passes through the oxide, and the second contribution consists of light which has passed through the oxide two or more times before being emitted at the air-oxide interface. These two portions of light are in phase only at specific wavelengths because of path length differences and phase changes occurring on reflection. The intensity of light reflected is then a function of the wavelength of the incident light, the thickness of the oxide film, and the optical properties of both the oxide and the metal. If this intensity is plotted as a function of thickness, minima such as shown in Fig. 5-14 occur. For white light falling on an oxide film, intensity minima occur at several wavelengths and give rise to the observed interference colors. Since the reflected color is a function of the oxide film thickness, the thickness may be estimated by comparison of the interference color with a set of standard colors.

The intensity of the interference colors is a function of the relative reflectivities of the air-oxide interface and that of the oxide-metal interface. For example, the combination of aluminum oxide on aluminum produces very shallow minima in intensity, and the interference colors are barely visible because of the high reflectivity of aluminum.

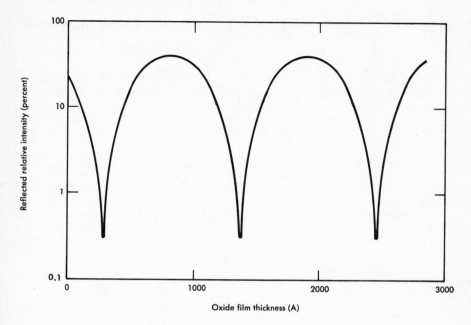

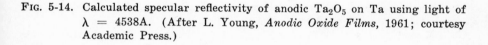

Oxide film thickness (A)

FIG. 5-14. Calculated specular reflectivity of anodic Ta_2O_5 on Ta using light of $\lambda = 4538A$. (After L. Young, *Anodic Oxide Films*, 1961; courtesy Academic Press.)

The combination of tantalum oxide on tantalum, however, produces brilliant interference colors since tantalum has a much lower reflectivity.

Current investigations with anodic thin films are concerned with the mechanisms of film growth, dielectric breakdown, and conduction. Industrial interest is centered on producing short-free capacitor dielectrics and controlling the temperature and humidity sensitivity of these capacitors.

REFERENCES

1. *Metals Handbook*, vol. II. American Society for Metals. Metals Park, Ohio, 1964.

2. Gutzeit, G. "An Outline of the Chemistry Involved in the Process of Catalytic Nickel Deposition from Aqueous Solution," *Plating*, 46, 1158-1164, 1275-1278, 1377-1378 (1959); 47, 63-70 (1960).

3. Randin, J.-P., and H. E. Hintermann. "Electroless Nickel Deposited at Controlled pH; Mechanical Properties as a Function of Phosphorus Content," *Plating*, 54, 523-532 (1967).

4. Gutzeit,G., and M. T. Mopp. "Kanigen Chemical Nickel Plating," *Corrosion Technology*, 3, 331-336 (1956).
5. Powell, C. F., J. H. Oxley, and J. M. Blocher, *Vapor Deposition*. New York: John Wiley and Sons, Inc., 1966.
6. Pourbaix, M., and G. Govaerts. "The Work of Cebelcor's 'Commission des Etudes Fondamentales et Applications'," *1st International Congress on Metallic Corrosion*. London: Butterworths, 1962. pp. 96-102.
7. Davies, J. A., B. Domeij, J. P. S. Pringle, and F. Brown. "The Migration of Metal and Oxygen during Anodic Film Formation," *J. Electrochem. Soc.*, 112, 675-680 (1965).
8. Güntherschulze, A., and H. Betz. "Movement of the Ionic Lattice of Insulators at Extreme Field Strengths," *Zeits f. Physik*, 92, 367-374 (1934).
9. Vermilyea, D. A. "The Kinetics of Formation and Structure of Anodic Oxide Films on Tantalum," *Acta. Met.*, 1, 282-294 (1953).
10. Klerer, J. "The Determination of the Density Anodization Factor and Dielectric Constant of Thin Ta_2O_5 Films," *J. Electrochem. Soc.*, 112, 896-899 (1965).
11. Berry, R. W., W. H. Jackson, G. I. Parisi, and A. H. Schaeffer. "A Critical Evalution of Tantalum Nitride Thin Film Resistors," *Proc. Electronics Comp. Conf.*, 86-96, 1964.
12. Berry, R. W., and D. J. Sloan. "Tantalum Printed Capacitors," *Proc. of the IRE*, 47, 1070-1075 (1959).
13. Schwartz, N., and R. Brown. "A Stylus Method for Evaluating the Thickness of Thin Films and Substrate Surface Roughness," *Trans. 8th Nat. Vac. Symposium, 1961*. New York: Pergamon Press, 1962. pp. 134-139.
14. Vermilyea, D. A. "Conduction and Rectification in Anodic Oxide Films," *J. Appl. Phys.*, 36, 3663-3671 (1965).
15. Schwartz, N., and M. Gresh. "Effect of Ambients and Contact Area on the Asymmetric Conduction of Anodic Tantalum Oxide Films," *J. Electrochem. Soc.*, 112, 295-300 (1965).
16. Gevers, M. "A New Theory on the Dielectric Losses, the Temperature Coefficient of the Dielectric Constant of Amorphous Solid Dielectrics, and the Relation between these Quantities," *Phillips Res. Reps.*, 1, 298-313 (1946).
17. McLean, D. A. "The A-C Properties of Tantalum Solid Electrolytic Capacitors," *J. Electrochem. Soc.*, 108, 48-56 (1961).

The following general texts cover the specialized fields noted in the title.
18. Gray, A. G. *Modern Electroplating*. New York: John Wiley and Sons, Inc., 1953.
19. Lowenbein, F. A. *Modern Electroplating*. New York: John Wiley and Sons, Inc., 1962.
20. Gorbunova, K. M., and A. A. Nikiforova. *Physical Chemical Process of Chemical Nickel Plating*. Moscow: Academy of Sciences, USSR, 1960.
21. *Symposium on Electroless Nickel Plating*, ASTM Special Technical Publication No. 265. Philadelphia: American Society for Testing and Materials, 1959.
22. Powell, C. F., J. H. Oxley, and J. M. Blocher. *Vapor Deposition*. New York: John Wiley and Sons, Inc., 1966.
23. Young, L. *Anodic Oxide Films*. New York: Academic Press, 1961.
24. Delahay, P., editor. *Advances in Electrochemistry and Electrochemical Engineering*, vol. 3. New York: John Wiley and Sons, Inc., 1963.
25. Bockris, J. O'M., editor. *Modern Aspects of Electrochemistry*, no. 2. London: Butterworths, 1959.

PRINCIPAL SYMBOLS FOR CHAPTER 5

Symbol	Definition
a	Constant
b	Constant
d	Film thickness
D	Density
D'	Electric displacement
e	Electronic charge
E	Electric field strength
E_d	Differential field strength
G	Equivalent weight
j	Current per unit area
k	Boltzmann constant
n	Constant
Q	Energy barrier height
t	Time
z	Constant
α	Constant
β	Constant
δ	Loss angle
$\tan \delta$	Dissipation factor
ϵ	Permittivity
ν	Frequency
λ	Half the separation distance between ions
ϕ	Potential

Chapter 6

Electrical Conduction in Metals

Thin films of metals have found many applications as resistors and conductors for electronic circuits. Thus, it is desirable to review some of the mechanisms involved in the conduction of electricity by thin metal films. In this chapter, a review of the elementary theory of conduction in metals is given, beginning with a perfect single crystal metal and the effects of impurities, strains, and alloying, and concluding with a discussion of various thin film effects.* The analysis follows the free electron theory as given, for example, by Kittel [1]. For a detailed discussion of metallic conduction, the reader is referred to the book by Wilson [2], and review articles by Slater, Jones, and Gerritsen [3]. A recent survey of the experimental data available is given by Meaden [4].

6.1   CONDUCTION IN BULK METALS

An atom may be described as a positively charged nucleus surrounded by electrons in orbit. The electrons are permitted to have orbits of only certain definite energy levels (shells), and the number of electrons in each level is strictly limited. The allowed energy levels and their occupation are described by the laws of quantum mechanics. The electrons arrange themselves to minimize their energy, so the shells of lowest energy are the first to be filled. If the electrons that are needed to balance the positive charge on the nucleus require the outer shell to be less than half full, the element

---

*The discussion of specific materials currently being used for thin film resistors
  and conductors will be presented in Chap. 7.

will be a metal. The electrons in the outer shell are called valence electrons. They determine the chemical properties of the metal, and in the solid state they are responsible for the metallic bond which causes one metal atom to attract another, keeping the metal intact. In a solid metal, the valence electrons lose their identification with individual atoms and form a nonlocalized sea of negative charge, permeating the entire solid. In this state they are called conduction electrons because they are responsible for the electrical conductivity. The lattice sites are occupied by the atom cores, which consist of the nuclei and the electrons occupying the inner shells.

Consider the behavior of a conduction electron in an electric field, **E**. If there were nothing to impede its motion, its velocity, **v**, would have a constant time derivative, given by

$$\frac{d\mathbf{v}}{dt} = \frac{-e\mathbf{E}}{m}$$

where $-e$ is the electron charge ($e > 0$) and m is the mass of the electron. The current density, **j**, of any assembly of electrons is

$$\mathbf{j} = -ne\mathbf{v}_d \qquad (6\text{-}1)$$

where n is the density of electrons, and $\mathbf{v}_d$ is the drift (i.e., average) velocity of the assembly. It immediately follows that

$$\frac{d\mathbf{j}}{dt} = \frac{ne^2\,\mathbf{E}}{m}$$

This cannot be the case, however, because it violates Ohm's law. In order to satisfy Ohm's law, there must be an impeding (or frictional) force proportional to $-\mathbf{v}$ added to the original equation of motion. So

$$\frac{d\mathbf{v}}{dt} = \frac{-e\mathbf{E}}{m} - C\mathbf{v} \qquad (6\text{-}2)$$

where C is a constant of proportionality. (The impeding force is analogous to a viscous force.) In the absence of an electric field ($\mathbf{E} = 0$), the solution to Eq. (6-2) is an exponential:

$$\mathbf{v} = \mathbf{v}_0\, e^{-Ct}$$

Thus, if the electron is given a velocity, it will decay exponentially to zero with a time constant $1/C$. The quantity $1/C$ has the character of a relaxation time, so it is convenient to let $\tau = 1/C$, and write

$$\frac{d\mathbf{v}}{dt} = \frac{-e\mathbf{E}}{m} - \frac{\mathbf{v}}{\tau}$$

When the system reaches a steady state (e.g., under a steady field $\mathbf{E}$), $d\mathbf{v}/dt$ will be zero, and

$$\mathbf{v} = \frac{-e\tau\mathbf{E}}{m} \tag{6-3}$$

and

$$\mathbf{j} = \frac{ne^2\tau\mathbf{E}}{m} \tag{6-4}$$

which is clearly consistent with Ohm's law. Resistivity, ρ, is defined by the relation $\mathbf{E} = \rho\mathbf{j}$, and ρ is then identified from Eq. (6-4) as

$$\rho = \frac{m}{ne^2\tau} \tag{6-5}$$

We will see later that one can equally well think of the frictional force, $-m\mathbf{v}/\tau$, as arising from a continuous drag on all electrons or from collisions which destroy the added drift velocity of electrons. In either case, there will be a resistivity and a relaxation time. Electron relaxation times are generally very small, typically about 10^{-14} seconds.

Conduction Electron Energy Distribution

The conduction electrons are not independent. Together they form a "gas," with a specific distribution of velocities and energies even in zero applied field. This distribution is dependent on the temperature of the metal, but it is not the same distribution as is found for a gas from the kinetic theory of gases. Classical concepts are inadequate to describe the nature of this electron "gas," because of the wave nature of the electron. Thus, principles of quantum mechanics must be introduced before further description can be given. The Pauli exclusion principle states that no more than two* electrons may occupy the same energy state simultaneously. Fermi-Dirac

*One for each of the two possible spin orientations.

statistics were developed to describe the velocity (or energy) distribution taking this into account. Thus electrons obey the Fermi-Dirac statistics rather than the classical Maxwell-Boltzmann statistics.

Particles obeying Fermi-Dirac statistics have the property that at 0°K they fill all available energy states up to a certain characteristic energy, called the Fermi energy, with all states of higher energy being empty. At finite temperatures, there is a very small transition region of width approximately $2kT$, centered about the Fermi energy, where the probability of being filled goes smoothly from one to zero. The Fermi energy is about 100 times kT at room temperature. Below the Fermi energy, the electrons are uniformly distributed in velocity space; that is, every velocity is as probable as every other one. By adding up all the states below the Fermi energy and setting this equal to the density of conduction electrons, it is possible to calculate the Fermi energy*

$$E_F = \frac{h^2}{8m}\left(\frac{3n}{\pi}\right)^{2/3} \qquad (6\text{-}6)$$

where h is Planck's constant. From this, v_F, the speed of an electron at the Fermi energy (by assuming that the Fermi surface is spherical), is

$$v_F = \frac{h}{2m}\left(\frac{3n}{\pi}\right)^{1/3} \qquad (6\text{-}7)$$

A typical Fermi speed for a metal is about 10^6 meters per second. This is much larger than practical drift velocities. In fact, if a No. 14 gauge copper wire is operated at its maximum current permitted by the National Electrical Code (20 amps), the electron drift velocity will be only 5×10^{-4} meters per second, which is negligible when compared to the Fermi speed. Under equilibrium conditions (zero field and zero temperature gradient), the Fermi distribution function gives a completely random velocity distribution, so there is no net flow in any direction. Under the influence of an electric field, however, the distribution is slightly disturbed, as shown in Fig. 6-1, and the net flow is no longer zero. Since the electron carries a negative charge, electron velocities antiparallel to the field become more probable, and electron velocities parallel to the field become less probable. We can think of the electric field as causing some of the parallel electrons to move over to the opposite side of the distribution, thus

*See, for example, pp. 252-257 in Kittel [1].

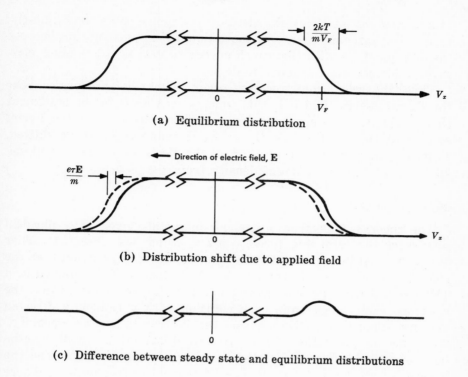

(a) Equilibrium distribution

(b) Distribution shift due to applied field

(c) Difference between steady state and equilibrium distributions

FIG. 6-1. Fermi distribution in velocity space, showing disturbance due to an electric field. The vertical axis is the probability of being filled. (After C. Kittel, *Introduction to Solid State Physics*, 2nd ed., 1956; courtesy John Wiley and Sons, Inc.)

affecting only those electrons with the Fermi speed. Alternatively, this may be considered as a shift of the entire distribution in the direction opposite to the field.

The amount of this shift is just equal to the drift velocity and is given by $-e\tau \mathbf{E}/m$, as in Eq. (6-3). The relaxation time can be thought of as being determined by collisions, with the allowed collisions governed by Fermi-Dirac statistics. The largest energy available to be given to an electron from the lattice in any one collision is only about kT. Then, an electron in the interior of the Fermi sea (i.e., with $E << E_F$) cannot be scattered at all, because all states within kT of it on all sides are completely filled, and the Pauli exclusion principle forbids a transition to a filled state. We also know that there are practically no electrons with energies more than kT greater than the Fermi energy, so there can be no loss of energy greater than about kT by *any* electron in any one collision. Similarly, no electron can gain energy by more than kT in one collision because the only particles with that much energy to give to it are other electrons, and they are forbidden to do so.

Thus, the relaxation processes occur primarily at energies near the Fermi energy, and it is this concept that leads to the statement that conduction processes involve only those electrons at the Fermi energy. Actually, of course, the entire distribution has been shifted, so it is proper to use the total of all the electrons in the Fermi sea when calculating the current density from $\mathbf{j} = -ne\mathbf{v}_d$.

Mean Free Path

The theory of electron scattering in metals gives us a physical picture of the retarding force, $-m\mathbf{v}/\tau$, which was inserted earlier and which led to agreement with Ohm's law. It is assumed that the electrons travel freely in space under the influence of the applied field (**E**) until they collide with something. It is further assumed that the electrons come away from each collision with a randomly directed velocity. That is, a collision causes the electron to lose all "memory" of its incoming direction. There will be a distribution of path lengths between collisions, just as in the case of molecules in a gas, and the average of all the path lengths is called the mean free path, λ. There will also be a mean time between collisions, τ', which will be given by

$$\tau' = \frac{\lambda}{v_F} \tag{6-8}$$

since the only electrons which may suffer collisions are those traveling with a speed v_F. The quantities λ and τ' are directly related to the resistivity, so they will now be calculated.

Consider a group of electrons, each of which has just made a collision. During the next small time interval, dt, some of them will be scattered again. Let the number remaining unscattered at time t be N. The rate of scattering, $-dN/dt$, will be proportional to N. That is,

$$-\frac{dN}{dt} = CN \tag{6-9}$$

The constant of proportionality, C, is assumed to be independent of time and the energy of the electron. This is a reasonable assumption, especially since all the electrons under consideration have very nearly the same (Fermi) energy over their entire path length. Under this assumption, Eq. (6-9) is valid for N for all times. Then Eq. (6-9) can be integrated and

$$N = N_0 e^{-Ct} \tag{6-10}$$

That is, the fraction left unscattered, after time t, is equal to e^{-Ct}. The average time between collisions is then

$$\tau' = \frac{1}{N_0} \int_0^{N_0} t\, dN$$

Thus, from Eqs. (6-9) and (6-10),

$$\tau' = C \int_0^\infty t e^{-Ct}\, dt$$

which becomes

$$\tau' = \frac{1}{C}$$

Thus, the constant of proportionality C in Eq. (6-9) is the reciprocal of the mean time between collisions.

The preceding discussion considered only those electrons which had just suffered a collision at time $t = 0$, but the same distribution of times-to-collision would result if all electrons had been considered, since the scattering probabilities are the same for an electron fresh from a collision as for one that is not so fresh. Thus, on the average, at any instant in time, the whole distribution of electrons will continue for another τ' seconds before colliding. Because of time-symmetry considerations, the distribution of times since the last collision must be the same as the distribution of times until the next collision, and the average time since the last collision is also τ'.

Thus, at any point in time, the average electron will have acquired a velocity

$$\mathbf{v}_d = \frac{-e\mathbf{E}\tau'}{m}$$

above and beyond its random velocity. The current density is given by

$$\mathbf{j} = -ne\mathbf{v}_d$$
$$= \frac{ne^2\tau'\mathbf{E}}{m}$$

By a comparison of this relation to Eq. (6-4), it is evident that the mean time to a collision is the same as the relaxation time, τ.

In Eq. (6-8), τ' can be replaced by τ, and the mean free path is given by

$$\lambda = \tau v_F$$

This relation allows the value of λ to be estimated. It is about 300 angstroms for typical metals at room temperatures. The resistivity from Eq. (6-5) may be expressed as a function of the mean free path:

$$\rho = \frac{mv_F}{ne^2}\left(\frac{1}{\lambda}\right) \tag{6-11}$$

or, by substituting for v_F from Eq. (6-7),

$$\rho = \frac{h}{2e^2}\left(\frac{3}{\pi n^2}\right)^{1/3}\left(\frac{1}{\lambda}\right) \tag{6-12}$$

The density of electrons, n, is essentially independent of temperature, so λ is the controlling factor in the temperature dependence of the electrical resistivity. The calculation of λ requires a knowledge of the scattering mechanisms.

If the crystallographic periodicity of a metal sample were perfect, there would be nothing for the electrons to collide with. That is, they could go through a perfectly periodic lattice without being scattered at all, and the resultant resistivity would be zero, as the mean free path would be infinite. This is a consequence of the wave nature of the electron. Anything that disrupts the periodicity of the lattice, however, shortens the mean free path and causes a finite resistivity. There are many things that can disrupt this periodicity, such as phonon scattering, impurity and defect scattering, and scattering from the surface of a thin sample, such as a thin film.

Phonon Scattering

One thing that can disrupt the normal periodicity of the lattice of atom cores is the thermal vibrations of the atomic cores from their equilibrium positions. These vibrations may be treated as standing waves, each of which can be resolved into two waves traveling in opposite directions. The energy contained in the waves is quantized in units of $h\nu$, where ν is the frequency. The waves travel at the speed of sound, and the quantum units are called phonons.* The phonons can collide with electrons, with an exchange of energy $h\nu$. That is, one phonon is created or annihilated in each collision. The total effect of the phonons on the resistivity is calculated by integrating over all phonon energies up to the maximum phonon energy, $h\nu_{max}$, which is given by $k\Theta_D$, where Θ_D is the Debye temperature obtained from the theory of specific heat for a solid. The form of the integral has been given by Grüneisen [5].

$$\rho_{ph} = \frac{C}{x_0^5} \int_0^{x_0} \frac{x^5 dx}{(e^x - 1)(1 - e^{-x})} \qquad (6\text{-}13)$$

where ρ_{ph} is the resistivity due to phonon scattering, x_0 is Θ_D/T, and C is a constant for a given metal, relatively independent of the temperature. The constant C depends on the atomic weight, the atomic volume, the effective mass of the electrons, and the Debye temperature of the metal. With the help of special curve-fitting techniques [4], Eq. (6-13) can be used to determine the value for Θ_D which best fits the resistivity data for any given metal. The value of Θ_D thus determined is designated Θ_R, and Fig. 6-2 shows values of Θ_R for some common metals. Also listed in Fig. 6-2 are values for Θ_D determined from specific heat data. The agreement between Θ_R and Θ_D is close enough that they are almost interchangeable when considering temperature-dependence of resistivity.

In the limit of $x_0 << 1$ (i.e., high temperatures), the integrand of Eq. (6-13) reduces to x^3 over the range of integration. It can then be integrated directly, with the result that

$$\rho_{ph} \cong \frac{C}{4x_0} = \frac{CT}{4\Theta_D} \qquad \text{for } T >> \Theta_D \qquad (6\text{-}14)$$

Thus, at high temperatures, the resistivity is directly proportional to the absolute temperature. At low temperatures, it is helpful to rephrase the Grüneisen integral by an integration by parts:

$$\rho_{ph} = \frac{C}{1 - e^{x_0}} + \frac{5C}{x_0^5} \int_0^{x_0} \frac{x^4 dx}{e^x - 1}$$

In this form, it is clear that the first term goes quickly to zero, as it reduces to $-Ce^{-x_0}$ at low temperatures. The integral in the second

*This is analogous to photons, the quantum units for electromagnetic waves which travel at the speed of light.

Element	Θ_R (°K)	Θ_D (°K)	Θ_D/Θ_R
Mg	340	325	0.96
Al	395	385	0.97
Ti	342	355	1.04
Cr	485	450	0.93
Co	401	380	0.95
Cu	320	320	1.00
Zn	175	245	1.40
Ga	215	240	1.12
Rb	55	63	1.14
Y	215	213	0.99
Rh	370	350	0.95
Pd	270	300	1.11
Ag	200	220	1.10
Cd	130	165	1.27
Cs	37	46	1.25
Ta	228	230	1.01
W	333	315	0.95
Re	294	290	0.99
Ir	305	285	0.94
Pt	240	225	0.94
Au	200	185	0.93
Tl	140	100	0.71

FIG. 6-2. Debye temperatures for some common metals. Θ_R is determined from the resistivity in the temperature range $\Theta_R/3$ to Θ_R, and Θ_D is determined from the specific heat in the temperature range $\Theta_D/2$ to Θ_D. (After G. T. Meaden, *Electrical Resistance of Metals*, 1965; courtesy Plenum Publishing Corp.)

term can be treated as a definite integral from zero to infinity, and it can be shown that

$$\int_0^\infty \frac{x^4 dx}{e^x - 1} = 4! \sum_{s=1}^\infty \frac{1}{s^5} = 24.88$$

Therefore, the expression for ρ_{ph} at low temperature becomes

$$\rho_{ph} \cong \frac{124.4 C T^5}{\Theta_D^5} \qquad \text{for } T << \Theta_D \qquad (6\text{-}15)$$

Thus, we see that at zero temperature the electrical resistivity becomes zero as, of course, it should, since both the lattice and the sea of electrons are at their zero-point energies, so neither can give up energy in a collision. At low but finite temperatures, the phonon resistivity is proportional to T^5. As shown earlier the resistivity is proportional to T at temperatures well above Θ_D. Figure 6-3(a) gives a normalized plot of the phonon resistivity* as a function of the temperature. Note that the curve approaches an asymptote which goes through the origin. If one looks at data in the temperature range from 0.5 Θ_D to 1.0 Θ_D, one could be misled into thinking that the curve is asymptotic to a straight line which intersects the temperature axis at a temperature of about 0.15 Θ_D. Debye temperatures are typically in the range of about 40°K to 400°K (see Fig. 6-2), so at room temperature and above, the phonon resistivity is fairly well described by a straight line, especially for those metals with low Θ_D.

The Grüneisen relation has had considerable success for many metals, and even in those cases where it does depart appreciably from experimental data, the reasons are usually clear. This relation has been replotted in Fig. 6-3(b) on a log-log scale, along with data for various metals. Because of the normalization, the agreement is best in the right-hand portion of the figure but is still good at low temperatures.

The temperature coefficient of resistivity is defined by

$$\alpha = \frac{1}{\rho}\frac{d\rho}{dT}$$

It can also be calculated from the Grüneisen equation [Eq. (6-13)]. From Eqs. (6-14) and (6-15),

$$\alpha_{\mathrm{ph}} = \frac{1}{T} \qquad \text{for } T >> \Theta_D$$

$$\alpha_{\mathrm{ph}} = \frac{5}{T} \qquad \text{for } T << \Theta_D$$

Figure 6-4 is a plot of the $\alpha_{\mathrm{ph}}T$ product as a function of T/Θ_D. The curve is the theoretical Grüneisen curve, and it indicates that at temperatures well above Θ_D, the temperature coefficient of resis-

---

*The resistivity is normalized by dividing by its value at $T = \Theta_D$ which turns out to be $\rho_\Theta = C/4(1.056)$. Thus, the slope of the asymptote is 1.056.

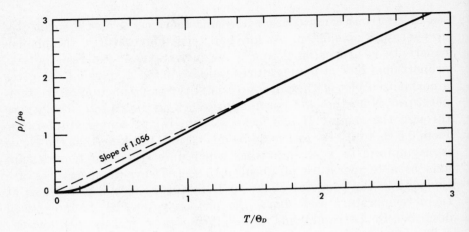

(a) Linear plot

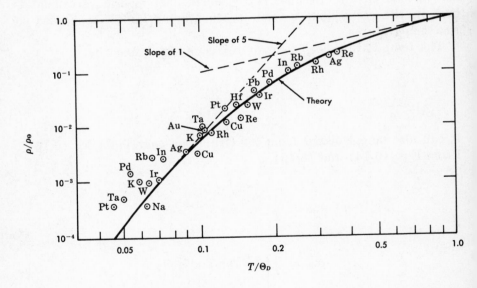

(b) Logarithmic plot. (After G. K. White, *Experimental Techniques in Low Temperature Physics*, 1959; courtesy Clarendon Press, Oxford.)

FIG. 6-3. Phonon resistivity as a function of temperature, according to the Grüneisen relation, Eq. (6-13).

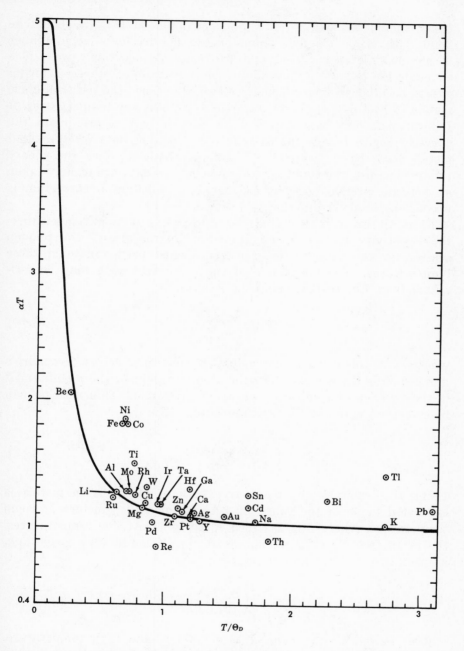

FIG. 6-4. The temperature coefficient of resistivity (α) times temperature (T) as a function of the reduced temperature, according to the Grüneisen relation. All data points are evaluated at $0\,°C$. (Data from Meaden [4].)

tivity should be equal to $1/T$. For example, at $0°C$, it should be $3660 \times 10^{-6}/°C$. The data points on Fig. 6-4 are for a large number of metals evaluated at $0°C$, and the relation holds fairly well. The elements Fe, Co, and Ni are off the curve because of other scattering effects peculiar to ferromagnetic materials. Since the data for pure metals in Fig. 6-3(a) follow a line of slope 5 at low temperatures, it follows that low temperature data for αT will approximate the curve of Fig. 6-4 up to the value of $\alpha T = 5$. The data for the alkali metals have been corrected for volume changes, since the theory relates to constant volume conditions, whereas data are usually taken at constant pressure. For these metals, a substantial correction is needed in the determination of ρ and α.

From all this, it is clear that, for pure metals at room temperature, the resistivity is fairly well described by the theory for phonon scattering, and thus phonon scattering must predominate at these temperatures. The magnitude of the mean free path may be estimated from Eq. (6-12), using the relation

$$n = \frac{f N_A D}{M} \qquad (6-16)$$

where n is the density of conduction electrons, N_A is Avogadro's number, D is the density, M is the atomic weight of the metal, and f is the number of conduction electrons per atom. This relation can be combined with Eq. (6-12) and the result is

$$\lambda = \frac{1}{2} \left(\frac{3}{\pi} \right)^{1/3} \left(\frac{h}{e^2 \rho} \right) \left(\frac{M}{f N_A D} \right)^{2/3} \qquad (6-17)$$

Using this relation, the mean free path of electrons in metals is estimated to be in the range of 1 to 6 times 10^{-8} meters at room temperature, where f is assumed equal to 1. At low temperatures, the mean free path for phonon scattering becomes very large, and, according to the Grüneisen relation,

$$\lambda_l = \lambda_h \left(\frac{\Theta_D{}^4 T_h}{497.6} \right) (T_l)^{-5}$$

where the subscripts l and h refer to low and high temperature. For example, if $\Theta_D = 100°K$, then the mean free path at $1°K$ is longer than the mean free path at room temperature by a factor of 6×10^7, which means that the electron mean free path for phonon collisions at 1°K is about 1 meter. Mean free paths of this length have not been observed because of other effects. Mean free paths longer than 10^{-3} meters have been observed but only in high-purity, strain-free single crystals at very low temperatures.

Impurity and Defect Scattering

The perfect periodicity of the metal lattice may also be interrupted by an impurity (interstitial or substitutional) or by any defect in the lattice such as a vacancy, dislocation, or grain boundary. It is assumed that the scattering due to these defects is independent of the phonon scattering, and thus the resistivities may be added together:

$$\rho = \rho_{ph} + \rho_r = \frac{m}{ne^2}\left(\frac{1}{\tau_{ph}} + \frac{1}{\tau_r}\right) \qquad (6\text{-}18)$$

where ρ_r is the resistivity due to impurities and defects. It is reasonable to assume that ρ_r is essentially independent of temperature (provided there is no annealing). This implies that $d\rho/dT$ is independent of impurity and defect content, thus making α inversely proportional to ρ as impurity or defect level is varied. This implication is known as Matthiessen's rule.* It is only an approximation but it holds quite well as long as ρ_r is much smaller than ρ_{ph}. At very low temperatures, as we have just seen, the phonon resistivity goes to zero, so

$$\lim_{T \to 0} \rho = \rho_r$$

Thus, ρ_r is called the residual resistivity, and it is an easy matter to measure it. For small impurity concentrations, ρ_r increases linearly with impurity content or defect density and so does ρ. Figure 6-5 shows the resistivity at 273°K for several dilute alloys of copper as a function of concentration. The linearity holds quite well, even up to concentrations where the resistivity has been increased by a factor of 5 or more.

---

*Matthiessen's rule is sometimes quoted as saying that $d\rho/dx$ is independent of temperature, where x is the impurity (or defect) content. The two statements are equivalent, since

$$\frac{\partial}{\partial x}\left(\frac{\partial \rho}{\partial T}\right) = \frac{\partial}{\partial T}\left(\frac{\partial \rho}{\partial x}\right)$$

and Matthiessen's rule says that this quantity is zero.

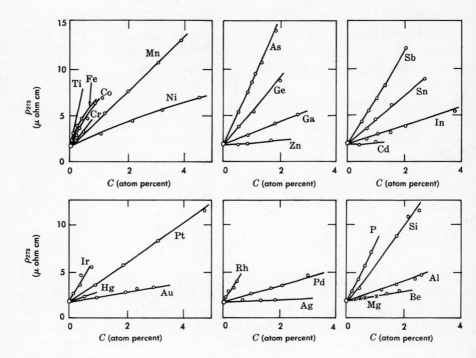

FIG. 6-5. Resistivity at 273°K for copper base binary alloys as a function of the concentration of the second component [7].

Effect of Stress

The effect of stress on the resistance of a metal sample may be divided into two terms: the change in dimensions and the change in resistivity. The resistance of a metal sample (in the shape of a rectangular parallelepiped) is given by

$$R = \frac{\rho l}{wd}$$

where l is the length (direction of the current), w is the width, and d the thickness of the sample. If a stress is applied to the sample, the resistance change will be expressed by

$$d \ln R = d \ln \rho + d \ln l - d \ln w - d \ln d \qquad (6\text{-}19)$$

For example, if the stress is a hydrostatic pressure, dP, and the sample is isotropic,

$$\frac{d \ln R}{dP} = \frac{d \ln \rho}{dP} + \frac{\kappa}{3}$$

where κ is the volume compressibility $(- \partial \ln V/\partial P)_T$ of the metal. Figure 6-6 is a table of $\kappa/3$ and $d \ln R/dP$ at $0°C$ for several metallic elements. It is seen that the effect due to the compressibility is of the order from 6 to 20 percent of the total change in resistance due to hydrostatic pressures. The table is not valid for pressures above 1000 atmospheres because the resistance is then no longer a linear function of the pressure.

Metal	$\kappa/3$ (Ref. [1])	$d \ln R/dP$ (Ref. [7])	$d \ln \rho/dP$
Li	2.90	$+7.00$	$+4.10$
Mg	0.98	-5.40	-6.38
Ca	1.90	$+9.48$	$+7.58$
Al	0.45	-4.29	-4.74
Fe	0.20	-2.42	-2.62
Ni	0.18	-1.77	-1.95
Nb	0.19	-1.40	-1.59
Mo	0.12	-1.31	-1.43
Rh	0.12	-1.65	-1.77
Pd	0.18	-2.10	-2.28
Ta	0.16	-1.62	-1.78
W	0.11	-1.33	-1.44
Pt	0.12	-1.92	-2.04
Cu	0.24	-1.92	-2.16
Ag	0.33	-3.48	-3.81
Au	0.19	-3.02	-3.21
Pb	0.79	-13.7	-14.5

FIG. 6-6. Effect of hydrostatic pressure on volume, resistance, and resistivity, all in units of 10^{-10} meters^2/kg.

The main reason that ρ is a function of stress is that the lattice vibrations decrease as the sample is compressed, resulting in a lowered resistivity. (In some cases, other more subtle effects enter.)

In a completely isotropic solid, such as a random, polycrystalline sample, one expects the effect of strain on the phonon spectrum to be independent of the relative direction between the strain and the current flow. Thus, it is possible to consider the problem in terms of the volume change produced by any given stress, so that

$$\frac{\partial \ln \rho}{\partial \ln d} = \frac{\partial \ln \rho}{\partial \ln w} = \frac{\partial \ln \rho}{\partial \ln l} = \frac{d \ln \rho}{d \ln V}$$

Given any particular type of stress, it will be possible to find the effect on the resistivity as long as the change in total volume is known. This will be useful in calculating effects of stress on thin films (Sec. 6.2).

Alloys

So far, this chapter has been concerned primarily with pure elemental metals. Some of the important materials for thin film conductors and resistors, however, are alloys. In discussing conduction in alloys, it is convenient to discuss first those systems which form solid solutions, and then those which are actually two-phase systems.

Solid Solutions. If a small amount of one metal is dissolved in another, the resistivity should increase, according to Eq. (6-18), because the solute atoms disrupt the lattice periodicity and act as scattering centers for the electrons. In this region, Matthiessen's rule holds fairly well, and the resistivity increase is roughly temperature-independent and proportional to the amount of solute added. As more of the solute is added to the solvent metal, however, the linearity disappears, and the curve of resistivity as a function of concentration for a solid solution is shown in Fig. 6-7. This curve is for the Au-Ag system, which is a rather ideal case because both elements have the same atomic volume and crystal structure and are miscible in all proportions. The curves are drawn for two different temperatures. Matthiessen's rule can be generalized to apply to the case of concentrated alloys by setting

$$\left(\frac{d\rho}{dT}\right)_{\text{alloy}} = a_1 \left(\frac{d\rho_1}{dT}\right) + a_2 \left(\frac{d\rho_2}{dT}\right) \tag{6-20}$$

where a_1 and a_2 are the volume percent concentrations of the two components. That is to say that the temperature derivative of the resistivity of the alloy is the average of the temperature derivatives

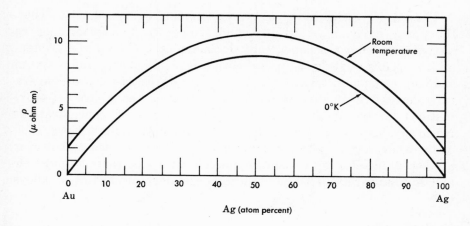

FIG. 6-7. Resistivity of gold-silver alloys at room temperature and 0°K [2].

of the two constituents, weighted by their relative concentrations. In this case, $d\rho/dT$ is 6.16 for Ag and 8.11 for Au, in units of 10^{-9} ohm cm per °C. Thus, the alloy of 50 volume percent has a temperature derivative of about 7×10^{-9} ohm cm per °C. Since the resistivity of this alloy is about 5 times that of either metal alone, the temperature coefficient of resistance (TCR) of the alloy will be about a factor of 5 smaller than the TCR of either of the two bulk metals. This is characteristic of alloys, especially for those that form solid solutions. In fact, it has been stated as a general rule [7] that, for binary solid solutions, the resistivity is higher and the TCR lower than for either of the components, and that the resistivity versus concentration curve will resemble the one shown in Fig. 6-7.

Heterogenous Mixtures. If the two components of a binary alloy are not miscible, they will coexist as a two-phase system. In this case, one expects the resistivity to be a kind of weighted average of the resistivities of the two components. Thus,

$$\rho_{\text{mix}} = a_1\rho_1 + a_2\rho_2 \tag{6-21}$$

This has been stated as a general rule [7] for immiscible mixtures. It holds fairly well, not only for systems which are never miscible but also for the immiscible region of a system where there is a miscibility gap in the phase diagram. If Eq. (6-21) is differentiated with respect to temperature, the result is the same as Eq. (6-20), so this formulation of Matthiessen's rule [Eq. (6-20)] holds for both solid solutions and immiscible mixtures of two phases.

The general behavior of alloys is summarized in Fig. 6-8, where the phase diagram, resistivity, and temperature derivative of resistivity are shown schematically for a typical solid solution system, an immiscible mixture, a system with a miscibility gap, and a system showing compound formation. In the first three cases, the temperature derivative of the resistivity is expected (as a first approximation) to be linear across the whole range of concentrations. Naturally, such simplified considerations have their limitations, especially when it comes to magnetic alloys. These diagrams, however, are useful in selecting resistance materials for electronic applications and for interpreting the composition-dependent properties of alloys.

Magnetic Alloys

Some of the alloys of interest in thin film technology contain atoms of transition metals which have a magnetic moment and which contribute to the resistivity of the alloy in a different fashion. This is sometimes called "magnon" scattering, since the magnetic entropy (disorder) can be resolved into spin-waves in a fashion similar to the way in which the lattice disorder is resolved into phonons. For dilute alloys, the resistivity approximately follows Matthiessen's rule (that is, $d\rho/dT$ is independent of impurity concentration), and the increase in resistivity is roughly proportional to the concentration of solute. Concentrated magnetic alloys, however, may become ferromagnetic (or antiferromagnetic) below a certain temperature, called the Curie (or Néel) temperature. Below the Curie temperature, the resistivity falls off rapidly with decreasing temperature as the magnon density rapidly decreases. Above the Curie temperature, the magnon contribution to the resistivity is independent of temperature, since complete magnetic disorder is achieved at the Curie point, and the alloy is paramagnetic in this region. In many cases, the magnon scattering is much larger than the phonon scattering. Thus, in the region above the Curie point, the material has a high resistivity but a small temperature coefficient of resistance.

In many cases, there is even a region of temperature where $d\rho/dT$ is negative, usually just above the Curie temperature. The reason for this is not clear, but it may be involved with some short-range ordering which is found only in the neighborhood of the Curie point. It is quite difficult, in general, to explain a negative $d\rho/dT$ in a metal, since the number of carriers is expected to be

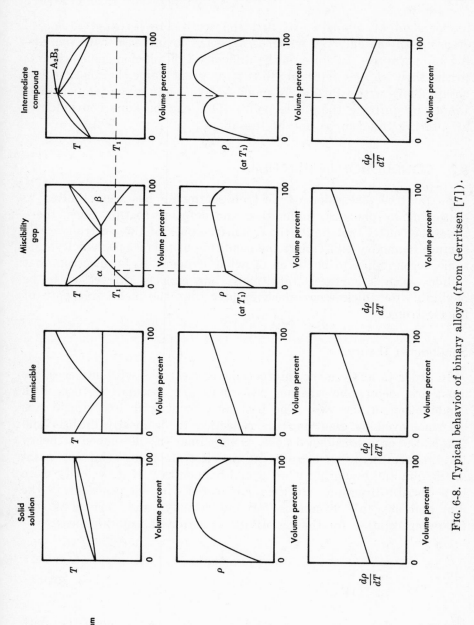

FIG. 6-8. Typical behavior of binary alloys (from Gerritsen [7]).

independent of temperature, and the scattering is expected to increase with the entropy, which is expected to increase with increasing temperature. In spite of the difficulty this phenomenon presents to the theorist, the engineer finds it of great value, because he can obtain alloys with vanishingly small $d\rho/dT$ at any given temperature, simply by shifting the Curie point. This can be done conveniently by altering the composition of the alloy.

6.2   CONDUCTION IN THIN FILMS

So far, the discussion of electron scattering has been limited to scattering by phonons, impurities, and defects. In thin films, there is another effect that limits the mean free path of the electrons. Any surface boundary will violate the condition of perfect periodicity, and some account has to be taken of what happens to an electron that collides with a surface. Intuitively, one expects this effect to be negligible for thicknesses much greater than the mean free path of the electrons.

Sondheimer Theory

In order to analyze the interactions of electrons with surfaces, we must first determine whether reflections are specular, diffuse, or a combination of the two. The best agreement with experiment has been achieved by assuming the reflection to be mostly diffuse, and that is the case considered here. Sondheimer [8] developed a theory for this scattering by considering an electron at an arbitrary point in the film and by noting what would happen to it if its velocity were in a certain direction. He then integrated over all positions in the film and over all directions for the velocity, and arrived at the following relation for the resistivity as a function of thickness:

$$\frac{\rho}{\rho_\infty} = \frac{1}{1 - \dfrac{3}{8x} + \dfrac{e^{-x}}{16x}(x^3 - x^2 - 10x + 6) - \dfrac{x}{16}(x^2 - 12)\displaystyle\int_x^\infty \frac{e^{-y}}{y}\,dy}$$

where $x = d/\lambda_\infty$, d is the film thickness, and λ_∞ is the mean free path that the electrons would have if the thickness were infinite. This relation, known as the Sondheimer relation, is shown as Fig. 6-9. Its validity is independent of whether the resistivity in the bulk is caused primarily by phonons, impurities, grain boundaries, or anything else.

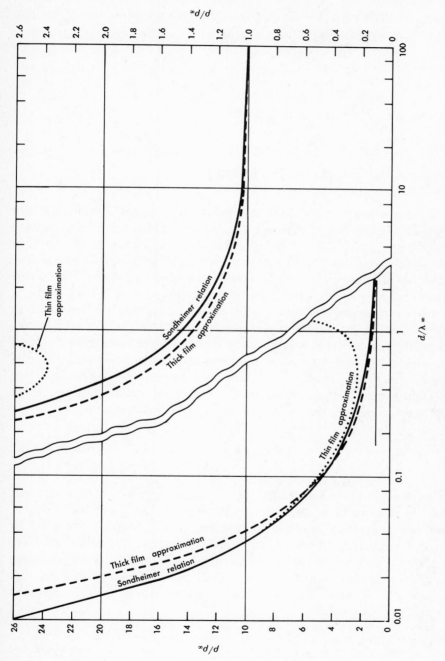

FIG. 6-9. Sondheimer relation for resistivity as a function of thickness.

The expression has been approximated at both large and small thicknesses, as shown by the dashed and dotted curves in Fig. 6-9. The thick* film approximation is

$$\frac{\rho}{\rho_\infty} = 1 + \frac{3\lambda_\infty}{8d} \qquad \text{for } \frac{d}{\lambda_\infty} > 0.1 \qquad\qquad (6\text{-}22)$$

and the thin film approximation is

$$\frac{\rho}{\rho_\infty} = \frac{4\lambda_\infty}{3d \left(\ln \dfrac{\lambda_\infty}{d} + 0.4228 \right)} \qquad \text{for } \frac{d}{\lambda_\infty} < 0.1 \qquad (6\text{-}23)$$

It is interesting to note that the thick film approximation actually crosses the exact curve at about $d/\lambda_\infty = 0.1$. Thus, for all thicknesses above $0.08\lambda_\infty$, the thick film approximation is better than the thin film approximation. We have seen earlier that electron mean free paths for phonon scattering in metals are of the order of 100 to 600 angstroms at room temperature, so even in the case where there is no other scattering mechanism, the region of validity of the thin film approximation is limited to thicknesses below about 50 angstroms. Films of this thickness are normally quite different in structure from the bulk metal and may be so agglomerated as to be electrically discontinuous. Thus, most of the data used to validate the Sondheimer relation were taken for films in the region of validity of the thick film approximation. Equation (6-22) can be rewritten as

$$\rho d = \rho_\infty \left(d + \frac{3\lambda_\infty}{8} \right)$$

Thus, if ρd is plotted as a function of d, the result will be a straight line whose slope is ρ_∞ and whose intercept on the d-axis is $- 3\lambda_\infty/8$. This is a convenient method for determining λ_∞ and ρ_∞ for a set of films of various thicknesses, provided, of course, that they all have smooth surfaces with no agglomeration. Figure 6-10 shows some plots of data taken at 60°K on alkali metal films plotted in this manner. It is seen that the data conform fairly well to straight

---

*The term "thick film," as used in this section refers to films thicker than λ_∞, and includes films normally classified as thin films, since λ_∞ may be only a few angstroms.

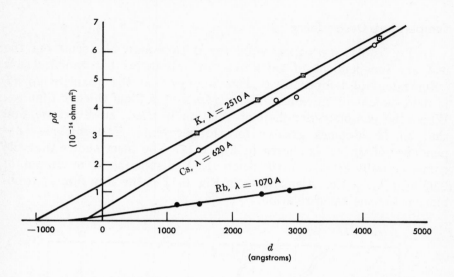

Fig. 6-10. Resistivity times thickness as a function of thickness for thin films of alkali metals at 60°K [9].

lines, and that the intercept is always on the predicted side of the origin. The estimated values for the mean free path are indicated on each set of data.

If the thick film approximation is used for the surface scattering, the resistivity can be written from Eq. (6-11):

$$\rho = \frac{mv_F}{ne^2}\left(\frac{1}{\lambda_\infty}\right)\left(1+\frac{3\lambda_\infty}{8d}\right) \qquad \text{for } d > 0.08\lambda_\infty$$

$$= \frac{mv_F}{ne^2}\left(\frac{1}{\lambda_\infty}+\frac{3}{8d}\right)$$

Thus, it can be seen that the thickness affects the resistivity as if it were a limitation on the mean free path equal to $8d/3$. The entire expression for the resistivity will look something like

$$\rho = \frac{mv_F}{ne^2}\left(\frac{1}{\lambda_{\mathrm{ph}}}+\frac{1}{\lambda_{\mathrm{imp}}}+\frac{1}{\lambda_{\mathrm{def}}}+\frac{1}{\lambda_{\mathrm{bdy}}}+\cdots+\frac{3}{8d}\right) \qquad (6\text{-}24)$$

where each of the effects that limit the mean free path is considered to act independently of the others. We have explicitly indicated only the phonon, impurity, defect, grain boundary, and surface scattering.

Temperature Dependence

In Eq. (6-24) (which is valid for thick films), all terms but the last are independent of thickness, and all terms but the first are independent of temperature. This implies that the quantity $d\rho/dT$ is independent of thickness, which is true for films thicker than λ_∞. When the complete Sondheimer relation is used, however, we find that $d\rho/dT$ becomes greater for thinner films. The thickness dependence of $d\rho/dT$ is shown in Fig. 6-11. The inset shows that the curve actually goes a little below 1.00, displaying a minimum of 0.98 at $d/\lambda_\infty = 1.5$. Also shown in Fig. 6-11 is the thin film approximation for the same relation.

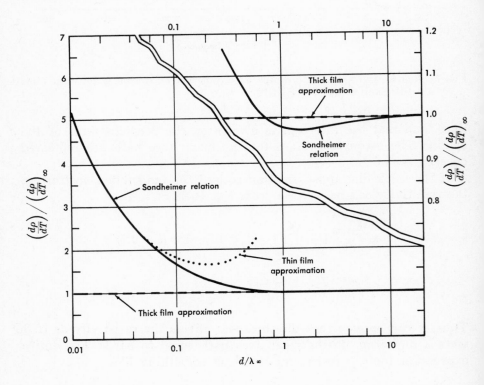

FIG. 6-11. Sondheimer relation for temperature derivative of resistivity as a function of temperature.

The temperature coefficient of resistivity (α) is also a function of thickness, and it can be found from Figs. 6-9 and 6-11. It is plotted in Fig. 6-12. Using the thick film approximation [Eq. (6-22)], we find that

$$\alpha = \alpha_\infty \left(\frac{1}{1 + \dfrac{3\lambda_\infty}{8d}} \right) \qquad (6\text{-}25)$$

In this case, the thick film approximation crosses the exact curve at $d = 0.4\lambda_\infty$, so its region of validity is not as great as the thick film approximation for ρ/ρ_∞. It does, however, give the general idea of the dependence of α on the thickness. The points shown in Fig. 6-12 are experimental data for some of the alkali metals measured in the range 60°K to 90°K [9]. The agreement is satisfactory in view of the experimental accuracies involved.

The temperature dependence of the resistance of a thin film can also be used to determine the residual mean free path, λ_r, of the material (that is, the mean free path at very low temperatures corrected for the surface scattering). Let the room temperature resistance be R_1 and the residual resistance be R_r (the residual resistance is the limit of the resistance as the temperature goes to zero). The difference between the two resistances will be inversely proportional to the room temperature phonon mean free path, λ_{ph}, if we assume that

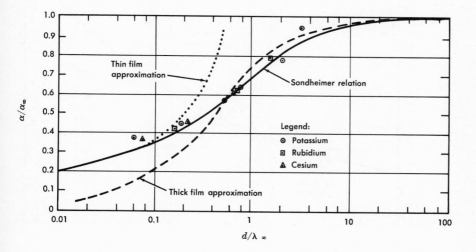

FIG. 6-12. Normalized temperature coefficient of resistivity as a function of thickness.

Matthiessen's rule holds. That is,

$$R_1 - R_r = \frac{C}{\lambda_{ph}}$$

where C is the constant of proportionality. This relation is exactly true only for the thick film approximation, but even in the thin film approximation, it is usually satisfactory. The residual resistance is

$$R_r = \frac{C}{\lambda_r} F\left(\frac{d}{\lambda_r}\right)$$

where $F(d/\lambda_r)$ is the Sondheimer function, given before as ρ/ρ_∞ (Fig. 6-9). Combining these two equations, we have

$$\frac{d}{\lambda_{ph}}\left(\frac{R_r}{R_1 - R_r}\right) = \frac{d}{\lambda_r} F\left(\frac{d}{\lambda_r}\right) \qquad (6\text{-}26)$$

The right-hand side of this equation is a function of d/λ_r only, and it has been plotted in Fig. 6-13. Thus, in order to find λ_r, it is neces-

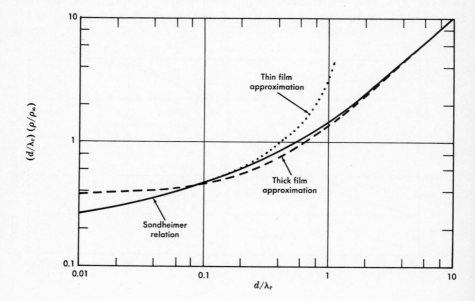

FIG. 6-13. Graph for finding the residual mean free path from the Sondheimer relation.

sary only to know the thickness, the resistance ratio, and λ_{ph} for the material given. Then, by consulting Fig. 6-13, we can find d/λ_r directly, and from this we can get λ_r. The mean free path at low temperatures is a significant parameter because it contains the effects of impurities, defects, and grain boundaries.

In the region where the thick film approximation applies, we can use Eq. (6-22) for $F(d/\lambda_r)$ and then solve for λ_r. The result is

$$\lambda_r = \frac{d}{\dfrac{d}{\lambda_{ph}}\left(\dfrac{R_r}{R_1 - R_r}\right) - \dfrac{3}{8}} \qquad (6\text{-}27)$$

An example of the use of Eq. (6-26) is now given for a certain thin film of potassium which has a ratio $(R_1 - R_r)/R_r$ of 30, and a thickness, d, of 5000A. The phonon mean free path of potassium at $0°C$ is 370A, so

$$\frac{d}{\lambda_{ph}}\left(\frac{R_r}{R_1 - R_r}\right) = \frac{5000A}{370A}\left(\frac{1}{30}\right) = 0.45$$

From Fig. 6-13, it is found that $d/\lambda_r = 0.085$, so $\lambda_r = 59,000A$. If we had used the approximate formula [Eq. (6-27)], we would have arrived at 67,000A. In any case, it is quite possible to have a residual mean free path that is much larger than the thickness. This means that the film is one of fairly high purity and large crystallite size. One would have to say that the crystallite size in this film was at least of the order of 50,000A, which is 10 times the film thickness. Thus, the film is made up of plate-like crystallites, whose diameter is at least 10 times the thickness.

Stress Effects in Thin Films

In this analysis, the film is assumed to be isotropic. In a noncubic metal, this assumption may not be valid since the resistivity and elastic constants can be a function of crystallographic orientation, and there may be a preferred orientation of these crystallites in the film, causing the resistivity to depend on the growth habit of the film. The following analysis assumes that there is no such preferred orientation, and that the film is completely isotropic with regard to its resistivity and elastic properties. In general, the stress applied to the sample is represented by a tensor of second rank, but if the metal is isotropic, the stress tensor can be diagonalized by choosing proper coordinates, which is done here by considering the stresses in the three dimensions, l, w, and d. If Y is the Young's modulus of the film and σ_f the Poisson's ratio, the strains will be given in terms of the stresses (P) as follows [10]:

$$\frac{\Delta l}{l} = \frac{1}{Y} \left[P_l - \sigma_f \left(P_w + P_d \right) \right] \qquad (6\text{-}28a)$$

$$\frac{\Delta w}{w} = \frac{1}{Y} \left[P_w - \sigma_f \left(P_d + P_l \right) \right] \qquad (6\text{-}28b)$$

$$\frac{\Delta d}{d} = \frac{1}{Y} \left[P_d - \sigma_f \left(P_l + P_w \right) \right] \qquad (6\text{-}28c)$$

It is next assumed that the film is so thin compared to the substrate that the presence of the film does not affect the stresses in the substrate, and the film conforms to the strains of the substrate in the dimensions l and w. That is, the strains $\Delta l/l$ and $\Delta w/w$ in the film are the same as in the surface of the substrate, and these are the same as if there were no film present. The other dimension (the thickness of the film) is not constrained. That is, $P_d = 0$. This does not mean, however, that $\Delta d/d$ is also zero. Let us consider a particular example.

Let the substrate be bent in such a way that the radius of curvature of its bending is constant along its length, and let the strain in this direction be ϵ. In this situation, the strain in the w-direction is $-\sigma_s\epsilon$, where σ_s is the Poisson's ratio for the substrate. Using these two conditions plus the fact that $P_d = 0$, the strain equations [Eqs. (6-28)] can be solved giving

$$\frac{\Delta l}{l} = \epsilon$$

$$\frac{\Delta w}{w} = -\sigma_s\epsilon$$

and

$$\frac{\Delta d}{d} = -\sigma_f \left(\frac{1 - \sigma_s}{1 - \sigma_f} \right) \epsilon$$

The fractional volume change is then

$$\frac{\Delta V}{V} = \frac{\Delta l}{l} + \frac{\Delta w}{w} + \frac{\Delta d}{d} = \left(\frac{1 - \sigma_s}{1 - \sigma_f} \right) (1 - 2\sigma_f) \, \epsilon$$

Since the film has been assumed to be isotropic, by the principle of superposition,* we can put

$$\frac{d \ln \rho}{d\epsilon} = \left(\frac{d \ln \rho}{dP}\right)\left(\frac{dP}{d \ln V}\right)\left(\frac{d \ln V}{d\epsilon}\right)$$

$$= \left(\frac{d \ln \rho}{dP}\right)\left(\frac{-1}{\kappa}\right)\left(\frac{1 - \sigma_s}{1 - \sigma_f}\right)(1 - 2\sigma_f)$$

Now, using Eq. (6-19), the fractional change in resistance can be found for a given change in the bending stress:

$$\frac{d \ln R}{d\epsilon} = \left(\frac{d \ln \rho}{dP}\right)\left(\frac{-1}{\kappa}\right)\left(\frac{1 - \sigma_s}{1 - \sigma_f}\right)(1 - 2\sigma_f) + \frac{1 + \sigma_s - 2\sigma_s\sigma_f}{1 - \sigma_f}$$

$$(6\text{-}29)$$

This is the basic equation for the change in resistance due to longitudinal stresses on the substrate. It is equally valid for the case where the substrate is bent with a uniform radius of curvature, and for the case where the substrate is stretched in the direction of current flow.

A very similar analysis leads to the formula for the case when the substrate is stretched (or bent) in a direction transverse to the current flow:

$$\frac{d \ln R}{d\epsilon} = \left(\frac{d \ln \rho}{dP}\right)\left(\frac{-1}{\kappa}\right)\left(\frac{1 - \sigma_s}{1 - \sigma_f}\right)(1 - 2\sigma_f) - \left(\frac{1 + \sigma_s - 2\sigma_f}{1 - \sigma_f}\right)$$

$$(6\text{-}30)$$

These two equations are the basis for the technology of foil strain gauges. The quantity $d \ln R/d\epsilon$ as evaluated from Eq. (6-29) is called the strain gauge factor. It must be evaluated for a certain value of σ_s, however, and this value is usually taken as 0.285, because that is the value of Poisson's ratio for the steel samples used in their calibration. As a matter of fact, it is possible to measure the Poisson ratio of a material by measuring the resistance change for longitudinal and transverse strain. The difference between the longitudinal and transverse gauge factors as given by Eqs. (6-29) and

*This principle states that each component stress produces the same effect if acting alone, as it contributes to the final result when acting in conjunction with the others [10].

(6-30) is simply $2(1 + \sigma_s)$. There is no reason that evaporated and sputtered films could not be used as strain gauges as well as the commercial thin foil strain gauges.

The second term in Eq. (6-29) for the strain gauge factor is seen to be of the order of magnitude of 1. If the Poisson ratios of the film and substrate are equal, then it is equal to $(1 + 2\sigma)$. Poisson ratios vary from about 0.15 to 0.40 (there is a theoretical upper limit of 0.5). Thus, the last term will contribute about $+1.6$ to the strain gauge factor. The first term will also be positive, because $d \ln \rho/dP$ is usually negative. Now let the Poisson's ratio be $\sigma_s = \sigma_f = 0.3$ and $d \ln \rho/dP$ be about 10 times $\kappa/3$ (see Fig. 6-6). Then the first term of Eq. (6-29) is about 1.3, so the total gauge factor is 2.9. Longitudinal gauge factors vary from about 2.0 to 4.0, which is consistent with these order-of-magnitude estimates. These estimates are also expected to be valid for thin metal films. For example, for a thin film of sputtered tantalum nitride, the longitudinal gauge factor was measured to be 2.1, as shown in Fig. 6-14. Transverse gauge factors can be either positive or negative, depending on whether the first or second term in Eq. (6-30) is larger. The transverse gauge factor of the tantalum nitride film mentioned above was -0.17.

The significance of the gauge factor for precision thin film resistors is apparent when it is considered that a 1-percent change in resistance

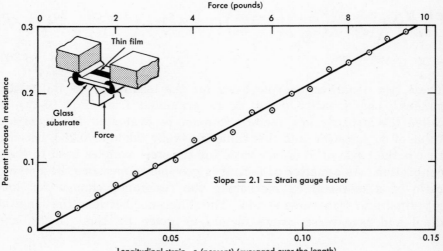

FIG. 6-14. Effect of strain on a tantalum nitride resistor.

can be induced by a strain of only 0.5 percent. If the substrate is glass, strains larger than 0.5 percent will probably break the glass. In practice, substrates are mounted to minimize the strain such that this condition is not approached. In normal use, the variation in resistance due to strain can be kept below 0.01 percent.

Effect of Expansion Mismatch on Temperature Coefficient of Resistance. The fact that thin metal film resistors are strain-sensitive also means that their temperature coefficients of resistance (TCR) may be a function of the mismatch between the thermal expansion coefficients of the film and the substrate. The temperature coefficient of resistivity (α) is not necessarily equal to TCR, even in the bulk material. From Eq. (6-19),

$$\frac{d \ln R}{dT} = \frac{d \ln \rho}{dT} - \frac{d \ln l}{dT}$$

for an isotropic bulk metal. In other words,

$$\text{TCR} = \alpha - \frac{\beta}{3} \qquad (6\text{-}31)$$

where β is the temperature coefficient of volume expansion $[\beta = (\partial \ln V / \partial T)_P]$. At room temperature, α is about 4000×10^{-6} per °C for metals and $\beta/3$ is typically 20×10^{-6} per °C. Thus, the effect of expansion is negligible for bulk metals, and TCR approximates α.

Both TCR and α are defined at constant pressure (i.e., zero stress) and not at constant volume. The quantity α_V at constant volume is related to α by

$$\alpha = \alpha_V + \beta \left(\frac{\partial \ln \rho}{\partial \ln V} \right)_T \qquad (6\text{-}32)$$

From here on, it is necessary to use partial derivatives because ρ is considered to be a function of $T, w, d,$ and l. It is assumed that

$$\left(\frac{\partial \ln \rho}{\partial \ln d} \right)_{T,w,l} = \left(\frac{\partial \ln \rho}{\partial \ln w} \right)_{T,l,d} = \left(\frac{\partial \ln \rho}{\partial \ln l} \right)_{T,d,w} = \gamma$$

where

$$\gamma = \left(\frac{\partial \ln \rho}{\partial \ln V} \right)_T$$

Then the TCR for a thin film, as derived from Eq. (6-19), is

$$TCR_f = \frac{d \ln R}{dT} = \frac{d \ln \rho}{dT} + \frac{d \ln l}{dT} - \frac{d \ln w}{dT} - \frac{d \ln d}{dT}$$

or

$$TCR_f = \left(\frac{\partial \ln \rho}{dT}\right)_{l,w,d} + \gamma \left(\frac{d \ln l}{dT} + \frac{d \ln w}{dT} + \frac{d \ln d}{dT}\right)$$

$$+ \frac{d \ln l}{dT} - \frac{d \ln w}{dT} - \frac{d \ln d}{dT}$$

Now it is assumed that the expansion of the film in the dimensions l and w is equal to the expansion of the substrate

$$\frac{d \ln l}{dT} = \frac{d \ln w}{dT} = \frac{\beta_s}{3}$$

where β_s is the bulk coefficient of thermal expansion of the substrate. Using this condition with Eqs. (6-28), we can find the expansion coefficient for the third dimension, d:

$$\frac{d \ln d}{dT} = \frac{\beta_f}{3} - \frac{2\sigma_f (\beta_s - \beta_f)}{3(1 - \sigma_f)}$$

Then

$$TCR_f = \alpha_V - \frac{\beta_f}{3} + \frac{2(\beta_s - \beta_f)}{3(1 - \sigma_f)} [\sigma_f + (1 - 2\sigma_f) \gamma] + \gamma\beta_f$$

With the use of Eqs. (6-31) and (6-32) this can be rewritten as

$$TCR_f = TCR + \frac{2(\beta_s - \beta_f)}{3(1 - \sigma_f)} [\sigma_f + (1 - 2\sigma_f) \gamma] \qquad (6\text{-}33)$$

It is thus clear that TCR_f (the temperature coefficient of resistance for a thin film) is not the same as that for a bulk material (TCR) of the same structure and purity unless $\beta_s = \beta_f$. Typical numbers that might be employed in Eq. (6-33) are $\sigma_f = 0.3$ and $\gamma = 3$. With these values,

$$TCR_f = TCR + 1.4 (\beta_s - \beta_f)$$

Then if β_f is taken as 60×10^{-6} per °C, and β_s as 15×10^{-6} per °C, the temperature coefficient of resistance of the film will be

$$\mathrm{TCR_f} = \mathrm{TCR} - 65 \times 10^{-6} \text{ per °C}$$

This is a significant correction to the TCR for thin film resistors, where $\mathrm{TCR_f}$ is typically 100×10^{-6} per °C, and may be as low as 10^{-6} per °C. It also gives one reason that $\mathrm{TCR_f}$ depends on the substrate. For example, changing from soda lime (soft) glass substrates to fused quartz substrates would change β_s from 28×10^{-6} to 2×10^{-6} per °C. This would change $\mathrm{TCR_f}$ by 36×10^{-6} per °C (again assuming $\sigma_f = 0.3$ and $\gamma = 3$).

Real Films

This chapter so far has considered thin films as differing from bulk material only in that they have one small dimension. Actually, real films can be markedly different in structure, purity, and strain, resulting in quite different properties, especially for extremely thin films. It is characteristic for very thin films that the $\mathrm{TCR_f}$ decreases (with decreasing thickness) until it becomes negative, and the resistivity increases faster than expected from the Sondheimer relation. Attempts have been made to explain this by postulating various kinds of energy barriers for the electrons which generally result in a resistivity proportional to $\exp(E/kT)$, where E is the barrier height. This gives a semiconductor-like behavior, with a $\mathrm{TCR_f}$ of $-E/kT^2$.

In the initial stages of film growth (say, less than 20 angstroms average thickness), the film consists of electrically and physically isolated nuclei (see Chap. 3). As the thickness increases, electrical continuity can be observed for films which are still physically discontinuous. Conductivity in this case is limited by electron transfer from one nucleus to the next. This process is not completely understood, as yet, but it is expected to cause nonohmic conduction, large negative temperature coefficients of resistance, and a strong dependence on nucleus size.

As the average film thickness increases, the interstices fill in, the resistivity rapidly decreases, and the temperature coefficient of resistivity becomes positive. The result is interpreted as a smooth change from essentially semiconducting behavior to metallic behavior. The details of this transition are not understood at all, and may vary considerably from one metal to another, especially in the case of alloys, or "doped" films.

Deviations from theoretical predictions are also caused by relatively high impurity concentrations in the films. In evaporated films, it is difficult to maintain a high ratio of deposition rate to residual gas arrival rate. In sputtered films, the presence of the sputtering gas can influence the properties of the film, and an analysis of sputtered films has shown as much as 2 atom percent of argon incorporated into the film. Attempts to correlate theory with experiments have been primarily concerned with films of the noble metals such as gold and silver, since it is much easier to deposit pure films of these metals than of refractory metals like tantalum. There is also less interaction between the substrate and the film.

Partly because of the high impurity concentration, thin films often have crystallites smaller than the film thickness. They are also likely to be highly strained due to (among other factors) distortions of the film lattice by the substrate lattice and to the mismatch in the thermal coefficient of expansion between the film and substrate, since many films are deposited at elevated temperatures. For these reasons, thin films commonly exhibit resistivities several times greater than the bulk resistivity of the same material, but they may still obey Matthiessen's rule, having approximately bulk values for $d\rho/dT$ at room temperature. In many cases, however, new structures are found, and no correlation with bulk properties is possible, as in beta tantalum, discussed in Chap. 4.

Normally, sputtered films have crystallite dimensions considerably smaller than the thickness, probably less than 100A. Thus, the effects of surface scattering are obscured by grain boundary scattering, and the Sondheimer theory cannot be applied. On films such as tantalum, it is possible to verify the small mean free path by measuring the resistance as a function of thickness. The thickness may be reduced by anodization, and the anodization voltage provides a measure of the remaining thickness (assuming the original thickness is uniform). Figure 6-15 shows a plot of conductance as a function of thickness for a tantalum nitride film, which was 435A before anodization (including an allowance of 75A for original oxide). The circles are data points, and the solid curve is the Sondheimer relation evaluated with $\lambda_\infty = 70A$, and $N_s\rho = 6.85 \times 10^4$ micro-ohm centimeter squares, where N_s is the number of "squares" of the resistor.* For thicknesses greater than λ_∞, the Sondheimer relation gives

$$\frac{1}{R} = \frac{1}{N_s\rho_\infty}\left(d - \frac{3\lambda_\infty}{8}\right)$$

*See Chap. 7 for the interpretation of the term "squares."

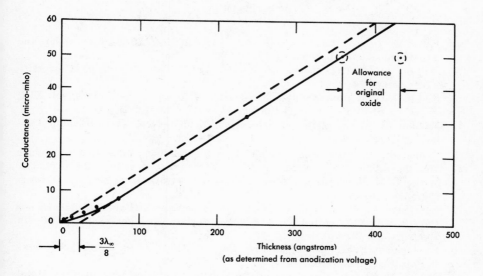

FIG. 6-15. Conductance as a function of thickness. The solid curve is from the Sondheimer theory for $\lambda_\infty = 70$ angstroms.

so the intercept of the asymptote is equal to $3\lambda_\infty/8$. In this case, the measured intercept gives λ_∞ as 70A, but it is almost certainly somewhat smaller because the first few layers are expected to have a higher resistivity due to a higher impurity content, and this would tend to make the line curve in the same direction. At any rate, it is clear from this type of plot that the mean free path in these films cannot be greater than about 100A, so for thicknesses greater than a few hundred angstroms, surface scattering is a negligible factor.

Reactive sputtering causes further deviations from the theory of metallic conduction. Tantalum nitride has a resistivity of a few hundred micro-ohm centimeters and a small, negative TCR (-50 to $-150 \times$ ppm/°C). Thus, it does not obey Matthiessen's rule at all and does not follow the theory of metallic conduction. Neither is it a well behaved semiconductor as its resistivity is too low and depends linearly on temperature.

The very properties which make these films difficult to understand in terms of the simple theories are often the same ones that make them useful as resistor materials, especially the low TCR's and the resistivity values. A variety of materials have been considered for use as thin film resistors, and the next chapter considers the material properties of some of the more important of them.

REFERENCES

1. Kittel, C. *Introduction to Solid State Physics.* New York: John Wiley and Sons, Inc., second ed., 1956.
2. Wilson, A. H. *The Theory of Metals.* Cambridge: Cambridge University Press, second ed., 1954.
3. Slater, J. C., H. Jones, and A. N. Gerritsen. *Handbuch der Physik,* vol. XIX, edited by S. Flügge. Berlin: Springer-Verlag, 1956 (in English).
4. Meaden, G. T. *Electrical Resistance of Metals.* New York: Plenum Publishing Corp., 1965.
5. Grüneisen, E. "Die Abhangigkeit des Elektrischen Widerstandes Reiner Metalle von der Temperatur," *Annalen der Physik Ser. 5,* 16, 530-540 (1933).
6. White, G. K. *Experimental Techniques in Low Temperature Physics.* Oxford: Clarendon Press, 1959.
7. Gerritsen, A. N. "Metallic Conductivity, Experimental Part," *Handbuch der Physik,* vol. XIX, edited by S. Flügge. Berlin: Springer-Verlag, 1956 (in English). pp. 137-226.
8. Sondheimer, E. H. "The Mean Free Path of Electrons in Metals," *Advances in Physics,* 1, 1-42 (1952).
9. Mayer, H. "Recent Developments in Conduction Phenomena in Thin Metal Films," *Structure and Properties of Thin Films,* edited by Neugebauer, Newkirk, and Vermilyea. New York: John Wiley and Sons, Inc., 1959. pp. 225-252.
10. Newman, F. H., and V. H. L. Searle. *The General Properties of Matter,* London: Ernest Benn, Limited, 1928. p. 139.

PRINCIPAL SYMBOLS FOR CHAPTER 6

SYMBOL	DEFINITION
d	Film thickness
D	Density
e	Electronic charge ($e > 0$)
$\mathbf{E}$	Electric field
E_F	Fermi energy
f	Number of conduction electrons per atom
$F\left(\dfrac{d}{\lambda_r}\right)$	Sondheimer function
h	Planck's constant
$\mathbf{j}$	Current density
k	Boltzmann's constant
l	Length of a sample (direction of current)
m	Electron mass
n	Electron density
N_A	Avogadro's number
P	Hydrostatic pressure
R	Resistance
R_r	Residual resistance
t	Time
T	Temperature
TCR	Temperature coefficient of resistance
TCR_f	Temperature coefficient of resistance for a film

SYMBOL	DEFINITION
$\mathbf{v}$	Electron velocity
$\mathbf{v}_d$	Electron drift velocity
v_F	Fermi speed
V	Volume
w	Width of a sample
x_0	Θ_D/T
Y	Young's modulus
α	Temperature coefficient of resistivity
β	Temperature coefficient of volume expansion $\left(\dfrac{\partial \ln V}{\partial T}\right)_P$
ϵ	Strain along the length
Θ_D	Debye temperature
Θ_R	Debye temperature determined from resistivity measurements
κ	Volume compressibility $\left(\dfrac{-\partial \ln V}{\partial P}\right)_T$
λ	Mean free path
λ_r	Residual mean free path
ν	Phonon frequency
ρ	Resistivity
ρ_r	Residual resistivity
ρ_Θ	Resistivity at $T = \Theta_D$
σ_f	Poisson's ratio of a film
σ_s	Poisson's ratio of a substrate
τ	Relaxation time
τ'	Mean time between collisions

Chapter 7

Resistor and Conductor Materials

Some of the basic mechanisms of conduction in metals were introduced in Chap. 6. This chapter deals with the various specific materials currently used for thin film resistors and conductors. The actual layout and design of components and circuits will be discussed in Chap. 12. More detailed information about thin film material properties is available in references 1, 2, and 3.

7.1 THIN FILM RESISTOR PARAMETERS

Before specific materials are considered, some important parameters used to describe their properties when the materials are used as thin film resistors are enumerated.

Sheet Resistance

The resistance of a thin film resistor is directly proportional to the resistivity, ρ, and inversely proportional to the thickness, d. It is therefore convenient to define a quantity, R_s, which is equal to ρ/d. The quantity R_s is called the sheet resistance and may be thought of as a material property since the film is essentially two-dimensional. A thin film resistor consisting of a simple rectangle of length l (in the direction of the current) and width w has a resistance of

$$R = \left(\frac{\rho}{d}\right)\left(\frac{l}{w}\right)$$

$$= R_s\left(\frac{l}{w}\right)$$

The ratio l/w is sometimes called the number of squares in the resistor, since it is equal to the number of squares of side w that can be superimposed on the resistor without overlapping. The term "squares" is a pure number, having no dimensions. The sheet resistance, R_s, has the units of ohms, but it is convenient to refer to it in ohms per square since the sheet resistance produces the resistance of the resistor when multiplied by the number of squares. The concept can then be broadened to include any arbitrarily shaped resistor by calling the quantity Rd/ρ the effective number of squares.

Since the sheet resistance is a basic material property, it is important to be able to measure it, preferably without the need of etching a pattern. This is commonly done with the "four-point probe" technique. In this method, four contacts are made to the film; current is injected through one contact and out another, and the resultant voltage drop across the two remaining points is measured. The sheet resistance is proportional to the ratio of the voltage, V, to the current, I. That is,

$$R_s = C\left(\frac{V}{I}\right)$$

where C is a constant of proportionality that depends on the configuration, position, and orientation of the probes and on the shape and size of the thin film sample. The four probes are commonly equally spaced in a straight line, with the two outer probes being the current probes. The spacing is usually made small compared to the planar dimensions of the film so that the film may be assumed to be infinite in extent (unless the probe is placed near a boundary). If the film were infinite in all directions, the constant C would be equal to $\pi/\ln 2$, or

$$C = 4.5324 \text{ (for an infinite film)}$$

When using the four-point probe technique, it is convenient to adjust the current until it is numerically equal to C (for example, 4.53 milliamps) so that the voltage reading will numerically equal the sheet resistance.

The constant C may have to be corrected for the size of the substrate. Figure 7-1(a) shows the amount of correction required for a four-point probe whose interprobe spacing (s) is 0.050 inch. The probe is centered on a rectangular film of dimensions a and b and is parallel to the dimension a. Two curves are shown, one for $a = b$

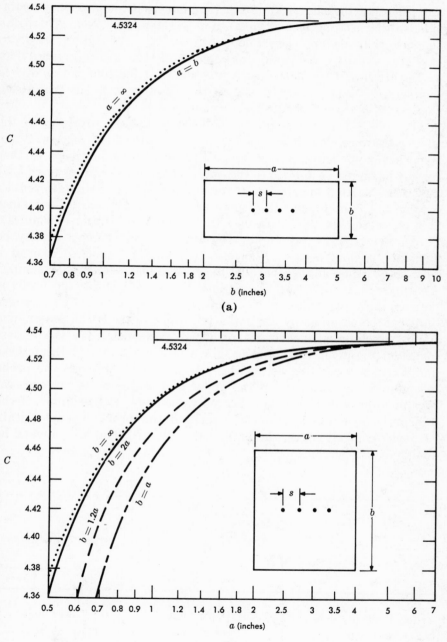

FIG. 7-1. Four-point probe (s = 0.050 inch) proportionality constant for rectangular films.

and the other for $a = \infty$. For practical purposes the two curves coincide, so that as long as a is equal to or greater than b (in the region shown), the constant C is independent of a.

Figure 7-1(b) gives values for C when b is equal to or greater than a. In this case, C is plotted as a function of a for four ratios of b/a (namely, 1, 1.2, 2, ∞). It is seen that if $b/a \geqq 2$ (in the region shown), C is essentially independent of b.

The four-point probe can also be used to check uniformity of the sheet resistance since it is sensitive to the region of its location. When the probe is not centered with respect to the sample, however, further corrections must be applied to the readings. Computer programs [4, 5] have been written to calculate C for any given case to any desired accuracy, but it is difficult to display results of such calculations since C is a function of four variables (a, b, and two position variables).

A simpler method of finding the approximate value of C may be adequate in many cases. Figure 7-2(a) shows a general case of a four-point probe aligned parallel to an edge of a rectangular sample. An upper limit for C can be obtained by assuming that the probe is positioned in the center of the sample, as in Fig. 7-2(b). A lower limit can be obtained by assuming the probe to be in the center of a smaller rectangle, one corner of which coincides with the closest corner of the given rectangle as shown in Fig. 7-2(c). Both of these limits can be evaluated using Fig. 7-1, since in both cases the probe is centered. The average of the upper and lower limits gives a reasonable approximation to the desired value of C. For example, if the interprobe spacing, s, is 0.050 inch and the probe is kept within 0.5 inch from all edges, the error using this method will always be less than 0.2 percent.

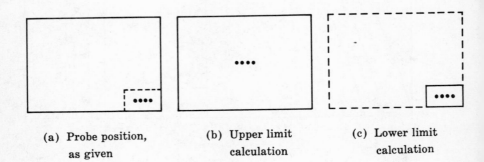

(a) Probe position,
as given

(b) Upper limit
calculation

(c) Lower limit
calculation

FIG. 7-2. Noncentered four-point probe approximation.

A problem is involved in four-point probe measurements if there are predrilled holes in a substrate. Many circuits have so many holes required for mounting components and wires that a location sufficiently far from all holes cannot be found. Calculations have also been made for this case but, again, too many variables are involved for complete data presentation. For a hole whose center is several spacings (s) from the center of the probe and whose diameter is not much larger than s, the value of C is approximated by

$$C = C_0 \left[1 - \frac{3\delta^2 s^2}{4 (\ln 4) (r^4)} \right]$$

where r is the distance from the center of the hole to the center of the probe array, δ is the hole diameter, and C_0 is the value of C when there is no hole [6].

Temperature Coefficient of Resistance

The temperature coefficient of resistance (TCR) is defined as $(1/R)(dR/dT)$. The TCR can be useful for film characterization, and it can be important in circuit functions where ambient temperature changes could cause unwanted changes in circuit performance. Thin films typically have very low TCR's compared to bulk metals, primarily because of the high resistivity of thin films. In many cases, it is possible to control the TCR of the thin film by varying the composition (e.g., by reactive sputtering) until the desired TCR is obtained. This may be useful, for example, in making resistance-capacitance circuits. In order to minimize the temperature coefficient of the resistance-capacitance product, the TCR may be made to compensate for the temperature coefficient of the capacitor.

For films thicker than a few hundred angstroms, dR/dT is essentially independent of temperature and

$$\text{TCR} \approx \frac{1}{R_{rt}} \frac{(R_1 - R_{rt})}{(T_1 - T_{rt})} \tag{7-1}$$

where R_{rt} is the resistance at room temperature (T_{rt}), and R_1 is the resistance at some other temperature, T_1. It has been found convenient to set T_1 at $-196°C$, the temperature of liquid nitrogen. This permits a rapid measurement over a wide temperature span ($220°C$), which may be necessary to get enough resistance change to measure accurately. Sometimes T_1 is made $+100°C$ or higher, but this procedure involves the complication that irreversible aging may take place at the temperature required to obtain accurate data on the TCR.

In spite of its low TCR value, a thin film can be used as a thermometer to measure the temperature of the film. Figure 7-3 shows data taken from a "meander" pattern thin film (tantalum nitride) resistor. The voltage drop across each loop was measured while a constant current (d.c.) was passed through the entire sample. The percent change in resistance of each section from its value at negligible current is divided by the TCR (assumed uniform over the resistor), and the complete temperature profile is obtained. The TCR of this film is -95×10^{-6} per $°C$ and, judging from the scatter of the points, the reproducibility in temperature measurement is better than $10°C$. When temperatures above room temperature are measured, the resistance change must be reversible and not due to aging. For this experiment, the resistance change was reversible as long as the resistor temperature never exceeded about $120°C$. The films had previously been heat treated at $250°C$ for 5 hours.

The TCR may be a function of thickness, not only for the reasons mentioned in Chap. 6, but also because the film composition and structure may vary. Thus, the TCR of a film may change upon trimming to value by either anodization or abrasion.

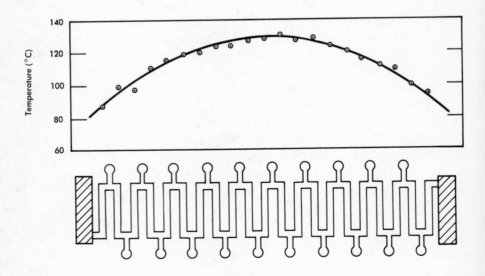

FIG. 7-3. Temperature profile of a tantalum resistor under a d.c. load, as measured by the reversible resistance change in each loop.

Voltage Coefficient of Resistance

Ohm's law states that the voltage across a resistor is proportional to the current passing through it, or that the resistance is independent of the applied voltage. This law loses applicability as the power dissipated becomes large enough to cause the resistor to heat appreciably. There can be, however, an additional effect that is noticeable in very thin films, where even at constant temperature the resistance is voltage-dependent. This effect can be measured by using pulsed d.c. and varying the duty cycle as the voltage is increased such that the average power dissipation is constant. The change in resistance measured in this way is described by a voltage coefficient of resistance, VCR:

$$\text{VCR} = \frac{1}{R}\left(\frac{\partial R}{\partial V}\right)_T$$

In films of practical thicknesses, this coefficient is not a major factor since it is of the order of 5 ppm/volt which is usually smaller than the effect due to the heating.

Noise Coefficient

Any resistor at a finite temperature, T, has a certain noise voltage even if no current is applied; this is called thermal noise. The square of the voltage thus produced is proportional to the resistance and the temperature, and is the same for all frequencies; i.e., it is pure, white noise. In addition to thermal noise, there may be a "current noise" voltage, which is only observed when a current is passed through the resistance. Current noise is often associated with incomplete contact, such as is found in granular structures, and poorly soldered joints. The voltage due to current noise is proportional to the resistance and the applied voltage. In this case, however, the frequency spectrum is not flat. The noise voltage is proportional to $(f)^{-1/2}$. Thus,

$$e^2(f) = \frac{BR^2V^2}{f}$$

where $e(f)$ is the noise voltage observed at frequency f, V is the applied d.c. voltage, and R is the resistance. The constant, B, depends on the material, the construction methods, and the configuration of the resistor. It is independent of applied voltage and of frequency; therefore, B can be used as a figure of merit for a given type of resistor. Materials used for thin film resistors have noise coefficients comparable to those of wire-wound resistors. These noise coefficients are considerably better than those of composition resistors and are too small to be measured on standard noise-measuring equipment. The subjects of noise in resistors and noise-measuring equipment are discussed in detail by Conrad et al [7].

High-Frequency Performance

At low frequencies, most resistors act as pure resistances with no measurable reactance. As the frequency is increased, however, the resistor eventually becomes reactive. In thin films, the frequency response is primarily a function of the pattern configuration of the resistor and its leads, so frequency response is not considered a material property; this subject is discussed in Chap. 12. It is sufficient to say here that thin films have an inherent advantage for high-frequency use because of their two-dimensional geometry.

Stability

In time, a resistor may drift from its initial resistance value, especially if operated at high temperatures. This effect is usually expressed as percent resistance change per unit time. Actually, the percent drift is not linear with time for several reasons. A thin film resistor usually becomes more stable as it ages, partly due to self-limiting processes such as surface oxidation, which forms protective layers. Thus, preaging of thin film resistors (by thermal treatment) can greatly increase their stability.

Humidity Effects. Some types of resistors are sensitive to the relative humidity. This is especially true of granular structures and composition resistors. With tantalum thin films, however, this has not been a significant problem as long as the leakage across the surface of the substrate is negligible. With less inert material such as nickel-chromium alloys, protection from humidity must be provided to prevent electrolytic corrosion.

Power Density

Another important parameter in thin film resistor design is power density. This is the ratio of power applied to the resistor to the area of a simple geometric figure that encompasses the entire resistor. It must be kept in mind that, even though the temperature increase may be proportional to the power applied for a given resistor, it may not be proportional to the power density. Approximate power density values, however, are helpful in deciding how large the resistor must be since the greater the power required, the greater the area needed to dissipate it. This topic is further discussed in Chap. 12.

7.2 THIN FILM RESISTOR MATERIALS

Many specific materials, each with special material properties, have been used for thin film resistors. Several of these materials are discussed in this section. Because of the importance of tantalum-based resistors to the communications industry, they are discussed in more detail in a separate section.

Chromium

Chromium was among the first metals to be investigated as a thin film resistor material [8]. It was originally selected because of its high vapor pressure (Fig. 3-5), high resistivity (13 μ ohm cm at 20°C), and excellent chemical stability. Aside from a small anomaly* around 39°C, the temperature coefficient of resistivity of chromium is about 3000×10^{-6} per °C in the region $-100°C$ to $+ 100°C$. Chromium is still used as an underlayer to provide adherence for other films, but nickel-chromium alloys seem to be superior to chromium as a practical resistor material.

Nickel-Chromium Alloys

One alloy commonly used for thin film resistors contains 80-percent nickel and 20-percent chromium. It is available as Nichrome V® from the Driver-Harris Company of Harrison, New Jersey. In bulk form, it has a resistivity of 108 μ ohm cm and a "nominal" TCR of 110×10^{-6} per °C. Thus, the bulk material is superior to chromium by a factor of 8 in the resistivity, and a factor of 25 in the TCR. In thin film form, the resistivity is the same as in the bulk even for films as thin as 50 angstroms, provided the composition and density of the film are kept the same as that of the bulk [9].† This implies, incidentally, that the mean free path in bulk Nichrome® is less than 50 angstroms; otherwise, the Sondheimer scattering would increase the resistivity. When Nichrome® is evaporated (typically at 1450° to 1600°C) from the melt, the first film formed is rich in chromium

*Chromium has a Néel point at 39°C below which it is antiferromagnetic.
†Somewhat different results have recently been obtained by Swanson and Campbell [10].

because of the higher vapor pressure of chromium (see Chap. 3). This has the advantage of increasing the adherence, since chromium-rich films are strongly adherent. If the film is to be used as a resistor, however, it may be preferable to obtain a uniform composition by the special methods described in Chap. 3, or by sputtering.

One problem when Nichrome® is evaporated from tungsten wires is that the nickel in the original charge can alloy with the tungsten. This further decreases the nickel content of the film and shortens the life of the filament.

Nichrome® films are not stable in high humidity, especially under a d.c. load, because of electrolysis. For this reason, discrete Nichrome® film resistors are usually hermetically sealed, and resistors that are part of a circuit on a substrate are coated with a thick layer of silicon monoxide. Under these conditions, the resistors have good stability, low moisture sensitivity, low voltage coefficients, and low noise coefficients. Their high-frequency performance depends primarily on the resistor pattern and is typical of the fine performance to be expected from thin film geometry.

Ni-Cr alloy films "not deposited by conventional techniques" are advertised [11] with temperature coefficients of resistance of $0 \pm 1 \times 10^{-6}$ per °C at room temperature. Such resistors are made by causing the maximum of the resistance-versus-temperature curve to coincide with room temperature by adjusting the composition (additives such as copper and aluminum are used for this purpose). Thus, even though the TCR may be exactly zero at room temperature, the curvature of the resistance-versus-temperature curve may produce a resistance deviation of -0.05 percent at $-60°C$ and also at $+150°C$. This is still a very small temperature dependence, and these Nichrome resistors are widely used for applications requiring low temperature dependences. The TCR can be controlled by adjusting the thickness of a coating on the back of the substrate, thus controlling the thermal expansion of the device which, in turn, affects the TCR (see Chap. 6). The usual technique is to make the films thick enough that the bulk resistivity and TCR are assured, thus requiring more film and more space for a given resistor. A typical thickness is 10,000 angstroms, which gives a sheet resistance of the order of 1 ohm per square. Since resistors made of Ni-Cr alloys cannot be anodically trimmed to value, they are generally trimmed mechanically or by etching away part of the pattern. The initial tolerances of these resistors are variously reported as ± 0.05 to ± 0.5 percent. They are available from 10 to 100,000 ohms.

Tin Dioxide

Tin dioxide films have been used both for discrete resistors [3] and, by at least one company, for integrated thin film circuits. Pure tin dioxide is an insulator, but can be made to behave like a semiconductor by making it deficient in oxygen. It is deposited by spraying tin tetrachloride (dissolved in water) onto substrates which are held at 500° to 800°C. The chloride is hydrolized to produce tin dioxide and hydrochloric acid. If this is done in the presence of a reducing atmosphere (which may be produced by organics in the solution), the oxide formed will be oxygen-deficient, making it an n-type semiconductor with a resistivity of the order of 2000 μ ohm cm and a temperature coefficient of resistance of the order of -500×10^{-6} per °C. These films can be doped by donors or acceptors. Acceptors raise the resistivity and make the TCR even more negative; the addition of donors causes the resistivity first to decrease and then increase after displaying a minimum at about 1000 μ ohm cm. Donors cause the TCR to go through zero and become positive, and then become negative again. The maximum positive TCR is about 500×10^{-6} per °C. Antimony (a donor) is the most common dopant, because it seems to improve the stability of the films and because it permits low TCR's. Zero TCR occurs where the resistivity is about 1400 μ ohm cm. Typical resistors made from tin dioxide have a TCR of $0 \pm 250 \times 10^{-6}$ per °C. The noise level is not as good as that of typical thin metal films, but is comparable to that of carbon film resistors. Variations in the deposition conditions can cause initial tolerances to be as high as ± 5 to ± 20 percent, but they can be abrasion-trimmed to value to within about ± 1 percent. These films are quite stable, especially after hermetic encapsulation, and can be made for precision- as well as power-resistor applications.

Chromium Silicon Monoxide

Chromium silicon monoxide belongs to that class of materials called cermets (from CERamics plus METals) and is deposited by flash evaporation of a powder mixture of chromium and silicon monoxide. The substrate is normally heated to temperatures of 160°C or greater. The resultant film is composed of metal grains embedded in a silicon monoxide matrix. These grains may be as small as 50 angstroms and need not necessarily touch in order for electronic conduction to be possible. The effective resistivity of this structure can be varied from 30 to 10^{-3} ohm cm by varying the atomic percent of chromium from 60 to 90 percent [1]. The size of the metal grains is controlled by adjusting the substrate temperature, which also affects the resistivity. The main advantages of the cermet films are their stability at high temperatures and their mechanical strength. Normal processing of these resistors includes covering them with about 10,000 angstroms of evaporated silicon monoxide. The completed resistor is then trimmed to value by heating in air or a reducing atmosphere. This process is capable of trimming the resistance values to ±2 percent. The final film may have a sheet resistance of between 30 and 1000 ohms per square and a resistivity of the order of 1000 μ ohm cm. The TCR of these films is of the order of -50×10^{-6} per °C. Their noise, voltage coefficient, and high-frequency behavior are reported to be comparable to those of Nichrome® resistors. More recent work on chromium silicon monoxide is described by K. E. G. Pitt [12].

Palladium-Silver Glaze

Palladium-silver is fired onto a substrate after having been applied as a mixture of metal and glass powders dispersed in an organic liquid [13]. During firing at about 800°C, the organic material burns off, the glass melts, and the metals alloy in a variety of compositions, along with some palladium oxide. The original alloy has about 40 percent (by weight) of silver because that particular alloy has the smallest TCR of all the palladium-silver alloys. The sheet resistance can be varied from 10 to 10^4 ohms per square by varying the metal-to-glass ratio of the glaze from 60:40 to 30:70.

The films are initially applied by silk-screen techniques onto the ceramic substrates to a thickness of as much as 0.005 cm. Thus, the effective resistivity of the glaze is of the order of 1 ohm cm. The controlling factor in the conduction is the resistivity of palladium oxide, which is formed with an excess of oxygen and acts like a p-type semiconductor. The TCR of the glaze is in the range of from -200 to -500×10^{-6} per °C. The sheet resistance of the glaze has a reproducibility of about ±20 percent, but the resistors can be trimmed to value by an abrasive air stream. The composition of the glass frit used in the original dispersion is critical in this type of resistor. The thermal coefficient of expansion of the glass should match that of the substrate to minimize strains which would otherwise develop upon cooling. When properly applied, the glaze is very adherent and strong. Because it is extremely resistant to wiper contact abrasion, it is well suited for use as resistor elements for potentiometers.

7.3 TANTALUM-BASED THIN FILM RESISTORS

Because of its high melting point, tantalum is classed as a refractory metal, along with tungsten, niobium, molybdenum, and rhenium. All of these metals have been considered for use in thin film resistors. Their primary advantage is their stability. High melting points are associated with high recrystallization temperatures, so these metals do not normally anneal at temperatures below a few hundred degrees centigrade. On surfaces exposed to air, they grow an oxide layer which is adherent and protective, so at moderately elevated temperatures they are chemically stable in air. Another advantage of some of the refractory metals is their ability to form an adherent oxide layer which can be used as a dielectric, thus permitting resistors and capacitors to be made from the same thin film material.

Of the refractory metals, by far the most important for thin film production is tantalum [14]. The refractory nature of tantalum makes it difficult to evaporate in the usual fashion from a boat or filament. Tantalum may be evaporated by electron-beam techniques and, if this is done carefully, the properties of the resultant film can be quite similar to those of bulk material. Resistivities as low as 20 μ ohm cm have been reported, as compared to 14 μ ohm cm for bulk tantalum. Figure 7-4 illustrates that the evaporated films have the same slope $(d\rho/dT)$ as bulk tantalum and that they have nearly the same superconducting transition temperature (4°K). Thus, these films seem to behave much like the bulk material (at least down to 1000A thickness), except for an additional residual resistivity which obeys Matthiessen's rule. The additional resistivity is generally an advantage for thin film resistors because it means greater sheet resistance for a given thickness, and this permits smaller components. An associated advantage is that the temperature coefficient of resistance of the film is somewhat smaller than that of the bulk.

Sputtering techniques for producing thin films of tantalum have been extensively developed in the last decade. The sputtered films have been shown to have many desirable characteristics, including higher resistivities, smaller temperature coefficients of resistivity, better stability, and process control by reactive sputtering. Figure 7-4 shows how the resistivity of a sputtered (bcc) tantalum film depends on temperature. Even though the room temperature resistivity has been increased by a factor of 3, Matthiessen's rule still holds and the

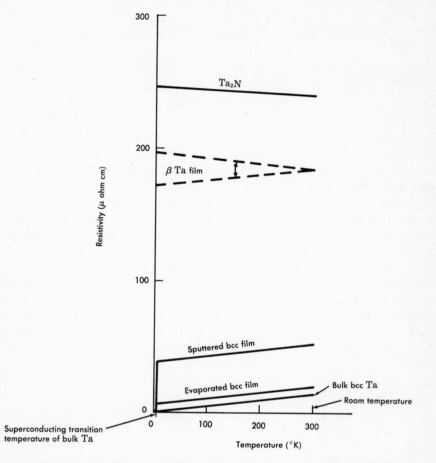

FIG. 7-4. Resistivity of several kinds of tantalum as a function of temperature.

film still becomes superconducting (although at a slightly lower temperature). Increasing the resistivity by a factor of 3 decreases the temperature coefficient of resistivity by a factor of 3 for this film.

As pointed out in Sec. 4.5, under most conditions, undoped sputtered tantalum film does not have the bcc structure; it is tetragonal (beta phase), has a much higher resistivity, and a lower temperature coefficient of resistivity. Figure 7-4 also shows the behavior of beta-tantalum films. These films have a resistivity of about 180 μ ohm cm and a temperature coefficient of resistance that ranges between $+100$ and -100×10^{-6} per °C. Thus, beta tantalum is better yet as a resistor material.

It appears entirely coincidental that beta tantalum has about the same resistivity and temperature coefficient of resistance as does Ta$_2$N, which is formed by adding a small percentage of nitrogen gas to the argon sputtering gas. One difference between the two is illustrated in Fig. 4-18. The nitrided film is much more stable upon exposure to air, and it is because of this excellent stability that the composition Ta$_2$N is used for resistors.

There are some applications where a certain temperature coefficient of resistance is specified and, in these cases, other compositions are considered. In particular, it may become necessary to match the temperature coefficient of capacitance of a thin film capacitor, which for tantalum oxide capacitors is about $+200 \times 10^{-6}$ per °C. In a resistance-capacitance circuit, if the temperature coefficient of capacitance is the negative of the temperature coefficient of resistance, the temperature coefficient of any characteristic frequency or time constant of the circuit is zero. Thus, it is desirable to be able to produce films with a TCR of -200×10^{-6} per °C. From Fig. 4-17, this could be done by nitrogen doping, with a resistivity of about 300 μ ohm cm and, from Fig. 4-18, this would only slightly deteriorate the stability. Unfortunately, however, films which are this heavily nitrided are also very difficult to anodize, so they cannot be easily trimmed to value. Oxygen doping is an alternate approach to obtaining a TCR of -200×10^{-6} per °C. The TCR-versus-partial-pressure curve for oxygen is quite similar to that for nitrogen, and the resistivity curve (Fig. 4-19) is also similar. The difference is that oxygen containing films are more easily anodize-trimmed to value, so they may be utilized at higher resistivities with more negative TCR's.

Another feature of practical importance is the ability to obtain a near-zero temperature coefficient of resistance. Low-density tantalum films produced by sputtering at low voltages (see Chap. 4) show high resistivities and near-zero temperature coefficients of resistance simultaneously. From Fig. 4-21, films sputtered at 2300 volts have a resistivity of about 1100 μ ohm cm and a near-zero TCR. These films are not as advantageous as one might expect, however, since they show rapid resistance shifts on aging. This rapid increase in resistance is the result of the porous nature of the films (Fig. 4-22), which allows for rapid oxidation. This resistance shift may be minimized by preoxidizing the films in an oven (which itself will increase the resistance several times), but this shifts the TCR to negative values of around -300×10^{-6} per °C. The final result is that the electrical properties of these films are similar to oxygen-containing films, except that high-resistivity films are more easily controlled with low-density tantalum.

Adjustment of Resistor Value

In order to take advantage of the excellent stability of tantalum thin film resistors, it is important to be able to initially adjust them to a precise value. This can be accomplished either by electrochemical techniques (anodizing), by thermal oxidation, or by mechanical techniques such as abrading. All three methods have been successfully used, but they all suffer from the primary disadvantage that they (normally) can only increase and not decrease the resistance. For this reason, sheet resistances are initially produced below the desired value so that all of the resistors can be trimmed to value, including those with an initial sheet resistance somewhat above the average for a given batch. This is not a major problem in manufacture, with the only difficulty being that these methods of adjusting to value do not allow for any "overshoot." If a film is inadvertently anodized too far, it must be discarded. Thus, the trimming process is one for which automation is helpful. If manually operated, there should be some electrical interlock that senses the resistor value and prevents the operator from overshooting the desired resistance.

Anodization Trimming. Tantalum resistors can be trimmed with extreme precision (0.01 percent in some cases) by anodization. Section 5.4 describes the process of anodization for the formation of capacitors, and there is little difference in the basic technique when forming anodic films for resistor trimming. Of course, the resistance of the film must be monitored either simultaneously or alternately with the anodization. Alternating current is sometimes used to monitor the resistance while the film is being anodized by direct current. Much higher current densities are permissible when resistor trimming than when forming a capacitor dielectric because the requirements on the capacitor dielectric are much more severe; current densities as high as 0.1 amp/cm^2 have been successfully used. With such a current density, the voltage rises at a rate of about 40 volts/sec. Care must be taken to insure that metals that do not form anodic oxides (such as gold or Nichrome®) are not exposed to the electrolyte, since such exposure will effectively short out the anodizing current. In circuit production, all termination areas electrically connected to the resistor being trimmed must be protected from the electrolyte. This is sometimes accomplished by making dams of a maskant around the area to be anodized. A few drops of electrolyte are then put inside the dam for the anodization. For mass trimming of circuits, it has also been found possible to coat, with an insulating resist (e.g., grease), all those areas not to be anodized, and then to immerse the entire slide in the solution. It is also possible to use a viscous electrolyte that can be applied selectively to the anodization areas, using silk screens or metal masks. Felt wicks and solid (gel) electrolytes have also been used for restricting the contact area of the electrolyte.

Anodization trimming follows Faraday's law, which states that the remaining metal thickness, d_m, will decrease at a rate proportional to the current density, j:

$$\frac{dd_m}{dt} = -\frac{j}{K_m} \tag{7-2}$$

and

$$\Delta d_m = -\frac{Q}{AK_m} \tag{7-3}$$

where Q is the total charge passed, A is the area* being anodized, and K_m is the number of coulombs required to oxidize 1 cubic centimeter of the metal.

$$K_m = \frac{eN_A D_m Z_m}{M_m}$$

where e is the charge on an electron, N_A is Avogadro's number, D_m is the density of the metal, Z_m is the number of electrons that pass for each atom of the metal oxidized (the valence), and M_m is the atomic weight of the metal. For undoped tantalum, $K_m = 44,200$ coulombs/cm³, and for tantalum nitride resistor material, $K_m \approx 40,000$ coulombs/cm³.

Similar equations are valid for d_o, the oxide thickness:

$$\frac{dd_o}{dt} = \frac{j}{K_o} \tag{7-4}$$

and

$$d_o = \frac{Q}{AK_o} \tag{7-5}$$

where

$$K_o = \frac{eN_A D_o Z_o}{M_o}$$

*Note that $A \approx w^2(R/R_s)$ where R is the desired resistance value, R_s is the desired sheet resistance, and w is the line width.

and D_o is the oxide density, Z_o is the number of electrons that pass for each molecule of the oxide formed, and M_o is the molecular weight of the oxide. The number of coulombs, K_o, required to form 1 cubic centimeter of the oxide is 17,500 coulombs/cm^3 for Ta$_2$O$_5$, and $K_o \approx 11,000$ coulombs/cm^3 for oxide formed from tantalum-nitride resistor material. The rate of voltage rise is given by

$$\frac{dV}{dt} = E_d \left(\frac{dd_o}{dt}\right) = \frac{E_d\, j}{K_o} \tag{7-6}$$

and

$$V = \frac{E_d Q}{K_o A} \tag{7-7}$$

where E_d is the differential field strength, which has been assumed constant in order to integrate. For tantalum and tantalum nitride, E_d is about 60 mv/A (E_d is a slowly varying function of j, as discussed in Chap. 5). If the metal thickness is uniform, the rate of resistance rise is

$$\frac{dR}{dt} = -\left(\frac{dd_m}{dt}\right)\left(\frac{R}{d_m}\right) = \frac{jR}{K_m d_m} = \left(\frac{1}{E_d}\right)\left(\frac{dV}{dt}\right)\left(\frac{K_o}{K_m}\right)\left(\frac{R}{d_m}\right) \tag{7-8}$$

The total voltage required to raise the resistance from R_1 to R_2 is given by

$$V = \left(\frac{E_d \rho N_s K_m}{R_2 K_o}\right)\left(\frac{R_2 - R_1}{R_1}\right) \tag{7-9}$$

where ρ is the resistivity and N_s is the number of squares of the resistor. If the metal film being anodized has nonuniform thickness, this relation does not hold because it has been assumed that the resistance is inversely proportional to the thickness. The invalidity is especially noticeable if a scratch has produced a thin place in the film. In fact, this is one way to detect irregularities of this sort.

Two points of practical importance should be mentioned about anodization trimming. The first is that the (interference) color of the film changes as the anodization proceeds; thus, observation will show whether the oxide layer is being uniformly formed. If the resistor is to have a very high resistance value, it may anodize to a thicker oxide layer near the termination connected to the anodizing power supply, especially if the current density is high. Thus, lower current densities and longer times may be required to trim to a uniform thickness on a large-value resistor. Non-uniformity of the oxide thickness may be detected by visual inspection.

The second point of practical importance is that the terminations must be masked, which means that in laying out the circuit, the terminations must be kept a certain distance away from resistor areas. It is difficult to silk-screen grease for anodization barriers to tolerances less than 0.015 inch. Also, each resistor track should be kept at least 0.015 inch away from every other resistor. If possible, it is desirable to have enough space between two resistors that a dam wall can be erected, so that both resistors can be trimmed with only one application of the grease. All the resistors on a substrate could be trimmed together while monitoring only one of them, but this method is limited in precision by the uniformity of the sheet resistance of the original material, which has been about ±3 percent in production of reasonable-size (0.75- by 1.8-inch) substrates.

Thermal Trimming. The sheet resistance of a tantalum or tantalum nitride film can be increased by heating in air at temperatures of several hundred degrees centigrade. This type of treatment has been used for stabilization of the resistance values, but it can also be used for trimming. The precision achieved is not as good as with anodization, because of errors involved in measuring resistance inside an oven and also because the temperature coefficient of resistance causes the resistance to be appreciably different from that at room temperature. Even with a TCR of -100×10^{-6} per °C, the resistance will change by four percent in a 400°C oven and, of course, there will be some variation of TCR values in any set of resistors. Also, heating is a process that is difficult to terminate instantaneously because a rapid quench would put severe strains in the substrate. Thus, with this method of trimming, it is not normally possible to trim to better than a few tenths of a percent. For resistors where this is sufficient, however, this process can be used to good advantage.

Mechanical Trimming. It is, of course, possible to change the resistance of a thin film resistor by mechanical methods. The objective in this type of trimming is to remove part of the film without causing any weak spots where local heating might occur. Any resistor could conceivably be trimmed to any higher value by making a scratch part of the distance across it. This is unsatisfactory, however, both because of the lack of precision and the loss in stability, especially under high current densities. The ideal mechanical trim technique would be to reduce the line width of the resistor uniformly along its length, and the Western Electric Company [15] has devised a design (Fig. 7-5) for discrete thin film resistors that effectively makes this possible. Instead of the film being deposited on a flat substrate, a ceramic substrate with grooves embossed in it is used; these grooves are more-or-less v-shaped in cross-section and run along the length of the resistor track. The resistor is trimmed by grinding layers off the top of the substrate, thereby effectively reducing the width of the resistor track until the resistance is trimmed to value. This type of trimming can be used to make inexpensive discrete resistors that are trimmed to better than ±5 percent of value.

Various other patterns have been developed to make use of mechanical trimming. These patterns usually employ parallel paths for part of the resistance so that one of the paths can be opened by scribing a line, and the other paths can be used to keep the current density from becoming too high. This is generally a quantized trimming technique, but it is helpful in some applications, particularly if the pattern can be laid out with many such paths in a digital fashion so that a wide range of resistance change can be achieved. With a scribing technique, there is always the possibility of damaging the substrate, especially if it is glass. Conduction paths may be severed by using an abrasive-containing air stream. The resistance may also be adjusted by abrading the entire surface of the resistor with, for example, a rubber ink-eraser. This is not recommended, however, because of the lack of control.

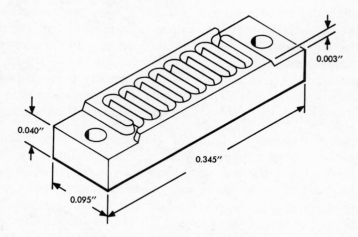

FIG. 7-5. Embossed ceramic substrate.

Aging and Stabilization

One of the most attractive features of tantalum film resistors is their excellent stability. The impervious oxide film formed by anodization or heat treatment inhibits chemical changes, and the high recrystallization temperature inhibits mechanical annealing that might change the resistance. A typical overload aging curve for a tantalum-nitride resistor on Corning Code 7059 glass is shown in Fig. 7-6. It is apparent that the film gains stability as time passes. Thus, an accelerated aging brought about by heating in air can make the resistors much more stable.

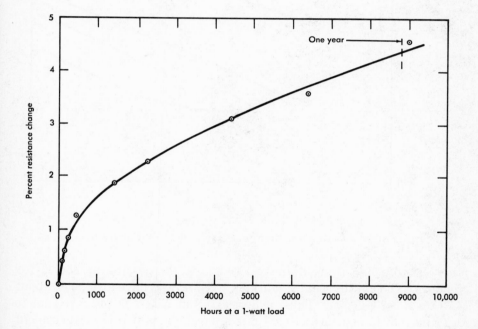

Fig. 7-6. Aging curve of a typical tantalum-nitride resistor at four times the rated power.

Oven Aging. The aging of these resistors is so slow that in order to study the relative stability of different processing sequences, their aging must be accelerated to obtain meaningful data in a reasonable time. This can be done by oven aging or by overload power dissipation in the resistor. Figure 7-7 shows some room-temperature data taken from resistors aged at several different temperatures, with each curve corresponding to the average of 20 different resistors. All of

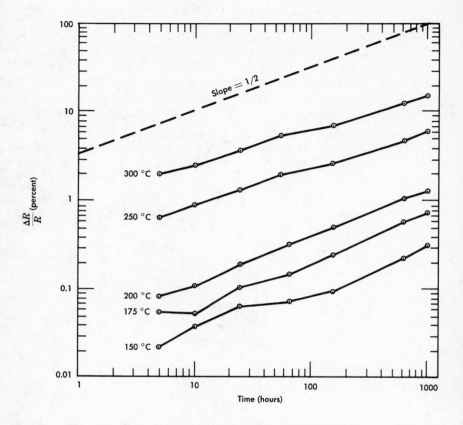

FIG. 7-7. Oven-aging data for "standard process" resistors.

the resistors had a preliminary "standard" processing at 250°C for 5 hours. The curves seem essentially parallel and have a slope of about 1/2 on a log-log plot. Thus, it appears that $\Delta R/R$ is proportional to $\sqrt{t}$ and that the curve of Fig. 7-6 is a parabola, which is in agreement with the parabolic oxidation law observed by Miller [16]. The aging is strongly temperature dependent. Aging mechanisms that are thermally activated generally produce aging that is proportional to $\exp -(E/kT)$, where E is a thermal activation energy and k is Boltzmann's constant. Thus, if log $(\Delta R/R)$ is plotted as a function of $1/T$ for a certain time of aging, the result is a straight line with a slope corresponding to an energy of 0.6 ev, or 14 kilocalories per mole, which again is in fair agreement with the oxidation energy of tantalum. Figure 7-8 shows such a plot obtained from the data in Fig. 7-7. All of these data are typical for tantalum-nitride films of the

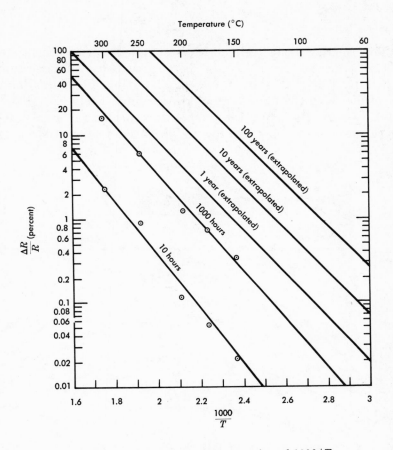

FIG. 7-8. Oven-aging data as a function of $1000/T$.

type currently being used for resistors. The time and temperature dependence can be combined to give

$$\frac{\Delta R}{R} = \sqrt{\frac{t}{t_0}} \exp\left(\frac{-T_0}{T}\right)$$

where $T_0 = 6,600°\mathrm{K}$ and $t_0 = 15$ msec for standard tantalum-nitride resistors. This expression is valid for films that have had the "standard process" of 5 hours at 250°C. For a given style of resistor, the constants of T_0 and t_0 are independent of temperature, at least from 150°C to 300°C, and are independent of time for aging periods of 10 to 10,000 hours. This relation can be extremely useful in evaluating the effects of aging and the benefits of preaging, which are discussed below.

Direct Current Load Aging. Resistor aging can be accelerated by applying a d.c. voltage to obtain the desired testing temperature by I^2R heating. Unfortunately, this type of testing introduces temperature gradients in the film, especially under high loads with glass substrates. Figure 7-9 shows the temperature profile for a certain testpattern resistor as observed by an infrared temperature scanner,

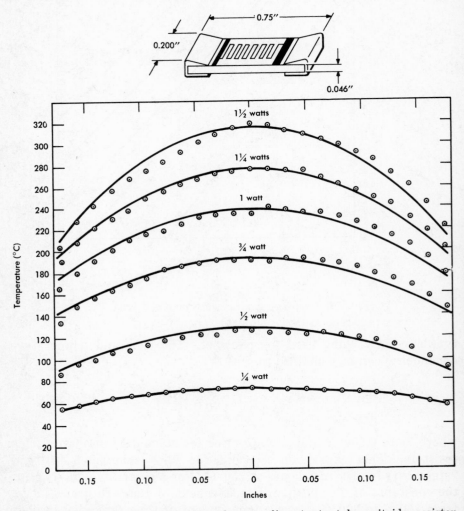

FIG. 7-9. Temperature distribution along a discrete tantalum-nitride resistor. The curves are symmetrical parabolas.

where the resistor was operated in free air but with a minimum of draft. Also shown in Fig. 7-9 are symmetrical parabolas that have been fitted to the data. In the approximation that the heat is produced uniformly over the surface of the resistor and conducted away to the leads with only longitudinal temperature gradients, the temperature profile is expected to be a parabola. Actually, these conditions do not exist because an appreciable fraction of the heat is transferred by radiation and convection in such an arrangement.

Figure 7-10 shows some long-term data obtained on tantalumnitride resistors on (0.2 by 0.75 inch) glass substrates under various

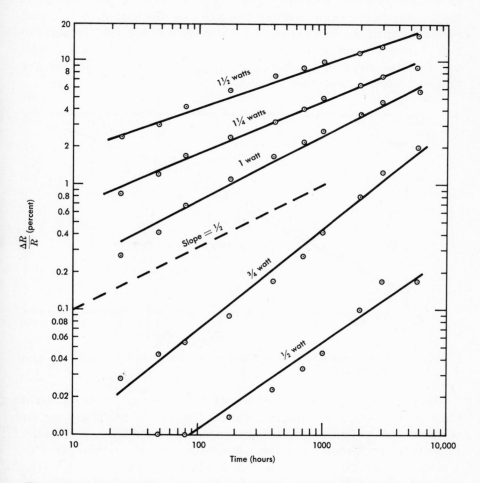

FIG. 7-10. Aging curves of tantalum-nitride resistors under d.c. overload conditions.

amounts of d.c. overloading. The $\sqrt{t}$ dependence is indicated at all power levels by the slope of 1/2, just as in the oven-aging curves. Thus, it is possible to assign to each power level an effective temperature by finding the time required to obtain a given percent change in the resistance. A comparison of Fig. 7-7 with Fig. 7-10 at a value of two percent for $\Delta R/R$ yields an effective temperature for four power levels; these data are plotted in Fig. 7-11 as solid circles. Also plotted in Fig. 7-11 is the average temperature (and the hot-spot temperature) at each power level as measured by the thermal plotter (see Fig. 7-9). The average temperature corresponds closely to the effective temperature during aging, again showing that the aging mechanism is the same whether the samples are heated by d.c. loading or by oven aging. Figure 7-11 also shows that the temperature rise is proportional to the applied power, at least up to about 1 watt, and that the slope of the line is about 200°C per watt for the particular pattern under test.

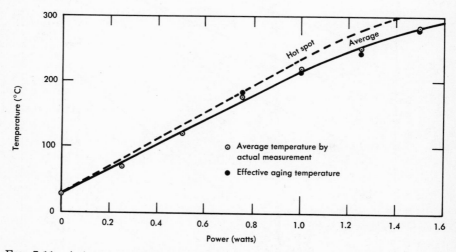

FIG. 7-11. Aging temperature as a function of power (0.2 by 0.75 inch glass substrates).

The previous data (Figs. 7-6 through 7-11) are for tantalum-nitride films of the normal composition (about Ta_2N). If the nitrogen level is decreased or increased, deviations occur from the $\sqrt{t}$ behavior. Especially troublesome are the low-nitrogen films, as they often exhibit a "negative dip." This is depicted in Fig. 7-12, which is a semi-log plot of $\Delta R/R$ versus time for typical films of high-nitrogen, normal-nitrogen, and low-nitrogen content. The negative dip may

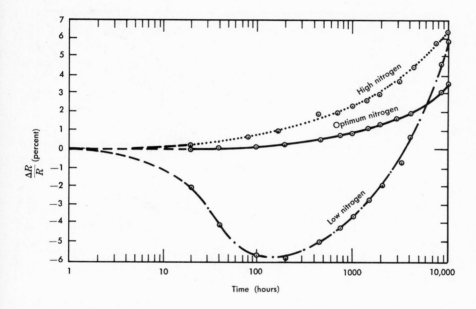

FIG. 7-12. Accelerated aging curves with different amounts of nitrogen doping.

be associated with some beta tantalum being transformed to the more stable, lower-resistivity bcc phase (see Fig. 4-17). Because of this negative dip, data obtained at the end of one particular test period, or for only one power level, cannot be trusted. For example, at 3500 hours, the low-nitrogen film looks very stable but actually has already made a large excursion from its initial resistance. The amount of negative dip also depends on the power dissipation; a film that shows a fairly flat curve under one power level may show a pronounced negative dip under a lower power level. This could cause considerable trouble if the resistor is tested at a higher level and operated at a lower one. Thus, it is best to avoid this low-nitrogen area altogether.

The temperature gradients produced by d.c. loading cause non-uniform aging. Figure 7-13 is a graph of the irreversible change in resistance of each "loop" of a meander pattern resistor (on Corning Code 7059 glass) due to aging at 1.25 watts for 10,600 hours (1.2 years) in air. The effect of the temperature profile is evident. During testing, the center loop temperature was estimated to be about 265°C, whereas loop No. 1 was about 210°C. A measurement of the total resistance change of the sample yields a value of about 6-percent change, but gives no information about the distribution of aging along

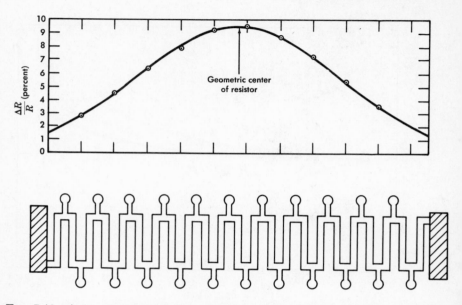

FIG. 7-13. Aging distribution for a tantalum-nitride resistor (0.2 by 0.75 inch)
after aging for 1.2 years at 1-1/4 watts.

the length of the sample. Figure 7-13 confirms that the aging pattern
is symmetrical about the center in spite of the fact that d.c. voltage is
applied. This may not be the case, however, for glasses containing
mobile alkali ions.

If the center section of the resistor ages more than a few percent,
it can change the temperature profile by changing the heat dissipation. If the resistor is connected to a constant-voltage power supply, the total current drawn will decrease; however, the power dissipation per unit area of surface will decrease faster at the edges than
in the middle of the resistor because the sheet resistance has become
higher in the center. Thus, although the hot-spot temperature will
probably (though not necessarily) decrease in time, the temperature
profile will become more sharply peaked in the center, causing the
aging to be even less uniform.

One important question that arises is whether a resistor section
heated to a certain temperature by I^2R heating will age the same as
if it were heated to the same temperature by being placed in an
oven; that is, does the flow of electrons through the film have any
effect on the aging other than to cause I^2R heating? An experiment
devised to answer this question consisted of fabricating three inter-
twined (but not connected) meander resistors (Fig. 7-14). The

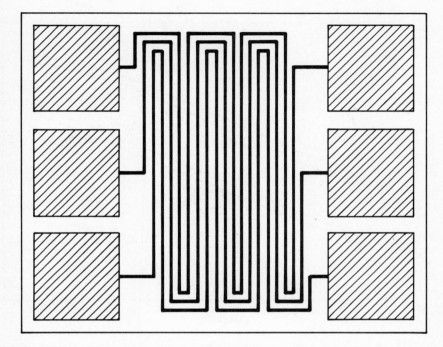

FIG. 7-14. Resistor pattern used to show equivalence of d.c. load aging and oven
aging.

two outer resistors were aged by applying power, while the middle
resistor was heated only by thermal conduction from the other two
resistors. Thus, the middle resistor had almost the same temperature
distribution along its length as the outer two did during the test,
but it had no current flowing through it. Figure 7-15 shows the results of the test for both glazed ceramic and Corning Code 7059 glass
substrates, with no thermal stabilization of the films. From the figure, it is seen that the loaded resistors consistently aged slightly more
than the unloaded ones. The resistors on the ceramic were heated to
an average* temperature of 144.2°C and a maximum temperature of
151°C by applying 3.8 watts. The resistors on glass were heated to
an average temperature of 129.4°C and a maximum temperature of
146°C by applying 1.8 watts. The small difference between the
loaded and unloaded resistors is completely attributable to the fact
that the loaded resistors were actually a few degrees hotter than the
no-load resistors.

*Averaged over the length of the resistor.

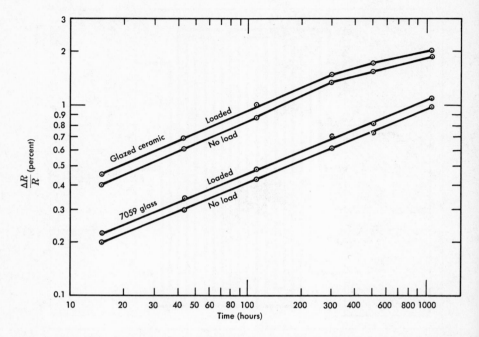

F<small>IG.</small> 7-15. Aging curves of resistors shown in Fig. 7-14. Only the outer two resistors had a d.c. load applied.

Thus, the aging due to I^2R heating seems equivalent to the aging due to external heating; therefore, in theory at least, one should be able to predict the aging characteristics of a circuit by measuring the operating temperature distribution.

In spite of the nonuniformity of aging in d.c.-load-tested resistors, the overall effect of d.c. loading can be compared with oven aging. Figure 7-16 shows accelerated-aging data for some resistors on Corning Code 7059 glass that were aged under 1-watt load and others aged at 290°C. The time dependence appears the same, but at 290°C the aging is accelerated by a factor of 320. Thus, in 27 hours at 290°C, the amount of resistor aging in 1 year at 1 watt can be determined. Note that this conversion factor is only valid for a particular pattern of resistor and for a particular substrate material (Corning Code 7059 glass).

Overload Test. When fine-line resistors are used, the problem of pinholes can be serious. Weak spots in a resistor may not appear until a hot spot develops when the resistor is being aged under load.

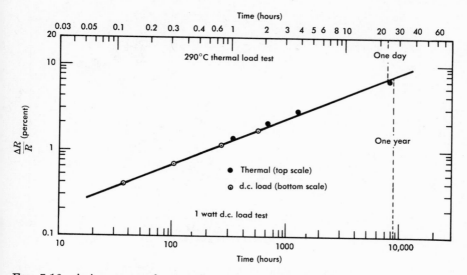

FIG. 7-16. Aging curves for tantalum nitride showing equivalence of 290°C and 1 watt with a time acceleration factor of about 320.

Many weak-spot resistors may be eliminated by giving them a strong pulse of current for a short time; this procedure is called the "overload test." A very thin spot in the resistor track becomes a hot spot causing a large change in the resistance value and perhaps even opening the resistor completely, like a fuse. A test pulse is established at a level somewhat below that which would appreciably change the resistance of a defect-free resistor, and any resistor which contains weak spots will be spotted and rejected. This process can save some visual inspection time that may otherwise be expended in critical circuits.

Preaging. The $\sqrt{t}$ dependence of the resistance shift during aging means that aged resistors are more stable than unaged resistors. Thus, accelerated aging can produce better stability and is a common procedure. A "standard" heat treatment consists of placing tantalum-nitride resistors in air at 250°C for 5 hours. This is normally applied after a preliminary anodization trim, and increases the resistance by 0.5 to 2 percent. The final anodization trim is performed after the thermal stabilization because it is much better controlled than the thermal stabilization.

More recent studies [17] have shown that higher preaging temperatures can dramatically increase the stability even above the

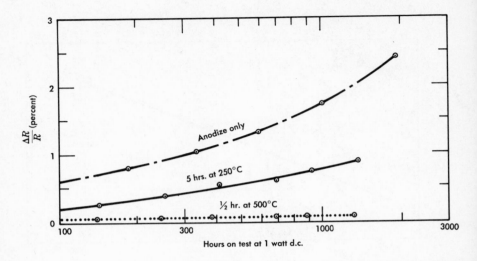

FIG. 7-17. Aging curves of tantalum nitride showing advantage of preaging.

5-hour-at-250°C aging process. Figure 7-17 shows aging curves for resistors prepared by anodization, anodization plus 5 hours at 250°C, and anodization plus 1/2 hour at 500°C. The stability has been improved by a factor of 10 by using the high-temperature aging. The amount of preaging can be measured by the change in resistance upon preaging. If the resistance is increased by a given amount, it seems of little importance whether this was accomplished by a high temperature for a short time, or a lower temperature for a long time. The 1-watt operation on 7059 glass for 1 year has been taken as a standard test condition and, from Fig. 7-16, 24 hours at 290°C is approximately equivalent. Thus, the change after 24 hours at 290°C can be plotted as a function of the change during the preaging process to see what resistance change is needed to obtain optimum stability. Figure 7-18 is an example of such a plot for a given initial film thickness (1440A). The stability improves rapidly with increased preaging up to a preaging level of 15 percent. As long as the resistor is increased by about 15 percent or more by the thermal process, it will have the optimum stability.

Preaging phenomena can be analyzed by using the formula that is valid for "standard process" resistors [18]:

$$\frac{\Delta R}{R} = \sqrt{\frac{t}{t_0}} \exp\left(\frac{-T_0}{T}\right) \qquad (7\text{-}10)$$

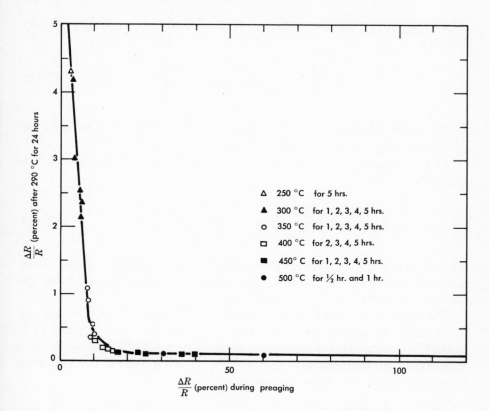

FIG. 7-18. Resistance change after a standard test (at 290°C for 24 hours) as a function of the amount of preaging as measured by the resistance change during the preaging (1440-angstrom films).

where t_0 is 15 milliseconds and T_0 is 6,600°K. The value for t_0 depends on the thickness of the sample, but T_0 is characteristic of the material and is related to the activation energy of the aging process. A resistor that is aged at a specific temperature for a given number of hours will change a certain percent in value. If the clock is then rezeroed after time t_1, a new aging relation is observed for the resistance change $\Delta R'/R'$ after that point, relative to the new time, t'. The origin translation gives

$$\frac{\Delta R'}{R'} = \sqrt{\frac{t'}{t'_0}}\left[\exp\left(\frac{-T_0}{T}\right)\right]\left(\sqrt{1 + \frac{t_1}{t'}} - \sqrt{\frac{t_1}{t'}}\right) \quad (7\text{-}11)$$

where $t'_0 = t_0\,[1 + (\Delta R/R)_{t_1}]^2$. The benefit of preaging is determined primarily by the ratio of t_1/t'. If $t_1/t' \gg 1$, Eq. (7-11) becomes

$$\frac{\Delta R'}{R'} = \frac{t'}{2\sqrt{t'_0 t_1}} \exp\left(\frac{-T_0}{T}\right) \qquad (7\text{-}12)$$

In this limit (short term), the preaging has reduced $\Delta R/R$ by a factor of $2\sqrt{t_1 t'_0/t' t_0}$, which can be a considerable help.

After considerable time, however, when $t_1/t' \ll 1$, Eq. (7-11) becomes

$$\frac{\Delta R'}{R'} = \sqrt{\frac{t'}{t'_0}} \exp\left(\frac{-T_0}{T}\right) \qquad (7\text{-}13)$$

This is the same functional dependence as Eq. (7-10). Thus, there is no benefit from the preaging for times (t') much longer than the effective preaging time (t_1), provided $t_0 \approx t'_0$.

This does not mean that a resistor must be aged for a time period comparable to its expected life, because it can always be aged at a higher temperature for a shorter time. From Eq. (7-10), it follows that aging at temperature T_2 for time t_2 is equivalent to aging at temperature T_1 for time t_1, where

$$\ln t_1 = \ln t_2 + 2T_0\left(\frac{T_2 - T_1}{T_1 T_2}\right) \qquad (7\text{-}14)$$

Figure 7-19 shows some generalized curves for this process; $\Delta R/R$ is normalized by multiplying by $\exp\left(T_0/T\right)$, and t is normalized by dividing by t_0. Then all unaged resistors will follow the straight line marked "standard process." The scale at the top is for $t_0 = 15$ msec, and the scales at the right are for the particular testing temperatures of 200° and 300°C. If the resistor is aged at a temperature for t_1 hours and then a new start is made using a new zero-time and a new zero-resistance, the curve traced out will be one of the family of curves shown, depending on the value of t_1/t_0. If, for example, the first test is stopped after $t_1/t_0 = 10^7$, the second test will follow the curve marked $10^7 \ t_0$. This preaging has made the resistor more stable, especially for short-term testing where it obeys Eq. (7-11). For long-term testing, the curve will asymptotically approach the "standard process" curve in accordance with Eq. (7-13).

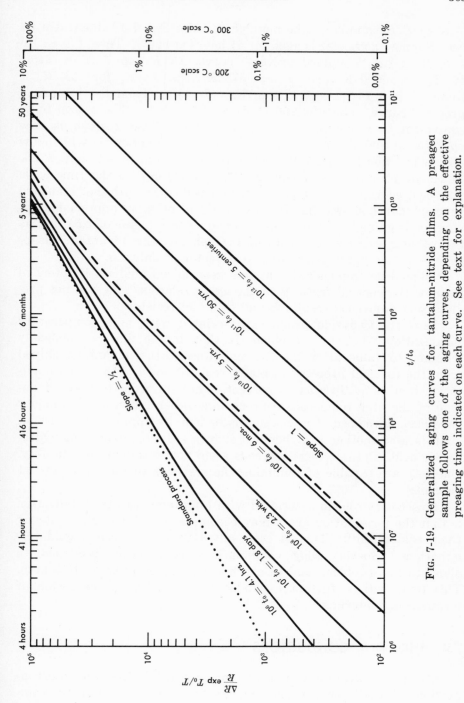

FIG. 7-19. Generalized aging curves for tantalum-nitride films. A preaged sample follows one of the aging curves, depending on the effective preaging time indicated on each curve. See text for explanation.

A specific example of the method of using Fig. 7-19 illustrates its use. Suppose a resistor is aged for 41 hours at 300°C. Then, $t/t_0 = 10^7$ and, from the "standard process" curve, $\Delta R/R$ exp $T_0/T = 3160$. If $T_0 = 6600°K$ and $T = 573°K$, the value for $\Delta R/R$ is found to be 3.14 percent. From Eq. (7-14) or from the "standard process" curve, using the 200°C scale of Fig. 7-19, this is seen to be equivalent to 8 months at 200°C (473°K). If the resistor is now operated at 200°C for the rest of its life, the aging curve will follow the dashed line in Fig. 7-19, where the horizontal axis is now t'/t'_0. Since $\Delta R/R$ is only 3.14 percent, $t'_0 = t_0$ (1.06) so the time scale may be used as given. Comparing the dashed curve with the "standard process" curve shows that for the first 8 months, an appreciable advantage is gained (at 8 months, it amounts to a factor of 2.5) but, as the resistor experiences years of further use, the advantage disappears. At five years, it amounts to a factor of only 1.5.

It should be emphasized that this generalized analysis is intended to show the general features of the aging relations and will not predict exact numerical results for all thin film resistors.

There are two side problems associated with high temperature stabilization. The first problem is that if the film is anodically trimmed to value after the thermal process, the advantage gained from the thermal process is greatly reduced, even if the final anodization trim is as little as 10 percent. Thus, in order to take advantage of the high temperature processing, other trimming techniques have been developed. One method is to trim by application of a heat lamp in air; another is a chemical etch in a dilute solution of hydrofluoric acid. Both of these methods seem not to impair the stability, and both are capable of trimming the resistors to tolerances of ±1 percent.

The second problem associated with high temperature stabilization is that the termination areas may become oxidized enough to impair their solderability. This is especially true of Nichrome®-gold terminations. Special alloy combinations have been developed to alleviate this problem, and research is still in progress in this area. This topic will be further expanded in the following discussion of termination materials.

7.4   THIN FILM CONDUCTORS

One of the advantages of thin film circuitry is that the interconnections between the various components are integral with the components, as opposed to being made individually later in the manufacture. This makes possible increased reliability and decreased cost. The interconnections then become part of the thin film technology and must be considered as part of the device, as must the areas where connections are made to other circuitry off the substrate. There are really two separate problems: interconnections among the thin film components on the substrate, and termination areas where wires or external components are to be joined to the substrate. The same thin film materials are usually used for both purposes, which results in a large saving of time and expense. Actually, the requirements of a termination area are such that any good termination area material would also serve as an adequate interconnection material. In addition to the factors of economy, the important properties of these films are adherence, sheet resistance, joinability, and the ability to withstand the heat treatment required to stabilize the resistors. The films should also be easily etchible in etchants that will not attack the resistor films of tantalum nitride. Unfortunately, no one material will adequately fulfill all the requirements. Therefore, various layered structures have been developed, with each layer satisfying one or more of the requirements.

Adherence

Most metals which have a low resistivity also have a low adherence to glass substrates, but will adhere if an adherent layer of Nichrome® is deposited first. The deposition of Nichrome® as a resistor material was discussed earlier in this chapter, and some of the problems associated with its evaporation were mentioned. One point mentioned is that when standard evaporation techniques are used, Nichrome® fractionates and produces an initial deposit that is chromium-rich. The chromium-rich film has a better adherence so, for this purpose, it is an advantage. Nichrome®, however, has a high resistivity, poor solderability, and can easily become oxidized. Rosin flux usually is not sufficient to remove the oxide from Nichrome®, so it is not a good material to use by itself for a thin film termination. The maximum practical thickness of Nichrome® that can be easily evaporated in one charge from a filament coil gives a sheet resistance of about 1.0 ohm per square, which is not low enough to be negligible compared to the resistor material (perhaps 50 ohms per square). Nichrome® works well as an adherent underlayer but it has a disadvantage when soldering is used for lead attachment. At high temperatures and high current densities, whiskers and other growths may be formed from the solder, with the Nichrome® layer being instrumental in their formation [19]. This phenomenon is not observed when titanium is used as the adherent layer, and thus titanium is now preferred to Nichrome® for this application.

Sheet Resistance

The low sheet resistance required can be obtained by an evaporation of copper or gold over the adherent layer, and either of these can bring the sheet resistance down to less than 0.1 ohm per square. Gold has the advantage over copper in that it does not corrode, but it has a higher resistivity and presents a problem in joining with soft solder because it alloys with the solder and makes it brittle, causing a mechanically poor connection. If the amount of gold is kept below that which would exceed the solubility limit of the solder, this embrittlement is minimized, but then the sheet resistance is not as low as may be desired. Thus, copper is normally used to provide low sheet resistance in combination with another layer to provide corrosion resistance.

Joinability

Copper is not an adequate conductor material because of oxidation, so it is desirable to add a protective layer of gold or palladium. A gold top layer has the disadvantage that it embrittles solder, whereas palladium apparently does not exhibit this effect. The combination of copper and palladium can be made with sheet resistances as low as 0.03 ohm per square. This combination is fairly resistant to thermal treatment (at 250°C for 5 hours), can be soldered with rosin flux after the standard thermal treatment, can be resoldered, and does not embrittle the solder. The problems associated with palladium are connected with its evaporation. Unfortunately, palladium alloys with tungsten, which is commonly used as a filament material. Palladium also has a relatively high melting temperature (1550°C) and evaporation temperature (1566°C), making it more difficult to evaporate than materials like gold and copper. A more attractive material for the top layer may be a 50-50 alloy of palladium-gold, which is considerably easier to evaporate since it melts at 1450°C and wets the tungsten filament without alloying with it. This alloy does not embrittle the solder, does not oxidize during the standard heat treatment, and can be resoldered many times using only rosin flux. It can be readily etched in the same reagents used to etch palladium.

Heat Treatment Resistance

High-temperature thermal processing (above 250°C) causes further difficulties. At these elevated temperatures, the chromium from the Nichrome® diffuses through the entire structure, arrives at the surface, and forms an oxide which is difficult to remove. If this type of heat treatment is planned, something other than Nichrome® should be used for the initial layer. Recent preliminary results indicate that a very thin (10 A) layer of titanium or nickel followed by 2000 A of palladium, and topped by 2000 A of gold, produces a very satisfactory termination material. In one test, a resistor was aged at a temperature approaching the softening point of the glass (800°C), during which time the resistor aged by a factor of 10^3, but the solderability of the termination was not impaired. In another test, the sheet resistivity of this termination increased from 0.05 ohm per square to 0.25 ohm per square after a few hours at 500°C, or 1/2 hour at 550°C. Thus, this combination seems to serve quite well for purposes of high-temperature preaging.

REFERENCES

1. Schwartz, N., and R. W. Berry. "Thin Film Components and Circuits," *The Physics of Thin Films*, 2, 363-425 (1964).
2. Neugebauer, C. A., J. B. Newkirk, and D. A. Vermilyea. *The Structure and Properties of Thin Films*. New York: John Wiley and Sons, Inc., 1959.
3. Halaby, S. A., L. V. Gregor, and S. M. Rubens. "The Materials of Thin Film Devices," *Electro-Technology*, 72 (3), 95-122 (Sept., 1963).
4. Logan, M. A. "An AC Bridge for Semiconductor Resistivity Measurements Using a Four-Point Probe," *Bell System Technical Journal*, 40, 885-919 (1961).
5. French, J. C. "Bibliography on the Measurement of Bulk Resistivity of Semiconductor Materials for Electronic Devices," Technical Note 232, National Bureau of Standards, Oct. 21, 1964.
6. Hall, P. M., and J. T. Koo. "Sheet Resistance Measurement on a Sample With a Circular Hole or a Circular Sample," *J. Appl. Phys.*, 38, 3112-3116 (1967).
7. Conrad, G. T., N. Newman, and A. P. Stansbury. "A Recommended Standard Resistor-Noise Test System," *IRE Transactions on Component Parts*, CP7 (3), 71-88 (Sept., 1960).
8. Graham, C. T. "Metallic Film Resistors," *Proceedings of the 1952 Electronic Components Conference of the IRE-AIEE*. pp. 61-63.
9. Alderson, R. H., and F. Ashworth. "Vacuum-Deposited Films of Nickel-Chromium Alloy," *British Journal of Applied Physics*, 8, 205-210 (1957).
10. Swanson, J. G., and D. S. Campbell. "The Structural and the Electrical Properties of 80:20 Ni-Cr Thin Films," *Thin Solid Films* 1, 183-202 (1967).
11. Vishay Resistor Products, Malvern, Pa. Design Data/Catalog R-401-D (1966).

12. Pitt, K. E. G. "Evaporated Cermet Resistors," *Thin Solid Films*, 1, 173-182 (1967).
13. Melan, E. H., and A. H. Mones. "The Glaze Resistor—Its Structure and Reliability," *Proceedings of the 1964 Electronic Components Conference of the IEEE*. pp. 76-85.
14. McLean, D. A., N. Schwartz, and E. D. Tidd. "Tantalum Film Technology," *Proceedings of the IEEE*, 52, 1450-1462 (1964).
15. Owens, J. L. "A Discrete Tantalum Nitride Thin Film Resistor on a Flat Embossed Ceramic Substrate," *Proceedings of the 1965 Electronic Components Conference of the IEEE*. pp. 129-135.
16. Miller, G. L. *Tantalum and Niobium*. Washington, D. C.: Butterworths Scientific Publications, 1959.
17. Kuo, C. Y., J. C. King, and J. S. Fisher. "Thermal Processing of Tantalum Nitride Resistors," *Proceedings of the 1965 Electronic Components Conference of the IEEE*. pp. 123-128.
18. Kuo, C. Y., and J. S. Fisher. To be published.
19. Berry, R. W., G. M. Bouton, W. C. Ellis, and D. E. Engling. "Growth of Whisker Crystals and Related Morphologies by Electrotransport," *Applied Physics Letters*, 9, 263-265 (1966).

PRINCIPAL SYMBOLS FOR CHAPTER 7

Symbol	Definition
a	Substrate dimension parallel to four-point probe
A	Area
b	Substrate dimension perpendicular to four-point probe
B	Material constant for current-noise voltage
C	Four-point probe constant
d	Thickness of film
D	Density
e	Electron charge $(e > 0)$
$e(f)$	Noise voltage
E	Activation energy
E_d	Differential field strength
f	Frequency
I	Current
j	Current density
k	Boltzmann's constant
K	Anodization constant (coulombs/cm^3)
l	Resistor length
M	Atomic weight
N_A	Avogadro's number
N_s	Number of squares
Q	Charge passed

Symbol	Definition
r	Distance between hole center and four-point probe center
R	Resistance
R_s	Sheet resistance (ohms per square)
s	Four-point probe spacing
t	Time
T	Temperature
TCR	Temperature coefficient of resistance
V	Voltage
w	Resistor width
δ	Hole diameter
ρ	Resistivity

Chapter 8

Capacitor Materials

The ability of thin film circuits to meet the needs of circuit designers is increased by the availability of thin film capacitors, in addition to resistors and conductors. The fabrication of capacitors using techniques compatible with those employed to form resistors and conductors has been one factor responsible for the rapid expansion of tantalum film technology. This chapter introduces some general capacitor properties and provides detailed information on certain thin film capacitor materials.

8.1 FUNDAMENTAL CAPACITOR PROPERTIES

A capacitor is a device capable of storing electric charge and consists of two electrodes (conductors) separated by a dielectric. The quantity of charge stored by a capacitor is proportional to the potential difference between the two electrodes, and the constant of proportionality is defined as the capacitance. The charge, q (usually expressed in coulombs), is $q = CV$ where the potential difference, V, is in volts and the capacitance, C, is in farads.

Capacitors block the passage of direct current but, because of their ability to store charge, allow alternating current to flow while altering the phase relation between potential and current.

Capacitance

The magnitude of the capacitance is a function of the electrode area, electrode separation, and the nature of the intervening dielectric, and

$$C = (8.85 \times 10^{-12}) \frac{KA}{d} \qquad (8\text{-}1)$$

371

where K is the dielectric constant (ratio of the permittivity of the dielectric to that of free space), 8.85×10^{-12} is the permittivity of free space in farads per meter, A is the area of coincidence of the metal electrodes in square meters, and d is the dielectric thickness in meters.

Energy Stored

As a capacitor is charged, the work done by the external source of power appears as the energy stored in the capacitor. This energy, W (in newton-meters), is given by

$$W = \frac{1}{2} C V^2 \tag{8-2}$$

The energy density in the dielectric (in newtons per square meter) is

$$U = (8.85 \times 10^{-12}) \frac{K V^2}{2d^2} \tag{8-3}$$

For example, a tantalum oxide capacitor with a 2000A dielectric ($K = 22$) charged to a potential of 30 volts has an energy density of about 2×10^6 newtons per square meter.

Leakage Current

The current flowing after a capacitor has been charged is the leakage current and is used as a measure of quality. Leakage current is determined by applying a d.c. voltage for a fixed time and measuring the current. This leakage current increases rapidly with increasing voltage and temperature. From the leakage current at a given test voltage, a (voltage-dependent) resistance may be calculated. The product of this resistance and the capacitance is called the insulation resistance of the capacitor and is expressed in ohm-farads as

$$\text{Insulation resistance} = \frac{V_{\text{test}}}{I_{\text{leakage}}} \times C \tag{8-4}$$

Care must be exercised in interpreting these measurements, since the time required to charge a capacitor varies with the dielectric.

This is due to the finite time required to displace charges in the dielectric. These times depend on the nature of the dielectric and may vary from a few seconds to months. This phenomenon is known as *dielectric absorption*.

Dielectric Breakdown

The capacitance may not be increased indefinitely by decreasing the thickness of the dielectric layer between the electrodes, as Eq. (8-1) might imply, because the dielectric breakdown strength must also be considered. If too high a voltage stress is applied to a capacitor, an electric arc will pierce the dielectric and destroy its insulating properties. The voltage stress that results in this destruction is known as the breakdown strength and is a characteristic of the dielectric material. Breakdown strengths range from a few hundred volts per centimeter to 10^7 volts per centimeter.

Maximum Working Stress

For practical capacitors, the maximum working stress is a more important parameter than the dielectric strength. The maximum working stress may be defined as the level of voltage stress for which the failure rate of capacitors is equal to the maximum allowable for a given application. This stress level is thus difficult to measure (or even estimate) since relatively long-time accelerated-aging tests are required. Accurate estimation of the maximum working stress would require a detailed knowledge of all possible failure mechanisms, and this knowledge is generally incomplete. Accelerated-aging tests are used, however, in an attempt to predict capacitor reliability quantitatively, and thus arrive at an acceptable stress level. Acceleration factors for the stress levels used in the tests are required, and it is generally necessary to test at more than one level of stress. The object of these tests is to obtain a plot of mean lifetime as a function of the applied stress, and then to extrapolate this to predict the stress for which the mean lifetime is acceptable. If the dominant failure mechanism is affected by more than one stress (such as voltage and temperature), similar plots must be obtained for each and, similarly, if more than one failure mechanism exists, the dominant one at operating levels must be determined.

There are two general methods used to obtain the necessary data. In one method, groups of capacitors are subjected to a wide range of stress, and at each stress level the time of each failure is recorded until at least half the capacitors fail. "Step-stress" testing is the other method used to obtain accelerated-aging data, and has been adopted because of its success in the semiconductor device field [1]. In this method, the capacitors are subjected to stresses which are increased step-wise with a constant time interval at each stress level, and the number of failures at each stress level is recorded. The latter method is obviously faster, but extrapolation is usually less accurate. Both techniques are depicted schematically in Fig. 8-1, which is a plot of stress versus mean time to failure but which also shows the distribution of failures under constant stress at A and those at constant time intervals at B. Either set of distributions permits the same extrapolation to C, which represents the working stress which would give the minimum acceptable mean lifetime, but both testing methods are usually used to obtain the extrapolation. Both voltage and temperature are important stresses for capacitors, and the effects of each should be determined.

Catastrophic failure implies either a short or open circuit with complete loss of the capacitor function. If the leakage current increases sufficiently (until the insulation resistance decreases to a few tenths of an ohm-farad) or if the capacitance changes by more than a few percent, the capacitor may no longer perform its function and thus may be considered a failure. In some instances (such as resistance-capacitance timing circuits), changes in capacitance as small as one tenth of one percent may be considered a failure.

A convenient factor for comparing various thin film capacitor materials has been suggested by McLean [2]. This factor arises because it is usually desirable to minimize the area required for a given capacitance and working stress. The capacitance per unit area is obtained by dividing Eq. (8-1) by the area:

$$\frac{C}{A} = (8.85 \times 10^{-12}) \frac{K}{d} \qquad (8\text{-}5)$$

Ideally, d is determined by the voltage that the capacitor must sustain, so that

$$d = \frac{V_w}{S} \qquad (8\text{-}6)$$

where V_w is the design (or working) voltage of the capacitor, and S is the maximum allowable stress in volts per meter. If Eqs. (8-5)

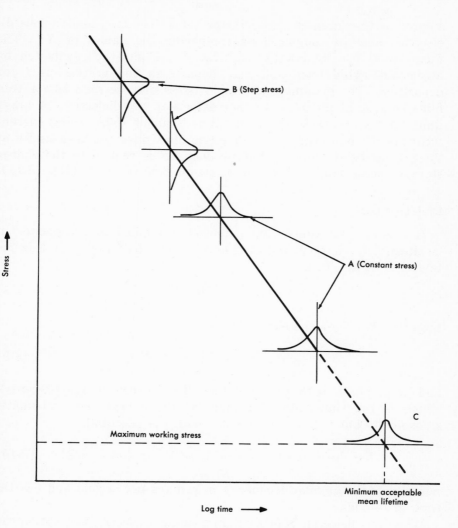

FIG. 8-1. Stress versus mean time to failure.

and (8-6) are combined,

$$\frac{C}{A} = (8.85 \times 10^{-12}) \frac{KS}{V_w} \tag{8-7}$$

This relationship is significant since both K and S are characteristics of a dielectric.

Therefore, KS, the product of the dielectric constant and the maximum allowable stress, is a material characteristic and may be considered as the "area efficiency factor" of a dielectric, and various dielectrics may be compared by comparing the values of KS. The factor K is usually readily available, but S must be evaluated by accelerated-aging tests and may depend on the process used for deposition. The maximum allowable stress may be reduced for thin films because of pinholes or other defects in the dielectric. A lower limit for the thickness of anodic films is about 100A, using current technology, while evaporated dielectric films have not been useful at thicknesses below 3000A. The product KS is related to the charge storage factor (CV_w/A) by the constant 8.85×10^{-12} (Eq. (8-7)).

Dielectric Loss

If a sinusoidal alternating potential is applied to a capacitor, it is alternately charged and discharged every half cycle. If this potential is of the form

$$V = V_0 \sin \omega t \tag{8-8}$$

then the charge on the electrodes is

$$q = CV = CV_0 \sin \omega t \tag{8-9}$$

and is in phase with the potential. The current I (in amperes), which is the time rate of change of the charge, also alternates sinusoidally but is 90° out of phase with the potential.

$$I = dq/dt = \omega CV_0 \cos \omega t = \omega CV_0 \sin (\omega t + \pi/2) \tag{8-10}$$

where ω is the angular frequency in radians per second and t is the time in seconds.

When the dielectric is other than vacuum, dielectric absorption can result in two effects: The finite time involved in moving charges results in a decrease in measured capacitance with increasing frequency (this decrease is usually quite small) and also results in a slight shift in the phase difference between current and voltage. The magnitude of this shift in phase is known as the loss angle, δ. The tangent of the loss angle is often used to express the phase relationship and the a.c. losses in the capacitor, and is commonly referred to either as "tan δ" or "dissipation factor." The reciprocal of this factor is another common term known as the quality factor, Q.

The loss is often expressed in terms of an equivalent circuit of a hypothetical resistor in series with the capacitor:

$$\tan \delta = \omega C_s R_s \qquad (8\text{-}11)$$

where C_s is the equivalent series capacitance and R_s is the equivalent series resistance.

Thin film capacitors also have a real resistance associated with their electrodes, so these capacitors are more conveniently represented by two resistors in series with the capacitor as shown in Fig. 8-2. In

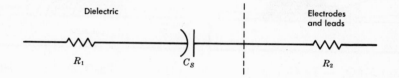

FIG. 8-2. Equivalent circuit for thin film capacitor.

this figure, R_1 is the hypothetical resistance which represents the dielectric loss of the capacitor dielectric, and R_2 is a true ohmic resistance in the capacitor electrodes and leads. With this representation, the dissipation factor becomes

$$\tan \delta = \omega C_s R_1 + \omega C_s R_2 = \tan \delta' + \omega C_s R_2 \qquad (8\text{-}12)$$

If the capacitance (C_s) and dissipation factor $(\tan \delta)$ are measured at two different frequencies, Eq. (8-12) may be solved for $\tan \delta'$ and R_2 [3]. This is only true, however, if $\tan \delta'$ is not a function of frequency. This condition occurs to a fair approximation for amorphous oxide dielectrics.

8.2 THIN FILM CAPACITOR MATERIALS

Dielectrics for thin film capacitors have been made using vacuum evaporated metal oxides, metal fluorides, reactively sputtered metal oxides, anodized valve metal oxides, and organic polymer films. Evaporated silicon monoxide, anodized tantalum oxide, and one of the polymer films are the most commonly used dielectrics and are discussed in this section. Figure 8-3 depicts schematically the structure of several thin film capacitors. These capacitors and others are the topics for the following discussion.

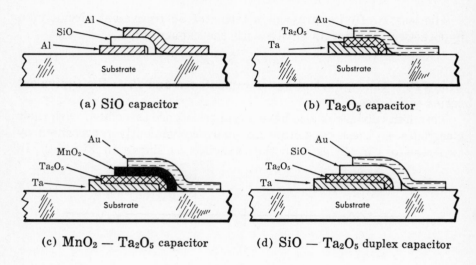

(a) SiO capacitor (b) Ta_2O_5 capacitor

(c) $MnO_2 - Ta_2O_5$ capacitor (d) SiO $- Ta_2O_5$ duplex capacitor

Fig. 8-3. Thin film capacitors.

Silicon Monoxide Capacitors

Silicon monoxide is probably the most frequently used evaporated dielectric since it is available in high purity and has a vapor pressure that allows evaporation at reasonable temperatures. Vacuum evaporation of silicon monoxide dielectric films, however, requires careful control of the deposition parameters (see Chap. 3) to attain mechanical stability and reproducible electrical properties. The degree of oxidation during both evaporation and subsequent annealing determines the relative amount of silicon, silicon monoxide, and silicon dioxide present in the film, and thus has a marked effect on film properties. Insufficient oxidation leads to dissipation factors as high as 0.1 and temperature coefficients of capacitance of about $+1000$ ppm/$^\circ$C; excess oxidation produces films with poor mechanical stability which peel off the substrate [4]. The degree of oxidation during deposition is controlled by the rate of deposition and the partial pressure of oxygen in the vacuum system. After deposition, annealing by heating in the presence of air is often used to reduce the temperature coefficient of capacitance, the dissipation factor, and the leakage current.

The yield of nonshorted silicon monoxide capacitors with aluminum electrodes has been improved by discharging an auxiliary capacitor across the thin film capacitor. This discharge vaporizes the metal film around pinholes in the dielectric, and thus averts shorts without allowing sufficient current to destroy the capacitor. This is the same method used to "clear" metallized paper and metallized polyester capacitors. Pinholes are a deterrent to producing capacitor dielectrics thinner than 5000A. In addition to this problem, a loss of yield results from sharp steps in the structure which apparently cause weak spots. This can be avoided by spacing the evaporation mask off the substrate so that shadowing occurs at pattern edges.

A smooth base electrode is essential to the reproducibility of capacitor properties. Thin films of both aluminum and copper have been shown to be satisfactory as base electrodes for silicon monoxide capacitors. Aluminum, copper, and gold have been used as upper or counterelectrodes. Aluminum is usually preferred for both electrodes because of its smoothness, lack of penetration of the dielectric, and ease of clearing by capacitive discharge.

Properties. Silicon monoxide is generally evaporated to thicknesses of from 5000 to 30,000A for capacitors, which results in a range of capacitance density from 0.01 to 0.0018 μf/cm^2. The capacitance precision depends on area and thickness control; under the best conditions a tolerance of ± 2 percent is possible, but ± 10 percent is more common.

The following properties were observed for capacitors prepared by depositing 1500A of aluminum as a base electrode, 5000A of silicon monoxide as a dielectric, and 4000A of aluminum as a counterelectrode. The finished capacitors were annealed at 400°C in air before their properties were measured. These capacitors had a capacitance density of 0.01 μf/cm^2; the silicon monoxide had a dielectric constant of 5.1, a dielectric strength of 2×10^6 volts/cm, and an allowable working stress of 6×10^5 volts/cm. The leakage current for a 0.001-μf capacitor after a 1-minute application of 30 volts was less than 1 nanoampere, which corresponds to an insulation resistance greater than 30 ohm-farads. The temperature coefficient of capacitance at 40-percent relative humidity ranged between +100 and +200 ppm/°C, and the change in capacitance between 0- and 87-percent relative humidity at 25°C was less than 0.5 percent. The dissipation factor at 100 kHz was 0.001, and the capacitance decreased by less than one percent when the frequency was increased from 0.1 to 100 kHz. A d.c. life stability test was conducted on 14 of these capacitors at 30 volts and 85°C for 32 weeks with no failure.

Tantalum Oxide Capacitors

Tantalum oxide thin film capacitors were first described by Berry and Sloan in 1959 [5]. Prior to this time, tantalum oxide prepared by anodizing sheet or sintered bulk tantalum was used as a dielectric in electrolytic and in tantalum solid capacitors. Berry and Sloan, however, provided the first indication that it might be practical to produce capacitors with an anodic Ta_2O_5 dielectric without either a liquid or a solid electrolyte interposed between the dielectric and a second electrode. Tantalum oxide films have a high dielectric constant and are formed in a way which yields a uniform, reproducible, and well controlled dielectric thickness.

Tantalum oxide thin film capacitor preparation begins with sputtering a tantalum film of about 4000A onto an insulating substrate. The capacitor electrode area may be defined either by sputtering through a mask or by photoetching. The tantalum film is then anodized to form the capacitor dielectric, as shown in Fig. 8-4. A 0.01-percent citric acid solution is a common electrolyte and is used with an initial current density of from 0.1 to 1 milliampere per square centimeter. Anodization is normally performed at constant

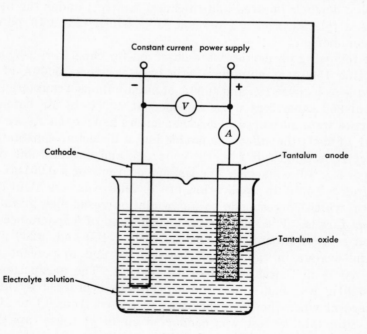

FIG. 8-4. Schematic representation of anodization of tantalum.

current until the desired anodization voltage is reached, and the anodization is continued for approximately 30 minutes at constant voltage. At this point, it is often desirable to anodically "back-etch" the capacitor for about 5 seconds at a constant voltage of approximately two-thirds the initial anodization voltage, using an electrolyte in which the reaction products are soluble. Current flows only at points where the oxide film is conducting; the tantalum film behind these points is dissolved in the electrolyte, and thus is removed from the capacitor structure. This can improve the characteristics of the final capacitor since there is no electrode left at these weak spots in the dielectric. The electrolyte commonly used for this step is a 0.01-percent solution of aluminum chloride in methanol. Following a distilled water rinse, the dielectric is returned to the anodization cell for an additional half-hour anodization at the formation voltage. If precise thickness control is required, the anodization time must be controlled, as mentioned in Chap. 5.

When anodization is completed, a thorough rinsing in distilled water is necessary to remove all traces of the electrolyte before the top or counterelectrode is deposited. The counterelectrode area may be defined either by using a metal mask during vacuum evaporation or by photoetching after deposition.

Properties. The dielectric constant of anodic tantalum oxide is 21.2, the density of the oxide is 8.0 gm/cm^3, and the voltage rate of anodization (or the anodization constant) is 16.0 A/volt [6]. Formation voltages range from 30 to 250 volts with corresponding capacitance densities of 0.4 to 0.05 $\mu f/cm^2$. A maximum in reliability appears to be obtained at a formation voltage of about 200 volts with a capacitance density of 0.06 $\mu f/cm^2$ [7].

Most of the data, however, have been collected from capacitors anodized at 130 volts. This provides an oxide thickness of 2000A and a capacitance density of 0.1 $\mu f/cm^2$. If the working voltage of the "130-volt" capacitor is assumed to be 30 volts, its charge storage factor (a figure of merit equal to the working voltage times the capacitance density) is 3 μf volts/cm^2.

The temperature coefficient of capacitance of tantalum oxide capacitors with gold counterelectrodes ranges from +170 to +250 ppm/°C. There is a small decrease in capacitance with increasing frequency, and this decrease is a characteristic of the dielectric. A large apparent decrease may be observed at high frequency if capacitance is measured on a "parallel" bridge. This decrease is not caused by the dielectric but rather by the series resistance of the capacitor, and the apparent decrease becomes insignificant if the electrode resistance is small or if the measurements are made on a "series" bridge. With gold counterelectrodes, the capacitance also increases with relative humidity; the increase is about 2 percent as the humidity rises from 0 to 86 percent. The effect of humidity is most likely due to the adsorption of a moisture layer between the dielectric and the counterelectrode. Thus, the use of a more adherent counterelectrode reduces the effect of humidity changes on capacitance so that a capacitor with an aluminum-Nichrome counterelectrode changes less than 0.1 percent between 0- and 86-percent relative humidity.

Dissipation factors reported for tantalum oxide capacitors range from 0.002 to 0.01 at frequencies less than 10 kHz. The dissipation factor increases rapidly with increasing frequency if the series resistance of the lower tantalum electrode is not negligible. A low series resistance may be assured by using a highly conducting film of aluminum beneath the tantalum electrode. A valve metal must be used since any other conductor beneath the tantalum would prevent anodization of the tantalum, and aluminum is chosen because it has the highest conductivity of all the valve metals.

Leakage currents are usually measured after some standard period of time following application of voltage in order to minimize the effects of charging currents and dielectric absorption. A period of 1 minute is common since this is long enough to eliminate charging currents (except for polarizations of very long relaxation time), while still allowing for a reasonable measurement time. Seventy-five volts has been chosen as the measurement potential for all tantalum capacitors anodized to 130 volts or greater [8]. Typical leakage currents are in the range of 0.01 to 0.1 ampere per farad, and capacitors are rejected if the leakage is greater than 1 ampere per farad. The nondestructive breakdown voltage is measured by applying a steadily increasing potential of 1 volt per second while simultaneously recording the current and the potential on an x-y recorder. The nondestructive breakdown voltage has been arbitrarily defined as the potential required to increase the current to three times the charging current [8]. This charging current is given by the product of the capacitance and the time rate of change of potential (1 volt per second in this case) so that the charging current in amperes is numerically equal to the capacitance in farads, and the breakdown voltage is then the stress required to cause a leakage of 2 amperes per farad. This test is generally conducted twice, once with the tantalum electrode positive (anodic breakdown), and then with this electrode negative (cathodic breakdown). Figure 8-5 shows a typical current-voltage envelope, and it can be seen that anodic breakdown voltages range from 65 to 90 percent of the formation voltage, while cathodic breakdown voltages range from 10 to 30 percent of the formation voltage for these units. This polar behavior of anodic films was previously described in Chap. 5.

Capacitors with an area less than 0.01 mm^2 exhibit symmetrical breakdown at high breakdown voltages and no influence of moisture on leakage, while capacitors with an area larger than 0.5 cm^2 have a significant probability of containing sufficient numbers of flaws to reduce the yield. This effect (at least until some means of avoiding these flaws is developed) imposes a practical upper limit on the amount of capacitance which may be placed on one substrate without liberalizing breakdown and leakage requirements. These flaws are also responsible for the dependence of some properties on the materials used for the counterelectrode. Adherent counterelectrodes, for example, usually reduce the anodic breakdown voltage with little effect on the cathodic breakdown voltage [9]. This is illustrated in Fig. 8-6, which gives breakdown voltages for a number of different

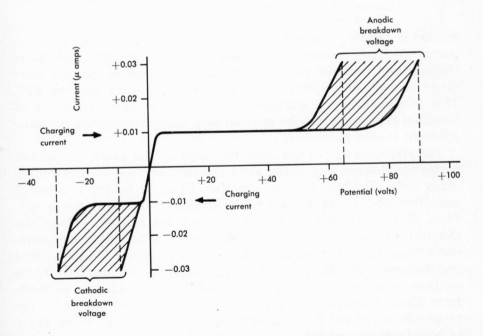

FIG. 8-5. Current-voltage characteristics of 0.01-μf tantalum oxide capacitors formed at 100 volts.

Metal*	Deposition	Breakdown voltage (anodic) †	Breakdown voltage (cathodic) †
Gold	Evaporated	91	15 to 20
Palladium	Evaporated	90	15 to 20
Copper	Evaporated	80	15 to 20
Antimony	Evaporated	75	15 to 20
Cadmium	Evaporated	71	15 to 20
Iron	Evaporated	45	15 to 20
Indium	Evaporated	41	15 to 20
Aluminum	Evaporated	24	15 to 20
Tantalum	Sputtered	15	15 to 20

*Listed in order of increasing adherence.
†Value of voltage for substantial current increase; catastrophic breakdown may be 5 to 30 volts higher.

Fig. 8-6. Effect of counterelectrode on anodic and cathodic breakdown voltages. (After McLean, Schwartz, and Tidd, "Tantalum Film Technology," in *Proceedings of IEEE*, 1964; courtesy The Institute of Electrical and Electronics Engineers, Inc.)

counterelectrodes. The variation in anodic breakdown voltage with counterelectrode material is probably caused by the ability of the adherent material to either penetrate the flaw to make a more direct contact, or prevent the admission of moisture into the flaw. This is qualitatively confirmed in that the humidity sensitivity of tantalum capacitors may be reduced to a negligible amount by using adherent counterelectrodes.

The structure of the tantalum film used as the anode can be important in determining initial yield and performance on accelerated-aging tests. The bcc tantalum films (see Chap. 4) tend to be stressed and often peel or crack upon being anodized. In addition, the surfaces of bcc film are often soft and are not smooth. These mechanical instabilities of bcc tantalum films result in generally unreliable capacitors. On the other hand, beta tantalum films are hard and adherent, and have a smooth and reproducible surface. These qualities allow capacitors of good yield and reliability to be made from films of beta tantalum and, except where noted otherwise, the properties described for tantalum capacitors are those which apply to beta tantalum anodes.

Some "doped" tantalum films with small additions of carbon, oxygen, or nitrogen have shown good performance when fabricated into capacitors. With nitrogen, insulating anodic films may even be grown from tantalum that shows the presence of Ta_2N. At these concentrations of nitrogen, however, the capacitance density is decreased. The relation of capacitance to nitrogen content is shown in Fig. 8-7 [10]; at low nitrogen levels the effect is small, but at higher levels the capacitance falls off rapidly. Once sufficient nitrogen is added to make the TaN structure evident (by X-ray diffraction), the maximum anodizing voltage attainable is greatly reduced. In addition, the anodic films no longer exhibit their usual clear interference colors. It is thus desirable to keep the nitrogen content of the films below the level at which the TaN structure is observed.

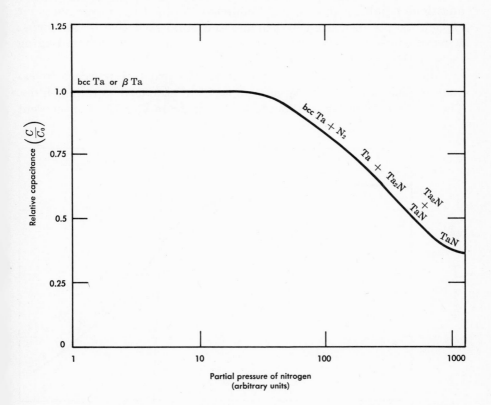

FIG. 8-7. Relative capacitance of anodic films formed on nitrogen-containing tantalum as a function of nitrogen pressure during sputtering.

The oxide growth factor (thickness of oxide per volt of anodization) does not change appreciably with the addition of nitrogen to the film, but the metal reduction factor (thickness of metal consumed per volt) does. Figure 8-8 shows this latter change as a function of nitrogen pressure during sputtering [10]. At low nitrogen levels there is good agreement with the anodization factor of bulk tantalum but, as the nitrogen content reaches the level at which stable resistors are made, the anodization factors decreases to about 4.5A of tantalum consumed per volt.

The reduction in capacitance observed with an increase in additive level arises from a change in the dielectric constant of the anodic film, since there is little effect on the oxide thickness. The addition of small amounts of carbon, nitrogen, or oxygen may increase the initial yield and improve the cathodic breakdown voltage, with oxygen additions resulting in a nearly "nonpolar" capacitor. These nonpolar capacitors (as measured by leakage and nondestructive breakdown), however, usually show much poorer performance on accelerated-aging tests if the tantalum is made negative.

Films containing carbon have shown the greatest increase in capacitor yield and reliability as compared to standard tantalum films. The films that exhibit this increase contain from 7 to 15 atom percent

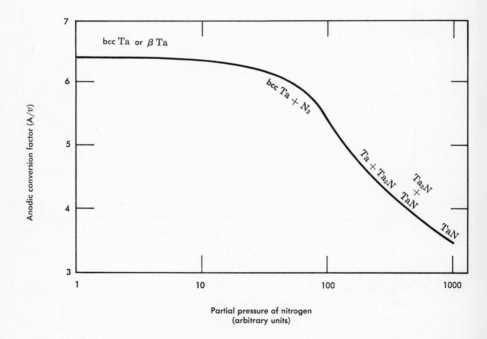

FIG. 8-8. Voltage rate of conversion on anodization for nitrogen-containing tantalum films as a function of nitrogen pressure during sputtering.

carbon, have a slightly distorted bcc tantalum structure, and are normally produced by reactive sputtering in a methane-argon atmosphere. The anodized film has a capacitance density between 70 and 90 percent of that obtained with beta tantalum, depending on the anodization voltage. The cathodic breakdown voltage is about 50 percent of the anodic value, and the dissipation factor is slightly less than for normal tantalum electrodes. There are, however, some practical problems remaining to be solved before capacitors may be manufactured using tantalum films containing carbon.

It is also possible to influence the character of anodically formed capacitors by using metallic additions to the tantalum. For example, small amounts of molybdenum, vanadium, or iron seem to increase the cathodic breakdown voltage with no harmful side effects. Thus, it may also be possible to create nonpolar capacitors or to increase reliability by proper alloying.

Another modification of the tantalum film is the low-density modification obtained by sputtering at reduced voltages. These films appear to contain many pores which give rise to a long induction period during anodization before the voltage begins to rise, since the initial anodization merely fills the pores with oxide. Capacitors formed on these low-density films have shown good performance with respect to catastrophic failure on accelerated-aging tests. The low-density tantalum has a high resistivity, however, which causes a high dissipation factor unless a conducting film of evaporated aluminum is used beneath the tantalum.

Manganese-Oxide Tantalum-Oxide Capacitors

A film of manganese oxide between the anodic tantalum oxide and the counterelectrode metal has been found to give a healing property analogous to that obtained with solid electrolytic capacitors [11]. Since the manganese oxide has a low resistivity, it acts electrically as an extension of the upper electrode so that the same capacitance is observed whether or not manganese oxide is present. In the healing process, the manganese oxide is apparently converted from a semiconductor to a dielectric at a point of incipient failure. This reduction is caused by the high leakage current and the temperature rise at such a point. A protective layer of manganese oxide permits larger-area capacitors to be made with satisfactory yield, and it also allows capacitor fabrication on rougher substrates.

There are several alternative methods of depositing manganese oxide on the anodized tantalum. These include pyrolysis of manganous nitrate, chemical reduction of permanganate, and reactive sputtering or reactive evaporation. In the thermal decomposition method, the substrate carrying the anodized tantalum is heated to 200°C and an aqueous solution of manganous nitrate is applied by spraying. The chemical reduction technique is analogous to electroless plating in that substrates suspended in a mixture of potassium permanganate and a reducing agent receive a coating of manganese dioxide [12]. A thickness of 4000A of manganese oxide is adequate. Manganese oxide may be removed from areas where it is not wanted by etching either in hydrochloric acid or a mixture of nitric acid and hydrogen peroxide. Except for somewhat higher dissipation factors and temperature coefficients, the properties of this type of capacitor are similar to those of the simple tantalum oxide capacitors since the dielectric is the same. The self-healing feature of the manganese oxide increases initial yield, life stability, working voltage, and capacitance density (films as thin as 100A may be used). The addition of a manganese oxide layer to silicon monoxide capacitors has also been demonstrated to improve their quality.

Silicon-Monoxide Tantalum-Oxide Duplex Capacitors

The high dielectric constant of tantalum oxide with its consequent high capacitance density may become a disadvantage if very low-valued capacitors are required. The area required for a capacitor of 100 pf is only 10^{-3} cm^2 (160 square mils) so that a small error in defining the capacitor area results in a large relative error in capacitance. If a capacitance precision of 5 percent or less is desired for 100 pf or lower, a lower capacitance density is desirable. Silicon monoxide is an obvious choice for this requirement because of its lower dielectric constant and because it may be deposited as a much thicker film (25,000A) than tantalum oxide (4000 to 5000A). If silicon monoxide alone were used, however, the other advantages of tantalum oxide capacitors would be lost. Capacitors that combine both dielectrics have thus been prepared in order to retain the advantages of each. Tantalum is first sputtered and anodized, and then a film of silicon monoxide is evaporated on top of the anodic oxide to form the "duplex dielectric."

The duplex capacitor behaves as two capacitors in series, with the capacitance determined principally by that of the low-valued layer; thus, the range of capacitance per unit area is 0.002 to 0.02 μf/cm^2, which is essentially the same as that for the silicon monoxide dielectric alone.

The duplex dielectric structure has additional advantages. The incidence of pinholes in silicon monoxide films is such that this dielectric must ordinarily be used as a fairly thick film to obtain a reasonable yield. Since the two-layer deposition reduces to insignificance the probability of pinholes extending through the entire dielectric, a much thinner silicon monoxide film (to 2000A) may be used. This provides a wider range of capacitance values available with good precision. The initial yield and stability on accelerated-aging tests of these capacitors are also greater than those of either type alone.

Another advantage of the duplex dielectrics is realized in the fabrication of distributed resistance-capacitance components. For these devices, both capacitance and resistance must be controlled. With tantalum oxide alone these two properties are interdependent, but with duplex films they may be independently controlled. The duplex structure also serves to produce more reliable crossover insulation than does silicon monoxide alone.

Duplex dielectric capacitors consisting of sputtered tantalum anodized to 100 volts and overlaid with 4500A of silicon monoxide with chromium-gold upper electrodes have been successfully utilized. These capacitors have a capacitance density of 0.01 $\mu f/cm^2$. The initial yield was 98 percent based on a leakage current of less than 10 amperes per farad at 20 volts. Step stress data [13] have been used to predict a failure rate of less than ten failures in 10^9 component hours at a stress level of 10 volts and 85°C.

Organic Polymer Capacitors

Most organic materials cannot be deposited by evaporation because they decompose below their vaporization temperatures. The spreading of a solution of a polymer with subsequent solvent evaporation is used to make lacquer film capacitors with dielectric thicknesses of the order of 0.1 mil. Another technique for depositing polymer films is to spread a monomer and polymerize it *in situ* with heat or ultraviolet light.

Films may also be produced by irradiating substrates with ultraviolet light in the presence of butadiene gas and have shown breakdown voltages of 5 to 6 $\times$ 10^6 volts per centimeter, low dissipation factors, and a dielectric constant of 2.65 [14]. The deposition is effected by irradiating only that portion of the substrate where film deposition is desired, so that no masking is required. Polymerization of gaseous monomers may also be initiated in selected areas by use of an electron beam. The rate of film formation is proportional to the pressure of the monomer vapor up to about 10^{-4} torr and is also proportional to the density of the electron beam. Another method of activating polymerization of organic monomers from the vapor phase employs an electrical discharge. Gas pressures of from 10^{-2} to 10^{-1} torr, discharge voltages of 100 to 300 volts, and currents of from 50 to 100 milliamperes are typical. Poorer definition of deposition areas is obtained than with irradiation.

Of the several organic thin film dielectrics, parylene (poly-para-xylylene) is chosen as an example for discussion.

Fabrication of Parylene Capacitors [15]. Parylene is a linear (non-crosslinked) polymer with average chain lengths of about 5000 units. A crude vacuum may be used for the parylene deposition with the background pressure reduced below the vapor pressure of 10^{-2} torr. The starting material is heated to 150° to 200°C to provide a reasonable vapor concentration. This vapor is heated to 650°C (pyrolized) to convert the dimer to the monomer. The monomer condenses, polymerizes on the substrate (which must be held below 50°C), and forms a continuous film.

The rate of deposition depends on substrate temperature and is increased by a factor of 10 if the substrate temperature is reduced from $+25°$ to $-25°$C. The process is complex and difficult to control since it depends on reaction time, pressure, temperature, concentration of monomer, and reaction chamber geometry. The use of a standard evaporation system for deposition would appear to be an advantage since the metal electrodes could be vacuum-deposited in the same chamber. The diffusion of parylene throughout the system, however, makes this impractical. The major problem associated with the fabrication of parylene capacitors has been the lack of adherence of the upper electrode. Difficulty has also been encountered in masking during deposition but, in principle, this may be overcome by photoetching since parylene becomes soluble in a 2-percent sodium carbonate solution after irradiation with ultraviolet light in the presence of air.

Properties of Parylene Capacitors. The dielectric constant of parylene is 2.65, and films with capacitance densities from 0.00025 to 0.025 $\mu f/cm^2$ and film thicknesses from 1000 to 100,000A have been produced. Thickness control is difficult and results in poor capacitance precision. The temperature coefficient of capacitance is -200 ppm/°C from $-50°$ to 100°C, and the effect of frequency on capacitance is very small from 0.1 to 1000 kHz. Dielectric strength is a function of thickness and ranges from 3 to 4 $\times$ 10^6 volts/cm.

The dissipation factor is low, with values of 0.0001 to 0.0006 reported, and the variation of dissipation factor with frequency is also small. Relative-humidity changes cause small changes in capacitance but very large changes in the dissipation factor.

Some properties of the thin film capacitor materials discussed in this chapter are summarized in Fig. 8-9. Of these materials, tantalum oxide is most often used because of its high area efficiency. When very low-valued capacitors are needed, the duplex silicon-monoxide tantalum-oxide dielectric film is preferred to silicon monoxide alone. Parylene, on the other hand, promises to provide reliable crossover insulation with only a small capacitance. The major advantages offered by thin film capacitors are small size, low cost, and their production as integral parts of thin film circuits.

REFERENCES

1. Howard, B. T., and G. A. Dodson. "Accelerated Aging of Semiconductors," *Bell Laboratories Record*, **40**, 8-11 (1962).
2. McLean, D. A. Whitehead Memorial Lecture, Conference on Insulation (NAS), Oct., 1966 (unpublished).
3. McLean, D. A. "The AC Properties of Tantalum Solid Electrolytic Capacitors," *J. Electrochem. Soc.*, **108**, 48-56 (1961).
4. Siddall, G. "Vacuum Deposition of Dielectric Films for Capacitors," *Vacuum*, **9**, 274-287 (1960).
5. Berry, R. W., and D. J. Sloan. "Tantalum Printed Capacitors," *Proc. of the IRE*, **47**, 1070-1075 (1959).
6. Klerer, J. "The Determination of the Density, Anodization Factor, and Dielectric Constant of Thin Ta_2O_5 Films," *J. Electrochem. Soc.*, **112**, 896-899 (1965).
7. Klerer, J. Personal communication.
8. Schwartz, N., and M. Gresh. "Effect of Ambients and Contact Area on the Asymmetric Conduction of Anodic Tantalum Oxide Films," *J. Electrochem. Soc.*, **112**, 295-300 (1965).
9. McLean, D. A., N. Schwartz, and E. D. Tidd. "Tantalum Film Technology," *Proc. of IEEE*, **52**, 1450-1462 (1964).
10. Gerstenberg, D. "Properties of Anodic Films Formed on Reactively Sputtered Tantalum," *J. Electrochem. Soc.* **113**, 542-547 (1966).
11. McLean, D. A., and F. E. Rosztoczy. "The Use of Manganese Oxide Counter Electrodes in Thin Film Capacitors. The TMM Capacitor," *Electrochem. Tech.*, **4**, 523-525 (1966).
12. Sharp, D. Personal communication.
13. Fuls, E. N. Personal communication.
14. White, P. "Thin Film Dielectrics," *Insulation*, **9**, 57-66 (1963).
15. Cariou, F. E., D. J. Valley, and W. E. Loeb. "Poly-para-xylylene in Thin Film Applications," *IEEE Trans. Parts, Materials, and Packaging*, **PMP-1**, 554-562 (1965).

Dielectric material	Base electrode	Counter-electrode	Dielectric constant	Allowable stress,* S (V/cm)	Area efficiency factor, KS (V/cm)	Charge storage factor (μ coul/cm^2)	Minimum useful thickness	Max. cap. density (μf/cm^2)
Ta$_2$O$_5$	Ta	MnO$_2$ + Au	22	2.0×10^6	44×10^6	4	100A	1
Ta$_2$O$_5$	Ta	Au	22	1.5×10^6	33×10^6	3	500A	0.2
SiO + Ta$_2$O$_5$	Ta	Al	6	2.5×10^6	15×10^6	1	2000A (SiO)	0.02
SiO	Al	Al	6	0.2×10^6	1.2×10^6	0.1	5000A	0.008
Parylene	Al	Al	2.65	1.3×10^6	3.4×10^6	0.3	1000A	0.025

FIG. 8-9. Thin film capacitor properties.

*These values are estimates based on available accelerated-aging data.

PRINCIPAL SYMBOLS FOR CHAPTER 8

SYMBOL	DEFINITION
A	Area
C	Capacitance
d	Film thickness
I	Current
$I_{leakage}$	Leakage current
K	Dielectric constant
q	Charge
R	Resistance
S	Maximum allowable stress
t	Time
U	Energy density
V	Voltage
W	Energy
δ	Loss angle
$\tan \delta$	Dissipation factor
$\tan \delta'$	Frequency-independent dissipation factor
ω	Angular frequency

Chapter 9

Substrates

Thin films cannot generally support themselves, and thus some form of carrier must be provided. This carrier, or "substrate," would ideally have no interaction with the thin film except for sufficient adhesion to provide support. If the film is to be part of a circuit, the substrate may also be called upon to conduct any heat generated in the film to a heat sink. In addition, the substrate must be compatible with the deposition processes and all subsequent processing or handling necessary for the use of the films. Another consideration is cost, particularly if the films are to be utilized commercially. Using these criteria, a list of the properties desired in an ideal substrate may be formulated as in Fig. 9-1. (In some circuit applications, particularly at microwave frequencies, the dielectric constant of the substrate can also be important.)

Once the ideal substrate has been defined, pertinent data on possible substrate materials may be compared with the ideal, and with one another. A substrate material that is ideal for all applications does not exist, since some of the desired properties depend on the intended use. The properties listed in Fig. 9-1 cannot be given an order of importance, so the film circuit designer must assess the application and assign the relative importance of the properties to arrive at the best choice. The intent of this chapter is to provide the information necessary for comparison of the various substrate materials with regard to each important property, and to indicate some of the techniques used to clean these substrates prior to thin film deposition.

Some materials may be eliminated from the list as not meeting some common requirement. For example, metals with resistivities of 10^{-6} ohm cm and semiconductors with values of 10^{-2} to 10^9 ohm cm are both very much inferior to insulators, with resistivities ranging

Desired property	Reason
Atomically smooth surface	Provide film uniformity
Perfect flatness	Provide mask definition
No porosity	Prevent excessive outgassing
Mechanical strength	Prevent breakage
Thermal coefficient of expansion equal to that of deposited film	Prevent film stress
High thermal conductivity	Prevent heating of circuit components
Resistance to thermal shock	Prevent damage during processing
Thermal stability	Permit heating during processing
Chemical stability	Permit the unlimited use of process reagents
High electrical resistance	Provide insulation of circuit components
Low cost	Permit commercial application

FIG. 9-1. Ideal substrate properties.

from 10^{14} to 10^{22} ohm cm. Metals and semiconductors may be coated with insulating films, but these are generally discounted because of parasitic capacitance. Another requirement is mechanical strength, and on this basis all the nonmetallic elements may be eliminated. The requirement of thermal stability essentially eliminates the organic plastics, which deform or decompose below 250°C. The inorganic compounds which remain include amorphous, polycrystal, and single crystal structures such as glass, ceramics, and sapphire. These examples are also the most commonly used substrate materials, and this discussion is limited to them.

9.1 SUBSTRATE MANUFACTURE

The methods used to manufacture substrates may affect their properties, and these methods are briefly reviewed for glass, ceramics, and sapphire.

Glass

Flat glass is presently manufactured using three different techniques to produce sheet glass, plate glass, and float glass. Sheet glass is made by melting the ingredients in a tank-type furnace, with raw materials added and melted at one end and a ribbon of glass drawn out at the other end. Plate glass is formed by allowing the molten glass to flow out of the bottom of the tank by gravity, passing it through forming rolls, annealing, and then grinding and polishing the surfaces. The float process involves moving a ribbon of molten glass across a pool of molten tin. Tin has a lower melting point than glass, and thus the glass may be allowed to cool while uniformly supported on the molten tin. This latter technique has the advantage that planar parallel surfaces are obtained without grinding or polishing. Float glass was first produced in the U. S. in 1964 and is currently available only in 1/4-inch and 3/16-inch thicknesses, and only in certain compositions.

The glass in most demand for substrates at present is Corning Code 7059 alumino borosilicate, and this is made as sheet glass. The molten glass is drawn as a ribbon from the tank, and the thickness is controlled by the rate of drawing and the viscosity of the melt. The viscosity is controlled by electrical resistance heating, but the high resistivity and viscosity of this glass make the operation difficult and may cause waviness, strain, and devitrification. The drawn glass sheet is cut to size by scribing with a diamond point and breaking. Cutting tolerances are ±0.015 inch in length and width, and 1 mil per inch is the maximum departure permitted from parallelism. The ease of breaking glass is a great advantage if circuits are made in multiple, but it becomes a potential cause for problems in circuit mounting.

Ceramics

Ceramics generally consist of sintered grains of the major component, with glassy phases as a binding matrix; their resistance to fracture is due to the interlocking nature and the intrinsic strength of the grains. "Green" sheet ceramic may be formed by die stamping, extrusion, or "doctor blade" techniques. All of these techniques have limitations on the range of thickness for which they are applicable, but the latter technique allows the largest areas and thinnest sheets to be made. In doctor blading, the basic mixture of the ceramic and organic binder in a solvent is formed into a long continuous tape by passing it under a knife blade. The tape is air dried to remove the solvent, and the desired shape of the substrate (including unusual features such as holes, orientation points, and alignment edges) is then punched out of the tape. The shaped pieces are prefired to remove the organic binder and are finally fired at a high temperature to mature and sinter the ceramic. These firing steps result in a linear shrinkage of 18 to 25 percent, so that dimensional tolerances are difficult to control. If a glaze is to be applied to the ceramic, it is next sprayed or squeegeed on and another firing is required. Ceramics with no glaze are often referred to as "as-fired" and, of course, the surface properties of the two are quite different. The surface of unglazed ceramics may be improved by grinding and polishing, but this is normally an expensive procedure.

Synthetic Sapphire

Single crystal alumina (Al_2O_3) is known as α alumina, sapphire, or corundum. The starting material for its synthesis is alum (ammonium aluminum sulfate) that has been purified by recrystallization and then calcined in a furnace at 1000°C. The light friable cake obtained from this treatment is milled and screened to produce a fine powder. This powder is heated in an oxy-hydrogen flame to form a single crystal boule of α alumina which is sometimes annealed to relieve stress. The boule is finally sliced and polished to form a substrate.

9.2 MATERIAL PROPERTIES

The properties listed in Fig. 9-1 are discussed in this section, and various substrate materials are compared on the bases of these properties.

Surface Smoothness

It is difficult, if not impossible, to prepare a surface which is atomically smooth over an appreciable area. Dislocation lines and surface point defects limit the best surface obtained to one having monatomic steps. Thin films are used in the thickness range of 50 to 50,000A; thus, substrate surface roughness in this range may have a direct effect on many film properties. Sharp discontinuities in a surface can be replicated in a dielectric film, with consequent enhancement of an electric field at a point, and can cause dielectric failure. In resistors, the effective path length may depend upon the degree of surface roughness and, more importantly, may vary from point to point if the roughness is not uniform. Some understanding of the variation in surface roughness on typical materials can be obtained from the results illustrated in Fig. 9-2. These Talysurf®

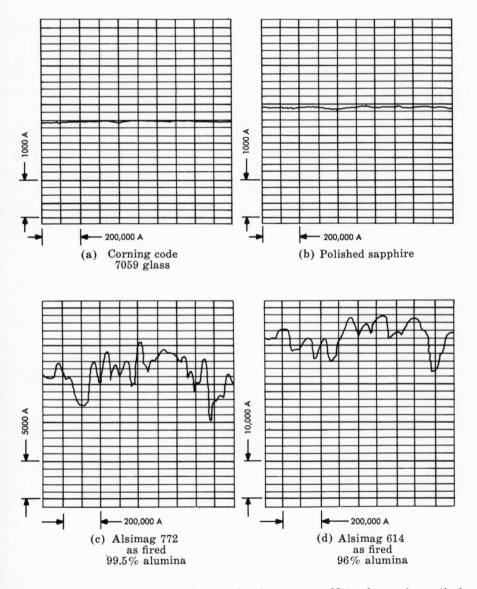

FIG. 9-2. Surface traces made by a stylus instrument. Note change in vertical scale.

recorder traces [1] were obtained by amplifying the motion of a
sensitive stylus as it traversed the material surface.* This instrument and its operation are described in Sec. 3.6. In Fig. 9-2(a), glass
exhibits the surface smoothness expected of a material drawn from
a melt. The occasional discontinuities caused by the drawing operation
are difficult to find. The curvature of the surface can be high, but the
microroughness is absent. The surface trace of a glazed ceramic
substrate has the same appearance as glass. The trace in Fig. 9-2(b)
shows polished sapphire to be only slightly rougher than glass. Small-
grain, high-density alumina has surface characteristics as exhibited
in Fig. 9-2(c) with peak-to-valley distances of as much as 10,000A.
Figure 9-2(d) illustrates the surface of an as-fired 96-percent alumina
ceramic with a surface showing peak-to-peak distances of about
15,000A. Comparing Figs. 9-2(c) and 9-2(d) reveals that the surface
trace indicates the grain structure of the material. The finer-grained
99.5-percent alumina thus shows smaller amplitude peaks but with
perhaps three peaks occurring in 200,000A of surface, whereas the
larger-grained 96-percent alumina exhibits about two peaks in this
distance. These ceramics may be mechanically polished and, when
the grain size is very small, the resulting surface is nearly as smooth
as that of glass. Such high-density, small-grain alumina substrates
are still difficult to obtain in large sizes, and the polishing operation
can be expensive.

Substrates may be listed with respect to their surface smoothness
as follows:

Substrate	Roughness (peak-to-valley distance)
Glass and glazes	<250A
Polished sapphire	<250A
Polished 99.94% alumina (ultrafine grain)	≈250A
As-fired 99.5% alumina (fine grain)	≈10,000A
As-fired 96% alumina	≈15,000A

Flatness

Surface flatness and dimensional tolerances are important in fitting
standard fixtures in processing. Line definition in the photolitho-

*Two limitations exist in interpreting these traces: (1) the stylus tip radius of
0.1 to 0.05 mil (25,000 to 12,500A) prevents fine resolution in the direction of
travel, but not in a direction normal to the surface; (2) the magnification is
not the same in these two directions, which pictorially emphasizes roughness.

graphic process is particularly affected by surface waviness. If the photoresist surface is not in intimate contact with the photographic mask during exposure, light will expose additional photoresist at the edge of opaque areas and definition will suffer. Both ceramics and glass are wavy, but the problem is much worse with ceramics because of warping during firing and because of glaze application problems. Surface deviations as great as 5 mils per inch can be tolerated if simultaneously associated with a large radius of curvature. Either the mask or the substrate will then bend enough to permit intimate contact during exposure. Deviations greater than 1 mil associated with a small radius of curvature are intolerable, especially when line widths in the range of 0.1 to 0.2 mil are desired.

Some approximate values of surface flatness are:

Material	Deviation from Flatness
Ground and polished surfaces	<0.1 mil/in.
Soda-lime glass	2 mil/in.
Corning Code 7059 glass	4 mil/in.
As-fired 99% or 96% alumina	5 mil/in.
Glazed 96% alumina	5 mil/in.

Porosity

The principal material sorbed by glass is water. At temperatures above 300°C, the water evolution [2] has been expressed by

$$m_w = at^{1/2} + b$$

where m_w is the mass of water desorbed in time t, and a and b are constants. The constant b arises from desorption of water by a hydrated layer on the surface of aged glass, but values of a depend on the type of glass and on temperature. Based on comparison of the a values, the substrates may be arranged in order of ease of water removal, beginning with the material from which water is most easily removed, as follows:

Soda-lime glass
Lead borosilicate glass (glaze)
Borosilicate glass
96% silica

High-density alumina ceramics and sapphire may be heated to higher temperatures than any of these glasses, and thus may be more thoroughly degassed.

Mechanical Strength

The resistance to mechanical shock of a brittle material, such as a ceramic or glass, is a function of its elastic (Young's) modulus and also depends on the presence of flaws [3]. For example, the theoretical tensile strength of glass has been estimated to be 10^{10} newtons/m² but because of surface flaws introduced during forming and handling, observed values of the tensile strength may be as low as 10^7 newtons/m². Polycrystalline ceramics are stronger than glass, as shown in Fig. 9-3.

Material	Modulus of elasticity (newton/m²)	Tensile strength (newton/m²)
Alumina ceramic	37×10^{10}	34×10^7
Beryllia	30×10^{10}	10×10^7
Glass	6×10^{10}	5.5 to 11×10^7
Silica	7×10^{10}	4.1 to 8.3×10^7

FIG. 9-3. Mechanical strength of substrates.

Thermal Expansion

The coefficients of thermal expansion of the common substrate materials are generally smaller than those of metals. Also, more than one metal is usually deposited when thin film circuits are made, so some compromise in matching the thermal coefficients of expansion is necessary. The coefficients of thermal expansion of principal substrate materials, along with those of some of the metals commonly deposited, are listed in Fig. 9-4.

Thermal Conductivity

Electrical insulators do not rely on a similar mechanism for both electrical and heat conduction, as do metals; thus, they may conduct heat well but electricity poorly. For example, of the materials considered, beryllia is the best electrical insulator and the best heat conductor, with a thermal conductivity better than that of aluminum. Thermal conductivities are listed in Fig. 9-5.

Metal	Substrate material	Coefficient of thermal expansion (ppm/°C)
Aluminum	—	20
Copper	—	14.5
Gold	—	14.2
Nickel	—	13
Palladium	—	12
—	Soda-lime glass	9.2
—	Synthetic sapphire	$\begin{cases} 6.66 & \parallel c \text{ axis} \\ 5.0 & \perp c \text{ axis} \end{cases}$
Tantalum	—	6.5
—	96% alumina	6.4
—	98% beryllia	6.1
—	99.5% alumina	6.0
—	Alkaline earth porcelain	5.0
—	7059 glass	4.5
—	7740 glass (Pyrex®)	3.25
—	96% silica (Vycor®)	0.8
—	Fused silica	0.56

®Registered trademark of Corning Glass Works.

FIG. 9-4. Thermal coefficient of linear expansion.

Material	Thermal conductivity at 25°C (watts/cm°C)
98% beryllia	2.10
Synthetic sapphire	0.42
99.5% alumina	0.37
96% alumina	0.35
Alkaline earth porcelain	<0.025
96% silica	0.0159
Fused silica	0.0142
7059 glass	0.0125
7740 glass (Pyrex)	0.0113
Soda-lime glass	0.0096

FIG. 9-5. Thermal conductivities of substrate materials.

Thermal conductivities are important for substrates carrying power-dissipating devices such as resistors. The surface temperature of resistors on substrates can be accurately determined by an infrared sensing device, which does not contact the surface to disturb the equilibrium thermal gradients. Temperature distributions for resistors on a variety of substrates are given in Fig. 9-6. These temperatures are specific to the condition of the test [4], which consists of applying power to a discrete resistor suspended by its axial leads in a still room ambient. In this configuration, the heat dissipation is from the film to the nearest heat sink, the substrate. The heat is dissipated from the substrate by radiation, convection to the ambient, and conduction to the axial leads. A glass substrate with a thermal conductivity of about 0.012 watt/cm°C results in the highest midpoint

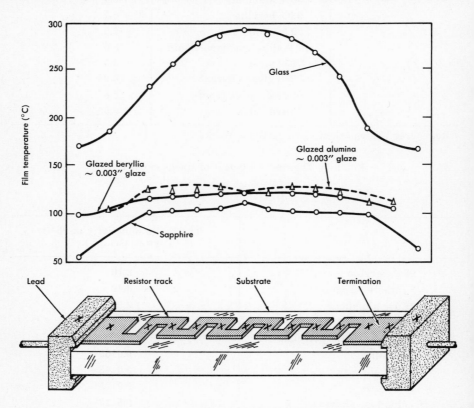

FIG. 9-6. Temperature rise in tantalum nitride resistors (0.16 inch by 0.49 inch) after 15 minutes at 1 watt (80 watts/in.² of film) at room ambient. Crosses on resistor pattern indicate where measurements were taken [4]. (After R. Brown, "Substrates for Tantalum Thin Film Circuits;" courtesy *American Ceramic Society Bulletin*.)

temperature with a very sharp thermal gradient toward the leads, which serve as heat sinks, indicating that the glass is the thermal barrier in this case. The use of an alumina substrate with thermal conductivity of 0.4 watt/cm°C and a thin glaze results in temperatures almost 160°C lower than that obtained on glass, and an essentially uniform temperature across the substrate. The substitution of glazed beryllia, with a thermal conductivity of 2.10, results in little decrease in surface temperature, indicating that the glaze is the limiting thermal barrier for both these structures. This is confirmed by measurement on sapphire (with a thermal conductivity essentially identical to alumina, but without a glaze) where a 20°C decrease in surface temperature is observed. These results are for powers eight times greater than the nominal rating of this resistor on a glass substrate. The decrease in surface temperature with high-thermal-conductivity substrates can allow two to four times higher power ratings for the same physical size, or a decrease in component size.

The consideration of power dissipation in thin film circuits is more complex since there may be many resistors with different power dissipations and heat sinks may be either remote or near to the power sources. The general conclusions about the influence of the thermal conductivity of the body and the glaze still apply.

Resistance to Thermal Shock

Thermal conductivity, specific heat, and density regulate the thermal gradient, while parameters such as elastic moduli and coefficient of thermal expansion determine the stress developed in substrates when their temperature is rapidly changed. A coefficient of thermal endurance F has been defined by Winkelmann and Schott [5] as

$$F = \frac{P}{\alpha Y} \sqrt{\frac{K}{Dc}}$$

where P is the tensile strength, α is the linear coefficient of thermal expansion, Y is Young's modulus, K is thermal conductivity, D is density, and c is specific heat. Although quantitative agreement of this equation with experiment is not good, its use does allow a numerical comparison of substrate materials to be made as shown in Fig. 9-7. This figure also includes the expansion coefficients since they are the most important factors in determining the ability of these substrates to withstand thermal shock.

Material	Relative thermal endurance factor	Thermal coefficient of linear expansion (ppm/°C)
Silica	13.0	0.56
Alumina	3.7	6.0
Beryllia	3.0	6.1
Glass	0.9	9

FIG. 9-7. Relative resistance to thermal shock.

Thermal Stability

None of the common substrate materials decompose rapidly at elevated temperatures, so the order of their thermal stability is the same as the order of their softening or melting points, shown in Fig. 9-8.

Material	Softening point (°C)
Synthetic sapphire	2040
99.9% alumina	2040
96% alumina	1650
98% beryllia	1600
Fused silica	1580
96% silica	1500
Alkaline earth porcelain	1250
7059 glass	872
7740 glass (Pyrex®)	820
Lead borosilicate (glaze)	725
Soda-lime glass	696

FIG. 9-8. Softening points for various substrate materials.

Chemical Stability

The chemical stability of a substrate is important in all phases of processing as well as in the performance of a completed thin film device. Soda-lime glass, with its high percentage of Na_2O (Fig. 9-9), is susceptible to surface degradation (weathering) during long-term exposure to moisture. Almost all glasses have a silicate network and are susceptible to hydrofluoric acid attack during the etching process

Material	Constituent									
	SiO_2	Na_2O	CaO	BaO	Al_2O_3	B_2O_3	PbO	MgO	As_2O_3	Fe_2O_3
Soda-lime glass	70.5	14.6	7.2	—	2.0	—	—	5.0	—	—
Pyrex® glass	79.1	4.6	—	—	trace	14.8	—	—	—	—
Corning Code 7059 glass	48.5	—	—	21.8	13.3	17.3	—	—	trace	—
Glazed alumina ceramic:										
Glaze	69.0	3.0	—	3.0	—	10.0	15.0	—	—	—
Ceramic	6.3	0.04	—	—	92	0.09	1.6	—	—	0.08

FIG. 9-9. Chemical composition of substrates (percent by weight).

used for refractory metals. The lead borosilicate used for ceramic glazes is also rapidly attacked by hydrofluoric acid solutions. Because of its high lead content, it is also susceptible to attack by some electrolytes used for anodization. This attack does not usually interfere extensively with processing, and the attack can be controlled to some extent by variations in the firing cycle of the glaze, which controls the chemical state of the lead species. Alumina, beryllia, and sapphire are outstanding because they are not attacked by hydrofluoric acid solutions, by anodizing solutions, or by long-term exposure to humidity. Figure 9-10 lists the etch rates of several glass substrate materials in various reagents.

Material	Etchant		
	5% HCl	5% NaOH	0.1% Na_2CO_3
Fused silica	0.001	2.8	0.12
96% silica	0.001	4.4	0.12
7740 glass (Pyrex®)	0.005	4.4	0.4
Soda-lime glass	0.02	2.0	0.4
7059 glass	5.5	14.8	1.2

FIG. 9-10. Etch rate of glass substrates at 95°C (mg/cm² per 24 hrs).

Another way to achieve chemical stability is to coat a glass or glazed ceramic substrate with a film of tantalum (100 to 300A), subsequently oxidized in a furnace. The etch rate of this tantalum oxide "underlay" is some one hundred times less than that of tantalum in the hydrofluoric nitric acid tantalum etchant, and has been especially useful when relatively thick films of tantalum must be etched.

Thin film resistor performance is directly related to the chemical composition of the substrate. In resistors at moderate d.c. power, the surface temperature rises and, simultaneously, a field exists along the surface. Alkali metal oxides may become ionized, and the ions may migrate through the substrate and electrochemically react with the film. The substrate becomes a high-temperature solid electrolyte, with reactions occurring at both the positive and the negative ends of the component. The magnitude of this effect, in terms of a change in resistance of tantalum nitride resistors on various substrates under accelerated aging, is given in Fig. 9-11. The stability of resistors on the glass substrates is inversely related to the Na_2O content, with alumino borosilicate (7059) glass resulting in the highest stability. For the soft glass (soda lime) and Pyrex (alkali borosilicate) samples, obvious color changes appear at the negative end of the resistor within a few hours after starting the test.

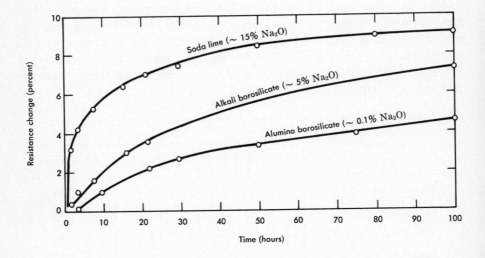

FIG. 9-11. Overload testing of nominal 0.5-watt tantalum nitride resistors (0.2 by 1.0 inch) on various substrates at room temperature. Power stress was 1.5 watts or 60 watts/in.^{2} of films [4]. (After R. Brown, "Substrates for Tantalum Thin Film Circuits;" courtesy *American Ceramic Society Bulletin*.)

The mechanism of this effect was studied by separating the simultaneous temperature and field effects associated with applying power to a given resistor [7]. The experiment used a pattern as shown in Fig. 9-12, and was performed as follows: The substrate was kept in an oven at 200°C with a potential of +90 volts maintained on resistor R_1 with respect to R_2, while R_3 and R_4 were maintained at the same potential as R_2 to act as controls. The negatively biased resistor, R_2, showed a change in color at the end of the test, and this gives rise to the mottled appearance of R_2 visible in Fig. 9-12. This dis-

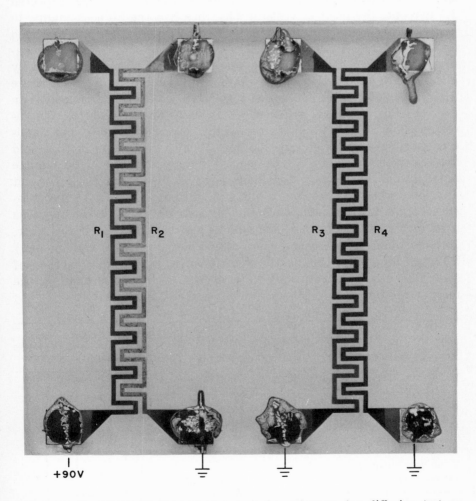

FIG. 9-12.   Tantalum nitride resistors used for substrate ion diffusion test on Pyrex after 1000 hours at 200°C.

colored material, however, is a water soluble compound of sodium and, when it is washed off, resistor R_2 has the same appearance as the other three resistors. After 1000 hours, the resistance of each track was measured, and Fig. 9-13 gives the changes in resistance observed for each of these resistors on a variety of different substrates. The change in resistance caused by thermal stress alone is in the range of 2.4 ±0.4 percent, as shown for resistors R_3 and R_4 for the last three substrates. (The low changes for the soft-glass substrates are not understood.) For soft glass, which has the highest Na_2O content, the changes in R_1 and R_2 are the greatest compared to the expected 2.4 percent; the positive resistor (R_1) has had the highest change in resistance, which was surprising since the discoloration was observed on the negative resistor (R_2). The large change in R_1 for the soda-lime substrate is attributed to a decrease in film thickness caused by a solid-state anodization of the tantalum nitride at the film-substrate interface. A smaller increase in resistance was observed for R_2 than for the controls. This difference may have been caused by the incorporation of sodium in the tantalum nitride film. For low-alkali-content glasses, such as alumino borosilicate, the change with a field applied is equivalent to that caused by aging alone. The change noted in the lead containing glaze is small, despite a Na_2O content approximating that of Pyrex. This is attributed mainly to the presence of the lead, which tends to decrease the mobility of the ions present. (Note the much higher volume resistivity of lead borosilicate, 10^{10} ohm cm, as compared with alkali borosilicate, $10^{7.2}$ ohm cm.) This ion migration test has proven useful in providing a method of screening new substrate materials, as well as in understanding the mechanism involved.

Substrate	Na_2O content (%)	Percent resistance change			
		Field present		No field	
		$R_1(+)$	$R_2(-)$	R_3	R_4
Soda lime	~15	10.1	0.6	1.2	1.2
Alkali borosilicate	~ 5	5.4	1.4	2.5	2.8
Alumino borosilicate (7059)	~ 0.1	2.4	2.4	2.4	2.4
Lead borosilicate glaze on alumina	~ 3	2.3	1.6	2.1	2.0

FIG. 9-13. Results of ion diffusion test on various substrates.

Electrical Conductivity

The glasses and ceramics considered for substrate materials are all good insulators at room temperature. The presence of variable-valence cations may reduce the band gap so that electronic conduction can occur, and the presence of mobile ions may permit ionic conduction, especially at elevated temperatures. Electrical resistivities have been reported for both surface and volume conduction. Surface resistivities are more a function of the surface condition (adsorbed water, etc.) than of the substrate material, however. Figure 9-14 lists only volume resistivities.

Material	Volume resistivity (ohm cm)		
	300°C	350°C	500°C
98% beryllia	$10^{13.8}$	$10^{13.3}$	$10^{12.2}$
7059 glass	$10^{12.4}$	$10^{11.7}$	$10^{8.8}$
Fused silica	10^{11}	$10^{10.2}$	$10^{8.4*}$
Synthetic sapphire	$>10^{10}$	——	——
99.9% alumina	$>10^{10}$	——	——
Lead borosilicate glaze	$>10^{10}$	——	——
96% alumina	10^{10}	$10^{9.4}$	$10^{7.9}$
96% silica	$10^{8.9}$	$10^{8.1}$	$10^{6.5*}$
Alkali earth porcelain	10^{8}	——	——
7740 glass (Pyrex®)	$10^{7.2}$	$10^{6.6}$	$10^{5.4*}$
Soda-lime glass	$10^{5.6}$	$10^{5.1}$	$10^{3.8*}$

*Obtained by extrapolation

FIG. 9-14. Electrical resistivities of substrate materials.

Cost

If the costs of substrate materials are normalized by assigning a cost of unity to unglazed 99.5-percent alumina ceramics, the relative costs of the various materials are:

Material	Cost
Synthetic sapphire	(400)
Beryllia	4.0
Glazed 96% alumina	1.66
Unglazed 99.5% alumina	1.0
Unglazed 96% alumina	0.9
7059 glass	0.65
Soda-lime glass	0.04

This comparison is based on the cost per square inch of 3-3/4 inch by 4-1/2 inch substrates purchased in large quantities (except for sapphire, which is based on a 1-inch by 1-inch substrate). The cost of separating a large substrate into individual circuits must eventually be added to these purchase figures, and this gives an additional advantage to the glass substrate materials.

In summary, several substrates exist for thin film applications. No one of these is best for all applications. Glazed alumina and Corning Code 7059 glass are presently the most popular materials, but the unglazed 99.5-percent alumina is gaining in popularity. Glazed alumina is equal to or better than 7059 glass in all respects except flatness and cost, although much of its thermal conductivity is sacrificed to gain the smoothness of the glaze. Ceramics also have an advantage in that they may be formed with holes or in irregular shapes. The more important properties of substrate materials are summarized in Figs. 9-15 and 9-16.

Substrate	Soda lime*	Alkali borosilicate*	Alumino borosilicate†	Lead borosilicate (ceramic glaze)‡	96% silica*	Fused silica*
Code number	0080	7740	7059	743	7900	7940
Annealing point, °C	512	565	650	—	910	1050
Softening point, °C	696	820	872	725	1500	1580
Thermal expansion coefficient, ppm/°C	9.2	3.25	4.5	—	0.8	0.56
Thermal conductivity, 25°C, watt/cm°C	0.0096	0.0113	0.0125	—	0.0159	0.0142
Density, g/cm³	2.47	2.23	2.76	—	2.18	2.20
Refractive index	1.51	1.47	1.53	1.58	1.46	1.46
Log volume resistivity, 300°C, ohm cm	5.6	7.2	12.4	>10	8.9	11
Dielectric constant, 1 MHz, 25°C	6.9	4.6	5.8	—	3.9	3.9
Loss tangent, 1 MHz, 25°C	0.01	0.0062	0.0011	—	0.0006	0.00002

*Bulletin B-83, Corning Glass Works, Corning, N. Y.
†Glass Substrates, Corning Glass Works, Corning, N. Y.
‡Bulletin 652, American Lava Corporation, Chattanooga, Tennessee.

FIG. 9-15. Properties of vitreous substrate materials.

Substrate	Alkaline earth porcelain*	96% alumina*	99.94% alumina†	Synthetic sapphire‡	98% beryllia*
Code number	531	614	E-37	—	735
Softening point, °C	1250	1650	2040§	2040§	1600
Thermal expansion coefficient, ppm/°C	5.0	6.4	6.7	6.66¶ 5.0 **	6.1
Thermal conductivity 25°C, watt/cm°C	<0.025	0.35	0.42	0.42	2.10
Density, g/cm³	—	3.70	3.99	3.98	2.90
Refractive index	—	—	—	1.8	—
Log volume resistivity, 300° C, ohm cm	8	10	>10	>10	13.8
Dielectric constant, 1 MHz at 25°C	5.4	9.3	9.1	11.5 ¶ 9.4 **	6.3
Loss tangent, 1 MHz at 25°C	0.026	0.0028	<0.0001	<0.0001	0.0006

*Bulletin 631, American Lava Corporation, Chattanooga, Tennessee.
†*Hypalox Ceramics*, American Feldmühle Corporation, Stamford, Connecticut.
‡*Physical Properties of Synthetic Sapphire*, A. Mellor Company, Providence, Rhode Island.
§Melting point.
¶Parallel to *c* axis.
**Perpendicular to *c* axis.

FIG. 9-16. Properties of crystalline substrate materials.

9.3 SUBSTRATE CLEANING

Before any substrate can be used, it must be adequately cleaned. The proper cleaning technique depends on the nature of the substrate, the nature of the contaminants, and the degree of cleanness required. Expected contaminants include those from manufacturing procedures, human contact (such as protein), and airborne dust, lint, and oil particles. If the substrates have received any prior processing, a host of new contaminants may be present; a knowledge of the processing steps may provide the list of those to be expected. As a minimum, the substrate surface must be made clean enough that contaminants do not interfere with the adherence of a vacuum deposited film. The ability to pass the water-break test is usually a necessary but not a sufficient condition to insure adherence. On visual inspection, using dark field illumination, there should be no lint or other particles and no film visible on any area.

Cleaning involves the breaking of adsorption bonds between the substrate and the contaminants without damaging the substrate surface itself. The energy to perform this step may be provided directly as heat or ion bombardment, by chemical reaction or solvation, or by mechanical scrubbing. Chemical cleaning methods ultimately depend on solvation and then removal of solvents. Mechanical action is best conducted in the presence of solvents and is most effective for removing gross contamination. The most logical order for combining the three techniques is to provide mechanical scrubbing in the presence of solvents, chemical reaction and solvation, and finally heat or ion bombardment (usually at reduced pressure).

Mechanical scrubbing has been performed by rubbing the substrate with chalk or cotton swabs, but a more effective procedure is to suspend the substrate in an ultrasonically agitated detergent solution. In this method, ultrasonic waves are transmitted to the solvent from a transducer and cause cavitation or bubble collapse with corresponding local surges of hydrostatic pressure. The stress created when a bubble cavity implodes on the surface of the substrate allows the solvent to penetrate between the contaminant and the substrate. The variable parameters of ultrasonic cleaning are the frequency of vibration and the power applied, as well as the temperature, vapor pressure, surface tension, and viscosity of the cleaning solution. The lower the frequency, the more violent the cavitation, but practical application is limited to frequencies above the audible range. An increase in power increases cavitation, but too much power may limit cavitation to the face of the transducer and reduce the cleaning action at the substrate surface. The effect of temperature on cavitation depends on the physical properties of the solvent used. Optimum results using water occur between 60° and 70°C. Liquids of low vapor pressure and viscosity produce better cavitation. Surface tension is the most important physical property of the cleaning solution, and greater surface tension leads to a greater energy release on bubble collapse. A quick but subjective test for ultrasonic action is to suspend a piece of aluminum foil in the solution (or in a beaker of water suspended in the solution) for about 15 seconds and then to visually inspect the foil for deformation.

Chemical treatment to remove contaminants is often used but is likely to attack the substrate surface. Acid cleaners react with contaminants such as grease and some oxides to convert them to more soluble compounds, and alkaline cleaners saponify fats and decrease surface tension. The effectiveness of solvents is probably more dependent on their ability to wet the substrate than their solvent action on the contaminants. The two main considerations of a solvent are that it be available in high purity and that any traces left on the substrate be easily removed in the vacuum system by heating to reasonable temperatures. Two effective final solvent rinses in common use are deionized distilled water and isopropanol. (The latter is usually used in a vapor degreaser.)

With unglazed ceramic substrates, a final cleaning by firing to a high temperature in air is beneficial. Glass or glazed ceramics may be "flamed," or they may be bombarded with ions in a gas discharge. A shielded electrode system should be used for this bombardment so that high energy electrons do not strike the substrate since they would cause polymerization rather than removal of any organic contaminants.

Substrates may become contaminated during and after cleaning. Cleaning tanks and substrate baskets must be immaculate and the surrounding atmosphere should be filtered. Substrate storage after cleaning is risky and should be minimized. Storage overnight beneath boiling, recirculating, deionized water is apparently safe for 7059 glass, and dry storage is possible for a few hours in a closed, clean glass dish or desiccator. Inside a vacuum system, substrates may be further cleaned by heating at reduced pressures and, with modern vacuum techniques, there is little chance for further contamination other than particulate matter.

The following substrate cleaning procedure has been used with good results on substrates of glass, glazed ceramic, unglazed ceramic, and sapphire. The first step in this system uses room-temperature ultrasonic agitation in a dilute solution of detergent to dislodge gross dirt. The second step is the same as the first, but is carried out at 70°C in a separate tank. The room-temperature bath is necessary first to remove protein material since heat denatures such contaminants and makes them more difficult to remove. The third operation is to replace the film of detergent solution on the substrate with a film of water. A hot-water spray rinse followed by an overflow rinse is usually effective. The objective of the fourth step is the conversion of organic materials (not previously removed) to water soluble compounds. A boiling solution of hydrogen peroxide is used to perform this function, since it is a strong oxidizing agent and is volatile. The fifth operation replaces the film of peroxide with one of water, and it is carried out in two steps: The first is a quick dip in hot distilled water; the second is a 15-minute soak in overflowing, boiling, deionized, distilled water.

Before substrates are loaded into a vacuum system, they are dried with hot (110°C) clean nitrogen for 15 minutes. A visual examination of glass substrate surfaces using oblique illumination will show any remaining particulate matter. No substrate ever appears perfect in these respects, and such a test is uncomfortably subjective. Some residue will almost always be observed at the lower substrate edge, especially where there has been contact with the substrate container during final drying. If the atmosphere used for the inspection is not well filtered, particulate matter will be observed to land on the substrate during examination and further confuse the test. The lack of confidence in this visual test and in other tests usually results in their being used only after trouble has been encountered with the appearance or adherence of deposited films.

Testing for Cleanness

It was observed and recorded nearly 75 years ago that the water vapor condensed on flamed glass after breathing on it had a different appearance from unflamed glass. Where glass has been flamed, a "black breath figure" is observed, which is a result of the water film forming an antireflection coating on the glass as its film thickness is uniformly reduced by drying. If the glass was not flamed, a "grey breath figure" occurs because the water does not wet the glass. Later, it was demonstrated that black breath figures could be obtained by other methods of cleaning, and also that removal of hydrophobic films from the surface produced this effect.

A more quantitative measure of the degree of wettability of a substrate surface is obtained by measuring the angle of contact between a water droplet and the surface. If the surface is easily wetted, this angle will be close to 0°; if not wetted, the water droplet will approach a spherical shape with a contact angle approaching 180°. There are two wettability tests which are more common than the direct measurement of contact angle and which allow classification into three degrees of cleanness, but the measurement of contact angle remains the only method which provides a quantitative measurement.

The two wettability tests are known as the water-break test and the atomizer test, and have ASTM designations of F22-62T and F21-62T. The water-break test is performed by withdrawing the surface to be tested, in a vertical position, from a container overflowing with pure water. The atomizer test is conducted by subjecting the dry surface to be tested to a fine spray of pure water. In both tests, the interpretation is based on the pattern of wetting. In the absence of hydrophobic films, the water-break test results in a continuous film of water over the surface, and the atomizer test results in the formation of a continuous film. The presence of a tenth of a monolayer of a hydrophobic film will cause the sprayed droplets of the atomizer test to remain as droplets; greater than one monolayer will cause the draining layer of the water-break test to break up into a discontinuous film. Thus, a surface may be classified as passing both, one, or neither test. Five degrees of cleanness have been defined by standard patterns of wetting by the atomizer test, but the interpretation is subjective.

Any wettability test may be misleading if there is any detergent remaining on the surface to be tested, and it must be remembered that only hydrophobic contaminants are detected by these methods. These methods are not readily applicable to unglazed ceramic surfaces since surface roughness interferes with the observations.

The resistance to sliding of a metal or glass surface over a glass substrate surface is related to cleanness, and the measurement of the coefficient of friction has been used to compare cleaning techniques. This test is destructive in that grooves are produced in the substrate.

The degree of adhesion of a metal film to a substrate is another indicator of cleanness of the substrate, although not all adhesion problems are caused by contaminated surfaces. Since the adhesion cannot be quantitatively measured, it indicates only that the cleaning was either adequate or inadequate. Another technique for judging glass substrate cleanness is to observe the substrate while it is edge-illuminated. By proper orientation of the substrate, the reflected light will show both particulate contamination and discontinuous films. This test is obviously subjective, but with a modicum of experience it becomes quite useful.

Once a cleaning technique has been developed, metal film adhesion is routinely tested and, if the adhesion is great enough that Scotch®

tape cannot be used to pull the film off the substrate, the cleaning is judged adequate. Particulate matter on a substrate will become obvious after metal deposition since it causes pinholes in the film. Particulate matter may also be detected by examination just prior to metal deposition.

REFERENCES

1. Schwartz, N., and R. Brown. "A Stylus Method for Evaluating the Thickness of Thin Films and Substrate Surface Roughness," *Trans. 8th Nat. Vac. Symposium, 1961*. New York: Pergamon Press, 1962. pp. 836-845.
2. Todd, B. J. "Outgassing of Glass," *J. Appl. Phys.*, 26, 1238-1243 (1955).
3. Kurkian, C. R., and J. C. Williams. "Ceramics and Glass," *Physical Design of Communication Equipment*, edited by W. O. Fleckenstein (in press).
4. Brown, R. "Substrates for Tantalum Thin Film Circuits," *Am. Ceramic Soc. Bulletin*, 45, 720-726 (1966).
5. Winkelmann, A., and O. Schott. "On the Thermal Endurance Coefficients of Various Glasses and Their Dependence on Chemical Composition," *Ann. Physik Chem.*, 51, 730-746 (1894).
6. Schwartz, N., and R. W. Berry. "Thin Film Components and Circuits," *Physics of Thin Films*, vol. 2, edited by G. Hass and R. E. Thun. New York: Academic Press, 1964.
7. Orr, W. H., and A. J. Masessa. "The Effect of Substrate Ion Migration on Thin Film Tantalum Nitride Resistors," *Electrochemical Society Extended Abstracts of Electrodeposition Division*, 8, 52 (Oct., 1965).

The following general texts cover the specialized fields noted in the title:
8. Morey, G. W. *The Properties of Glass*. New York: Reinhold Publishing Corporation, 1954.
9. Holland, L. *The Properties of Glass Surfaces*. New York: John Wiley and Sons, Inc., 1964.

Chapter 10
Pattern Generation

The production of electrical components and circuits from films requires the generation of geometrical patterns in the films. Pattern generation is accomplished by either masking during deposition to permit only selected substrate areas to receive the film or by uniformly coating the substrate with film and then selectively removing it to leave the desired film pattern. Each of these procedures requires a mask containing the image which is to be transferred to the substrate. Other means of processing films, such as laser-beam shaping of thin films and direct photochemical etching of substrates, do not require masks, but these procedures will not be discussed in this chapter.

Several masks are ordinarily needed to fabricate a complete thin film circuit. These include the aperture types, such as etched mechanical masks and silk-screen stencils, and the photomasks, such as high-resolution photographic emulsions on glass or polyester bases and chromium on glass masks. An introduction to their uses is provided in the example below.

10.1 GENERAL EXAMPLE

The shaping of a thin film circuit can involve any or all deposition and masking techniques in a variety of sequences. One example, the preamplifier shown in Fig. 10-1, is described to demonstrate the use of some of the shaping techniques. The circuit layout, shown in Fig. 10-2, includes information for making two photoetch masks, three silk screens, and a combination metal mask, which are shown in Fig. 10-3. A machined mask, Fig. 10-3(d), was also required for sputtering capacitor electrode areas, and the drawing used to fabricate the mask is shown in Fig. 10-4.

The sequence followed to fabricate the circuit was:

1. The substrate was coated with tantalum nitride resistor film.
2. Nichrome®, gold, and palladium were evaporated on the tantalum.

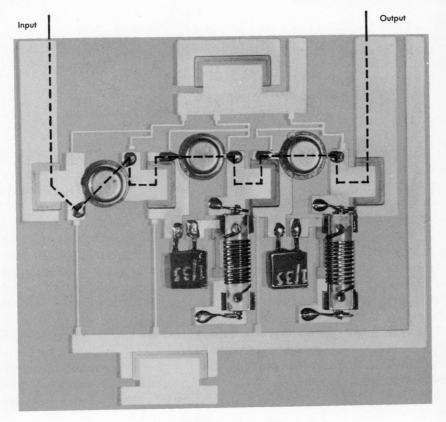

FIG. 10-1. Thin film preamplifier circuit.

3. The conductors and resistors were photoetched.
4. The capacitor areas were defined by sputtering their tantalum anodes in place through the machined mask.
5. The silk screen was used to apply grease to all areas except the capacitors and to anodize these areas.
6. A second silk screen was used to define the resistor anodization areas which were subsequently anodized to adjust their resistance values.
7. The substrates were baked at 250°C for 5 hours.
8. A third silk screen was used to protect the remaining circuit while the capacitor areas were back-etched and reanodized.
9. The gold capacitor counterelectrode film was evaporated through a combination metal mask.

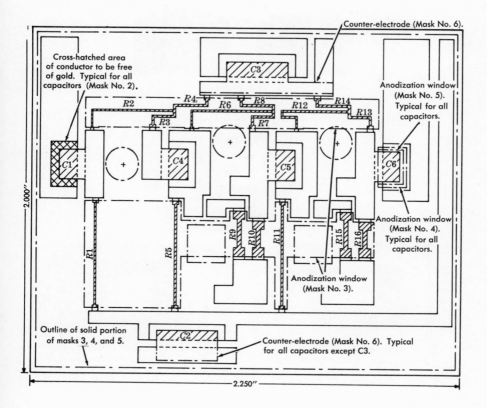

FIG. 10-2. Thin film preamplifier circuit layout.

10.2 PHOTOLITHOGRAPHIC PROCESSING

Photolithographic techniques for area-controlling the chemical reactions utilized in thin film and semiconductor processing have been
used for at least a decade. Most of the planar technology now employed in the fabrication of semiconductor devices, semiconductor
integrated circuits, and thin film circuits depends upon this art-
cum-science labeled photolithography. The general practice has been
to generate the primary pattern at from 10X to 1000X the final size,
then to reduce it to final size by photographic means, to use a contact
printing procedure to transfer the image to the photoresist-coated
substrate, and finally to etch or plate the pattern into the substrate.

(a) Photoetch masks

(b) Silk screens

(d) Machined mask

(c) Combination metal mask

FIG. 10-3. Masks used to generate thin film preamplifier circuit pattern.

Photomasks

The masks used in exposing the photoresists are usually precise photographic reductions made by copying an accurate large-scale facsimile of the circuit pattern called an "art master." The art

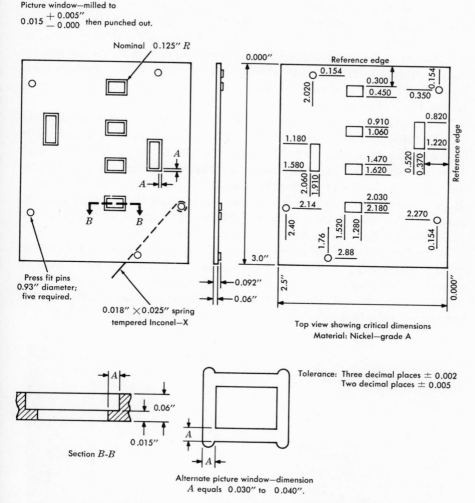

FIG. 10-4. Engineering drawing of machined metal sputter mask.

masters are "cut" with the aid of a "coordinatograph" by scribing lines through a red plastic overlay on clear Mylar® film. The red layer is stripped from the unwanted areas, and the pattern remains red on a transparent background. Art masters are cut with a nominal tolerance of ±0.5 mil and are generally made 10 times the actual circuit size. For circuit linewidths of less than 5 mils, the art master may be prepared at greater magnifications. It is at this point that all the necessary compensations for limitations of the lenses and photographic recording media must be introduced so that the final device dimensions will correspond to the designer's intentions. The art master is photographically copied and reduced to make the master mask, which is the size of the actual circuit. After photographic reduction in size is accomplished, contact prints are prepared for use as working photomasks.

The contact-printing process consists of holding the master mask in intimate emulsion-to-emulsion contact with an unexposed plate and exposing the pair to light. Care must be taken to provide a high level of cleanness; dirt particles introduced between the plates prevent good contact and often print out as unwanted images. The result is poor copies, and the dust remains to damage the emulsion of the master. For this reason, contact-printing operations are carried out in a laminar-flow clean hood where possible.

A contact printer which was designed and constructed for this purpose is illustrated in Fig. 10-5. It has several features not found on commercial units. For example, the photographic master is used to form a vacuum seal, thus eliminating additional glass surfaces that introduce interference fringes, which degrade copies. Provision is

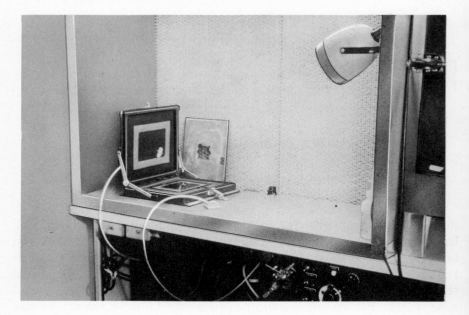

FIG. 10-5. Contact printer.

also made for locating the image with respect to the plate edges. The master can, without being removed from the frame, be inspected for dirt or damage between printings. The first copies, which are made directly from the master, are called submasters and are in turn contact-printed to produce the production copies. Where masks other than the photographic emulsion-on-glass type are required, the image is transferred to the desired material by contact printing with photoresist chemicals.

Photomask fabrication has also been performed with the aid of a computer. One technique is to supply the computer with the coordinates of the circuit components and have it control the cutting of an art master on the coordinatograph. In another technique, the computer is made to trace the geometrical pattern directly on photographic film to produce a mask. This procedure facilitates the rapid production of "bread-board" masks during circuit development. Another procedure, similar in principle, is to utilize computer program control of a scanning electron beam to generate the primary pattern on the phosphor-coated face of a cathode ray tube and record this image photographically. This technique suffers from defects in the phosphor raster, but these may be compensated for by appropriate feedback from test patterns into the program control. Another technique promising high recording speed and resolution is the direct exposure of photographic master plates at 10X final size in an electron beam machine. As pattern complexity increases and finest dimensions decrease, this type of versatile high resolution recorder will become essential.

In the discussion of the photographic procedures used to produce reduced and step-and-repeated images, we shall assume that the art master is a red plastic overlay on a clear polyester base.

Photographic Techniques

The making of photographic masks is, in principle, not different from the taking of ordinary photographs, but the scale of the images is drastically reduced. The technology has thus been termed microphotography. The loss of sharpness that would normally result from this reduction of image size is combated by using very fine grain emulsions (which are very slow) and by employing only the center of the camera lens. These measures result in restrictions in processing that must be accepted as inevitable inconveniences. For example, the slow emulsions require relatively long exposures and high levels of illumination, and only stationary originals can be readily used.

The camera must also be fixed firmly so that the aerial image does not move with respect to the emulsion during exposure.

The art master previously described is usually supported on glass and illuminated from behind; it is located at the "object plane," as shown in Fig. 10-6. Both the art master and the photographic plate are centered on the optical axis of the camera lens system, with the photographic plate located at the "image plane." The tolerance in the position of the art master along the optical axis, within which no deterioration of image sharpness can be seen, is known as the "depth of field." The corresponding, but much smaller, tolerance in the position of the emulsion is known as the "depth of focus."

The object of the photographic process is to precisely reproduce, with a specific reduction, the pattern cut in the art master on the emulsion of the photographic plate. The procedures used in this step tax the capabilities of both the lens system and the photographic materials; thus, all of the critical elements in this process are discussed.

Cameras. A camera system consists of three basic components: a screen or art master support, a lens transport, and a photographic plate holder. The construction of the conventional graphic arts

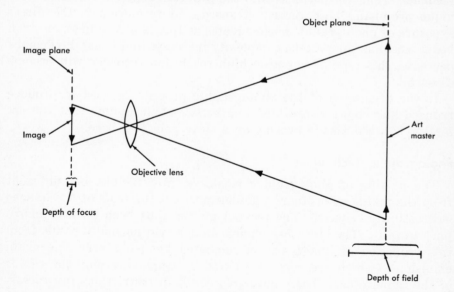

FIG. 10-6. Geometry of optical system for microphotography. (Reproduced with permission from G. W. W. Stevens, *Microphotography*, 1957, Chapman and Hall, London.)

camera allows for movement of both the screen and the lens along the optical axis, and this is satisfactory only when there is liberal tolerance in image formation. For photoshaping applications, the most desirable approach to guarantee repeatable results is a fixed-reduction system, but the lack of flexibility can be a drawback. If several different reduction ratios are required for process flexibility, the cost of having many separate fixed-reduction systems could be excessive. The next best approach is a camera design which provides the reliability of a fixed-reduction system but allows for the different reduction ratios by the use of a series of selected lenses. This implies that the system be designed to be capable of accepting the lenses and be ready for immediate use without needing further checks for focus, reduction, collimation, and exposure.

Regardless of how nearly perfect other arrangements may be, sharp images will never be produced if vibration causes any relative motion of the three components of the system. Vibration is extremely important since exposure times for the high-resolution emulsions are long, seldom less than several seconds and, on occasion, as long as several minutes. An effective method of minimizing the effects of vibrations is to mount all camera components on a common rigid base supported on isolation mounts. Several commercial mounts allow better than 98 percent isolation from transient vibration for frequencies down to 1 or 2 Hz.

In a variable-reduction camera system, two of the three components (screen, lens, and plate) must be capable of being moved. In most process cameras, the screen and lens are the movable components in order that the negative plate can remain as a fixed work station in a darkroom area. This arrangement may be convenient to work with, but alignment is difficult to maintain. Moving a structure as large and as massive as the screen makes it difficult to ensure repeatability of position and perpendicularity of the plane of the screen to the optical axis of the system. Since a transport for the photographic plate is considerably smaller, it is more practical for it to be movable with the screen being permanently fixed. At least two of the components of the system should be provided with a means of collimation. The screen must be made parallel to the photographic plate, and the optical axis of the lens must be placed on a line perpendicular to the two parallel planes. These adjustments are essential for photoshaping applications to prevent image distortion.

Camera Lenses. An optical system is subject to problems of spherical aberration, coma, astigmatism, curvature of field, and distortion. These subjects are treated in optics textbooks [1], and are not discussed here except to say that these problems are all aggravated by increasing the aperture. This section treats only those effects that are uniquely critical for photoshaping microcircuits.

The resolving power of a lens system is measured in terms of its ability to produce separate images of closely spaced objects. This power is limited by the ratio of the focal length to the lens aperture. This ratio is called the f-number of the lens. The limit on lens resolution is imposed by diffraction (because of the wave nature of light) and is given by the relation $S = f\lambda$, where S is the image separation distance, f is the ratio of focal length to aperture for the lens, and λ is the wavelength of light used. The resolving power (RP) of photographic objectives is more often quoted as the reciprocal of separation in lines, or cycles, per millimeter, with each cycle being comprised of a line and a space. Thus,

$$RP = \frac{10^7}{\lambda f} \text{ lines per millimeter}$$

where λ is in angstroms.

Photoshaping tools present an optical system with a challenge not normally encountered in photography; i.e., the simultaneous achievement of a resolution approaching the limit of light optics for the contrast required, a uniform image quality over a selected field size, and a distortion-free field. The inherent characteristics of any optical system are such that the highest degree of resolution can be achieved only at the expense of image uniformity. It has already been shown that high resolution requires a large lens aperture. But increasing the aperture also increases the geometric aberrations, particularly the curvature of field. The reciprocal relationship between resolution and flatness of field points to the major compromise in the selection of optics for photoshaping. This compromise is illustrated in Fig. 10-7. One curve represents diffraction-limited resolution, and the other represents the decrease caused by aberrations which increase the radial distance from the lens axis. It is seen from these two curves that maximum resolution is achieved at the point where they cross; the resolution decreases to the right because of aberrations and to the left because of diffraction. A lens should be set at the optimum aperture for both photographing and focusing because the plane of best focus changes with aperture if the predominant aberration is spherical aberration.

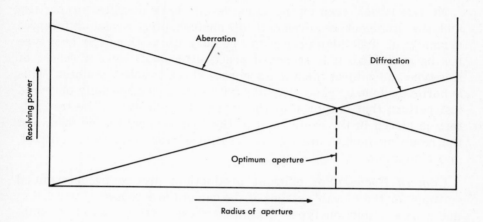

FIG. 10-7. Resolving power versus radius of aperture, as limited by diffraction
and aberration.

The size of the flat field over which a lens will give satisfactory
resolution is limited. Lenses compounded from spherical elements
normally reproduce a flat original as a curved surface which is con-
cave toward the lens. The useful area of the field is the portion of
the curved surface that deviates from the paraxial focal plane by no
more than the depth of focus of the lens. The size of the useful field
is thus related to the focal length of the lens. This can be easily understood by comparing the curvature in a fixed arc length for circles of
increasing radii. The larger the radius, the less is the curvature and,
in general, the larger the focal length of the lens, the less is the
curvature of field. Lenses designed for the graphic arts are considered to be well corrected for curvature of field and, in normal
applications, are intended to cover a field diagonal equal to the
focal length.

For the most part, lenses for photoshaping have been selected from
optical systems intended for other applications. The alternatives are:
1. The use of a very high-power lens, such as a microscopic ob-
    jective, for very high resolution over a small field.
2. The use of a long focal length lens, such as a graphic arts lens,
    and the acceptance of its lower resolution to gain the increased
    field.
To obtain optimum results on the most sophisticated microcircuitry,
however, lenses should be specifically designed for these applications.

No two lenses, even of the same design, have identical properties, and the stringent requirements of photoshaping reveal differences normally of little significance in ordinary uses. Once the lens type has been decided, it is standard practice to obtain several lenses of this type and subject them to a rigorous check to select the best. This is normally done by photographing with each lens a specially prepared test pattern that is typical of the expected application. The test pattern is placed in the center and in the four corners of the field, and the resulting images are closely scrutinized for resolution, contrast, and distortion.

Camera Focus. For efficient production, precise predetermined settings for the movable stages of the camera are necessary when it is not a fixed-reduction type. Obtaining these settings involves establishing sharp focus and the desired reduction ratio simultaneously. Once this has been achieved, the mechanical problem of maintaining it must still be solved.

The plane of sharpest image formation should be placed at or just below the emulsion surface. For the low *f*-number systems employed in photoshaping, this placement can be made satisfactorily only if the means of moving the image plane and of sensing its location are sufficiently sensitive. The normal technique of using a ground-glass screen in the image plane to scatter the light and to make the image appear as if it were contained in the screen is convenient but not precise. A superior method, illustrated in Fig. 10-8, employs a photographic plate whose emulsion surface has been lightly scratched or dirtied with a fingerprint. The intention is to view the dirt particles on the surface of the emulsion at the same position that the plane of the aerial image is located. This is best accomplished by affixing

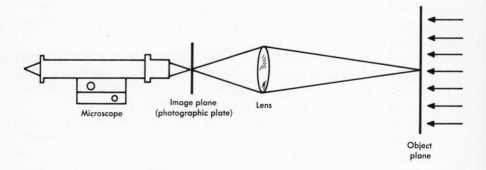

Fig. 10-8. Visual focusing technique.

the microscope to the base supporting the photographic plate so that the two may be moved as a pair. This visual method of focusing will permit observing the changes in the image, but it will not show the response of the emulsion to these changes. This can only be shown by actual photographic tests, and knowing the response is especially important when uniformity over a finite field area is desired. The most reliable method is to record the image in the emulsion for a series of focus settings close to the visually determined focus setting and to inspect the developed images at high magnification under a microscope to find the optimum focus.

The camera must be designed so that the lens-to-emulsion distance can be reproducibly altered by very small increments. The size of the increment should be sufficiently small to allow two or three settings around the correct focus to give almost equally sharp images, but large enough to enable observable changes in image quality on either side of the optimum focus without having to make too many images in a series. The aperture of the lens is a factor in determining the optimum size of the increment, as indicated by

$$\text{Increment size} = \lambda \frac{\sqrt{1 - (NA)^2}}{(NA)^2}$$

where NA is the numerical aperture of the lens and λ is the wavelength of the incident light.

Illumination of the Art Master. In photoshaping applications, uniform illumination of the art master at maximum contrast is essential in order to provide equal exposure over the entire field. For this reason, the art master is prepared in the form of a transparency rather than as an ink drawing. The two most common methods of illumination are the condenser lens system and the diffuse screen. Condenser systems provide efficient utilization of the available light and are employed when condenser optics are available that can illuminate the entire area of the transparency. The diffuse screen is used when it is necessary to illuminate transparencies larger than can be handled with condenser systems.

Photographic Emulsions and Plates. Many materials could conceivably form the basis of a photographic recording system, each with a different degree of sensitivity. Most, however, are highly insensitive, including many which have gained widespread application for image recording. For example, the prints or drawings which comprise part of an engineer's everyday life are made either by the blueprint process, which depends on the change in oxidation state of an iron salt when exposed to light, or by the diazo process, which depends on the decomposition of a diazonium salt by light. In either case, not more than one molecule of the compound can be affected by a single quantum of light, and seldom is a single quantum sufficient; hence, these materials are satisfactory only when used with light of high intensity. Many relatively low-sensitivity materials are used in photoshaping applications in the electronic industry today, such as the photoresist materials previously discussed.

Of all the processes available, the familiar photographic process involving silver halide emulsions* is considered to have the highest sensitivity. A silver halide emulsion consists of grains (crystals) of silver halide dispersed in a gelatin layer on some support such as a glass plate or a plastic film. This process has been in use for well over 100 years, but it still eludes a complete explanation. Until recently it has been classed as an empirical art and thought of as a chemical mystery. James and Hamilton [2] indicate that an explanation is dependent on solid-state physics as well as chemistry, and that the concepts of holes and electrons, crystal dislocations, and charge transport are as important to the explanation as are reduction, solubility, and catalysis. The photographic system appears to act as an amplifier with an amplification factor (in some instances) as high as 10^9 and with such sensitivity that it can respond to an energy input as small as 10^{-11} erg.

Silver halide itself is limited in the same sense that the iron and diazonium salts are limited, since only one molecule of silver halide can be affected by a single quantum of light. The greater sensitivity arises from an amplification during the developing process. The few silver atoms released upon the action of light form a nucleus around which other silver atoms collect during development until the number has been increased by as much as a factor of 10^9.

The basic photochemical reaction in a silver halide is the transfer of an electron from the halide ion to a silver ion:

$$Ag^+ + Br^- \xrightarrow{h\nu} Ag^0 + Br^0$$

*Although the use of the word "emulsion" to describe light-sensitive layers is not chemically correct, it has gained universal recognition in photography, and will be so used here.

The electron is released from the halide ion by the absorption of a quantum of light and wanders about in the crystal until it finally becomes trapped at some site. Here, it may be joined by a silver ion to form an atom of silver. A second electron released by another quantum of light may be trapped at the same site to form another silver atom. It is the formation of these silver clusters that makes it possible for the developer to act on the crystal to reduce more silver ions and produce one of the minute silver particles contained in the opaque regions recognized as images.

The prime concern in photoshaping applications is to establish a sharp, well defined image in the emulsion layer, and the properties of an emulsion must be such that this image is not degraded by any of the following conditions:

1. *Refraction at emulsion surface.* The different indices of refraction of air and the emulsion cause the entering rays to bend at the air-emulsion interface, which makes a cone of rays to be less convergent inside the emulsion than in air, as shown in Fig. 10-9. If the emulsion surface is not smooth, entering light

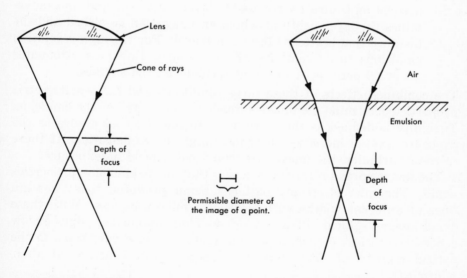

(a) Image formation in air   (b) Image formation in emulsion

Fig. 10-9. Refraction effects at the surface of a photographic emulsion. (Reproduced with permission from G. W. W. Stevens, *Microphotography*, 1957, Chapman and Hall, London.)

rays will be refracted at different angles, partially destroying the image. Hence, a smooth surface is essential to prevent such nonuniform refraction.

2. *Light scattering from grains.* The amount of light scattered by grains in an emulsion is related to grain size, and less scattering arises from smaller grains. Emulsions suitable for producing high-contrast images of the order of 1 micron in size must possess small grains (0.2 micron and less) distributed uniformly in the gelatin layer. The wavelength of the exposing light also affects scattering, but this factor is negligible in high-resolution emulsions as long as the grain size is a fraction of this wavelength. While diffraction effects are minimized by using shorter wavelengths, increasingly smaller images down to the grain size cannot be produced simply by using shorter wavelengths of exposing light. The unfortunate circumstance is that scattering in the emulsion increases very rapidly as the wavelength and grain size approach equality, and the gains expected from decreased diffraction are not attained.

3. *Absorption of light energy passing through the emulsion.* Light absorption is directly releated to layer thickness, and thus determines the layer depth at which an image can be placed and the distance that scattered light can travel. For the high-resolution emulsions in current use, the total layer thickness attenuates about 60 percent of the light incident on the surface.

The combined effects of these three conditions and the sensitometric properties determine the image spread or "turbidity" of the emulsion. Turbidity is defined as the increase in image size that occurs as the exposure level is increased to achieve high image densities. If there were no turbidity, the image size would not depend on exposure.

The image spread referred to here is that due to emulsion properties alone. The microelectronic patterns being produced today contain lines of different widths with some of 1 micron or less. With these small images, optical diffraction effects alone may cause edges of low contrast with loss of definition and may also cause variations in the optical density of developed line images as a function of their width [3].

A typical high-resolution emulsion layer is 6 microns thick before chemical processing and is reduced to 4 microns by the processing; it is very thin relative to its glass support. Although it has been generally accepted that the emulsion thickness must be small compared to the width of the finest black line required, this need not be a limitation. If the lens system has a small depth of focus, the image is confined to a small upper portion of the emulsion layer; thus, the depth of focus may be of more importance than the emulsion thickness.

The emulsions used for photoshaping applications are supported by transparent plates which can contribute internal reflections and degrade image quality. Oblique rays from a lens or light scattered by the emulsion can pass through both the emulsion layer and the support, reflect from the back surface of the support, and return along a different path to expose the emulsion layer. The effect of this scattered light is called halation and in its most recognizable form appears as a halo around the periphery of an image. Unfortunately, thick supports and the high-contrast photographic process required for photoshaping tend to enhance the conditions for halation. The thick supports increase the halo diameter, and the high-contrast process magnifies and records the low levels of reflected energy. To counteract this situation, the back surface of the support is coated with a thin layer of light-absorbing material (called an antihalation surface) which is removed in the developing process.

The spectral transmission of the support, while not important in photoshaping-tool manufacture, is important in its use. The spectral transmission characteristics for commercial photographic glass is shown in Fig. 10-10. The important portion of the range extends from 6000 A down to the sharp cutoff at 3000 A in the long ultraviolet region, since this is the spectrum of energy which can be used to expose resist materials in device manufacture.

Whereas the optical properties of the support affect image quality, the mechanical properties determine the ease or difficulty of obtaining focus. Glass is the most common support material for photoshaping applications because of its dimensional stability. Its humidity coefficient is negligible and its thermal coefficient of linear expansion is only 8 ppm/°C. Dimensional changes caused by temperature, when they occur, are reversible so that glass serves as a reliable support material where accuracy of image placement is of prime importance.

The emulsion-coating operation produces a layer of emulsion approximately 6 microns thick which conforms to the topography of the support material. If the surface of the support is rough, the emulsion surface will exhibit this same roughness and tend to have a disruptive effect on images, especially those of small size. To insure the smoothest surface possible, flat drawn or polished glass is used.

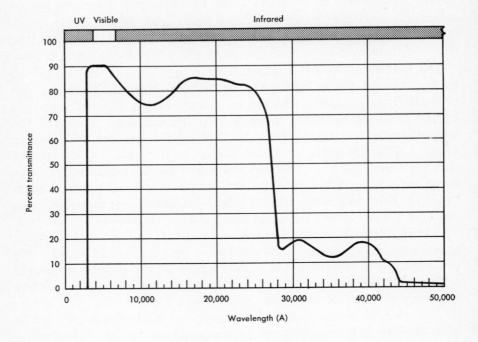

FIG. 10-10. Spectral transmission of photographic glass [4]. (Reprinted with permission of Eastman Kodak Co.)

In addition to being stable and smooth, a support material must meet certain dimensional requirements if uniform image quality is to be achieved. The emulsion surface of the photographic plate must not deviate from a true plane by more than the depth of focus of the lens if the focus in a camera system is to be maintained. The allowable tolerance is fixed by the numerical aperture (NA) of the specific lens system, with the acceptable deviation being smaller for lenses of higher numerical aperture. For example, a lens of 0.1 NA requires a flatness of 50 microns, while a lens of 0.5 NA requires a flatness of 2 microns over the entire image area. The types of commercially available photographic plates are given in Fig. 10-11.

Image Development. After the latent image has been established by exposing the photographic emulsion to light, it must be developed [5] to produce a visible silver image. In the conventional process there are two paths by which the mechanism can proceed. Generally

Plate type	Flatness (mil/in.)	Thickness (in.)	Remarks
Selected flat	1.0	0.030 to 0.130	Flat drawn fire polished
Ultra flat	0.5	0.060 to 0.250	Flat drawn fire polished
Ultra flat polished plate	0.5	0.190 Max.	Polished plate free from blisters and scratches
Micro flat	0.02	0.250 only	Polished plate or float

FIG. 10-11. Types of photographic plates [4]. (Reprinted with permission of Eastman Kodak Co.)

both are in operation, but one predominates. The first of these is called chemical or direct development, and in the case of photoshaping applications it is the one which occurs first and dominates the process. The second, for purely historical reasons, is called physical development, or more properly as it applies to photoshaping, "solution" physical development, and is responsible for the attainment of high density, high contrast images.

Chemical Development. In this process, the exposed grain is reduced very rapidly at one or more discrete points on the grain surface, corresponding to the latent image nuclei, and proceeds from these points until the entire grain is reduced. A catalytic reaction proceeds at the interface between the silver nucleus and the silver halide crystal, where silver atoms and silver ions are closely associated. The silver ions are reduced to silver, and the halide ions pass into solution. A new silver/silver-halide interface is thus formed, the reaction continues, and the silver nucleus grows in size. This growth has a preferential direction to form filaments of silver as shown by electron micrographs.

Physical Development. In this process, the developer contains silver ions (either as an intentional addition or as a result of the solvent action of the developer) in addition to the reducing agent. These ions are reduced to metallic silver by the same catalytic chemical action that takes place in the chemical development process, but the reduction can occur at any place the nucleus contacts the developer. The result is a total overall enlargement of the nucleus rather than the preferential filamentary growth.

It might be expected that the extremely small size and large surface area of grains in high-resolution emulsions would permit them to be more rapidly developed than coarse-grained emulsions. The reverse is true, however. The optical density given by high-resolution plates continues to increase long after a conventional emulsion would have completed its development. The density increase in such cases is accompanied by an increase in the size of silver particles until they are larger than the grains of the unexposed emulsion. The effect, particularly marked at lower densities, is the result of physical development. The explanation of this sensitivity behavior is connected with the enormous surface area resulting from the high density of minute silver halide particles in the emulsion layer. The greatest increase in grain size occurs where the number of nuclei is small and the local silver ion concentration is the highest, i.e., in the areas that have received the smallest exposure. Grain growth becomes obtrusive when line images are made with an optical system that fails to produce a really sharp line. This produces a low-density fringe whose edge is markedly affected by grain growth and results in a line with a central black core flanked by a ragged fringe of coarse grains.

Processing. Regardless of the process, the developer and development time are of paramount importance. A brief development time gives an image of fine structure with moderate contrast; lengthening the time improves contrast but increases grain size. Thus, a compromise of image sharpness and resolution is often necessary.

Undeveloped silver halides are normally removed from emulsions in a solution containing sodium or ammonium thiosulfate. This solution is called a hypo or fixing bath and it must:
1. Dissolve the undeveloped silver halides completely.
2. Form salts with the dissolved silver halides that are soluble in the fixing bath itself and are stable when diluted (to eliminate decomposition during subsequent washing).
3. Not attack the gelatin or the emulsion support.
4. Not seriously affect the silver grains of the developed image.

The chemistry of fixation is complex and is not completely understood. It is known, however, that the thiosulfate forms several types of stable, complex compounds with silver ions and thus reduces the concentration of free silver ions in solution. When the total quantity of dissolved silver (and hence the total quantity of silver-thiosulfate complexes) reaches a certain level, the fixing bath is said to be exhausted and should not be used beyond this point.

After the image is fixed, the thiosulfate must be thoroughly removed by a water-washing to prevent its oxidizing and destroying the image. So-called hypo-eliminators or cleaning agents are useful to remove the thiosulfate and complex silver salts rapidly from the gelatin layer. These solutions are intended to oxidize thiosulfate to tetrathionate, but do not remove all of the thiosulfates. They do, however, reduce the normal washing time from 20 minutes to 5 minutes.

The highest yield of quality silver-emulsion photographic masks can only be obtained by operating under clean room conditions and following explicitly the manufacturers' recommended handling and chemical processing techniques with the following additional step [6]. Between the first washing step to remove hypo and the final distilled water rinse, the plate should be scrubbed lightly with a 1% solution of sodium hydroxide, using a very soft sponge. This tends to remove particulate matter from the gelatin surface which could create pinholes in subsequent photoresist exposures.

When an exposed photographic emulsion layer is immersed in a developer, grains which have received more light are reduced to silver more rapidly than those which have received less. The developer thus functions as an amplifier, magnifying the effect of the exposure.

The quantity of silver formed in an image is recognized in terms of its ability to block the passage of light. This ability is generally measured as a transmittance, T, or opacity, $1/T$. Transmittance is defined as I_t/I_o, where I_t is the intensity of the transmitted light and I_o is the intensity of the incident light. In practice, the optical density, D, is commonly used in image evaluation and is defined as the logarithm of opacity. Thus, $D = -\log T$.

Hurter and Driffield suggested that a plot of optical density versus exposure would yield a curve characteristic of the emulsion layer for the chosen development condition [7]. This curve is commonly called the characteristic curve and is illustrated in Fig. 10-12. To obtain this curve, a group of intensity-scale exposures on a photographic plate is produced by a sensitometer. The sensitometer directs light from a source of known quality and quantity through a photographic step tablet on the photographic plate. The plate is processed for a given

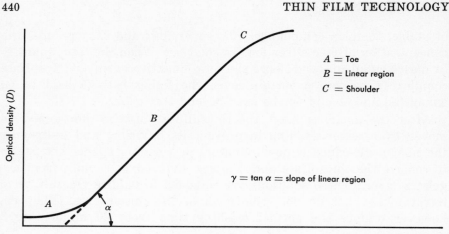

FIG. 10-12. Emulsion characteristic curve [8]. (Reprinted with permission of Eastman Kodak Co.)

time under conditions as uniform as possible. The effect of the intensity-scale exposures is then measured in units of optical density.

In the characteristic curve, any given increment, ΔD, corresponds to an exposure increment of $\Delta \log E$, and the ratio of these two is the slope, S, at any point on the curve. This ratio corresponds to the gain of the developer amplification process. If the slope is greater than 1, the contrast of an image will be increased; if the slope is less than 1, the contrast will be decreased. The slope in the linear region of the curve is commonly called gamma and is a major factor affecting the contrast and edge sharpness of a photographic tool. In the ideal case, gamma would be infinite, since exposures below a given level would result in absolutely clear areas and those above would result in maximum density, with the transition from clear to opaque occurring over an infinitesimal region. This is not possible in practice, but gammas of maximum value consistent with other pertinent image criteria (such as graininess) are sought. Typical D versus $\log E$ curves for high-resolution plates are shown in Fig. 10-13.

Density and Density Gradients. A representative plot of density as a function of distance is shown in Fig. 10-14, which indicates that the clear areas of a mask are not completely clear and the opaque areas are not completely opaque. This itself does not present a problem, but the transition between the "clear" and the "opaque" areas occurs over a finite distance and there is an uncertainty in the location

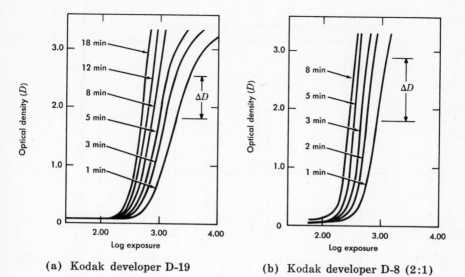

(a) Kodak developer D-19   (b) Kodak developer D-8 (2:1)

FIG. 10-13. Characteristic curves for high-resolution emulsions at various development times [9]. (Reprinted with permission of Eastman Kodak Co.)

of the image edge. An expansion of Fig. 10-14 is shown in Fig. 10-15. A comparison of measurements on a toolmaker's microscope with microdensitometer measurements has shown that the eye may locate the image edge at a density level of 0.4 to 0.8, depending on the operator, the magnification, the illumination, and the edge-density gradient. Since this is such an inexact quantity, the edge has been arbitrarily defined as the 0.3-density (50-percent transmission) level of an image. The location of the real edge depends on the application of the mask and must ultimately be determined in this manner.

The sharpest edges possible are required if reproducible image sizes are to be obtained. The edge-density gradient, a measure of image sharpness, is defined as the slope of the density-versus-distance curve between the 90- and 10-percent transmission points or the 0.04- to 1.0-density levels. (The slit width of the measuring instrument must be considered to obtain a true gradient.) It has been found that edges exhibiting high quality in device application possess edge-density gradients in the neighborhood of 2 density units per micron.

Latent Image Decay. After exposure of a photographic plate, the latent image "fades" with time, and this is especially noticeable with high-resolution plates because of the microscopic size of the grain.

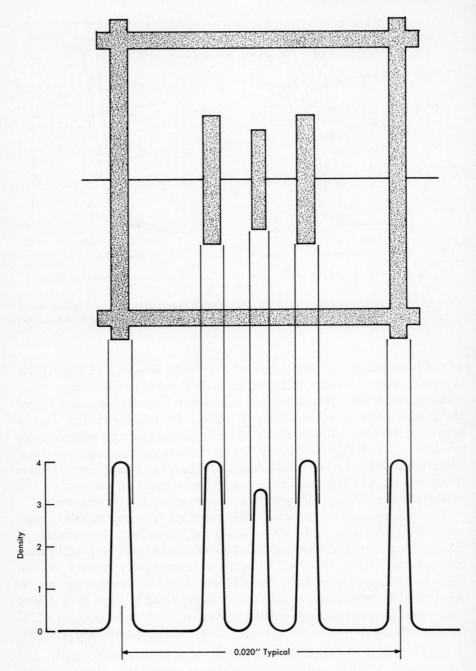

FIG. 10-14. Idealized plot of optical density versus distance for a photomask.

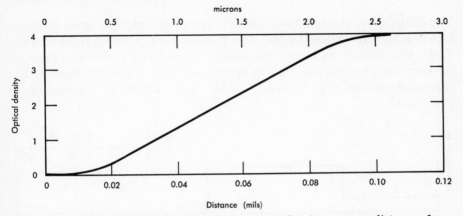

FIG. 10-15. Expanded idealized plot of optical density versus distance for a photomask.

If processing is delayed after exposure, the density becomes lower than if processing had not been delayed. As shown in Fig. 10-16, the density decreases rapidly during the first hour and somewhat less significantly thereafter. Thus, when constant density is important, a uniform time between exposure and processing must be adopted.

Adjacency Effect. This effect is of particular importance to photo-shaping applications and occurs when a high-density area is immediately adjacent to a low-density area. The influx of the relatively fresh developer into the heavily exposed area from the lightly exposed area accelerates the development at the edge of the exposed region.

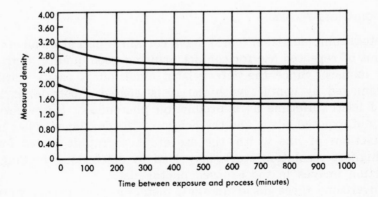

FIG. 10-16. Latent image decay curves for Kodak high-resolution plates. The curves were plotted for densities of 2.0 (lower curve) and 3.0 (upper curve) [6]. (Reprinted with permission of Eastman Kodak Co.)

The diffusion of the reaction products from the heavily exposed region into the lightly exposed area also retards development in this area. Thus, the density near the boundaries is reduced below the level of the surroundings on the "transparent" side and above the level of the surroundings on the "opaque" side. This effect is illustrated in Fig. 10-17.

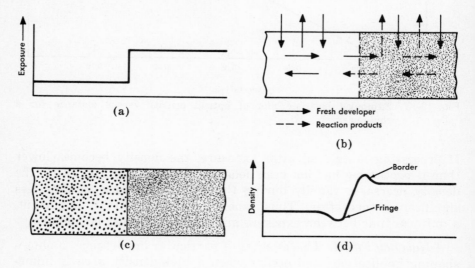

FIG. 10-17. Adjacency effect [5]. (After James and Higgins, *Fundamentals of Photographic Theory;* courtesy Morgan and Morgan.)

Metal-on-Glass Masks

Photographic emulsion masks have certain inherent disadvantages in terms of repeated use for contact printing good quality, high resolution images. Since the soft gelatin emulsion is easily scratched and damaged by contact with the resist-coated substrate, it has a limited useful lifetime as a production photomask. Since both the density of exposed silver and the height of the developed relief image are functions of line width, the use of photographic plates for the very highest resolution work may lead to deleterious fogging and diffraction images in the exposed photoresist.

To overcome these disadvantages, metal-on-glass masks were developed. Those in general use consist of a thin chromium film uniformly deposited on a flat glass substrate into which the image of the master mask has been etched using photoresist techniques. These chromium-on-glass masks are amenable to repeated cleaning procedures, are scratch resistant, and have provided up to ten times as many acceptable prints as the more conventional photographic emulsion masks. They also provide line images of uniform height and optical density.

Kodak photosensitive metal-clad plates consist of 2-inch by 2-inch by 0.06-inch glass plates coated on one side with 600 A of chromium, which, in turn, is coated with 6000 A of KTFR (Kodak Thin Film Resist), the whole being protected by a water-soluble polymeric coating. The resist is exposed to the image of a master mask by a contact printing procedure after removal of the protective coating. After development and drying, the pattern is etched into the chromium film using a ferricyanide etch. The most critical step in the successful processing of this type of mask is the development of the photoresist.

Photoresists

Photoresists are film-forming materials which undergo marked changes in solubility under the influence of ultraviolet light. For use in the planar technology of semiconductor and thin film circuit manufacture, resist systems must be capable of reproducing images with excellent acuity and high resolution, be insensitive to a wide variety of etchants, be readily and totally removable, and be safe to handle. Two classes of photoresists exist: negative and positive resist systems, differentiated by the nature of their photochemical change in solubility. Negative resists, on exposure to ultraviolet light, undergo closs-linking reactions which decrease their solubility in certain solvent systems. Positive photoresists undergo photochemical decomposition reactions which affect their dipole moments, increasing their solubility in aqueous solvent systems.

There are two distinct families of Kodak resists within the negative resist classification. These are (1) the cross-linkage poly(vinyl cinnamate) resin systems KPR, KPR-2, KPR-3, and KOR and (2) the polyisoprene plus diazido cross-linking agent systems KMER and KTFR. The positive resist systems AZ-1350 and AZ-111, sold by the Shipley Company, consist of polar phenolic type resins plus ortho-quinone diazides. An introduction to the chemistry of these systems is presented below so that the reader will have a greater appreciation of the chemical significance of each of the processing steps.

Negative Photoresists. The first family of resists mentioned above, referred to as the KPR family, consists of organic solvent solutions of poly(vinyl cinnamate) of intermediate molecular weight

($M_N \cong$ 200,000) plus a small percentage of added photosensitizer. This poly(vinyl cinnamate) structure is shown by:

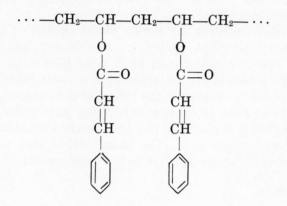

After application and drying of the photoresist film, the cinnamate side chains undergo a photochemical coupling reaction analogous to the well known dimerization of cinnamic acid [10].

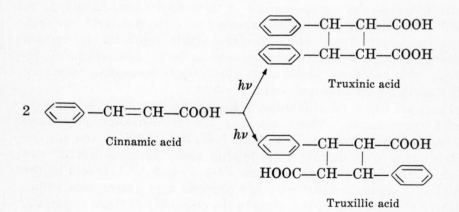

This reaction produces a cross-linked network in the exposed regions of the polymer film which markedly reduces its solubility in its prior solvents.

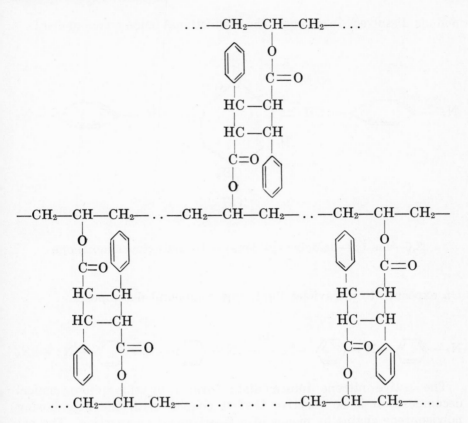

The advantages of poly(vinyl cinnamate) lie in its inert alkane backbone and in the strength of the two carbon-carbon bonds formed at each cross-link Its weakest point is the ester linkage between the side chain and backbone, which tends to be hydrolyzed in strongly basic solutions. Within the cinnamate family of photoresists, KPR-2 and KPR-3 have a higher percent solids content than the parent KPR; hence, they are capable of forming thicker, more chemically resistant films. KOR, while also forming thicker films, is unique in having sufficient added photosensitizer to give it a tenfold increase in photographic speed over KPR and to push its sensitivity well into the visible region of the spectrum. In general, KOR is too sensitive to be routinely handled in gold-fluorescent lighted rooms, and is incapable of the very highest resolution work because of the fluorescence properties of the added sensitizer.

The KMER and KTFR family of photoresists consists of low molecular weight polyisoprenes ($M_N \cong 60,000$) plus aromatic diazido compounds dissolved in xylene. The bifunctional azide compound is:

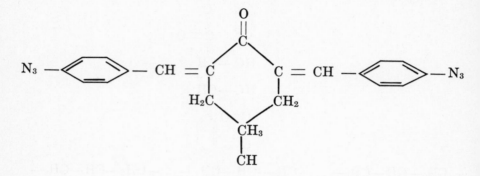

2,6—bis (p—azidobenzylidene)—4—methylcyclohexanone

On exposure to ultraviolet light, this compound decomposes

$$N_3 - \langle \rangle - R - \langle \rangle - N_3 \xrightarrow{h\nu} :N - \langle \rangle - R - \langle \rangle - N: + 2N_2$$

The active nitrene intermediate formed by this photochemical decomposition is so reactive that it couples with the neighboring polyisoprene chains by means of a C—H insertion reaction. The net result is once again a completely cross-linked polymeric network of reduced solubility. KTFR is designed for use in thinner films than KMER to provide higher resolution. It is formulated as a cleaner system than KMER and also has a higher optical density. KMER, on the other hand, is the leader in etch resistance among common photoresists.

Positive Photoresists. The AZ-1350 and AZ-111 positive-image forming photoresists sold by the Shipley Company consist of organic ester solutions of orthoquinone-diazides plus carboxy-methyl-ethers of polymeric phenol formaldehydes. These photoresists, soluble initially only in organic solvents, become soluble in aqueous alkaline solutions after exposure to light. This change in solubility depends upon the photochemical decomposition of the orthoquinone-diazide followed by rearrangement and a subsequent hydrolysis resulting in a carboxylic acid:

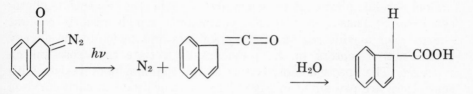

The relative change in solubility is not as great as in the case of the cross-linking polymers and, therefore, development conditions are much more critical. The polymeric component present, which is to serve the etch resist, is marginally soluble in both the organic solvent and the aqueous alkaline developer. This quality decreases its potential etch resistance, especially in basic aqueous systems, but it offers the important advantage of simplified post-operation removal of residue.

Photoresist Technology. The most critical step in the photoresist process is the exposure to ultraviolet light. It is during this step that the significant chemical change in the solubility of the system takes place. Next in order of chemical significance is the development step. In the case of negative photoresists, this involves the dissolution of nonexposed material by organic solvents. During such development, the polymeric network in the exposed region tends to absorb a large quantity of solvent and swell. Swelling and subsequent retraction during drying can detract from image quality and weaken adhesion to the substrate. Theoretically, swelling can be minimized by increasing the cross-linking in the network or by using a developing agent which has a large molal volume or is thermodynamically a poor solvent [11]. The first alternative is not possible in the case of photoresists since it would require overexposure with the inherent danger of fogging the image. Some developing systems which incorporate poorer solvents of large molal volume, such as dipropyl carbonate, have been investigated and have proven successful.

Because of the cross-linking reactions in the KPR and KPFR systems, it has been proposed to incorporate about two percent of a specific organo-silane coupling agent into these resists to improve substrate adherence. This material, γ-methacryloxypropyltrimethoxy silane, chemically couples to oxide substrates and to the cross-linking resins when heated in a prebake operation. In the case of pure photoresists, the prebake serves merely to dry the photoresist film and is optimally carried out in a flowing nitrogen ambient to minimize oxidation of the photoresist.

Commercial photoresists normally contain the essential photo-sensitizers. These are organic compounds which absorb radiant energy over a wide region of the spectrum and transfer the energy to the active centers in the polymers to initiate the cross-linking reaction. The spectral sensitivities of the common photoresists with sensitizers are presented in Fig. 10-18. The ordinate in each curve is a relative scale of sensitivity, which is the reciprocal of the integrated light intensity required for optimum exposure. The sensitizers in-crease the sensitivity of the basic photoresist system 200 to 2000 fold. The vertical lines in the illustration define a transmission window in silver emulsion photomasks near $\lambda = 3200$ A. Illumination in this band may thus serve as a source of background fog.

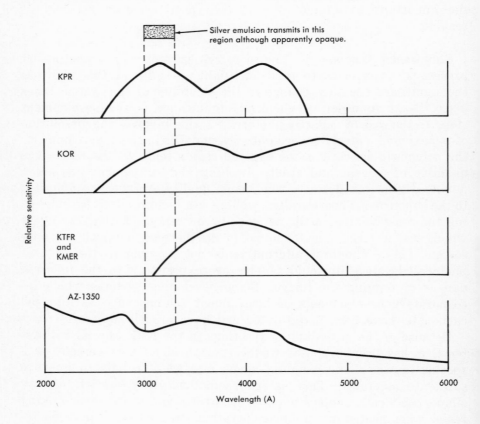

FIG. 10-18. Spectral sensitivity of photoresists. (KPR, KOR, KTFR, and KMER curves reprinted with permission of Eastman Kodak Co. [12]. AZ-1350 curve from product data sheet DS1350A, Shipley Co., Inc.)

In KTFR and KMER, both the sensitizer and the active nitrene intermediate in the cross-linking reaction are sensitive to oxygen. Therefore, unless oxygen is excluded during the exposure step, the photographic speed and resulting film thickness will be adversely affected. This effect is most pronounced in thin KTFR films used for high-resolution images.

Etching

Circuit areas protected with photoresist remain on the substrate while the unprotected areas are dissolved away in an etchant. A good etchant has a high rate of dissolution for the film to be removed and a negligible rate of attack on the photoresist and on the substrate. The pattern definition obtained in photoetching depends on mask quality, photoresist properties, photoresist thickness and uniformity, exposure techniques, development, etchant, film to be etched, and etching technique. With a given mask, the photoresist thickness, exposure, and development are of primary importance.

Some of the common films and their etchants are listed below:

Film	Common etchants
Tantalum	Mixture of hydrofluoric acid and nitric acid
Nichrome®	Hydrochloric acid, or a mixture of hydrochloric acid and copper chloride
Gold	Aqua regia, or a mixture of potassium iodide and iodine
Palladium	Aqua regia, or a mixture of potassium iodide and iodine
Aluminum	Sodium hydroxide or phosphoric acid
Titanium	Dilute hydrofluoric acid
Copper	Hydrochloric acid, ferric chloride, or nitric acid
Tantalum oxide	Hydrofluoric acid
Manganese oxide	Hydrochloric acid or hydrogen peroxide

10.3 MECHANICAL MASKING

A mechanical mask is used to block depositing atoms from reaching the substrate except in a desired pattern. Vacuum evaporation through a mechanical mask will generally provide a pattern within ±2 mils of the mask pattern, but tolerances less than ±1 mil may be obtained with special care. The precision obtained depends upon the separation of the mask and the substrate, which in turn depends on mask and substrate flatness. Masks with more open structure are less likely to remain flat in use. Sputtering imposes extra requirements on the masks since higher substrate temperatures (to 500°C) are used and masks tend to warp. This warping is especially serious since sputtering presents a wide area source, and penetration into a gap between the mask and the substrate is approximately equal to the size of the gap. Figure 10-4 is a drawing of the mask pictured in Fig. 10-3(d). This particular mask was formed by selectively thinning 60-mil nickel by milling and then precision-punching the required apertures in the thinned areas.

For mechanical masking during evaporation, a "combination" copper-nickel mask, Fig. 10-3(c), is often used, and it is constructed as shown in Fig. 10-19. Mask thicknesses to 30 mils are available with pattern tolerances of 0.1 mil. Since these masks are formed by etching and electroforming techniques, the minimum obtainable land area between openings depends on the mask thickness. Generally, the width of the narrowest land should be no less than the mask thickness. The beam-type structure under the land areas permits long thin openings with fine definition. With any mechanical mask, it is impossible to provide isolated land areas. The photoresist techniques used in making combination metal masks are similar to those used in photoetching.

10.4   OTHER METHODS OF PATTERN GENERATION

Problems arise in the use of photoetching when combinations of various films are to be etched and the etchant for one attacks an underlying film or the substrate itself. Sometimes the solution to this problem is to use a mechanical mask during film deposition, but indirect methods also exist and several examples are described.

Conformal Masks. A deposited film of an easily etched metal may be used as a mask. This technique is best described by citing a particular example. A copper or aluminum film is first deposited on the substrate, and the desired pattern is formed by photoetching openings in the film. Tantalum is sputtered over the entire surface. This combination of films is then immersed in an etchant for the masking metal, dissolving this metal and freeing the overlying tantalum. Thus, the tantalum remains only where openings had been made in the mask. This method provides a thin mask which is in intimate contact

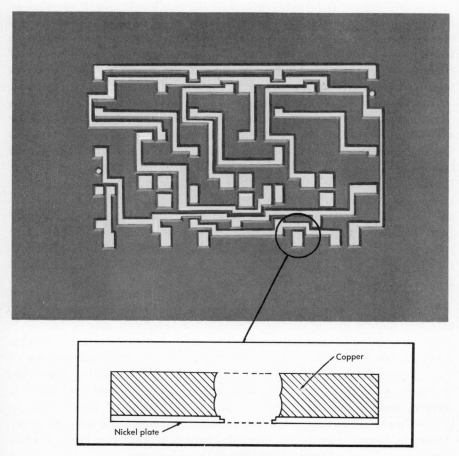

FIG. 10-19. Combination metal mask.

with the substrate, and thus provides sharp edge definition. Mild etchants are sufficient, so there is no attack of the substrate itself.

Anodic Conversion. If the tantalum film is thin enough that it may be anodized completely (this will be the case for many resistor applications), tantalum in unwanted areas may be converted completely to oxide by anodization instead of being etched away. Unless the initial film is of uniform thickness, however, islands of unconverted tantalum may be left in the "clear" areas. Of course, some method must be used to prevent the anodization of the desired tantalum, and there are several ways to accomplish this.

Anodization with Photoresist. The most obvious approach to confining the anodization to a desired area is to use the same resists used in the direct etch technique. There are several problems with this approach. Anodizing electrolytes may creep under the edge of the resist layer, and the resist itself may suffer electrical breakdown during anodization if too high a voltage is attempted.

Anodic Conversion with Aluminum Resist. Since aluminum may also be anodized, it may be used to define an anodization area. In this procedure, aluminum is deposited over the entire surface of the tantalum-coated substrate. The aluminum is then removed from the unwanted area by using a selective etch and photoresist. The entire substrate is anodized, using an electrolyte suitable for both aluminum and tantalum, until the tantalum is completely converted to oxide as just described. (Sufficient aluminum must be deposited so that it is not completely converted to oxide.) The remaining aluminum is removed with a mild etchant, leaving the bare tantalum in the desired pattern.

Etching with a Tantalum Oxide Resist. Since tantalum oxide is attacked slowly by the normal etchants for tantalum, anodic tantalum oxide may be used as a resist. The technique is quite similar to that just described, except that the desired areas are anodized instead of the unwanted areas. The tantalum must not be completely converted to oxide with this technique, and anodization is continued only to some moderate potential, such as 25 volts. Either of the resists described may be utilized to confine the anodization to the desired areas. After the resist is removed, the substrate is placed in a fluoride etch, and the oxide over the desired area acts as a resist, preventing attack on this area.

Silk Screen Stencils. Resist coatings such as grease and vinyl inks may be applied to thin film circuits for anodization masks using silk screens as shown in Fig. 10-3(b). Screen cloth is stainless steel mesh with a polymer film filling the holes in the screen except in the areas through which the resist is to be applied. The pattern in the polymer film on the screen is formed photographically. The definition obtained when a silk screen is used depends on squeegee speed and pressure, screen mesh size (from 165 to 325 mesh), and operator skill. Under the best conditions, lines may be obtained with edges defined to ± 3 mils, but registration tolerances of ± 15 mils are common and should be allowed for in circuit design.

10.5 MANUFACTURE OF PHOTOSHAPING TOOLS

Specific facilities for generating art masters, making master masks, and manufacturing aperture masks are described to illustrate the basic processes and equipment involved. The facilities described are those designed at the Allentown Works of the Western Electric Company.

Generating the Art Master

The initial step in making any photoshaping tool is the generation of an art master. The drawings supplied by the device designer provide all information to be contained in the art master from which the final mask is made. The prints specify the desired final dimensions and the allowable tolerances.

An accurate enlarged layout of the pattern is made with a precision drafting table, the coordinatograph, illustrated in Fig. 10-20. The accuracy of this instrument has been improved by providing precision engraved scales with an optical readout. The position of the cutting head may be held to an accuracy of 0.0001 inch. The combination of this with other variables provides an overall accuracy of about 0.0005

FIG. 10-20. Coordinatograph.

inch. The art master material consists of a 0.005-inch-thick clear Mylar® base covered with a 0.0015-inch-thick red plastic overlay. This overlay is easily stripped from the Mylar® after the pattern has been scribed. Figure 10-21 shows the cutting knife and roller in the correct position to cut the art master material. The roller rides on the material, enabling accurate vertical positioning of the knife. Precision adjustment of the knife is provided to limit the depth of cut to the thickness of the overlay. After the pattern has been cut, the overlay is stripped from the areas of the pattern that are to transmit light. The end product is called the art master, from which the master mask is made.

FIG. 10-21. Cutting head.

Making the Master Mask

The sequence of steps from art master to master mask depends on the type of device to be produced. Figure 10-22 is a flow chart of the steps required to make masks for both thin film and semiconductor devices.

Thin Film Mask Processing. There are some advantages in having several identical small circuit patterns appear on one mask, so that

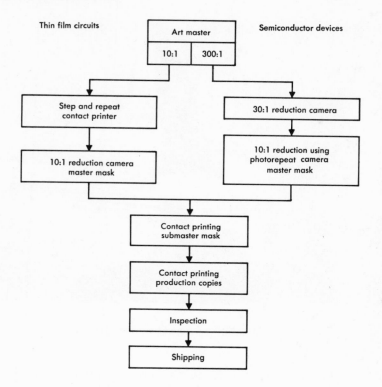

FIG. 10-22. Mask production flow chart.

they may be processed simultaneously on one substrate which is later
cut to separate the individual circuits. In some cases also there is a
repetition of the same circuit component within a single circuit. In
both cases, it is efficient to copy one art master and to generate the
multiple-image art master photographically. To obtain these multiple-
image arrays, a step-and-repeat contact printer, shown in Fig. 10-23,
is a convenient tool. During the operation, the art master of a single
circuit or component is held on a movable head. Below the head
a large sheet of unexposed photographic film is held on a flat bed.
The head, which may be positioned along two perpendicular axes,
is placed at the first array position and then lowered to bring the unit
image and the film into intimate contact. The film is exposed by a
light source mounted in the top section of the head. The head is
then moved to the next array position, and the entire sequence is
repeated. The result is an array of identical images ten times the
final size which, after developing, becomes the art master for the final

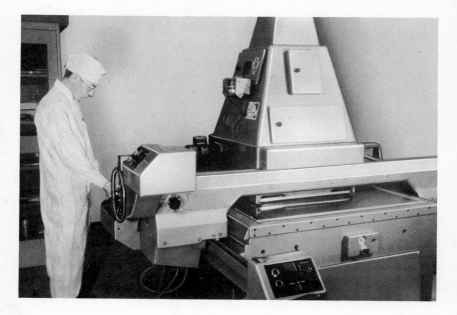

Fig. 10-23. Step-and-repeat contact printer.

reduction. An array may contain as many as 50 images. To accurately reduce this array to final size, a 10:1 reduction camera is used. This camera, like every piece of optical equipment used in the manufacture of photoshaping tools, is designed to perform its specific function and no other.

One such camera has a special base to isolate it from vibrations created by other plant equipment. This support structure is necessary to make possible the long exposures required for high-resolution plates. Figure 10-24 is a diagram of the fixed-reduction camera. The bed of this camera is constructed of 8-inch steel I-beams encased in concrete. The structure is 20 feet long, 5 feet wide, and weighs 12 tons. The complete bed rests on air mounts placed at 4-foot intervals under the structure. This construction limits the relative motion between corresponding points on the art master and negative planes to less than 10 millionths of an inch, and provides 97 percent isolation from all vibrations above 3 Hz.

Figure 10-25 shows the lens and plate holder of the 10:1 camera. This fixed-reduction camera is capable of photographing 0.001-inch lines over a 4.0- by 4.0-inch field within an accuracy of 0.0001 inch. The screen used with this camera is a single sheet of 0.25-inch plate

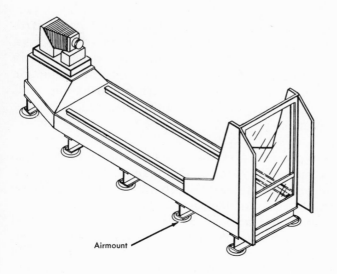

Sectional view

Airmount

FIG. 10-24. 10:1 camera diagram.

FIG. 10-25. 10:1 camera.

glass that is flat within 0.002 inch. The plate is fixed in a plane perpendicular to the optical axis within 0.50 second of arc. An electrostatic charge is used to hold the art master against the glass screen to avoid the lateral distortion of the Mylar® normally encountered in conventional methods of mounting. The product from this reduction camera is a glass master photographic mask. All photoshaping tools used in device production are second- or third-generation contact prints made from this master.

Semiconductor Device Mask Processing. The flow chart in Fig. 10-22 lists the sequence of steps followed in the manufacture of photoshaping tools for semiconductor integrated circuits and devices. These masks contain a multiple array of identical images that vary in size from 0.020 by 0.020 inch to about 0.100 by 0.100 inch. The final arrays are from 1- to 2-inch squares and may contain several thousand identical images. These masks start with an art master 300 times final size, and the first mask-processing step is a 30:1 photographic reduction.

The system used for this reduction is designed for various reduction factors, with a special lens being installed for each factor. When used for the 30:1 reduction, the camera is capable of producing 0.0001-inch lines over a 0.8-inch-square field with a position accuracy of 0.00005 inch. Like the 10:1 camera previously described, this camera has a special base to isolate it from vibration. Its bed is smaller than that of the 10:1 camera, but it is similar in construction. The exposed plate from the 30:1 camera contains the latent image of a unit detail at 10 times final size. After developing, the plate is contact-printed, and the print becomes the target (or reticle) for the final reduction. This final reduction for semiconductor device masks is made with a photorepeat camera. One such machine, shown in Fig. 10-26, consists of three units: a master reticle alignment instrument, an electronic control unit, and a reduction camera with a flash lamp and a precision coordinate stage. The reticle alignment instrument enables aligning the photographic image with the reticle holder.

The control of the flash intensity, and therefore exposure, is a function of the electronic unit. This unit also controls the spacing between exposures along one coordinate axis and the number of exposures to be made in one row.

The camera unit uses a 10-power objective lens to reduce the master reticle to final size. The photographic plate onto which the image is transferred is positioned on the coordinate table. A xenon flash

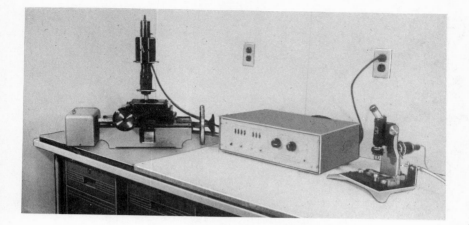

FIG. 10-26. A commercial photorepeat camera.

tube provides a high-intensity 10-microsecond light pulse that permits the reduction exposure to be made while the table is in motion. The resulting blur is only about 0.5 microinch.

This camera has had good success in producing semiconductor masks, but several limitations exist in the system. For example, the lead screws have periodic error of 1 micron. The single camera head requires that the series of complementary masks required for a device be exposed one at a time, making registration of the masks dependent upon the temperature-repeatability of the machine. A third limitation is the necessity of having an operator manually position the table in both the x and y axes before starting each row of an array; any error renders the master mask unusable. To overcome these limitations, a new system, known as the Multiple Array Photorepeat System (MAP), was designed and constructed. This system is shown in Fig. 10-27.

The MAP System can simultaneously expose an array of images or details on 10 photographic plates. Thus, up to 10 masks related to a single device may be made, with precise registration between them. The table in Fig. 10-27 summarizes the characteristics of the MAP System. The exposed plates from the single or multiple head system are the masters for semiconductor device masks. Submasters and production copies are then made by contact printing.

Mask production capability	10 masks simultaneously
Resolution	0.0001-inch lines
Registration within set of details	Approaches zero
Field coverage:	
3X	0.200-inch diameter
10X	0.070-inch diameter
Image spacing:	
X-axis	0.004-3 inches
Y-axis	0.000020-0.45 inch
Tolerance	0.000020 inch
Vibration isolation	98% above 3 Hz
Operator requirement	Make initial setup and start cycle

Fig. 10-27. Multiple array photorepeat system.

Manufacturing Aperture Masks

The processes used in making metal masks and silk screens are applications of photolithographic techniques. Precision photographic masters provide the starting point for all these tools.

Figure 10-28 offers a comparison of the relative cross-sectional thickness of various aperture masks with respect to each other. These tools have a basic similarity in that each has open areas through which a solid, a semisolid, or a liquid can be transferred or deposited upon a receiver. For example, vaporized metal can be deposited through a molybdenum foil mask onto a thin film substrate. The essential processing steps are described briefly for each type of tool.

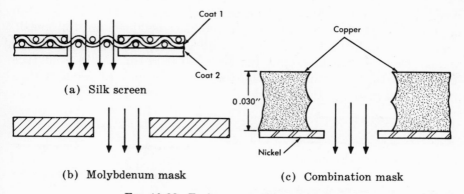

FIG. 10-28. Tool cross-section comparison.

Silk-Screen Stencils. The screen cloth is normally a 165- to 325-mesh stainless steel. The mesh, held by mounting screws, is stretched on an aluminum frame. The screens and frames are cleaned, using a hot peroxide wash, followed by a water rinse and a forced air dry. A sensitized poly (vinyl alcohol) (PVA) emulsion is squeegeed into the screen mesh, as in Fig. 10-29, to apply a smooth even coat. The screen is air-dried, and a second coat is applied to the contact surface. This second coat is from 0.4 to 1.0 mil thick and helps to cushion the steel mesh from abrasive action during use. When the second coat is dry, a pattern is exposed on the emulsion using an ultraviolet light source. The ultraviolet light makes the exposed emulsion water-insoluble, while unexposed portions of the pattern remain water-soluble. A gentle agitation of the screen in a water bath develops the pattern in the emulsion. After a thorough drying, the screen is ready to use. Some typical uses are for applying metalizing paste on ceramic headers and applying an anodization resist coating for thin film circuits.

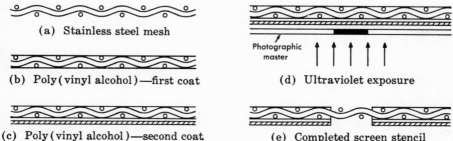

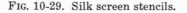

FIG. 10-29. Silk screen stencils.

Molybdenum Foil Masks. The second type of tool which is often used in device manufacture is the etched molybdenum foil mask. Figure 10-30 illustrates the various stages of processing for a "moly" mask. The foil ranges from 0.001 to 0.005 inch thick with a surface finish of 4 to 6 microinches. The foil is coated on both sides with a photoresist material by whirling.

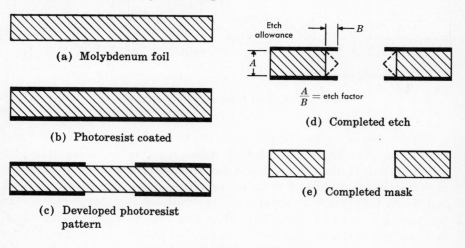

(a) Molybdenum foil

(b) Photoresist coated

(c) Developed photoresist pattern

Etch allowance

$\dfrac{A}{B}$ = etch factor

(d) Completed etch

(e) Completed mask

FIG. 10-30. Molybdenum masks.

The pattern is exposed on the photoresist coating on both sides, with alignment of both the front and back sides guaranteed by the exposure fixture. This fixture is unique in that no manual alignment is required in making the two sides of the pattern. A front-side pattern is mounted in one side of the fixture and an unexposed photographic plate in the other side of the fixture (under darkroom conditions). The second side of the pattern is exposed from the first master and developed (by reversal processing) while still mounted in the fixture. The only alignment variable in this method is the ability of the fixture to re-register on its pins.

The exposed image or pattern is dissolved off the foil with developer. The unprotected pattern on the foil is then electrolytically etched in a 4-4-1 mixture of phosphoric acid, water, and sulfuric acid controlled to a temperature of 130°F. Normally, 3 volts at a current density of a few tenths of an ampere per square inch (of etched area) is used for etching pattern arrays on 0.005 molybdenum. The etch allowance (Fig. 10-30) is the amount by which the radius of a hole must be reduced when the master pattern is prepared and is equal to half the foil thickness. Evaporation of gold patterns on ceramic headers and evaporation of contact patterns on thin film circuits are typical applications for these masks.

Combination Masks. A combination copper-nickel mask is used where precision patterns from a rigid mask are required. A beryllium-copper alloy sheet 0.010 to 0.030 inch thick is used as a base material for the mask illustrated in Fig. 10-31. Both sides are coated with a photoresist by whirling. A pattern is exposed on each side of the sheet with an ultraviolet light source. There is a positive pattern on one side and a negative pattern on the other side. A nickel layer, 0.0007 inch thick, is then electroplated on one surface of the copper sheet, with the positive photoresist image (shown by the dashed line in the illustration) used to define the unplated areas. The electroplating bath is usually nickel sulfamate maintained at 120°F, and the plating current density is approximately 0.2 ampere/in.^{2}. This laminate is then selectively etched from both sides, and a nickel flash is applied to produce the final mask. Typical applications of this type of mask include the evaporation of thin film contact material and the application of viscous anodizing solutions to thin film resistor patterns.

Summary of Characteristics of Aperture Masks. The tolerance on individual pattern size for a typical fine-mesh silk screen is ±0.003 inch. This is roughly ±1 mesh opening. The tolerance achievable on

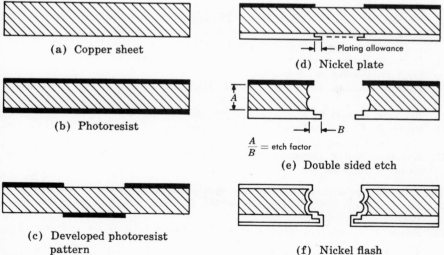

(a) Copper sheet

(b) Photoresist

(c) Developed photoresist pattern

(d) Nickel plate

Plating allowance

$\frac{A}{B}$ = etch factor

(e) Double sided etch

(f) Nickel flash

FIG. 10-31. Combination metal masks.

1666   THIN FILM TECHNOLOGY

"moly" masks is ±1/5 of the foil thickness. For example, on 0.005-inch molybdenum, a tolerance of ±0.001 inch would be expected. Combination masks can achieve ±0.001 inch on individual pattern size, regardless of the sheet thickness. An accuracy of ±0.002 inch on pattern location is easily attainable on all three types and can usually be improved if array size is reduced.

The smallest open area on an etched "moly" mask is twice the foil thickness and is a process limitation of the etch bath. On combination masks, the smallest space between open areas can be no less than the copper sheet thickness. This is necessary to ensure proper back support for the nickel plate.

REFERENCES

1. Jenkins, F. A., and H. E. White. *Fundamentals of Optics*. New York: McGraw-Hill Book Company, Inc., 1950.
2. James, T. H., and J. F. Hamilton. "The Photographic Process," *International Science and Technology*, no. 42, pp. 38-44 (June, 1965).
3. Altman, J. H. "Photography of Fine Slits Near the Diffraction Limit," *Photographic Sci. and Eng.*, 10, 140-143 (1966).
4. Eastman Kodak Company, "Physical Characteristics of Kodak Glass Plates," Publication no. Q-35.
5. James, T. H., and G. C. Higgins. *Fundamentals of Photographic Theory*, Chaps. 5 and 6. New York: Morgan and Morgan, 1960.
6. Eastman Kodak Company, "Techniques of Microphotography," Publication no. P-52.
7. Hurter, F., and V. C. Driffield. "Photochemical Investigations and a New Method of Determination of the Sensitiveness of Photographic Plates," *J. Soc. Chem. Ind., London*, 9, 455-469 (1890).
8. Eastman Kodak Company, "Kodak Plates and Films for Science and Industry," Publication no. P-9.
9. Eastman Kodak Company, "Kodak High Resolution Plates," Publication no. P-47.
10. Bertram, J., and K. Kursten. *J. Prakt. Chem.*, 51, 324 (1895).
11. Flory, P. J. *Principles of Polymer Chemistry*. Ithaca, New York: Cornell University Press, 1953.
12. Eastman Kodak Company, "Kodak Photosensitive Resists for Industry," Kodak Publication no. P-7.

The following general texts cover the specialized fields noted in the title:
13. Stevens, G. W. W. *Microphotography*. London: Chapman and Hall, Ltd., 1957.
14. Kosar, J. *Light Sensitive Systems*. New York: John Wiley and Sons, Inc., 1965.
15. Mees, C. E. K., and T. H. James. *The Theory of the Photographic Process*, New York: The Macmillan Company, 1966.

Chapter 11
Component and Circuit Design

Traditionally, the problems of component design, circuit design, and equipment design have been considered separate entities with well-defined interfaces. This has been drastically changed by use of thin films and semiconductor integrated circuits, since a complete circuit function may be included in a single device. The substrate is an important part of a thin film circuit package and is also an integral part of each film component. The components themselves are custom tailored to meet the circuit needs in the most advantageous manner.

In order to optimize the use of thin film technology, equipment and circuit designers need a working knowledge of the various factors involved in the fabrication and performance of thin film devices. Although a device design engineer can provide much of the required information, the system development engineers are best equipped to evaluate most of the complex interrelations that shape the design of a custom-made thin film device.

There are five major considerations relating to the device design problem: overall system requirements, required circuit performance, properties of film components, specific film properties, and film circuit fabrication methods. Since these elements are not independent, it is important that they all be considered as the design evolves. This chapter points out many of these interrelations and their importance in the realization of a good thin film design. Some examples of how these design inputs relate to the device problem are given.

The overall physical requirements of the system influence the choice of substrate size and shape, the location of terminations, and the packaging techniques to be employed. In deciding how much circuitry should go on a single substrate, it is important to consider the factors of convenient substrate size, opportunity for redundant modules, minimization of the number of interconnections, and favorable fabrication yield. The problems of power dissipation and high-frequency performance must be considered for the overall system design as well as in the design of individual circuits and components. Although some of the above considerations appear to go beyond the immediate problem of device design, they form important boundary conditions for the device design problem.

The close relationship between circuit performance and component characteristics makes it imperative that the properties, advantages, and limitations of thin films be considered even during the initial stages of circuit design. The component properties available are determined by the film configuration and method of fabrication, as well as by the physical, chemical, and electronic properties of the films themselves. The conventional component requirements of resistance, capacitance, voltage, current, power, and precision must be related to film device design factors such as the types of films used, their relative positions, thicknesses, and geometrical patterns, as well as the fabrication techniques employed. This chapter provides the necessary relationships between component properties and film properties for the various types of film components available, including resistors, capacitors, conductors, and distributed RC networks. Crossovers, land areas for external terminations and appliquéd devices, and ultrahigh frequency inductors are also considered.

The procedures of custom designing a complete film circuit are introduced by giving the various steps. Specific examples are given to illustrate the influence of fabrication procedures and circuit performance requirements on the film circuit layout. Circuit packaging, appliquéd components, and bonding techniques are considered with particular attention given to beam-lead devices.

The scope of the overall design problem is best illustrated by a specific example. A DATA-PHONE® data set, chosen as an exploratory development vehicle, is used as such an example. One of the objectives of the design was to achieve a drastic size reduction without changing the overall data transmission performance.

The original data set is shown in the center of Fig. 11-1, with its associated telephone to the right. The thin film version of this same data set is on the left, with its telephone being an integral part of the receiver so that the whole set could be placed conveniently on a desk. The overall volume reduction achieved was about 17 to 1.

Fig. 11-1. Size comparison of a conventional data set (402B) and experimental thin film model.

The original data set has eight data channels and one timing channel, all operating in parallel at different frequencies in the audio spectrum. Tuned LC filters establish the mark and space frequency of each channel. The inductors needed for these filters make a sizable contribution to the overall volume of the set. Audio-frequency inductors are not available in thin film form; thus, the design of the exploratory set had to start with new circuits that made optimum use of the components available in thin film form. These circuits used active RC filters instead of LC filters to allow almost all passive components to be realized as thin film.

The circuit diagram for the new filters of one data channel is shown in Fig. 11-2. The circuit consists of three narrow bandpass amplifiers. One amplifier rejects signals of the adjacent channels while the other two amplifiers form the mark and space bandpass filters of the desired channel.

The thin film substrate for this discriminator filter circuit is shown in Fig. 11-3. There are 33 film resistors and 12 film capacitors with the required film interconnections on the front side of the glazed ceramic substrate, which measures 2 by 3.3 by 0.03 inches. Mounted on the back of the substrate are nine transistors and three large-value capacitors. Additional conducting paths are provided in thin film form on the back of the substrate. Holes in the ceramic provide through-connections for conducting paths and wire leads to some of the transistors. This circuit was developed before the invention of beam-lead transistors.

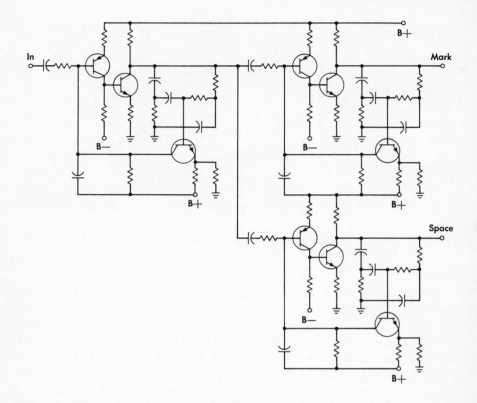

FIG. 11-2. Circuit diagram of data channel discriminator filter for one of the eight data channels in a thin film data set.

The entire data set, with the exception of the power supply, was made using thin film substrates of this size. Thus, 21 substrates were mounted vertically, side by side, and were plugged into a common wiring board. This packing arrangement can be seen in Fig. 11-4, which shows the thin film data set with the top of the housing removed. The height of the substrates was determined by the height of the equipment package. The length was chosen to provide enough area for the data channel discriminator filter circuit shown in Fig. 11-2. This circuit formed a convenient repetitive module since the circuitry is nearly the same for all eight data channels. The other circuits in the set are generally less complex and less critical in performance requirements. The remaining circuitry was subdivided to fit conveniently on the same size substrate. The thin film model contained about 600 film resistors, 200 film capacitors, 30 appliquéd capacitors, 180 transistors, and 180 diodes.

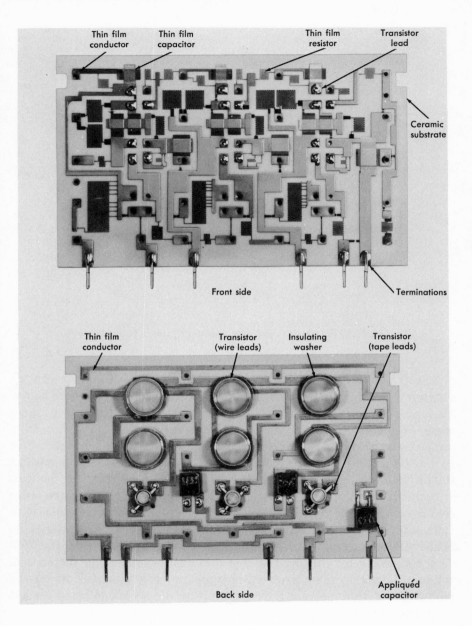

FIG. 11-3. Thin film data channel discriminator filter for one data channel in a thin film data set.

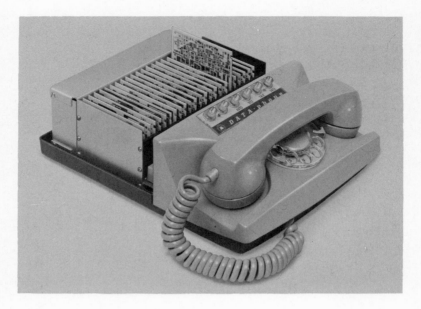

Fig. 11-4. Thin film data set with the top of the housing removed.

The feasibility model was successful in that the intended size reduction was achieved and the transmission characteristics were equivalent to the conventional set. As in any development project, there were problems in certain areas of the design that were not fully appreciated until the experimental set was fabricated. It is instructive to discuss two of these problems since they typify the problems that the physical designer may encounter when designing thin film devices.

Some of the first data channel bandpass amplifiers fabricated were found to oscillate. The cause was traced to excessive resistance in some of the film ground conductors that provided an undesirable positive feedback path in the amplifiers. The problem lent itself to easy correction by altering the thin film design to shorten the critical conducting paths. This problem is typical of the subtleties of circuit performance that normally do not appear on a conventional circuit diagram. One way to prevent this type of problem would be to include in the circuit diagram an upper limit of resistance on each of the wiring paths that could affect circuit performance.

There was also a problem of power dissipation in the thin film data set. The operating temperature of the top of the housing was excessive for equipment destined to go on top of a desk. The film components themselves were properly designed to prevent hot spots on the substrates. The total power dissipated in each substrate was not excessive and much of the heat was effectively carried away by convection in the air space between substrates. The top of the housing, however, was the major heat sink for this convection process, and further design effort would have been required to overcome this problem. Again, the need for close coordination of all aspects of the physical design is emphasized.

This chapter treats, in some detail, the problem of power dissipation in film components, but the designer must also be concerned with the transport of heat away from the substrate, the ultimate sink for this heat transport process, and the resulting operating temperatures that will exist in the equipment. The dissipation of power can become a critical factor in equipment designs that employ thin film devices to achieve high component densities, unless the total power requirements are reduced in proportion to the volume reduction. As the power dissipated per unit volume goes up, the operating temperature goes up, and the equipment performance and reliability can be affected adversely.

Many of the discussions in this chapter relate specifically to tantalum film technology [1, 2]. The design principles and procedures, however, should be applicable to other film materials that may be of interest in the future.

11.1 FILM RESISTORS

Choice of Materials

Before the design of a thin film resistor can begin, it is necessary to choose the most appropriate film and substrate material for the particular application under consideration. The properties, advantages, and limitations of several different types of resistor films and substrates are discussed in Chaps. 7 and 9. It is anticipated that the list of available materials will grow and change as thin film technology continues to develop.

The two important resistor film material properties that have a direct bearing on the resistor design are electrical resistivity and stability with time. Other material properties which may be of importance in some applications are temperature coefficient of resistance, voltage coefficient of resistance, and current noise. Although these latter three properties may influence the choice of resistor material, they do not have a direct bearing on other resistor design considerations.*

The choice of substrate material involves many factors other than resistor performance, but some substrate properties are directly involved in the resistor design. Resistor stability is intimately related to substrate inertness and thermal conductivity. The surface smoothness, or lack of it, can affect sheet resistance.

Resistor Properties

Once the resistor film and substrate materials have been chosen, the design problem consists of establishing a geometrical pattern of a given thickness which meets the resistor requirements of:
1. Stability with time
2. Power rating
3. Size
4. Resistance value
5. Initial precision
6. Parasitics (or frequency response)

Generally, the pattern that evolves will resemble one of the two shown in Fig. 11-5. The straight-line pattern on the left is normally used for low-value resistors, and the meandering-line pattern on the right is used for larger values. A major part of the resistor design problem is the specification of the dimensions indicated in this figure.

Stability, Power, and Size

The minimum size of a resistor is often determined by its power rating and stability requirements. The required stability tolerance can be used to establish a design factor for the maximum value of the average power density. This density is defined as the total power dissipated divided by the effective area of the resistor, or

$$q = \frac{P_R}{A_R} \tag{11-1}$$

The effective area of the resistor is defined as the area of a simple geometric figure that encompasses the entire resistor. For the

*In some film materials, the temperature coefficient of resistance is a function of film thickness. For some applications, this fact may influence the choice of film thickness. Other factors influencing the choice of film thickness will be discussed later.

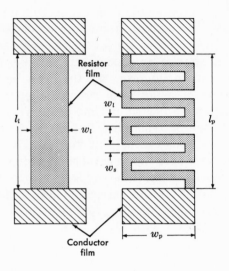

FIG. 11-5. Thin film resistor patterns.

straight-line resistor shown in Fig. 11-5, the resistor area is

$$A_R = w_l l_l \qquad (11\text{-}2)$$

where w_l is the width of the line and l_l is the length of the line. For the meandering-line pattern in Fig. 11-5, the effective resistor area is

$$A_R = w_p l_p \qquad (11\text{-}3)$$

where w_p is the width of the pattern and l_p is the length. Thus, the substrate area between meandering lines is considered a part of the resistor area. These definitions of A_R, for the two styles of resistors, result in areas that are comparable from the standpoint of heat dissipation, provided the line width for the straight-line resistor is sufficiently large and the space width of the meandering type is sufficiently narrow.

Establishing a design maximum for the average power density is a complex matter. The critical parameter is the temperature of the film. As the power dissipated goes up, the film temperature increases and the resistor becomes susceptible to thermal oxidation. The formation of an oxide on the surface of the film or diffusion of oxygen into the film increases its resistance. Both of these processes often involve a thermally activated diffusion, with the rate increasing exponentially with temperature (see Chap. 7). Other resistor-aging mechanisms, such as annealing of the film or electrolysis of the substrate, are also accelerated by elevating the temperature.

The percentage change in resistance produced by these aging processes is a film material property that can be measured experimentally. The effect of exposing a resistor film to an elevated temperature for an extended period of time is shown in Fig. 11-6, where the percentage change of resistance is plotted as a function of time for tantalum nitride resistors held at 200°C. The magnitude of the resistance change produced in this thermal aging experiment is insensitive to the substrate material as long as the material is chemically inert and smooth.* If an electrical field is present, the resistance change can be appreciably greater for substrates containing mobile ions such as sodium (see Chap. 9).

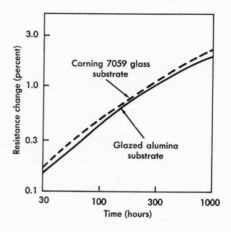

FIG. 11-6. Resistance change versus time for accelerated oven aging of tantalum nitride film resistors. Resistors are 44 ohms/square, prestabilized at 250°C for 5 hours and oven aged at 200°C.

The amount of power input needed to elevate the film temperature to some particular value depends on (a) the thermal conductivity of the substrate; (b) the substrate thickness, size, and shape; (c) the ambient conditions; (d) the size, shape, and location of the resistor; and (e) the location of heat sinks or other heat sources on the sub-

---

*Resistors on rough, unglazed alumina will change slightly more than those on glazed alumina or glass when they are oven aged at the same temperature.

strate. The situation is further complicated by the fact that the thermal conductivity is temperature dependent for most substrate materials. Also, the power dissipation in the film can be nonuniform, and thick conducting films in the vicinity can contribute to the thermal conductance.

A general solution to this problem is not available, and working out a particular solution for each application is impractical. If an accurate solution were obtained, it would be found that the temperature of the film is not uniform over the entire resistor path. For a substrate material with a low conductivity, such as glass, the center of the resistor is considerably hotter than the ends (see Fig. 7-9). In order to relate the film temperature to resistor stability, an integration over the resistor area is necessary.

An alternative to analytically determining film temperature is to relate resistor stability directly to the average density of dissipated power. This is done by aging resistors at different power levels and by measuring the percentage change in resistance as a function of time. Such a study was made of tantalum nitride resistors on Corning 7059 glass in a room temperature ambient [3]. The curves shown in Fig. 11-7 are based on the results of this accelerated aging experiment. Each curve represents the *maximum resistance change expected** for any resistor among a large group of resistors aged at the average power density shown.

These curves are extrapolated from experimental data obtained using a resistor pattern with an area of 0.05 in.^{2} (see Fig. 7-10). Changing the size and shape of the pattern changes the temperature distribution and, hence, the resistor stability. It has been found, however, that considerable change can be made in the pattern without greatly changing the relation between stability and average power density.

The temperature of the test specimens, and therefore the resistor stability, depends on the manner of dissipation of heat from the substrate. In general, this is by conduction through leads and any other physically contacting parts, by convection to the surrounding air, and by radiation. For these and other tests described in this section, the substrates were supported horizontally to minimize convection, and no parts, other than the leads, were allowed to contact the substrate. The results are somewhat conservative; nevertheless, they have been found to be quite useful for design purposes.

---

*The average resistance change is less.

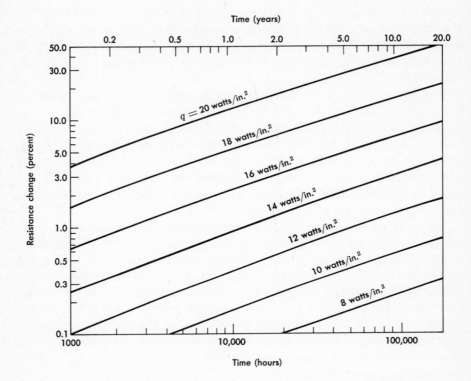

FIG. 11-7. Maximum expected resistance change versus time for accelerated power aging of tantalum nitride film resistors on Corning 7059 glass substrates 0.05 inch thick. Resistors are 50 ohms/square and pre-stabilized at 250°C for 5 hours.

The curves shown in Fig. 11-7 approach asymptotes that were described in Chap. 7 by

$$\frac{\Delta R}{R} = \sqrt{\frac{t}{t_0}} \exp\left(\frac{-T_0}{T}\right) \qquad (11\text{-}4)$$

where $\Delta R/R$ is the fractional increase in resistance, t is the time, T is the average temperature of the resistor, and both t_0 and T_0 are constants. For a tantalum nitride resistor film on 7059 glass (0.05 inch thick) and with an ambient temperature of 25°C:

$$t_0 = 0.015 \text{ sec}$$

$$T_0 = 6600°K$$

The $\sqrt{t}$ is indicative of the oxidation mechanism; t_0 and T_0 are the characteristic aging properties of the film material. They are independent of the substrate as long as the material is smooth and chemically inert.

Increasing the ambient temperature will merely shift the curves upward; therefore, a general relation for the average film temperature can be written as

$$T = T_A + \frac{q}{h} \tag{11-5}$$

where T_A is the ambient temperature, q is the average power density, and h is a heat transfer coefficient for the particular substrate material being considered. For the curve shown in Fig. 7-11, the measured value of h is 0.10 watt/in.^{2}°C for 0.05-inch 7059 glass. It is about 0.2 watt/in.^{2}°C for 0.03-inch glazed ceramic.

The measured value of h is independent of resistor film material but it depends on the size and shape of the resistor pattern, the location of the heat sinks, and the substrate thickness, so these values should be used with caution. The data for the curve in Fig. 7-11 were obtained using a resistor with an area of 0.050 in.^{2}.

Equation (11-4) can be combined with Eq. (11-5) to determine a maximum design value for q:

$$q \leq h \left[\frac{T_0}{\frac{1}{2}\ln\left(\frac{t}{t_0}\right) - \ln\left(\frac{\Delta R}{R}\right)} - T_A \right] \tag{11-6}$$

Once this upper limit on the average power density has been established for the time and stability requirements of the resistors, the minimum area for each resistor can be calculated using Eq. (11-1).

The curves in Fig. 11-7 and the resulting expressions [Eqs. (11-4) and (11-6)] apply to films with sheet resistance of about 50 ohms/square. These results can be adapted for use with films having other values of sheet resistance, as shown later in this section. According to Eq. (11-6), the maximum design value for q is proportional to h, and the power density on glazed ceramic can be twice that on glass.

The above discussion shows how a maximum average power density and a minimum size can be calculated for a tantalum nitride film resistor. A similar procedure can be used for any resistor film material. Before this is possible, however, it is necessary to perform accelerated aging studies in order to generate curves like those shown in Fig. 11-7.

The procedures given above are only approximations. The average power density does not completely define the resistor stability, particularly when the substrate thermal conductivity is high. It would be desirable to provide for the effect of resistor size (or total power) and shape as well as other design variations, but useful analytical solutions to the heat transfer problem have not yet been generated. The alternative is to make temperature measurements on preliminary models of the circuit. An important qualification to the foregoing analysis is that time-averaged power must be de-rated for small duty cycles unless the repetition frequency is very high. This is because of the time delay in transferring pulses of heat from the film to the substrate. The maximum temperature achieved by the film should not be greater than the temperature assumed for stability requirements.

Film Thickness

Film thickness may affect resistor stability and, since oxidation and oxygen diffusion more seriously affect the top layer of the film, it is natural to expect a larger percentage change in resistance in a thinner film. This effect is evident from the curve plotted in Fig. 11-8 [4]. The percentage change in resistance is plotted as a function of sheet resistance for tantalum nitride resistors oven-aged for 1850 hours at 150°C. It will be recalled from Chap. 7 that the sheet resistance is

$$R_S = \frac{\rho}{d} \tag{11-7}$$

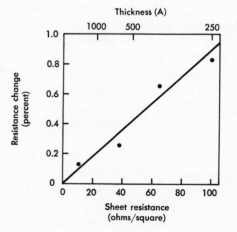

FIG. 11-8. Resistance change versus sheet resistance for tantalum nitride film resistors with resistivity $\rho = 250 \ \mu$ ohm cm, oven aged at 150°C for 1850 hours.

where ρ is the resistivity of the film material and d is the film thickness. Thus, the percentage change in resistance varies inversely with the resistor film thickness. On the other hand, for a resistor dissipating a given amount of power, changing the film thickness has a negligible effect on film temperature.

As previously mentioned, design curves given in Fig. 11-7. or calculations made using Eq. (11-6), are for a tantalum nitride film with a sheet resistance of about 50 ohms/square. For tantalum nitride films with different sheet resistances, it is necessary to scale the stability requirement accordingly. If the design sheet resistance is 100 ohms/square (twice that used to establish the curves), the value of $\Delta R/R$ used to determine q should be one-half that actually required.

Another way of accomplishing this same end is to add an R_S term to Eqs. (11-4) and (11-6). Thus,

$$\frac{\Delta R}{R} \leqq \frac{R_S}{R'_S} \sqrt{\frac{t}{t_0}} \, \exp\left(\frac{-T_0}{T}\right) \qquad (11\text{-}8)$$

and

$$q \leqq h\left[\frac{T_0}{\frac{1}{2}\ln\left(\frac{t}{t_0}\right) - \ln\left(\frac{\Delta R}{R}\right) + \ln\left(\frac{R_S}{R'_S}\right)} - T_A\right] \qquad (11\text{-}9)$$

where R'_S is 50 ohms for a tantalum nitride film.

Figure 11-8 shows that resistor stability requirements impose a lower limit on film thickness and, hence, an upper limit on the sheet resistance of a given material. In the case of tantalum nitride, 100 ohms/square is considered a practical upper limit. There are several factors that place an upper limit on film thickness. If the film is too thick, it may be mechanically unstable. Thick films take longer to deposit and, although the low sheet resistance of a thick film may be desirable for low-value resistors, it is inappropriate for large-value resistors.

In the case of tantalum films, the method of resistance adjustment also imposes an upper limit on film thickness. Tantalum resistors are adjusted to value by an anodization process that converts the top portion of the metal film to an insulating film of tantalum oxide (see Chap. 7). If the sheet resistance of a sputtered film is controlled in production to within ±10 percent, it is necessary that anodization be able to provide a 20 percent adjustment of film thickness. Thus, the upper limit on film thickness is five times (i.e., $1/0.2$) the thickness adjustment range available. Normally, the anodization voltage used for resistor adjustment ranges from 30 volts to about 130 volts.*

*A 30-volt anodization provides the minimum protective oxide coating. Above 130 volts, confining or masking of the electrolyte becomes a problem.

The anodization constant for conversion of tantalum nitride is about 4.5A/v (see Chap. 5); therefore, the 100-volt spread corresponds to a 450A change in the resistor film thickness. Thus, the original film thickness should not exceed 2250A (i.e., 5 $\times$ 450). The resistivity of tantalum nitride is about 250 μ ohm cm and, hence, the minimum sheet resistance is 11 ohms/square as calculated by Eq. (11-7). If the control of the sputtered-film sheet resistance is better than ±10 percent, a lower value of R_S is possible. Normally, 10 ohms/square is considered a practical lower limit of sheet resistance for a tantalum nitride film to be used as a resistor.

The exact choice of sheet resistance between the limits established above involves many considerations, the most important being the range of resistance values on a given circuit. For low-value, high-power resistors, a low value of R_S is better, since q can be higher for a given film material and, hence, less area is needed. Also, resistor patterns are more satisfactory if $R_S < R$.

For high-value, low-power resistors, a high value of R_S is better, since the minimum size in this case is usually controlled by sheet resistance rather than by power requirements (unless R_S is made correspondingly higher). One way of achieving a higher sheet resistance is to use a material with a higher resistivity, such as oxygen-doped or low-density tantalum (see Chaps. 4 and 7).

Resistor Line Width

Once the design value for the sheet resistance and the maximum value for the average power density have been determined, the delineation of the resistor pattern must be considered. Chapter 7 states that the sheet resistance and resistance value determine the number of squares required in the pattern. Thus,

$$n = \frac{R}{R_S} \qquad (11\text{-}10)$$

where n is the number of squares.

For the straight-line pattern shown on the left in Fig. 11-5,

$$n = \frac{l_l}{w_l} = \frac{A_R}{w_l^2} \qquad (11\text{-}11)$$

where A_R is the resistor area given in Eq. (11-2). For the meandering-line pattern shown on the right in Fig. 11-5, n is given approximately by the number of meanders (straight segments of length w_p) times the number of squares in each meander, or

$$n \cong \frac{l_p}{w_l + w_s} \times \frac{w_p}{w_l} = \frac{A_R}{w_l^2 \,(1 + w_s/w_l)} \qquad (11\text{-}12)$$

where A_R is the resistor area given by Eq. (11-3). A more exact determination of the number of squares is given later. Actually, Eq. (11-12) can be used for both cases since it reduces to Eq. (11-11) when $w_s = 0$

It is evident from Eq. (11-12) that the number of squares required and the minimum resistor area (based on power requirements) establish an approximate minimum line width for the resistor. Changing the space-width-to-line-width ratio (w_s/w_l) may influence the results but normally by only a slight amount. Equation (11-12) has been plotted in Fig. 11-9 for the case where $w_s/w_l = 1$.

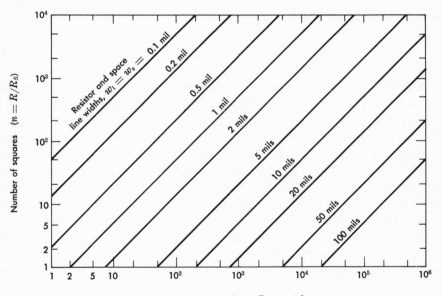

FIG. 11-9. Number of squares in a resistor path versus resistor area for a meandering-line film resistor.

Figure 11-9 shows that low-value, high-power resistors require wide lines, whereas high-value, low-power resistors can be made with very narrow lines. In fact, the minimum line width (based only on power requirements) may be so narrow that they cannot be produced.

The minimum line width achievable with chemical etching techniques, at present, is about a tenth of a mil. The exact limit depends on the film material and thickness. Lines this narrow, however, are sensitive to small defects in the film so that the fabrication yield may be low. The yield increases with line width and is satisfactory for lines more than 1 mil wide.

Another factor that places a lower limit on line width is resistor precision. In the case of tantalum films, it is possible to achieve design resistance values by anodizing each resistor individually to the required precision, the limit being about ± 0.01 percent under ideal conditions. In applications requiring less precision, it may be less expensive to anodize all the resistors at once, while monitoring only one resistance value. In this case, the pattern geometry and uniformity of sheet resistance determine the precision of the resistors not monitored. This can be expressed in differential form using Eqs. (11-10) and (11-11). Thus,

$$\frac{\Delta R}{R} = \frac{\Delta R_s}{R_s} + \frac{\Delta l_l}{l_l} - \frac{\Delta w_l}{w_l} \tag{11-13}$$

A sheet resistance uniformity of better than ± 1 percent can be achieved over large areas. Figure 11-10 illustrates the uniformity of a sputtered beta-tantalum film over an area of 36 in.^{2}. The sheet resistance shown was measured at the center of each 1-inch square.

The path length term in Eq. (11-13) is negligible compared to the other two terms if $n >> 1$, which is usually the case. The path-width term in this expression is often the controlling factor. In the chemical etching process, the magnitude of Δw_l is a function of the film material, the film thickness, the substrate material, the nature of the pattern, the etchant used, etc. (see Chap. 10). In the case of tantalum films, Δw_l can routinely be kept to less than a tenth of a mil. If a resistor tracking tolerance of ± 5 percent is required, a minimum line width of about 2 mils might be appropriate.

Fortunately, two resistors on the same substrate will have approximately the same $\Delta w_l/w_l$ if they have the same w_l and w_s. Thus, the best way to insure that the values of two resistors track each other is to use the same w_l and w_s and to place them physically close together so that R_s is nearly the same.

As discussed above, fabrication cost and fabrication yield can impose a lower limit on line width which may be larger than that established by power consideration alone. When this is the case, it is evident from Fig. 11-9 that the minimum resistor area is a function

6 in. × 6 in.

15.00 −0.72%	15.00 −0.72%	15.12 +0.07%	15.08 −0.19%	15.16 +0.34%	15.20 +0.60%
15.00 −0.72%	15.12 +0.07%	15.16 +0.34%	15.18 +0.47%	15.16 +0.34%	15.12 +0.07%
15.10 −0.06%	15.08 −0.19%	15.16 +0.34%	15.18 +0.47%	15.16 +0.34%	15.14 +0.20%
15.14 +0.20%	15.06 −0.32%	15.20 +0.60%	15.17 +0.40%	15.20 +0.60%	15.02 +0.41%
15.16 +0.34%	15.13 +0.14%	15.18 +0.47%	15.10 −0.06%	15.10 −0.06%	15.00 −0.72%
15.20 +0.60%	15.12 +0.07%	15.14 +0.20%	14.96 −0.99%	14.98 −0.85%	14.96 −0.99%

FIG. 11-10. A 6-inch by 6-inch sputtered beta-tantalum film with measurements of sheet resistance (ohms/square) and percent deviations from the average value of 15.11 ohms/square (sputtered in a bell jar system with a 14-inch diameter cathode).

of the minimum w_l available and R_S. Thus, rewriting Eq. (11-12),

$$A_R = \frac{R}{R_S} w_l^2 \left(1 + \frac{w_s}{w_l}\right) \tag{11-14}$$

On the other hand, if w_l is not a limiting factor, the minimum resistor area is a function of q, as established from resistor stability requirements using Eq. (11-9). Thus,

$$A_R = \frac{P_R}{q} \tag{11-15}$$

There are several ways in which new developments may provide even smaller resistors than are presently available. New methods of pattern generation, such as electron beam or laser beam techniques, may be developed to the point where lower values of w_l and Δw_l are achieved. More stable materials may be developed to allow higher power densities and, hence, allow less area to be used for a given amount of power. Higher sheet resistances will provide more resistance per unit area. From the circuit designer's point of view, it is clear that circuits employing low-value resistors at low power levels will require less space. Also, the technique of resistor adjustment must be considered before a resistor is made arbitrarily small.

Number of Squares

Thus far, no limits have been established for the number of squares in a pattern. An upper limit on n is needed to insure a high yield during fabrication. The probability that a film defect will damage a resistor is greater over a longer path and, at the same time, a narrower path is more vulnerable since smaller defects are effective in damaging the resistor. If the total number of squares in all the patterns on a substrate gets too large, the substrate fabrication yield decreases. Thus, the total resistance on a substrate is more important than the number of resistors or the value of any one resistor.

Substrate yield, along with line width and total area considerations, normally limits the total number of squares to something less than 50,000. This implies an upper limit on the total resistance per substrate such that $\Sigma R < 5 \times 10^4 \, R_S$.

There is also a practical lower limit on the number of squares in any one resistor pattern. Patterns consisting of less than one square may be difficult to fabricate within required tolerances. When path lengths become very small, resistor tolerances are affected since the Δl_l term in Eq. (11-13) can no longer be neglected. Also, the small margins available between the conducting pattern and the anodizing electrolyte make resistor adjustment more difficult.

There are various methods of circumventing this difficulty. One method of achieving a fractional square pattern in a tantalum film is shown in Fig. 11-11. An aluminum film placed above or below the tantalum film provides a series of shorting sections which reduce the resistance without reducing the area for heat dissipation. Aluminum is chosen because it is a good conductor while being compatible with the anodization process. If power is not a problem, a single resistive section may be used. As materials with high sheet resistivities are further developed, fractional square patterns will become more important.

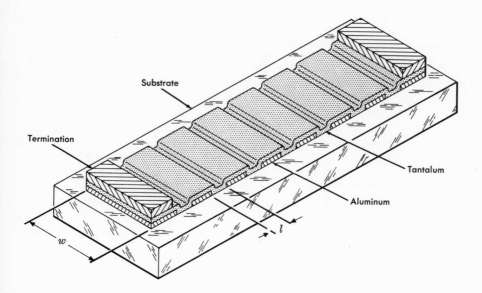

FIG. 11-11. Tantalum film resistor with low aspect ratio using aluminum shorting bars [1].

Counting Squares

Equation (11-12) was developed without considering the effect of the right-angle bends in the resistor path. The resistance of these bends must be accounted for if the proper resistance ratios are to be achieved in a circuit pattern.

In the region of a bend, the current density is nonuniform and the concept of a square of film material must be interpreted carefully. In order that Eq. (11-10) remain valid, the number of squares associated with bends, or other special geometries, must be determined by finding the resistance they contribute to the path. It is convenient to consider the straight sections in the normal manner (l_i/w_i) and to assign a special number to the "square" in the corner. Actually, the current density and equal potential lines are distorted in part of the straight section and, hence, their contribution is not l_i/w_i. This point should be kept in mind in the discussions that follow.

There are various ways of attacking this problem, but the simplest is probably conformal mapping. With one or more suitable coordinate transformations, the particular geometry of interest is mapped into the rectangular shape of the straight-line pattern shown on the left in Fig. 11-5. The results of some of these calculations are shown in Fig. 11-12 [5].

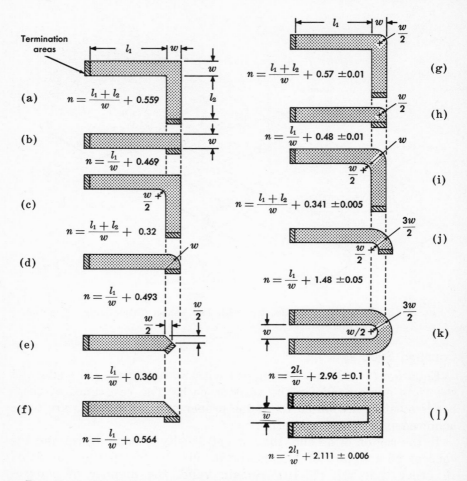

FIG. 11-12. Various film resistor paths with effective number of squares as indicated.

From Fig. 11-12(a) it appears that the corner of a simple right-angle bend contributes 0.559 square to the resistor path. This number is for a perfect corner, with a long, straight section leading into it. When $l_2 = w$, 0.559 is still a good approximation. When $l_2 = w/2$, 0.556 is a good approximation. When l_2 goes to zero, as shown in Fig. 11-12(b), the corner contribution is 0.469 square, a drop of only 0.090 square.

Any small radius on the inside corner of a right-angle bend will decrease the number of squares ascribed to the corner [6]. The decrease (Δn) can be approximated for a small corner radius, r, by

$$\Delta n = -0.6\left(\frac{r}{w}\right)^{4/3} \tag{11-16}$$

where $r < w/2$. An inside-corner radius is often provided intentionally to relieve the high current density which occurs near a sharp corner. In high-power applications, the high current density produces a hot spot which can cause an excessive resistance change at that point. Other types of bends and their solutions are shown in Fig. 11-12. Where approximations have been used, the magnitude of maximum error is indicated.

The conformal mapping technique can be used on more complex geometries by breaking them up into simpler shapes. This is done by making effective use of symmetry effects, equal potential lines, and current stream lines. An example of this, the lattice pattern in Fig. 11-13, is of interest for certain high-frequency applications. The objective is to achieve a low inductance with a wide, short, signal path, while still providing the required resistance. The interconnected pattern has the added advantage that a break or a pinhole in any one of the narrow lines produces only a small effect on the total resistance.

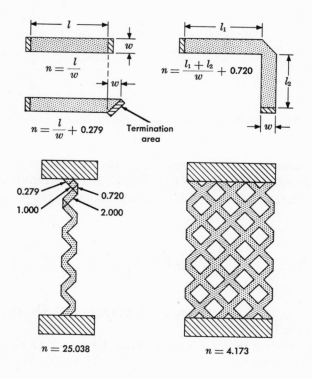

FIG. 11-13.   Lattice pattern for a film resistor broken down into its basic elements.

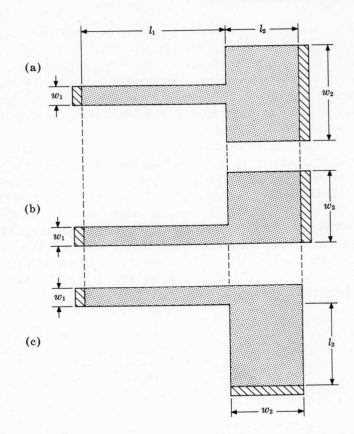

FIG. 11-14. Film resistor pattern with abrupt changes in resistor line width.

Another geometry of interest is an abrupt change in line width. Some typical examples are given in Fig. 11-14. The number of squares in the symmetric, straight-line segment shown in Fig. 11-14(a) is given by

$$n = \frac{l_1}{w_1} + \frac{l_2}{w_2} + \frac{1}{2\pi}\left[\frac{(S^2 + 1)}{S}\ln\left(\frac{S+1}{S-1}\right) - 2\ln\left(\frac{4S}{S^2 - 1}\right)\right] \qquad (11\text{-}17)$$

where $S = w_2/w_1$ and $l_2 > w_2/3$. A graphic plot for smaller values of l_2 is shown in Fig. 11-15.

If the symmetric pattern is split down the center (a current stream line), the nonsymmetric pattern in Fig. 11-14(b) results, and the

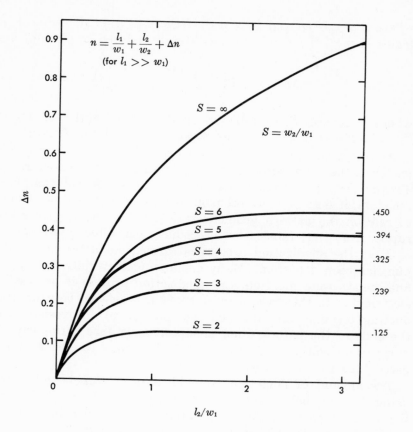

FIG. 11-15. Number of squares associated with an abrupt change in path width versus the length of the wide portion of the path [see Fig. 11-14(a)] [5].

number of squares is given by

$$n = \frac{l_1}{w_1} + \frac{l_2}{w_2} + \frac{1}{\pi}\left[\frac{(S^2+1)}{S}\ln\left(\frac{S+1}{S-1}\right) - 2\ln\left(\frac{4S}{S^2-1}\right)\right] \qquad (11\text{-}18)$$

where $S = w_2/w_1$ and $l_2 > 2w_2/3$.

The number of squares for a right-angle bend with a change of line width, Fig. 11-14(c), is given by

$$n = \frac{l_1}{w_1} + \frac{l_2}{w_2} + \frac{1}{S} - \frac{2}{\pi}\ln\left(\frac{4S}{S^2+1}\right)$$

$$+ \frac{(S^2-1)}{\pi S}\cos^{-1}\left(\frac{S^2-1}{S^2+1}\right) \qquad (11\text{-}19)$$

where $S = w_2/w_1$ and $l_2 > w_2$. If there is a small radius r on the inside corner, the number of squares is reduced by the amount

$$\Delta n = -0.4 \left(\frac{S^2 + 1}{S}\right)^{2/3} \left(\frac{r}{\sqrt{w_1 w_2}}\right)^{4/3} \qquad (11\text{-}20)$$

where r is less than half the smaller line width. If $w_1 = w_2$, Eq. (11-20) reduces to the expression given in Eq. (11-16).

Although changes in line width are more prevalent in conductor patterns, they can serve an important function for resistors. It has already been pointed out that in the adjustment of tantalum film resistors, it may be more economical to anodize them all at once and to rely on the pattern geometries to produce the proper resistance ratios. However, fabrication procedures can affect these ratios.

Often the anodization process is carried out after the conducting film has been deposited. Many conducting materials do not anodize and, hence, they short out the anodization process if exposed to the electrolyte. In this case, a mask margin is needed between the conducting terminal and the anodized portion of the resistors. Since the ends of the path are not anodized, resistance ratios can change during anodization and the amount of change can be a significant percentage for short resistors.

The amount of deviation from ideal ratios can be minimized by using the dumbbell-type pattern shown in Fig. 11-16. By broadening the resistor path at the termination, the number of squares not anodized is made smaller (for a fixed mask margin).

The situation is best understood by comparing the resistance before and after anodization. Before anodization,

$$R_O = \frac{\rho}{d_O}\, n \qquad (11\text{-}21)$$

where ρ is the resistivity, d_O is the original film thickness, and n is the total number of squares in the pattern. After anodization,

$$R_F = \frac{\rho}{d_F}\, n_a + \frac{\rho}{d_O}\, (2n_b) \qquad (11\text{-}22)$$

where d_F is the film thickness after anodization, n_a is the number of squares anodized, and n_b is the number of squares not anodized at each end of the path. Resistance ratios can be maintained either by making the ratio n_b/n_a the same for all the resistors or by making $n_b \ll n_a$. The dumbbell accomplishes the latter, even for short

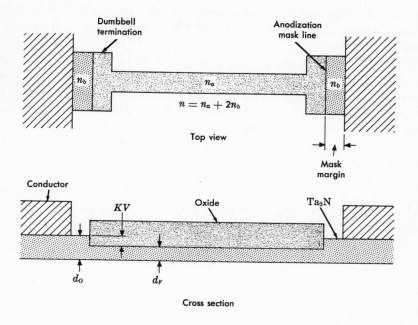

FIG. 11-16. Film resistor with dumbbell terminations for mass anodization.

resistor paths. If the uniform ratio approach is used, the results are sensitive to the position of the anodizing mask pattern which often is not located precisely.

Another way of obtaining the proper final resistance ratios is to correct the number of squares in each pattern so that the proper ratios are achieved after anodization rather than before. The correction consists of adding to each pattern the following number of squares:

$$\Delta n = (2n_b)\frac{KV}{d_O} \tag{11-23}$$

where V is the designed anodization voltage and K is the anodization constant of the metal film. Using the correction along with the dumbbell pattern produces the best results.

The values of $\Delta R_S/R_S$ over the substrate area are dependent on the details of the sputtering process. With proper control, variations such as those shown in Fig. 11-10 are readily achieved on large substrates. It should be noted that there are regions of high R_S and of low R_S with only moderate differences between adjacent blocks. Thus, control within any one small area improves as the area becomes smaller.

When all resistors on a substrate are made to a single line width, any overetching or underetching of the pattern affects all resistors alike and the resulting resistance changes can be compensated for by mass anodization. On the other hand, if various line widths are used, a given amount of overetch will reduce all line widths by the same amount, causing a larger proportionate increase in resistance for narrow-track resistors than for wide ones, and mass anodization is no longer completely compensating.

In this case, it is desirable to monitor a resistance with an intermediate line width. Optimum control is achieved when this intermediate width is

$$w_m = \frac{2w_w\, w_n}{w_w + w_n}$$

where w_w, w_n, and w_m are the widths of the widest, narrowest, and monitor lines, respectively. In this case, the $\Delta R/R$ (due to etching variations) for the widest lines should be the same magnitude as for the narrowest lines, but of the opposite sign.

The corresponding deviation in resistance due to line width control is

$$\frac{\Delta R}{R} = \frac{\delta w\,(w_w - w_n)}{w_m\,(w_w + w_n)}$$

where δw represents the average amount of overetching or underetching of the pattern.

Resistor Parasitics

Thin film components in integrated devices have several distinct advantages over conventional components in regard to parasitics. Much of the parasitic behavior of a conventional component is associated with the leads. In a thin film circuit, the leads can be made extremely short and, in some cases, they can be eliminated entirely. The interconnections on a thin film circuit have a fixed configuration. This is also true of the component itself. Thus, the parasitics can be determined analytically from the configuration of the film circuit and the electrical constants of the materials used. The electrical constants of interest are the resistivity (ρ), dielectric constant (ϵ_r), and relative permeability (μ_r). Since the thin film devices considered here do not employ magnetic materials, the material constants (hence the parasitics) are independent of the voltages and currents used. Once the parasitics are determined, they will be substantially the same for every device made. Therefore, it is possible to shape the configuration to achieve the optimum device response with respect to the critical parasitics.

The primary parasitics associated with a film resistor are series inductance, shunting capacitance, shunting resistance, and series resistance. The inductive and capacitive parasitics are similar to those found in conventional components, yet they can be considerably smaller for a film resistor. Usually they are negligible, except at very high frequencies. The shunting resistance refers to a leakage resistance across the substrate. The series resistance refers to the lead-in resistance of the conducting film. Clearly, the series parasitics are of more concern for low resistance values, whereas the shunting parasitics are more important for high resistances.

Series Inductance. Any conducting path has a self-inductance associated with it. This inductance is a function of the geometry and permeability $\mu_r\mu_0$ of the surrounding materials. In the absence of any magnetic material, $\mu_r = 1$ and the inductance becomes dependent only on the dimensions and the permeability of free space $(\mu_0 = 4\,\pi \times 10^{-7}\,\text{henry/m})$.

This section gives formulas that approximate the inductance of some of the geometries prevalent in thin film resistors [7]. These formulas are applicable only to cases where the relative permeabilities of the substrate and film materials are equal to 1. It is also assumed that the film thickness is small compared to the width of the path. The skin effect at very high frequencies can reduce the inductance slightly below these calculated values.

The self-inductance of a single, straight, resistor line shown in Fig. 11-17(a) is given by

$$L = \frac{\mu_0 l_l}{2\pi}\left[\ln\left(\frac{2l_l}{w_l}\right) + \frac{1}{2} + \frac{w_l}{3l_l}\right] \qquad (11\text{-}24)$$

where l_l is the length of the line, w_l is the width of the line, and $w_l \ll l_l$. The last term in Eq. (11-24) is usually much smaller than the rest and is often neglected.

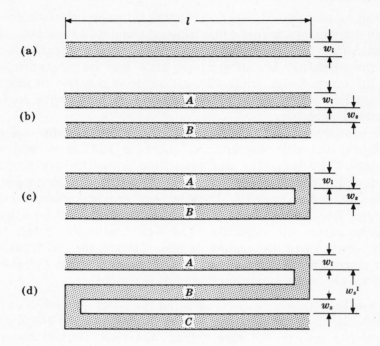

FIG. 11-17. Various resistor patterns used to calculate parasitic series inductance.

For a thin film resistor path 10 mils wide and 1 inch long, the inductance is 29.5 nanohenrys. For a path 10 mils wide and only 0.1 inch long, the inductance is 1.79 nanohenrys. Not only does the inductance decrease as the length decreases but, in fact, the inductance per unit length decreases with decreasing length because of the logarithm term in Eq. (11-24). As the line width is made narrow, however, the inductance increases because of the same term.

When there is a second resistor path next to the first, their mutual inductance must be considered. The mutual inductance between the two parallel resistor lines shown in Fig. 11-17(b) is given by

$$L_m = \frac{\mu_0 l_l}{2\pi}\left[\ln\left(\frac{2l_l}{w_l + w_s}\right) - 1 + \frac{w_l + w_s}{l_l} - \frac{1}{4}\left(\frac{w_l + w_s}{l_l}\right)^2 \right. $$
$$\left. + \frac{1}{12(1 + w_s/w_l)^2}\right] \quad (11\text{-}25)$$

where w_s is the width of the space between the lines and $w_l \ll l_l$ and $w_s \ll l_l$. The last three terms in Eq. (11-25) are usually small and are often neglected.

For two 10-mil lines 1 inch long with a 10-mil space between them, the mutual inductance is 18.5 nanohenrys. It is 0.77 nanohenry if the lines are 0.1 inch long. Thus, the mutual inductance decreases as the length decreases and the mutual inductance per unit length also decreases as the length decreases. Decreasing the line width and space width again increases the mutual inductance.

The total inductance of the pair of resistor lines can be found from the self-inductance of each line and the mutual inductance between them. If the lines are connected in parallel, so that the current travels in the same direction in each path, the mutual inductance adds to the self-inductance of each line and the inductance of the parallel combination is

$$L_p = \frac{(L_A + L_m)(L_B + L_m)}{L_A + L_B + 2L_m} = \frac{L + L_m}{2} \qquad (11\text{-}26)$$

where L_A, L_B, and L_m are given by Eqs. (11-24) and (11-25). If the lines are connected in series so that the current travels in opposite directions, the mutual inductance subtracts from the self-inductance of each line and

$$L_s = L_A + L_B - 2L_m = 2(L - L_m) \qquad (11\text{-}27)$$

In the meandering resistor pattern, the various resistor lines are connected in series and, hence, Eq. (11-27) would apply to the two-meander pattern shown in Fig. 11-17(c). If the small terms in Eqs. (11-24) and (11-25) are neglected, substitution into Eq. (11-27) gives the inductance of the two-meander pattern:

$$L_2 = \frac{\mu_0 l_l}{\pi} \left[\ln\left(1 + \frac{w_s}{w_l}\right) + \frac{3}{2} \right] \qquad (11\text{-}28)$$

where l_l is the length of each meander.

The logarithm term does not contain the line length and, hence, the inductance per unit length of the path is independent of the length. For a total path length of 1 inch ($l_l = 0.5$ inch), a 10-mil line width, and a 10-mil space width, the inductance is 11.1 nanohenrys. This should be compared with the 29.5 nanohenrys of a single straight line. The comparison illustrates that the meandering-line resistor is, in effect, a noninductive winding.

The same procedures can be used on the three-meander pattern shown in Fig. 11-17(d). Adding the various mutual inductances to the self-inductance of each line in the pattern gives

$$L_3 = L_A + L_B + L_C - 2L_{m(AB)} - 2L_{m(BC)} + 2L_{m(AC)}$$

$$= 3L - 4L_m + 2L_m{}^I = 3\left(L - \frac{4}{3}L_m + \frac{2}{3}L_m{}^I\right) \quad (11\text{-}29a)$$

where $L_m{}^I$ is given by Eq. (11-25) but using $w_s{}^I$ for w_s with

$$w_s{}^I = w_l + 2w_s \tag{11-29b}$$

The $L_m{}^I$ term has a positive sign since the currents in paths a and c are in the same direction.

Equations (11-24) and (11-25) may be substituted in Eq. (11-29a) and, if the small terms are neglected, the total inductance of the three-meander pattern in Fig. 11-17(d) can be expressed as

$$L_3 = \frac{3\mu_0 l_l}{2\pi}\left\{\ln\left[\frac{(2l_l)^{1/3}\left(\dfrac{w_l + w_s}{2}\right)^{2/3}}{w_l}\right] + \frac{7}{6}\right\} \quad (11\text{-}30)$$

For a total path length of 1 inch ($l_l = 1/3$ inch), a line width of 10 mils, and a space width of 10 mils, the total inductance of the three-meander pattern is 13.0 nanohenrys or slightly greater than the 11.1 nanohenrys of the equivalent two-meander pattern.

This general procedure can be used to find the inductance of any meandering-line pattern. As the number of meanders increases, there will be more mutual inductance terms to consider. The general relation for a pattern with p meanders, in series, can be written as

$$L_T = pL - 2(p-1)L_m + 2(p-2)L_m{}^I$$

$$- 2(p-3)L_m{}^{II} + \dots \quad (11\text{-}31)$$

where L is the self-inductance of each meander given by Eq. (11-24); L_m is the mutual inductance between nearest-neighbor lines given by Eq. (11-25); $L_m{}^I$ is the mutual inductance between next-nearest neighbors given by Eq. (11-25), using a space distance of $w_s{}^I = w_l + 2w_s$; and Lm^{II} is the mutual inductance between next-next-nearest neighbors given by Eq. (11-25), using a space distance of $w_s{}^{II} = 2w_l + 3w_s$.

A certain amount of caution must be exercised with this approach. It may be found that, as more terms are added, the smaller terms in

Eqs. (11-24) and (11-25) can no longer be neglected. As the mutual inductance of resistor lines farther and farther apart is considered, it may be found that the space distance is no longer small compared to the length and, hence, Eq. (11-25) is no longer accurate. When this is the case, a better approximation may be obtained by replacing the resistor lines with filaments at x distance apart. For nearest-neighbor lines, $x = w_l + w_s$; for next-nearest neighbors, $x = 2(w_l + w_s)$, etc. The exact expression for the mutual inductance of two filaments of any length is

$$L_m = \frac{\mu_0 l_l}{2\pi}\left[\ln\left(\frac{l_l}{x} + \sqrt{1 + \frac{l_l^2}{x^2}}\right) - \sqrt{1 + \frac{x^2}{l_l^2}} + \frac{x}{l_l}\right] \quad (11\text{-}32)$$

Since resistor lines have a finite width, Eq. (11-32) should only be used when $w_l << x$.

There are several general conclusions concerning the series inductance of thin film resistors:

1. The lead inductance of a film resistor can be made extremely small by placing components close to each other.
2. The meander pattern is basically a noninductive pattern so that the series inductance of thin film resistors is small. A straight-line resistor has more inductance per inch of path.
3. If a particular resistor pattern is given and all the dimensional ratios are maintained, the inductance of the pattern can be decreased by decreasing the size of the pattern. In fact, the inductance is directly proportional to a linear dimension of the pattern [see Eqs. (11-24) and (11-25)].
4. If a pattern is scaled down to reduce the inductance, the resistance will remain fixed since the number of squares in the pattern will not change during the scaling operation [see Eqs. (11-11) and (11-12)]. Thus, scaling down reduces the inductance without changing the resistance.
5. Increasing the sheet resistance of the film reduces the series inductance effects. A higher sheet resistance means a shorter path with less inductance for a given amount of resistance. Conversely, increasing the sheet resistance without changing the pattern will increase the resistance but will not change the inductance. Thus, the series inductive reactance has less effect on the impedance of the resistor.
6. Increasing the space-width-to-line-width ratio while keeping the length constant will increase the inductance of the path [see Eq. (11-28)].

Shunting Capacitance. The straight-line resistor pattern shown on the left in Fig. 11-5 has very little capacitance associated with the resistor path itself. In fact, the large termination areas provide considerably more shunting capacitance than the path does. In the meandering-line pattern shown on the right in Fig. 11-5 there is a shunting capacitance between adjacent meanders that can be significant, particularly when the substrate material has a high dielectric constant. In principle, this capacitance can be calculated if the dielectric constant is known and if the substrate thickness and pattern geometry are specified [8, 9, 10].

A detailed calculation of the capacitance is beyond the scope of this chapter. However, it is possible to simplify the geometry and to perform calculations that provide some insight into the magnitude of the capacitance and to show how it is related to resistor geometry.

A cross section of a substrate with two lines of a meandering resistor pattern on it is shown in Fig. 11-18(a). In order to simplify the calculation, it is assumed that the two resistor lines shown are

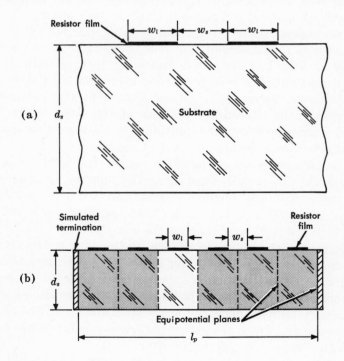

FIG. 11-18. Cross-sectional view of resistor pattern used to calculate parasitic shunting capacitance.

isolated and that the substrate extends a considerable distance to the right and left of the two lines. It is also assumed that the line length is large compared to the line and space widths.

Once the geometry is idealized in this manner, it is possible to conformally transform the configuration shown in Fig. 11-18(a) into a parallel-plate capacitor structure. This procedure is the same technique that was used to determine the number of squares in complicated resistor patterns. In making the transformation as indicated, the substrate material fills the region between the parallel plates so that the capacitance is readily calculated if the dielectric constant of the substrate is known.

Note that in Fig. 11-18(a), the space above the substrate will also make a contribution to the capacitance between the resistor lines. This contribution can be evaluated using a similar transformation. It might appear from this figure that the film thickness should enter into this calculation. The film thickness shown is, however, a great exaggeration and is actually negligible compared to the other dimensions. The resulting relation for the total capacitance between two parallel resistor lines on a substrate is given by

$$C = \frac{\epsilon_0 l_1}{2} \left[\frac{K(k'_1)}{K(k_1)} + \epsilon_r \frac{K(k'_2)}{K(k_2)} \right]$$ (11-33a)

where

$$k_1 = \frac{w_s}{2w_l + w_s}$$ (11-33b)

and

$$k_2 = \frac{\tanh\left(\dfrac{\pi w_s}{4d_S}\right)}{\tanh\left[\dfrac{\pi(2w_l + w_s)}{4d_S}\right]}$$ (11-33c)

The function $K(k)$ is the complete elliptical integral of the first kind, which is a tabulated function of k, and k' is $\sqrt{1 - k^2}$. The ϵ_0 and ϵ_r are the permittivity of free space and the dielectric constant of the substrate material. The l_l, w_l, w_s, and d_S refer to the line length, line width, space width, and substrate thickness, respectively. This expression neglects any contribution of the space beneath the substrate, but the contribution of this region is small compared to the total capacitance when $\epsilon_r >> 1$ or $d_S >> (2w_l + w_s)$. The quantity k_2 approaches k_1 when $d_S >> (2w_l + w_s)$.

For two 10-mil lines with a 10-mil space between them on a 50-mil Corning 7059 glass substrate, the capacitance per unit length is 1.2×10^{-12} farad/in. The dielectric constant of 7059 glass in 5.8, and ϵ_0 is 8.854×10^{-12} farad/m. The substrate contribution to the capacitance is 1.0×10^{-12} farad/in. whereas the space above the substrate contributes 0.18×10^{-12} farad/in.

The effect of the various dimensions on the substrate contribution to the capacitance is shown in Fig. 11-19. The normalized capacitance is plotted as a function of w_s/d_s for different values of w_l/d_s. As would be expected, decreasing the space width or increasing the line width increases the capacitance per unit length. The effect of substrate thickness can be seen by comparing points A and B in Fig. 11-19. Point A corresponds to $w_l/d_s = 0.1$ and $w_s/d_s = 0.2$. Decreasing the substrate thickness by a factor of 20 establishes point B, which corresponds to $w_l/d_s = 2$ and $w_s/d_s = 4$. The capacitance for point B is about one-third of that for point A. Thus, reducing the substrate thickness reduces the substrate contribution to the capacitance.

Figure 11-19 can also be used to determine the effect of scaling the size of a resistor pattern up or down without changing the relative dimensions. If the effect of substrate thickness is ignored, the shunting capacitance of the pattern will be directly proportional

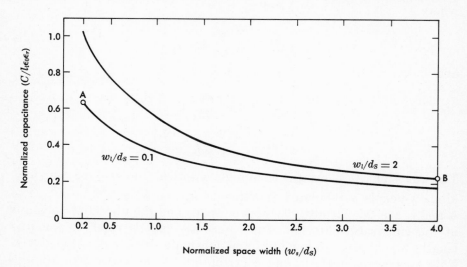

FIG. 11-19. Calculated shunting capacitance versus space between resistor lines.

to the linear dimensions of the pattern. This can be seen from Eq. (11-33). If d_S is made very large, $k_2 \cong k_1$ and C is proportional to l_l.

With a substrate of finite thickness, the effect of scaling down the pattern is the same as going from point B to point A in Fig. 11-19. The normalized capacitance is increased by about a factor of 3. The normalization factor contains a linear dimension (l_l), so that the net effect of reducing the size of the pattern by a factor of 20 is to reduce the shunting capacitance by a factor of 7. The above numbers apply to a particular range of w_l/d_S and w_s/d_S. A reduction of pattern size always reduces the shunting capacitance, but the effect of the size reduction is greater when the pattern dimensions are small compared to the substrate thickness.

Equation (11-33) can be used to estimate the shunting capacitance in a resistor pattern containing only two lines. If more lines are added, the capacitance between different pairs of adjacent lines are all connected in series. The total capacitance of this series combination, however, will be smaller than the total capacitance of the pattern because of the shunting effect of the capacitance between next-nearest neighbors as well as between lines even farther apart. As the number of lines grows, the network of series-parallel combinations becomes very complex. It also should be noted that calculations made using Eq. (11-33) are slightly in error when the additional conducting material of the other lines is present on the substrate surface.

An alternative solution to the problem is to consider the many-line pattern as a whole and then to break it up into small identical segments that can be treated analytically. An example of this procedure is shown in Fig. 11-18 (b). Termination areas have been simulated by coating the ends of the substrate with conducting material. While this simulation might seem artificial, it preserves the symmetry needed for analysis in a geometry that should be a satisfactory approximation for a many-line pattern. By use of symmetry arguments and equal potential planes, it is possible to establish the clear area in the drawing as a basic segment. The dashed lines represent equal potential planes. Since there are six such segments in series, the total capacitance of the path is one-sixth the capacitance of this segment. The capacitance between the planes can be found by conformal mapping. The equations are complex and will not be given in detail. Two limiting cases will provide some insight into the general nature of the solution. Consider the case where $w_l = w_s$. The capacitance of the clear segment in Fig. 11-18(b) is

$$C = \frac{\epsilon_0 \epsilon_r l_l d_S}{w_s + w_l} \tag{11-34}$$

when $w_s = w_l$ and $w_l << d_S$, and

$$C = \frac{\epsilon_0 \epsilon_r l_l d_S}{w_s + \dfrac{4 \ln 2}{\pi} d_S} \tag{11-35}$$

when $w_s = w_l$ and $w_l >> d_S$.

Using Eq. (11-34) for the case where $w_l = 10$ mils, $w_s = 10$ mils, and $d_S = 50$ mils, it is found that the equivalent substrate capacitance between two planes with Corning 7059 glass is about 4×10^{-12} farad/in. of line length. Although the geometries are different, this value should be compared with the capacitance of 1.0×10^{-12} farad/in. which was calculated using Eq. (11-33). The difference is due mostly to the shunting effects of neighboring lines in a many-line pattern.

Equation (11-34) suggests a method of estimating the capacitance of a resistor without looking at details of the pattern. When the total capacitance of the six segments shown in Fig. 11-18 is calculated, the denominator will be $6(w_s + w_l)$, which is the exact length of the pattern (l_p). The length of each line (l_l) is the exact width of the pattern (w_p). Thus, the total shunting capacitance of the resistor is

$$C_T \cong \frac{\epsilon_0 \epsilon_r w_p d_S}{l_p} \tag{11-36}$$

when $w_l = w_s$, $w_l << d_S$, and $d_S << l_p$. Although this neglects smaller capacitance terms associated with the air spaces, experiments have shown that Eq. (11-36) gives an estimate of the shunting capacitance sufficiently good for analysis of parasitics. The restriction $d_S << l_p$ is imposed because the simulated terminations shown in Fig. 11-18(b) are not realistic unless this inequality is satisfied.

There is little point in trying to carry detailed capacitance calculations much farther, for there are other features of the resistor path which complicate the picture. The resistor lines are connected at one end so that the capacitance of interest is actually distributed along the resistor path. This makes the lump-equivalent model only an approximation, particularly at high frequencies where the shunting capacitance effects are most important. An equivalent circuit of the meandering-line resistor, which includes the distributed nature of the capacitance, is shown in Fig. 11-20. This equivalent circuit also includes the series inductance effects and it was drawn to represent the meandering-line resistor shown in Fig. 11-5. Although this circuit should represent the resistor more accurately than the lumped model, it is still only an approximation.

Several general conclusions can be stated concerning the shunting capacitance of thin film resistors:

1. The shunting capacitance depends on the dielectric constant of the substrate material and on the substrate thickness. The

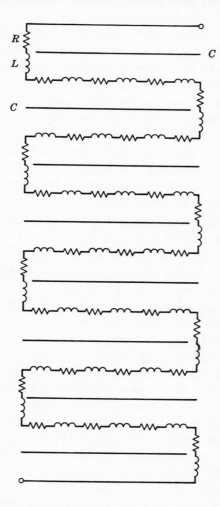

Fig. 11-20. Schematic diagram of thin film resistor including distributed parasitics.

capacitance is minimized by using a thin substrate with a low
dielectric constant [see Eq. (11-36)].

2. The meandering resistor pattern has more shunting capacitance than the single-line pattern. Also, the meandering pattern
   is usually needed for large resistances where the shunting effect
   is more critical. Many short meanders will give less shunting
   capacitance than a few long meanders, since the capacitances
   between meanders are in series [see Fig. 11-18(b)].

3. Decreasing the size of the pattern can decrease the shunting
   capacitance, but it will not change the resistance, since the
   number of squares stays the same [see Eqs. (11-12) and
   (11-33)]. The largest decrease in capacitance by scaling down
   will be realized only when the pattern is small compared to the
   substrate thickness [see Eq. 11-36)].

4. A higher sheet resistance will mean a shorter path and a
   narrower pattern for the same resistance value and, hence, there
   will be less shunting capacitance. For a given pattern, however,
   increasing the sheet resistance increases the resistance of the
   path without changing the capacitance. In this case, the reactance of the shunting capacitance forms a larger part of the
   total impedance of the resistor.

5. Decreasing the space-width-to-line-width ratio tends to increase
   the shunting capacitance (see Fig. 11-19).

6. The high-frequency performance is limited either by the series
   inductance or by the shunting capacitance, depending to a large
   extent on the magnitude of the resistance.

Measurements have been made on tantalum nitride film resistors
with different meandering patterns. In one case, the pattern details
were varied while the overall pattern dimensions were maintained
at 0.25 inch long and 0.1 inch wide. The total shunting capacitance
was found to be about 0.1 picofarads and the series inductance was
about 5 nanohenrys. Above approximately 1000 ohms, the shunting
capacitance limited performance; below 1000 ohms, the series inductance was limiting. In the region of 1000 ohms, the impedance
was essentially pure resistance to above 100 MHz.

To appreciably improve the high-frequency performance, it is
necessary to reduce the shunting capacitance. This may be done by
decreasing the w_p/l_p ratio. Although this increases the series inductance, a considerable increase can be made without significantly
affecting the impedance of the resistor.

Series and Shunting Resistances. The potential sources of parasitic series resistance are the resistance of the conducting film leading up to the resistor and any contact resistance between the conducting film and the resistor film underneath it. Experience has shown that, with the use of proper processing techniques, the contact resistance is well under 10 milliohms. The resistance of the conducting film depends on the number of squares of film in the conducting path and the sheet resistance of the conducting material. The sheet resistance of evaporated conducting films is normally between 0.05 and 0.1 ohm/square. Lower sheet resistances can be obtained by plating or solder coating the evaporated film (see Sec. 11.3). The number of squares in the conducting path leading up to a particular resistor is flexible. In laying out a circuit, it is usually possible to keep the lead-in path of a critical resistor to less than one square.

It might appear that any parasitic series resistance could be compensated for in the design value of the resistor. In some cases this may be true, but in precision resistor applications, a parasitic resistance can be of concern. If the series resistance is an appreciable percentage of the total, it can affect the temperature coefficient of resistance, the voltage coefficient of resistance, the current noise, and the resistor stability. These effects are not usually serious unless the total resistance value is less than 10 ohms.

A parasitic shunting resistance can occur if there is a leakage current flowing through the substrate material or on the substrate surface. The substrate materials of principal interest, such as alkali-free glass and ceramic, are good insulators. At 200°C the volume resistivities range from 10^{13} to 10^{16} ohm cm. At room temperature, they are about two orders of magnitude higher. Soda-lime glasses have considerably lower resistivities (see Chap. 9).

A rough approximation of the effective sheet resistance of these materials can be obtained by dividing the resistivity by the substrate thickness. For a substrate thickness of about 40 mils, the effective sheet resistance opposing bulk leakage currents is in the range of 10^{14} to 10^{17} ohms/square at 200°C.

The surface leakage current can also be characterized by an effective sheet resistance. In this case, the sheet resistance is more sensitive to surface cleanness and relative humidity than to the temperature. In the absence of gross contamination, the effective surface sheet resistance in different ambients ranges from 10^{10} to 10^{15} ohms/square for the substrate materials of interest. The surface and bulk leakage current paths are generally in parallel, so the smaller resistance will conduct most of the current. It is seen from the above figures that, for Corning 7059 glass (which is essentially alkali free) and glazed alumina, the surface leakage is the principal potential source of parasitic shunting resistance.

One of the most critical regions for a surface leakage current is across the space areas of the meandering resistor path. Consider the case of two meanders as they are shown in Fig. 11-17(c). The approximate resistance of this path is given by

$$R \cong (2R_s) \frac{l_l}{w_l} \tag{11-37}$$

where R_s is the sheet resistance of the film material, l_l is the length of one meander, and w_l is the line width. The shunting resistance of the surface leakage across the space between meanders is distributed along the path. It can be shown that the resistance of the path, with this leakage present, is given by

$$R' = (2R_s) \frac{l_l}{w_l} \frac{\tanh \theta}{\theta} \tag{11-38a}$$

where

$$\theta = \sqrt{\frac{2R_s l_l^2}{R'_s w_s w_l}} \tag{11-38b}$$

R'_s is the sheet resistance of the substrate surface and w_s is the space width. Equation (11-38) applies to the two-meander pattern shown in Fig. 11-17(c). A treatment of the many-line patterns is more complex, but it yields similar results.

As long as the surface leakage current is small, Eq. (11-38) can be approximated by

$$R' = (2R_s) \frac{l_l}{w_l} \left[1 - \frac{2}{3} \left(\frac{R_s}{R'_s} \right) \left(\frac{l_l}{w_l} \right)^2 \left(\frac{w_l}{w_s} \right) \right] \tag{11-39}$$

where the second term in the brackets represents the fractional change in resistance due to the parasitic shunting resistance. From Eq. (11-11) it is seen that (l_l/w_l) is the number of squares in each meander.

Consider the case where the film sheet resistance is 100 ohms/square and the sheet resistance of the substrate surface is as low as 10^{10} ohms/square. For $w_l = w_s$ and $l_l = 100 \, w_l$, the fractional change in resistance is $(2/3) \times 10^{-4}$ or about 0.01 percent. For higher film sheet resistances or greater aspect ratios, the shunting resistance could be troublesome unless the substrate sheet resistance were also increased. The above results indicate the importance of surface cleanness when meandering resistor patterns are employed.

It should be noted from Eq. (11-39) that it is the length-to-width aspect ratio used in the pattern that is important rather than the line and space widths themselves. Surface leakage does not prevent the use of very fine lines and spaces as long as the length of the line segments is also reduced so that the pattern-aspect ratios remain the same. Equation (11-39) also confirms the intuitive conclusion that, if l_l and w_l are maintained constant, increasing the space width decreases the effect of surface leakage.

When films with a very high sheet resistance are used ($R_S > 10^4$ ohms/square), it may be necessary to place a limit on the line-length-to-line-width ratio that is used in a pattern. It may also be necessary to coat the resistors with an insulating material to prevent inadvertent surface contamination. Fortunately, long meandering patterns are generally not needed when R_S is very large. There is an upper limit on a usable sheet resistance, even for a straight-line resistor pattern. This limit might be expressed as $R_S < 10^{-5} R'_S$.

Skin Effect. Electrical conduction in metals at high frequencies is characterized by a quantity called skin depth, defined [11] as

$$\delta_S = \frac{1}{\sqrt{\pi f \mu \sigma}} \tag{11-40}$$

where f is the frequency, μ is the permeability, and σ is the conductivity. For a solid conductor with a circular cross section, most of the current is conducted within one skin depth of the surface. If the frequency is high enough so that the skin depth is less than the conductor radius, the resistance of the conductor is a strong function of frequency.

The same phenomenon can occur in thin film resistor paths, but here the current distribution on the surface is not uniform. The current density is greatest at the edges of the film path, considerable on the surfaces, and least in the center of the body of the film. Although an analytical description of the current density is not readily available, it can be said that the skin effect on a film conductor is generally less than on a solid, round conductor with the same cross-sectional area. Thus, the skin effect will not be important for thin film conductors as long as

$$\delta_S >> \sqrt{\frac{d w_l}{\pi}} \tag{11-41}$$

where d is the film thickness and w_l is the width of the path. Solving Eqs. (11-40) and (11-41) for the line width gives

$$w_l << \frac{R_S}{f\mu} \qquad (11\text{-}42)$$

where R_S is the sheet resistance of the film in ohms/square. For non-magnetic materials, $\mu = \mu_0 = 4\pi \times 10^{-7}$ henrys/m where w_l is in meters. Equation (11-42) establishes the maximum resistor line width that should be used to avoid the resistance being frequency dependent. If w_l is made equal to $R_S/f\mu$ at the particular frequency of interest, the a.c. resistance will be a few percent higher than the d.c. resistance; if w_l is half this value, the resistance difference will be a few tenths of a percent. The resistance difference continues to fall off exponentially as w_l is made smaller.

As an example of what this means in a resistor design, consider the case where the sheet resistance is as low as 10 ohms/square. At 1 GHz, the quantity $R_S/f\mu = 0.313$ inch, or more than an order of magnitude larger than most resistor lines. Thus, for frequencies up to 1 GHz, the skin effect causes no resistance change in thin film resistors.

11.2   FILM CAPACITORS

Capacitor Properties and Materials

Conceptually, the design of a thin film capacitor is simpler than that of the film resistor, since the layers of different film materials naturally produce a parallel plate structure. There are, however, many subtleties to a capacitor design, most of which relate to the choice of materials and to how the material properties affect capacitor performance. This situation is in sharp contrast to the design of a film resistor. Although resistor film material properties are of vital importance, they tend to be overshadowed by geometric considerations when details of the design are worked out.

The material properties of primary importance for film capacitors are, to a large extent, the properties of the dielectric film. In many instances, however, the metal electrodes and the substrates affect the properties of the dielectric; thus, all the materials should be considered together.

Some of the more important material properties which affect capacitor design and performance are:

1. Dielectric constant
2. Dielectric strength
3. Dielectric loss
4. Stability
5. Temperature coefficient of capacitance
6. Voltage coefficient of capacitance
7. Insulation resistance
8. Surface smoothness

The properties, advantages, and disadvantages of the materials available for different capacitor structures were discussed in Chap. 8. As time goes on, new materials and structures with different or improved properties will become available. In many instances, the capacitor performance requirements will dictate the choice of materials. If a circuit requires extremely stable capacitance values, the choice would probably be a tantalum oxide dielectric with an adherent top electrode, such as gold or aluminum on top of Nichrome® or titanium. On the other hand, if the circuit needs several large-value capacitors, the choice would be a tantalum oxide dielectric, but with manganese oxide used as the bottom layer of the top electrode. For very small-valued capacitors, a duplex dielectric of silicon oxide on top of tantalum oxide might be preferred.

Dielectric Thickness

Once the best combination of capacitor materials has been decided upon, the design problem is to determine the thickness of each layer of material and the linear dimensions of a pattern for the structure. The first consideration is the thickness of the dielectric film. It will be recalled that the capacitance of a parallel-plate capacitor is

$$C = \epsilon_0 \epsilon_r \frac{A_C}{d_\epsilon} \qquad (11\text{-}43)$$

where ϵ_0 is the permittivity of free space, 8.854×10^{-12} farad/m; ϵ_r is the dielectric constant; A_C is the active area of the capacitor; and d_ϵ is the thickness of the dielectric film. The active area of the capacitor is that area of the dielectric film that has a metal electrode on both sides (i.e., top and bottom). This definition may seem obvious, but it is an important distinction when the total area of a capacitor pattern is considered. For a rectangular geometry, the active area is defined by its length and width:

$$A_C = l_C w_C \tag{11-44}$$

Equation (11-43) is a precise relation for a thin film capacitor, since fringing effects at the edges are negligible (since $d_\epsilon << l_C$ and $d_\epsilon << w_C$).

Equation (11-43) indicates that the dielectric constant and the dielectric thickness of the material together determine the capacitance density which is defined as the capacitance per unit area. Thus,

$$\frac{C}{A_C} = \frac{\epsilon_0 \epsilon_r}{d_\epsilon} \tag{11-45}$$

where C/A_C is the capacitance density. For a given material, the film thickness establishes the capacitance density which, in turn, determines the area needed for a particular capacitance value.

The appropriate choice of film thickness depends upon the required working voltage and the dielectric strength of the material. For an anodic dielectric, the field used to grow the oxide establishes an upper limit on the dielectric strength. This limit is the reciprocal of the anodization constant, or about 6.3×10^6 v/cm for tantalum oxide at room temperature. Because of defects in the dielectric film, the dielectric strength is actually less than this upper limit, determinable by measuring the voltage at which the capacitor breaks down. Thus,

$$\mathcal{E}_B = \frac{V_B}{d_\epsilon} \tag{11-46}$$

where $\mathcal{E}_B$ is the dielectric breakdown strength of the film, V_B is the breakdown voltage of the capacitor, and d_ϵ is the dielectric thickness.

Since anodic film capacitors tend to be polar in nature, the dielectric strength will depend on the polarity of V_B. This chapter will restrict the discussion to the anodic, or blocking, direction (base electrode positive) and will treat the anodic film capacitors as polar devices when it comes to biasing voltages. Signal voltages on unbiased capacitors are not usually a problem, since the anodic film capacitor can withstand some cathodic voltage. In applications where nonpolar capacitors are required, anodic film capacitors can be used back-to-back, as will be discussed later.

The dielectric strength is approximately independent of film thickness over a limited range of film thickness. For very thin dielectric films, however, the density of defects is higher so that the dielectric strength is lower. One advantage of using manganese oxide in the top electrode is the tendency of this semiconducting layer to heal or isolate the defects in the dielectric film so that the high dielectric strength is still available even in a very thin film (see Chap. 8).

There is also an upper limit on film thickness for some dielectrics. In the case of evaporated dielectric films, mechanical stress in the film can result in crazing or rupturing if the film is too thick. These effects are noted in SiO films that are more than 3 to 5 microns thick. The maximum thickness of an anodic tantalum oxide film is about 0.65 micron (6500 A), which corresponds to an anodization voltage of about 400 volts. At higher anodization voltages, the defect density increases sharply due to scintillations and field recrystallization in the oxide during anodization. These phenomena are due to the combined effects of both electrical and mechanical stresses that act upon the thicker films.

Between the thickness limits discussed above, the dielectric strength is a film material property that is nearly independent of the film thickness, although it does depend on the electrode materials. For a tantalum oxide capacitor with a gold top electrode, it is about 4×10^6 v/cm. With Nichrome® next to the tantalum oxide it is lower, but with manganese oxide it is higher.

The rated voltage, or working voltage, of a capacitor must obviously be less than the breakdown voltage. How much less depends upon the substrate temperature and the time at rated voltage required in actual use. The rated voltage seldom exceeds half the breakdown voltage and, for some organic films, it may be as low as 10 percent of the breakdown voltage. Thus,

$$V_R = kV_B = kd_\epsilon \, \mathcal{E}_B \qquad (11\text{-}47)$$

where V_R is the rated voltage and k is a number less than 1, depending on the capacitor materials and the conditions under which the capacitor will be used (time at voltage and temperature).

A lower limit on the dielectric film thickness can now be established from the voltage requirement of the capacitor and the dielectric strength of the material chosen. Thus,

$$d_\epsilon \geqq \frac{V}{k \, \mathcal{E}_B} \qquad (11\text{-}48)$$

where V is the voltage that the capacitor must withstand. Equations (11-45) and (11-48) can then be used to calculate the maximum capacitance density that can be used in the capacitor design.

Normally, all the capacitors on a substrate have the same dielectric thickness, since the dielectric is formed in a single operation. Thus, all the capacitors will have the same capacitance density, which must satisfy the voltage requirements of the capacitor experiencing the highest voltage. There are ways around this situation, such as putting capacitors in series or using different dielectric thicknesses. The former can take up more room and the latter requires extra processing steps. Still, if one small-valued capacitor has a significantly higher voltage requirement than all the rest, one of these solutions might be acceptable in order to conserve space.

At this point, it is of interest to define the capacitor charge storage factor, which is the working voltage times the capacitance density, or

$$V_R \frac{C}{A_C} = k\epsilon_0\epsilon_r \, \mathcal{E}_B \qquad\qquad (11\text{-}49)$$

where the units are farad v/m^2, or charge per unit area. The charge storage factor is a film material property which determines how much substrate area is needed to provide a given capacitance with a given voltage capability. The charge storage factor is independent of the film thickness as long as the dielectric strength is independent of film thickness.

For a tantalum oxide dielectric film, the charge storage factor is about 20 μfarad $v/in.^2$; for a silicon monoxide dielectric, it is about 0.7 μfarad $v/in.^2$ (see Fig. 8-9). Since the charge storage factor of tantalum oxide is about 30 times that of silicon monoxide, it takes only 3 percent of the area to make a capacitor using tantalum oxide. This difference is due partly to the higher dielectric constant of tantalum oxide and partly to its higher dielectric strength.

Capacitor Yield

In the discussions above, the breakdown voltage and dielectric strength were treated as fixed numbers for a particular capacitor structure. Actually, the breakdown of film capacitors occurs at point defects in the dielectric film, and each defect has its own breakdown voltage. The numbers quoted above represent average, or mean values, determined from testing large numbers of samples. As processing techniques are perfected, the measured breakdown voltage tends to increase and then level off. The more the processing is under control, the tighter the distribution is around the mean value. There will still be some capacitors with defects that have low breakdown voltages which are weeded out by applying a test voltage to every capacitor. The problem then becomes one of device yield, and in a film circuit with many components, the individual component yield must be very high in order for the circuit yield to be reasonable.

The above considerations have a direct bearing on capacitor design decisions. When Eq. (11-48) is used to determine the minimum dielectric thickness, the k factor used is based primarily on reliability of the devices which are accepted, not device yield. In other words, a capacitor designed with this thickness has a very low probability of failing during its design lifetime, *once it has passed the initial screening voltage*. The question of capacitor yield can dictate a thicker dielectric film and can limit the amount of capacitance that should go onto one substrate.

It has already been pointed out that both very thin and very thick films have high defect densities. It is natural to expect that there is a particular dielectric thickness range for which the defect density is at a minimum and for which the capacitor yield is at a maximum. Such a range has been found for tantalum oxide dielectrics to be from 3000 A to 3500 A. If the thickness, calculated by Eq. (11-48), is less than 3000 A, the decision is whether or not to increase the dielectric thickness. A thickness increase will increase the capacitor yield but, at the same time, will require more substrate area. The decision can be difficult since substrate area and capacitor yield both affect the cost. If the thickness calculated using Eq. (11-48) is greater than 3500 A, there is no decision to be made since the thicker dielectric is required to insure capacitor reliability.

The other decision that affects device yield is the total amount of capacitance per substrate. It is not the number of capacitors nor the value of any one capacitor that is important but, rather, it is the total capacitance per substrate, since a defect in the area of any capacitor will cause a defective device. The maximum amount of capacitance per substrate that still gives a reasonable yield is a function of the capacitor materials involved. A number that has been used for tantalum oxide capacitors with a plain metal counterelectrode is 0.05 μfarad per substrate. When manganese dioxide is used in the electrode, the capacitance per substrate may be increased to more than 1 μfarad. These numbers will probably increase as fabrication processes are improved.

Capacitor Losses

The three principal sources of loss (power dissipation) in thin film capacitors are dielectric loss, electrode resistance, and lead resistance. The planar geometry of the film capacitor makes the electrode and lead resistance important considerations in the design of a capacitor pattern. On the other hand, the dielectric loss is a property of the dielectric material and, hence, should be considered along with other properties when the capacitor materials are first chosen.

Capacitor losses are normally evaluated in terms of the dissipation factor of the capacitor (see Chap. 8). It is of interest to consider separately the contribution that the dielectric makes to the total capacitor loss. The dielectric loss (tan δ') is thus defined as the dissipation factor of the dielectric alone. The dielectric loss is independent of the dielectric area and thickness. For most of the dielectrics used in thin film capacitors, the dielectric loss is also independent of frequency to the first approximation. This is an important point since it means that the high-frequency performance of a thin film capacitor is normally not limited by the properties of the dielectric.

It is useful to represent the film capacitor by an equivalent circuit when analyzing its effect on circuit performance. Because the electrode resistance and lead resistance are, in effect, in series with the dielectric, it is convenient to represent the capacitor with a series equivalent circuit. Thus, the impedance of the thin film capacitor is

$$Z = R - \frac{j}{\omega C} \qquad (11\text{-}50)$$

where C is the equivalent series capacitance and R is the equivalent series resistance used to represent all the losses. The dissipation factor of the capacitor is

$$\tan \delta = \omega RC \qquad (11\text{-}51)$$

where ω is the angular frequency $(2\pi f)$.

The R in Eqs. (11-50) and (11-51) is not a simple ohmic resistance, since it includes the dielectric losses. These losses can be separated by rewriting Eq. (11-51):

$$\tan \delta = \tan \delta' + \omega R'C \qquad (11\text{-}52)$$

where tan δ' is the dielectric loss and R' is an ohmic resistance that represents the total effective series resistance provided by the electrodes and the lead paths. It is seen that at very high frequencies, the $\omega R'C$ term is the major source of loss. Equations (11-51) and (11-52) can be used to show the two distinct parts of R. Thus,

$$R = \frac{\tan \delta'}{\omega C} + R' \qquad (11\text{-}53)$$

It will be recalled that tan δ' is independent of frequency; therefore, the portion of R that represents the dielectric loss is frequency dependent.

If the R given by Eq. (11-53) is substituted back into Eq. (11-50), the impedance of the capacitor can be expressed as

$$Z = \frac{\tan \delta'}{\omega C} + R' - \frac{j}{\omega C} \qquad (11\text{-}54)$$

where the frequency dependence of Z is now expressed explicitly. It can be seen that at very low frequencies, the tan δ' term dominates the real part of the impedance, but at very high frequencies, the R' term dominates. A typical value of tan δ' for a tantalum oxide dielectric is 0.005.

The magnitude of the R' term is determined by the resistivities of the electrode materials, the thickness of the electrodes, and the shape of the capacitor pattern. When tantalum is used as the base electrode, its high resistivity makes the design of the pattern more critical. In this respect it is important to consider the distributed nature of a thin film capacitor when determining the magnitude of the R' term.

A cross-sectional view of a thin film capacitor is shown in Fig. 11-21. For purposes of this discussion, it is assumed that the sheet resistance of the top electrode is negligible compared to the base electrode. This is a valid assumption for the gold and tantalum combination shown in Fig. 11-21. It is also assumed that the cross section shown is uniform across the width of the capacitor, which has a rectangular shape. Some lead resistance of the type shown is unavoidable, since the gold top electrode cannot touch the tantalum base electrode nor the gold conducting film leading up to the capacitor. The exact margin required depends on the processing techniques used and the sequence of fabrication steps employed.

The fact that the capacitance is distributed along the base electrode has been indicated in the electrical schematic diagram also shown in Fig. 11-21. A section of the distributed structure has been enlarged to indicate how the dielectric loss can be taken into account. The series equivalent representation shown is a matter of choice; alternatively, a parallel representation could have been used.

The equation for the total impedance of the capacitor structure shown in Fig. 11-21 is

$$Z = R_L + R_E \frac{\operatorname{ctnh} \theta}{\theta} \qquad (11\text{-}55)$$

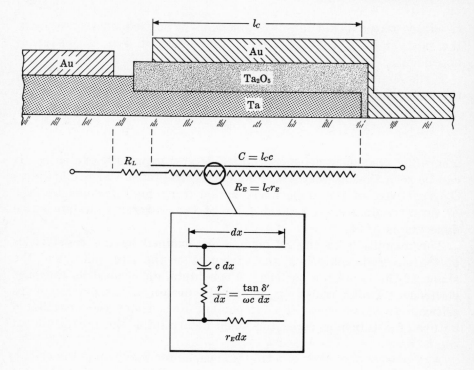

FIG. 11-21. Cross-sectional view and series equivalent schematic of thin film
capacitor showing parasitics.

where R_E is the total resistance of the base electrode. Thus,

$$R_E = R_S \frac{l_C}{w_C} \tag{11-56}$$

where R_S is the sheet resistance of the electrode (after anodization),
w_C is the capacitor width, and l_C is the capacitor length. The parameter θ is given by

$$\theta = \sqrt{\frac{j\omega R_E C}{1 + j \tan \delta'}} \tag{11-57}$$

where C is the equivalent series capacitance indicated in Fig. 11-21
and $\tan \delta'$ is the dielectric loss. Equation (11-55) was derived using
the usual methods of distributed parameter analysis [12, 13]. If the
frequency is low enough so that the total loss in the capacitor is small,

Eq. (11-55) can be approximated by

$$Z = \frac{\tan \delta'}{\omega C} + R_L + \frac{R_E}{3} - \frac{j}{\omega C} \qquad (11\text{-}58)$$

At higher frequencies, the original expression should be used.* Comparing Eqs. (11-54) and (11-58), it is seen that effective ohmic resistance, in series with the capacitor dielectric, is

$$R' = R_L + \frac{R_E}{3} \qquad (11\text{-}59)$$

This result applies specifically to the rectangular capacitor pattern shown in Fig. 11-21. Changing the shape, or bringing the base electrode out on more than one edge of the capacitor, will change R'. Capacitor patterns that minimize the effects of R_L and R_E are discussed later.

For a base electrode of tantalum, the minimum value of R' that can be obtained is less than 1 ohm. A value of 1 ohm corresponds to a loss of about 6 percent in a 1000-pf capacitor operated at 10 MHz. As evident from Eq. (11-52), decreasing either the capacitance or the frequency will reduce the loss factor. If a metal with a higher conductivity is used in the base electrode, the value of R' can be reduced by at least an order of magnitude, which will decrease the high-frequency loss by a similar amount. An aluminum film is used under the tantalum film to accomplish this purpose.

Capacitor Patterns

There are two capacitor patterns of primary interest, the crossed-electrode pattern and the minimum-resistance pattern, as shown in Fig. 11-22. Other patterns which may be of interest in some special applications are the nonpolar pattern and the adjustable capacitor pattern. The pattern to be used depends on the circuit requirements.

Crossed-Electrode Pattern. To control the value of a capacitor, it is necessary to control the dielectric constant, the dielectric thickness, and the capacitor area. The dielectric constant is a function of the structure and composition of the dielectric film and, hence, it is

*A series expansion of Eq. (11-55) shows that at higher frequencies (where the total loss is no longer small), the distributed nature of the film capacitor structure produces a decrease in the effective capacitance of the reactive impedance. On the other hand, the capacitance density of a tantalum-oxide dielectric itself has been found to be essentially independent of frequency up to frequencies in excess of 1 GHz.

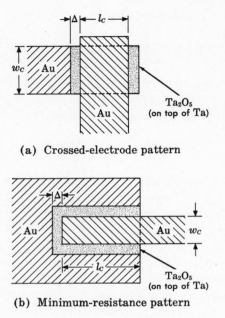

(a) Crossed-electrode pattern

(b) Minimum-resistance pattern

FIG. 11-22.   Crossed-electrode and minimum-resistance film capacitor patterns.

sensitive to the materials used and the conditions that exist during the formation process. The dielectric thickness must also be controlled when the film is formed. The dielectric constant and thickness together determine the capacitance density [see Eq. (11-45)]. For a tantalum oxide dielectric, the thickness is precisely determined by the anodization voltage and, thus, the capacitance density can be controlled to better than 1 percent. The control on the capacitance density of an evaporated dielectric can be almost as good if the deposition process is carefully monitored.

The area of a capacitor is determined by the overlap of the base-electrode and top-electrode patterns. Thus, the area accuracy is a function of the precision of each pattern and the pattern registration. If the electrode patterns are defined by photolithographic techniques, the line definition can be better than a tenth of a mil. Pattern registration, however, is generally not this good unless special, more costly techniques are employed.

The crossed-electrode pattern shown in Fig. 11-22 (a) is a means of achieving good area definition without precise mask alignment. It can be seen on inspection that, because of the overlap provided, a translational misalignment of patterns does not change the area of the capacitor. A rotational misalignment increases the area by the amount

$$\frac{\Delta A_C}{A_C} = \sec(\Delta\theta) - 1 \tag{11-60}$$

where $\Delta\theta$ is the amount of angular misalignment of the two electrode patterns. Rotational alignment of better than $1°$ is not difficult to achieve and a $1°$ misalignment corresponds to an area increase of less than 0.016 percent. This tolerance is independent of the capacitor area.

The area tolerance due to line definition errors is most serious in small-area capacitors. Consider, for example, a 30-pf capacitor. If a tantalum oxide dielectric is used, the minimum capacitance density available is about 0.3 μfarad/in.^{2}, due to practical limitations on dielectric thickness. Thus, the maximum size of the capacitor is a square 10 mils on a side. With this small an area, it might be difficult to control the capacitance to within ±5 percent. If the duplex dielectric structure (SiO on top of Ta_2O_5) is used, the capacitance density can be reduced by more than an order of magnitude. With the area increased by a corresponding amount, line definition will no longer be as serious a limitation and a 5-percent capacitor tolerance would be reasonable. Another alternative would be to put several larger area tantalum oxide capacitors in series. When tantalum is used for the base electrode, the crossed-electrode pattern has the same effective series resistance as the pattern depicted in Fig. 11-21 and, hence, the analysis developed in Eqs. (11-55) through (11-59) is applicable to this pattern.

Minimum Resistance Pattern. The pattern shown in Fig. 11-22(b) is used to reduce effective series resistance of the base electrode. By bringing the base electrode out on three sides of the capacitor, both the lead and effective electrode resistances are kept to a minimum. It can readily be shown that this puts an upper limit on the total effective resistance, in series with the dielectric, given by

$$R' < R_S \left(\frac{\Delta}{2l_C} + \frac{w_C}{12l_C} \right) \tag{11-61}$$

where R_S is the sheet resistance of the anodized tantalum, Δ is the total gold-to-gold mask margin, and w_C and l_C are the length and width of the capacitor as indicated in Fig. 11-22(b). The mask margin can range from 5 to 50 mils, depending on the process sequence and process techniques employed. The w_C/l_C ratio can be reduced as much as desired but, if it is made too small, the long path of the top electrode will start to contribute to the series resistance. It is clear from Eq. (11-61) that, for a capacitor of any appreciable size ($l_c \gg \Delta$), the effective ohmic series resistance can be a small fraction of the sheet resistance and, in fact, can be significantly less than 1 ohm, since R_S can be as low as 5 ohms/square. It must be noted here, however, that the value of the capacitor in the minimum-resistance pattern is affected by the precision available in mask alignment.

Nonpolar Pattern. Another pattern of interest is the nonpolar capacitor pattern. As mentioned earlier, anodic film capacitors are considered to be polar devices in their present state of development, even though they have some capability in the cathodic direction. When a circuit is laid out, it is important that the tantalum electrode be positive and that the top electrode be negative. Small signal voltages in the cathodic direction are not a problem, but a large bias can cause a voltage breakdown in the dielectric. In applications where a nonpolar capacitor is needed, two film capacitors can be placed back-to-back as shown in Fig. 11-23(a). The forward-biased capacitor limits the d.c. leakage current of the reverse-biased capacitor and,

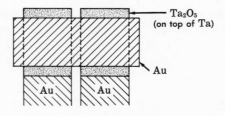

(a) Nonpolar pattern

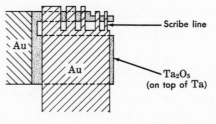

(b) Adjustable capacitor pattern

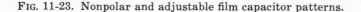

FIG. 11-23. Nonpolar and adjustable film capacitor patterns.

hence, protects it from voltage breakdown. Since the two capacitors are in series, this arrangement requires four times as much area to achieve the same capacitance value.

Adjustable Capacitor Pattern. There may be times when the capacitance precision achieved with the crossed-electrode pattern is not good enough. At present, there are no well controlled means of capacitor adjustment comparable to that used for resistors; however, there are means for some degree of adjustment. One method is shown in Fig. 11-23(b). The capacitor pattern consists of a main capacitor with several small fingers that serve as trimming capacitors. The total capacitance can be reduced by the desired amount by scribing the gold counterelectrode leading up to a particular trimming capacitor. The arrangement of the finger permits the scribing to be done in regions where only one electrode is present, thereby avoiding the possibility of shorting. A pattern of the type shown in Fig. 11-23(b) can be used to adjust the capacitance by as much as 10 percent and, at the same time, come within 0.2 percent of the design value. The areas of the trimming capacitors shown form a binary series.

When the capacitance value is very small, the parallel arrangement may become impractical because of the extremely small size of the trimming capacitors required and a series arrangement of trimming capacitors can be used. Unfortunately, considerable area is needed since the series trimming capacitors must all be much larger than the capacitance being adjusted.

Sometimes a precision capacitor requirement can be replaced by a resistor adjustment. If the circuit requirement is actually a precise frequency response, the resistor adjustments can be used to control the various RC products. This approach has been used for precise tuning of thin film RC filters (see Sec. 11.5).

Capacitor Parasitics

The major capacitor parasitics are the dielectric loss and the resistance in series with the dielectric. Both of these capacitor losses have already been discussed. Additional parasitics are the series inductance and the d.c. leakage current.

Series Inductance. The series inductance of a conducting path was discussed in the case of resistors in Sec. 11.1. To estimate the series inductance of a capacitor, it is usually sufficient to ignore the dielectric and to consider that both electrodes together form a single conduction path. Equation (11-24) can be used to estimate the inductance if the electrode path is appreciably longer than it is wide, an assumption on which this equation is based. If the length-to-width ratio is not large, an estimate of the inductance can be obtained by

$$L = \frac{\mu_0 l}{2\pi} \left[\ln\left(\frac{l}{x} + \sqrt{1 + \frac{l^2}{x^2}}\right) - \sqrt{1 + \frac{x^2}{l^2}} + \frac{x}{l} \right] \qquad (11\text{-}62\text{a})$$

where

$$x = 0.223w \qquad (11\text{-}62\text{b})$$

When the electrode length l and width w are in meters, the inductance is in henrys.

Consider the case of a 1000-pf tantalum oxide capacitor with a capacitance density of 0.1 μfarad/cm^2. For a square pattern, the square sides would measure 40 mils. Allowing for a suitable lead length, the overall length might be 80 mils and the width, 40 mils. The series inductance calculated from Eq. (11-24) or (11-62) is about 0.8 nanohenry.

The series inductance produces a series resonance with the capacitance at a frequency of

$$f = \frac{1}{2\pi \sqrt{LC}} \qquad (11\text{-}63)$$

The 1000-pf capacitor discussed above would go through a series resonance at a frequency of about 180 MHz. Above this frequency, the capacitor would act like an inductor; at an even higher frequency, there is a second resonance between the series inductance and the shunting capacitance of the terminations.

If the capacitance value is reduced by a factor of 10 and the pattern is scaled down without changing any dimensional ratios, the series inductance will be reduced by $10^{1/2}$ since the length term in Eq. (11-62) is reduced by this amount. Equation (11-63) shows that the frequency at which the series resonance occurs would be increased by a factor of $10^{3/4}$. Thus, for a 100-pf capacitor, the resonant frequency would be about 1 GHz. Measurements made at frequencies to about 1 GHz indicate that the capacitance and loss of the tantalum oxide dielectric are essentially independent of frequency.

In many applications, the series resonance phenomena may not be important. For instance, a capacitor used for a.c. coupling or d.c. blocking still can serve this function at frequencies above the series resonance, even though the a.c. performance is that of an inductor.

D.C. Leakage. Another form of capacitor parasitic is the d.c. leakage current, which is related to the insulation resistance of the dielectric. This resistance is usually extremely high and does not affect the a.c. properties. The dielectric loss usually controls the a.c. performance and, when a resistance is used to represent this loss, it should not be confused with the insulation resistance.

A 0.01-μfarad tantalum oxide capacitor anodized at 130 volts has a leakage current of less than 10^{-8} amp for 75 volts applied for 1 minute. A d.c. leakage current of this magnitude is usually of no importance as far as circuit performance is concerned.

The leakage current of a capacitor can be characterized by its insulation resistance, which is equal to the product of the d.c. resistance and the capacitance value and, hence, has the units of ohm farads. The 0.01-μfarad capacitor above has an insulation resistance of 75 ohm farads. The insulation resistance is actually a function of the measuring voltage and the time it is applied, so that caution must be used in interpreting quoted values. The insulation resistance value is utilized principally as a measure of the quality of the capacitor dielectric and, hence, the reliability of the device.

11.3 CONDUCTING FILMS

Uses and Materials

Conducting films (discussed in Chap. 7) are used for many different purposes in thin film circuits. One major use provides a conducting path between interconnected film components. The same conducting film often serves as land areas for bonding of appliquéd components and for attaching external leads. It has already been indicated in the discussion of capacitors that various conducting films may be used for the capacitor electrodes and that capacitor properties are sensitive to the particular material chosen. Conducting films can also be used to form low-value, low-Q inductors for high-frequency applications.

The various uses above help to establish a list of requirements for the conducting film materials. If some of the special requirements of film capacitors are excluded, the major requirements are:
1. Low sheet resistance
2. Resistance to corrosion
3. Bondable top surface
4. Adherence to the substrate surface and other films
5. Compatibility with processing of other film components

Generally, these requirements are not met by any one material. Thus, the conducting films used in thin film circuits consist of multiple layers of different metals. The layers are usually evaporated, one after the other, in a single evaporation process.

The low sheet resistance is provided by metals with low resistivity, such as copper or gold. Since these materials, by themselves, do not adhere well to the substrate, a thin film of more adherent material, such as Nichrome®, chromium, or titanium, is evaporated immediately before the copper or gold. Copper is susceptible to oxidation, or other forms of corrosion, so that it is covered by a more noble metal, such as gold or palladium. A gold film is the best choice when a thermal compression bond is to be formed between the conducting film and a gold lead. If a lead is to be soldered to the conducting film, palladium is better than gold, since gold forms a brittle alloy with most solder materials. On the other hand, pure palladium is difficult to evaporate from a conventional tungsten filament, for it tends to alloy with the tungsten at the evaporation temperature. Thus, a 50-50 palladium-gold alloy is often used as the top layer of the conducting film when solder bonds are required. The gold alleviates the palladium-tungsten alloying problem and the palladium prevents gold embrittlement of the solder joint.

There are other conducting materials available that fulfill some of the requirements but still are not in prevalent use for various reasons. Silver is an excellent conductor, but silver will migrate across the substrate surface under certain ambient conditions when an electric field is present. This migration can produce a filament of silver between adjacent conducting paths and thus create a short circuit. Some of the other noble metals, such as platinum and rhodium, have certain attractive properties for use as conducting films, but they are difficult to deposit by evaporation.

Some of the other possible conducting materials are not compatible with the fabrication of film resistors and capacitors because of their chemical activity or extreme inertness. On the other hand, a material like aluminum is attractive because it is compatible with the anodization process. One possible disadvantage of aluminum is that conventional soldering and welding techniques cannot be used to attach leads, although ultrasonic methods have proved successful.

Conducting Paths and Land Areas

Normally, the conducting paths between components and the land areas used for bonding leads are both formed with the same conducting film. This procedure usually imposes no design limitations and it minimizes the number of processing steps. If the need arises, however, it is clear that different conducting films can be used for these two functions.

The design of a land area is relatively simple. The top layer of the conducting film must be appropriate for the type of bond to be made. The size and shape of the land area is dictated by the lead to be attached. Generally, a flat tape lead is preferable to a round lead for surface attachment since the greater contact area of the flat lead provides greater bond strength. The various types of bonds are discussed in more detail in Chap. 12.

The primary objective in the design of a conducting path is to minimize the resistance of the path by keeping the number of squares in the path small and the sheet resistance of the material low. Many of the considerations discussed in the design of film resistors apply to the design of conducting paths. This is particularly true of the parasitic inductance and capacitance associated with the paths.

Figure 11-24 lists the bulk resistivity of several materials commonly used in conducting films. The resistivity of evaporated films is generally higher; thus, the table also shows the sheet resistance expected in evaporated films 1000 A thick. For evaporated films thicker than 1000 A, the sheet resistance will decrease roughly as $1/d$, where d is the thickness, since the resistivity remains approximately the same. However, if the films are much thinner than 1000 A, the effective resistivity may increase (Chap. 6) and, hence, the sheet resistance may increase more rapidly than expected. The relation between sheet resistance, thickness, and resistivity was given in Eq. (11-7).

Metal	Bulk resistivity (μ ohm cm)	Sheet resistance for 1000 A film (ohms/square)
Copper	1.7	0.20
Gold	2.4	0.27
Aluminum	2.8	0.33
Palladium	11	1.3
Palladium-gold (50-50 alloy)	21	3
Titanium	55	10
Nichrome®	100	15

FIG. 11-24. Bulk resistivity and film sheet resistance of various metals used for film conductors.

The Nichrome® or titanium films, evaporated to provide adherence, make a negligible contribution to the conductance of the path compared with the copper or gold layer that is evaporated next. The limitation to achieving a high sheet conductance (or low sheet resistance) is the limit placed on the thickness of the deposit. As already noted, a very thick film tends to be mechanically unstable. Actually, the evaporation process usually limits the thickness to values well below the unstable point, since it is inconvenient to evaporate films much thicker than 10,000 A. The parameters that become troublesome are the quantity of the evaporant source material, the length of time required, and the radiant heating of the substrate and parts of the evaporator fixture. From Fig. 11-24 it is seen that, in view of this limitation on thickness, minimum sheet resistances for evaporated copper and gold are approximately 0.02 and 0.03 ohm/square, respectively. An additional layer of pure palladium or palladium-gold alloy reduces the resistance only slightly, since these bonding layers are generally only about 2000 A thick. In finding the sheet resistance of a multiple-layer film, it is easiest to use the sheet conductance (the reciprocal of sheet resistance), since the various layers of film are in parallel and the sheet conductances are additive.

If the sheet resistances of evaporated films are not low enough to meet the needs of a circuit, the evaporated conducting films may be plated with copper or gold, either before or after the conductor pattern has been etched. Plated films tend to be less dense than either evaporated films or bulk material; hence, the resistivities tend to be higher. For instance, a 0.5-mil plated film of copper may have a resistivity of 2.3 μ ohm cm, 35 percent higher than the bulk value and 15 percent higher than evaporated. The exact value is a function of various parameters of the plating process. The plating thickness is again limited by mechanical stresses in the film and too thick a deposit will peel off when the substrate temperature is elevated. Using present techniques, a practical upper limit on the thickness is about 1 mil (250,000 A), and the lower limit on sheet resistance of a plated conducting film is about 0.001 ohm/square.

Another method used to reduce the sheet resistance of evaporated films is to coat them with solder by either solder-dipping or wave-soldering the substrate surface. The high resistivity of most solders makes the increase in conductivity only moderate, but the technique has the advantage that leads can be bonded in the same operation (see Chap. 12). One disadvantage is that, under certain conditions, whiskers of solder material will grow from the solder coating at the positive terminal of a film resistor (see Chap. 7). A solder coating also produces stresses in the substrate when the temperature is varied because of the mismatch in the thermal expansion coefficients. A thick solder coating on a glazed ceramic substrate can generate enough stress to fracture the glaze when the substrate is temperature-cycled.

In high-frequency applications, the series inductance, coupling capacitance, and skin effect may be important aspects of conducting films. The series inductance of most simple conducting-film patterns can be calculated using Eqs. (11-24) and (11-62). In cases where mutual inductance effects are important, Eqs. (11-25) and (11-32) can be used. The capacitance between conducting paths or land areas can be estimated using Eqs. (11-33) and (11-36).

The skin effect, however, is more critical for film conductors since the resistivities and sheet resistances are lower. It will be recalled from Eq. (11-40) that the skin depth is inversely proportional to the square root of the conductivity of the material. At a frequency of 1 GHz, the skin depth of bulk copper is about 20,000 A. It would appear that for this skin depth, the skin effect would be negligible at 1 GHz for evaporated film conductors under 10,000 A thick. Actually, this may not be true, since current distribution is not uniform on the surface of a conductor with a rectangular cross section. As was indicated in Eq. (11-41), the cross-sectional area of the film should be considered rather than the film thickness. Thus, a copper conducting path 10,000 A thick and more than 0.5 mil wide will have an a.c. resistance at 1 GHz which is higher than its d.c. resistance because of the skin effect. As the frequency is increased, the resistance increases.

Fortunately, the increase in resistance is not very large until the skin depth is less than the film thickness. For example, consider the case of a copper conducting path 10,000 A thick and 1 mil wide. It is seen from Eq. (11-42) that the first significant increase in resistance due to the skin effect will occur at a frequency of about 500 MHz. At 1 GHz, the a.c. resistance of the path is about 10 percent larger than the d.c. resistance; at 4 GHz, it is only about 35 percent larger. At 4 GHz, the skin depth is equal to the film thickness. Thus, for a conducting path where some resistance change is tolerable, the skin effect does not become a serious consideration until the skin depth is less than the film thickness. This is in contrast to resistor design considerations where small changes are important.

Crossovers

In designing a thin film circuit, it is often necessary that two film conducting paths cross somewhere on the substrate surface. The more complex the circuit, the more often this situation arises. In developing a circuit layout, the number of conducting-path crossovers are usually minimized by rearranging the various components. It is sometimes possible to eliminate crossovers in circuits having film capacitors by making appropriate use of the capacitor electrodes. Figure 11-25 shows a situation where a film capacitor can be used in this manner. A crossover problem is sometimes eliminated by providing an extra external lead. Thus, in some instances, two ground leads or two power supply leads may be provided to eliminate crossovers in the film circuit.

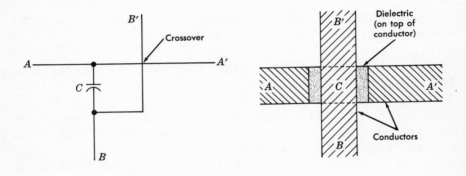

FIG. 11-25. Schematic showing a film capacitor used as a conducting path crossover.

There are several methods of providing for crossovers that cannot be eliminated in the circuit layout. One method is to use a thin film insulator crossover similar to the capacitor crossover shown in Fig. 11-25. This method introduces a parasitic capacitance between the conducting paths; thus, consideration should be given to which pair of conducting paths should be crossed. Normally, the crossover is designed to provide as little capacitance as possible and, hence, a thick dielectric film with a low dielectric constant is best suited for the crossover insulation. The capacitance can also be decreased by using narrower conducting paths. Making the paths too narrow can increase their path resistance to the point where it becomes a problem. Thus, the crossover design is a compromise between the coupling capacitance between the paths and the series resistance of each path.

In situations where the series resistance and coupling capacitance are not critical, a tantalum film conductor and a tantalum oxide dielectric may be used in the crossover structure. If the capacitance is critical but the resistance is not, a duplex dielectric of tantalum oxide and silicon oxide may be best. If both the resistance and capacitance are critical, the tantalum conducting film is not suitable and, hence, an evaporated dielectric such as silicon oxide might be used by itself. The minimum capacitance density in this case is about 0.002 μfarad/cm^2. Various organic films are also being developed to provide the insulating layer. If even lower values of coupling capacitance are required, it is possible to use an appliquéd lead to form a crossover by stitch-bonding a small lead to two land areas on either side of the conducting path to be crossed.

In the case of a ceramic substrate with holes in it, conducting paths on the back can provide for circuit crossovers (see Fig. 11-3, lower view). This method has two disadvantages. The back-side wiring requires additional circuit processing and may complicate the processing of the front side of the substrate. There is also a problem in providing a suitable conducting path through the holes.

Film Inductors

The development of thin film inductors has not been rapid for several reasons. An inductor is basically a volume element since inductance is achieved by filling a volume with a magnetic field, while for resistors, capacitors, and transistors, the volume requirement is a secondary rather than a fundamental need. Magnetic materials greatly increase the flux density to require less volume for a given amount of inductance, but the volume available per unit area in a thin film device is, by definition, very small. Also, the magnetic properties achieved in many bulk materials have proved difficult to reproduce in thin film form. While this situation should improve with time, this chapter is limited to a discussion of film inductors that do not employ magnetic materials.

The amount of inductance achieved with thin films is relatively small and the Q's of the inductors are low so that their properties are useful only at very high frequencies. The Q of an inductor is defined as

$$Q = \frac{2\pi f L}{R} \qquad (11\text{-}64)$$

where f is the frequency, L is the inductance, and R is the a.c. series resistance. The limitations of film inductors are best illustrated with some example calculations.

Consider a conducting strip of copper 10,000 A thick, 10 mils wide, and 1 inch long. From Fig. 11-24 it is concluded that the sheet resistance is about 0.02 ohm/square. The conducting path has 100 squares, so the series resistance is about 2 ohms. The self-inductance of this path was calculated before, using Eq. (11-24), and is about 29.5 nanohenrys. The Q of the inductor at a frequency of 10 MHz is found, from Eq. (11-64), to be about 1. As the frequency increases, the Q increases since it is proportional to f. At frequencies above 65 MHz, however, the skin effect starts to increase the a.c. series resistance [see Eq. (11-42)] which in turn limits the Q. At a frequency of 100 MHz, the Q is less than 10 and, at 1 GHz, it is considerably less than 100. In fact, at some frequency above 1 GHz, the skin effect becomes more dominant than the frequency term so that further increase in frequency causes a decrease in Q.

The low-frequency Q of the inductor can be improved by plating-up the conducting film to provide a lower d.c. resistance. This does not, however, prevent the skin effect from limiting the Q at higher frequencies. For a 1-mil plating, the d.c. series resistance is about 0.1 ohm and the 10-MHz Q is about 20.

To increase the amount of inductance, it is necessary to provide a longer path. One way to do this is to form a spiral which provides the additional benefit that the mutual inductance of adjacent paths adds to the self-inductance. There will also be, however, a distributed shunting capacitance between the adjacent paths.

Assume for the moment that the two paths shown in Fig. 11-17(b) are part of a square spiral inductor. If the rest of the inductor is neglected, the effect of the shunting capacitance for these two paths may be calculated. It was shown by Eq. (11-33) that if the paths are 10 mils wide and 10 mils apart, the capacitance between them is 1.2×10^{-12} farad/in. for a 7059 glass substrate, 50 mils thick. The mutual inductance between them, calculated using Eq. (11-25), is 18.5 nanohenrys if the paths are 1 inch long. Since the current is in the same direction in the case of a spiral, the mutual inductance is added to the self-inductance [see Eq. (11-26)] and the total inductance for the two lines is 96 nanohenrys. If the distributed nature of the capacitance is neglected, a calculation of a resonant frequency of these two paths may be made using Eq. (11-63). The frequency found in this manner is 470 MHz.

The calculation above indicates that the spiral structure is limited by shunting capacitance effects at high frequencies. To establish the true resonant frequency, the whole spiral path would have to be considered, but some conclusions may be drawn from the simple model used. If the lines are placed closer together to increase the inductance per unit area, the capacitance effects will increase. If the lines are made narrow to achieve the same end, the series resistance will increase.

Although spiral film inductors have been designed with inductances as high as 100 microhenrys, these are large-area devices (3 inches in diameter). Normally, with an inductance anywhere near this value, it is far better to appliqué a conventional component to the film substrate. In fact, area and performance requirements will normally limit the range of useful film inductors to very small values (no more than a few hundred nanohenrys). (An example of a 20-nanohenry film inductor is shown in Fig. 11-39.

11.4 DISTRIBUTED FILM COMPONENTS

The two-dimensional geometry of thin films is ideally suited for forming distributed components. By intentionally distributing the component function over a well-defined geometric area, unusual and useful electrical performance characteristics can be achieved. The two types of distributed film components which have received the most attention are area resistor networks and distributed RC structures. These distributed components provide a means of achieving monomorphic designs to perform a whole circuit function in a single structure.

Since the distributed nature of these structures is the essence of their performance, lumped-component equivalent circuits are often totally inadequate to describe their function. Exact analytical solutions of the electrical characteristics are available, however, for some of the simpler geometries used in distributed components. From these solutions, a structure that provides a particular circuit function can be synthesized. The equations often involve hyperbolic functions or elliptical integrals which make the circuit design problem more complex. This complexity, however, is not a serious handicap, since computers may be used for the more involved calculations.

Area-Resistor Networks

The area-resistor network is a means of replacing a network of lumped resistor elements with a single, continuous area of resistor film, with conducting electrodes placed at various points on the resistor film. The size and shape of the resistor area and the size, shape, and position of the electrodes determine the resistance between any two electrodes. The area-resistor pattern is synthesized to make the resistance between any two electrodes on the area resistor the same as the resistance between the corresponding two terminals of the lumped-resistor network. The advantages achieved with the area resistor are:

1. Smaller size
2. Smaller parasitics
3. Better high-frequency performance
4. Less sensitivity to point defects

The limitations of the area resistor are:

1. Film uniformity
2. Pattern alignment
3. Resistance adjustment

The area-resistor network is similar to a network of lumped film resistors where no individual resistor trimming is done. The resistance ratios depend upon the pattern definition and the uniformity of the sheet resistance of the film. The proper resistance level of the network is achieved by monitoring one resistance value during fabrication. The major difference arises from the fact that the area resistor is more dependent upon pattern alignment than the discrete resistor network and is less sensitive to point defects in the film and pattern edge definition. Also, the option of individual resistor adjustment is no longer available, except in the simplest networks. The elimination of the space between resistors and the conductor interconnections can make the area resistor smaller and eliminate some of the series inductance and shunting capacitance.

Some examples of area-resistor networks are shown in Fig. 11-26. The film layout of the area resistor is shown on the right and the lumped-resistor equivalent circuit is shown on the left. These networks could be used to provide a given amount of insertion loss in a signal path.

Analytical solutions are available for the examples shown in Fig. 11-26 [14]. For the three-terminal network, the solution provides the dimensional ratios for a, l, and w to achieve the desired resistance ratio between R_1 and R_3. It is clear from the symmetry of the thin film layout that $R_1 = R_2$ for this network. The solution for the four-terminal network at the bottom of Fig. 11-26 can be obtained by

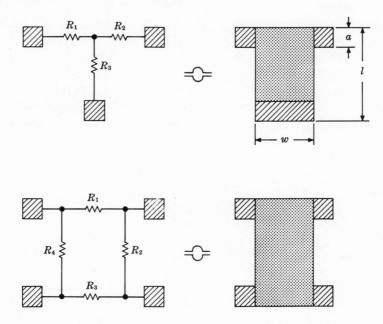

FIG. 11-26. Thin film area-resistor networks and lumped resistor equivalent circuits. (After Schwartz and Berry, "Thin Film Components and Circuits," in *Physics of Thin Films*, Vol. II, 1964; courtesy Academic Press.)

noting the symmetry and using the solution for the three-terminal network above.

As additional terminals are added, the area-resistor analysis becomes more complex and solutions to a particular geometry of interest may not be available. There are, however, formal methods of generating solutions that may be applicable to the geometry in question. One of the techniques that is widely used is the conformal mapping procedure which was discussed previously.

Distributed RC Structure

The thin film distributed RC structure is a logical extension of the structures used for individual film capacitors and resistors. In its simplest form, the structure can be thought of as a film capacitor in which one of the conducting electrodes has been replaced by a thin film resistor. Thus, the capacitance is distributed along the resistor path. The structure resembles a transmission line and, in fact, transmission line analysis techniques can be used to determine the elec-

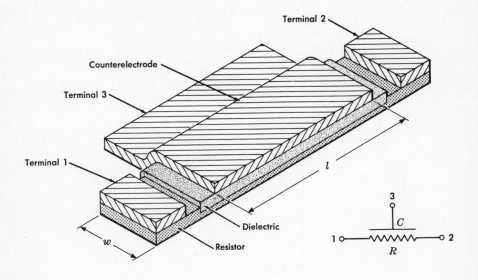

FIG. 11-27. Thin film distributed RC structure.

trical performance characteristics of the distributed RC structure
[12, 13, 15]. One obvious application of this structure is the simulation
of a transmission line with a three-terminal line build-out network.

A drawing of the three-terminal distributed RC structure described
above is shown in Fig. 11-27, along with the electrical symbol used
to represent it as a component in an electrical circuit diagram. It is
possible to replace both capacitor conducting electrodes with film
resistors and obtain a four-terminal distributed device. A drawing
of this structure and its electrical symbol is shown in Fig. 11-28.

The above two structures are commonly referred to as uniform
$\overline{RC}$ networks. The word "uniform" indicates that the resistance and
capacitance per unit length are constant over the length of the path,
with the symbol $\overline{RC}$ used to indicate a distributed-parameter RC
structure. It is, of course, possible to vary the resistance and capacitance per unit length by changing the width of the resistor path. This
technique is often used to enhance the properties of the distributed
structure. The most common types of path-width variation are linear
and exponential tapers. The structure is then called a tapered $\overline{RC}$
network. A drawing of an exponentially tapered $\overline{RC}$ network is shown
in Fig. 11-29, along with its electrical symbol. For a linear taper, the
capacitor electrode in the symbol would be straight instead of curved.

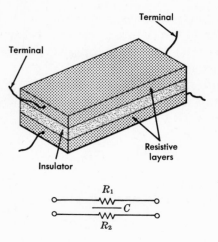

FIG. 11-28. Thin film distributed RC structure with two resistive layers.

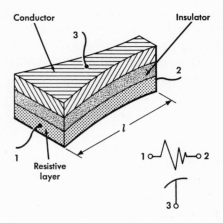

FIG. 11-29. Tapered thin film distributed RC structure.

Many other line-width variations could be used but, unless the geometry is kept relatively simple, the mathematical analysis becomes difficult.

The $\overline{RC}$ structure differs from the area-resistor in that there is no true lumped-component equivalent circuit for the distributed parameter network. This is not necessarily an advantage, but complex circuit functions can be synthesized with these distributed structures. The same circuit functions can usually be synthesized with lumped components, but it normally requires many more of them. In other words, the unique electrical characteristics of the $\overline{RC}$ structure lead to efficient synthesis procedures. For a better understanding of synthesized structures, the electrical characteristics of a uniform $\overline{RC}$ segment are examined.

Consider the uniform $\overline{RC}$ structure shown in Fig. 11-27. A transmission line analysis shows the complex impedance between terminals 1 and 3 of this network to be

$$Z_{13} = \frac{R \text{ ctnh } \theta}{\theta} \tag{11-65a}$$

where R is the total resistance of the path (i.e., the d.c. resistance between terminals 1 and 2) and θ is given by

$$\theta = \sqrt{j\omega RC} \tag{11-65b}$$

where ω is the angular frequency and C is the total capacitance [compare with Eq. (11-55)].

The voltage transmission coefficient of the structure may be of more interest. Consider that an input voltage V_{13} is placed between terminals 1 and 3 and an output voltage V_{23} is measured between terminals 2 and 3. The open-circuit voltage transfer function of the structure is then given by

$$T = \frac{V_{23}}{V_{13}} = \text{sech } \theta \tag{11-66}$$

where θ is defined by Eq. (11-65b).

This transfer function is plotted in Fig. 11-30 as a function of frequency using a polar diagram in the complex plane. When ωRC is approximately 20, it is seen that the output voltage is 180° out of phase with the input voltage. The polar plot is a spiral going around the origin in a clockwise direction as the frequency increases. The curve gets closer and closer to the origin with each turn but never actually reaches it. This continuously changing phase is a distinctive feature of the distributed structure that cannot be duplicated with lumped components.

A simple example of the component economy provided by $\overline{RC}$ structures is shown in Fig. 11-31. The desired circuit function in this case is a notch filter where the transmission coefficient goes to zero at a particular frequency, f_0, as shown at the top of Fig. 11-31. This transmission characteristic can be achieved with three lumped resistors and three lumped capacitors arranged in the twin-tee configuration as shown. It can also be achieved using an

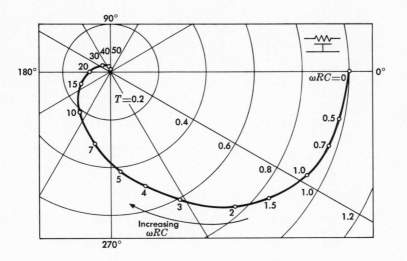

FIG. 11-30. Polar plot of the no-load transmission coefficient of a distributed
     parameter RC network.

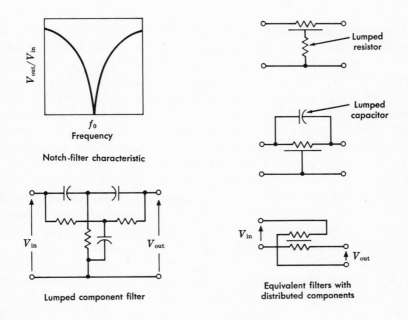

FIG. 11-31. Circuit diagrams for achieving notch-filter transmission
     characteristics.

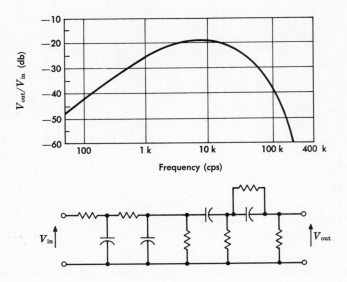

FIG. 11-32. Lumped element RC network and frequency characteristic [13].

$\overline{RC}$ component in combination with either one lumped resistor or one lumped capacitor, or it can be achieved with a single $\overline{RC}$ structure which has resistor paths on both sides of the dielectric. The circuit diagrams of these various configurations are all shown in Fig. 11-31.

An example of a more complex synthesis is shown in Fig. 11-32. In this case, the transfer function is the bandpass characteristic shown. This function can be approximated to the desired degree by the network of 10 lumped resistors and capacitors shown. The function is also achieved by the cascaded $\overline{RC}$ network shown in Fig. 11-33. This distributed structure was designed using a formal synthesis procedure to determine the geometry. The number of stubs, the positions of the stubs, and the changes in line width are all custom tailored to produce the proper frequency response. This structure was made using thin films, and it performed in the expected manner.

It can be seen from the above discussion that most of the physical dimensions, or at least the dimension ratios of an $\overline{RC}$ structure, come out of the circuit design synthesis procedures. The rest of the physical design problem involves the choice of materials and film thicknesses and the determination of the actual dimensions from the dimensional ratios.

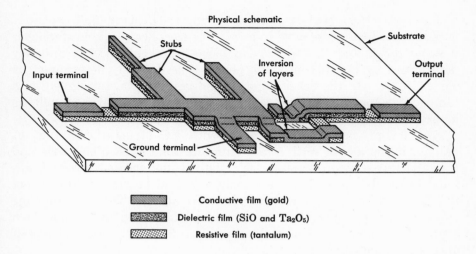

Conductive film (gold)

Dielectric film (SiO and Ta_2O_5)

Resistive film (tantalum)

FIG. 11-33. Synthesized thin film distributed RC structure with the bandpass transfer functions shown in Fig. 11-32 [13].

The considerations that are important in choosing film materials and in determining film thicknesses are the same as those that apply in the design of lumped resistors and capacitors. There is, however, less freedom of choice in the distributed structure, since requirements for the resistors and capacitors must be met simultaneously with the same materials and dimensions.

Consider, for example, the uniform $\overline{RC}$ network shown in Fig. 11-27. The resistance of this network is given by

$$R = R_S \frac{l}{w} \tag{11-67}$$

where R_S is the sheet resistance, l is the length of the path, and w is the path width. The capacitance of this network is given by

$$C = lw \left(\frac{C}{A_C} \right) \tag{11-68}$$

where C/A_C is the capacitance density. The RC product, an important performance parameter, is obtained from Eqs. (11-67) and (11-68):

$$RC = R_S l^2 \left(\frac{C}{A_C} \right) \tag{11-69}$$

It is interesting that the RC product does not depend on the line width, although the impedance level of the network does, as seen from Eqs. (11-65) and (11-67).

The restrictions that arise from having the resistor and capacitor superimposed can be seen as follows. The electrical design establishes a value for R and C. Once R_S and C/A_C are chosen, l is uniquely determined by Eq. (11-69) and, hence, w is uniquely determined by Eq. (11-67). On the other hand, if the resistor were designed as a lumped component, l and w could be scaled up and down at will, as long as the proper ratio was maintained.

Another problem that can arise with $\overline{RC}$ structures is that of resistor adjustment. With a tantalum film for the base electrode and the capacitor structure on top, the resistor is not accessible for adjustment in the usual fashion. Obviously, it would be desirable to invert the structure so that the resistor film were on top. This may be done using materials other than tantalum and tantalum oxide, but then some of the advantages of stable tantalum films are lost, and resistance adjustment may not be uniform.

One other point that should be made concerns power dissipation. The distributed structures are currently restricted in the amount of power that can be dissipated per unit area. The capacitor dielectric materials that have been developed are evaluated with essentially no power being dissipated in the component. Thus, the only thermal stress is that provided by the ambient conditions. If these same dielectrics are to be used in distributed structures, the maximum power density for the resistor should be restricted to a lower value than that which would be appropriate for the resistor by itself. Fortunately, most of the potential applications for $\overline{RC}$ structures are in low-power circuitry.

In spite of these shortcomings, $\overline{RC}$ structures should play a significant role in the circuit designs of the future. They offer unique electrical characteristics and an economy in the component count. They eliminate many parasitics and, hence, provide improved high-frequency performance. The shunting capacitance of the resistor and the series resistance of the capacitor, for example, are no longer limiting parasitics. The dielectric loss is still present, but it does not limit the high-frequency performance, as pointed out in Sec. 11.3.

As thin film technology continues to develop, it may be possible to build up a structure with many alternate layers of resistive and capacitive material. Such a multiple-terminal device will require very little surface area and could be extremely versatile from a circuit designer's point of view. The material problems and fabrication cost of a many-layered structure are, however, items that require considerable attention before such devices are proved feasible.

11.5 FILM CIRCUIT LAYOUT

The design of a film circuit layout consists of determining position, size, and shape of the film components, interconnections, and termination areas. The objective of the design is to meet the electrical and physical requirements of the circuit and, at the same time, minimize the fabrication cost.

For a circuit with many components, the layout design is a complex problem. It involves all the considerations of individual component design plus the problem of how various portions of the circuit may interrelate electrically. The whole circuit must be considered when fabrication processes and process sequences are selected, since each component on the substrate is involved, to some degree, in each step of the fabrication procedure.

Before the layout procedure can actually begin, a circuit module (the amount of circuitry on a single substrate) has to be determined. In many cases, the circuit function neatly defines the circuit module. In other cases, the choice of a suitable module may be a complex problem in itself. The decision may be based on the overall requirements of the system or a compromise based on fabrication yield and cost considerations. It should be remembered that, when the circuit is fabricated, there may be many circuit modules processed together on a single large substrate until this substrate is broken up into individual circuit modules.

Once a circuit module has been chosen, it is necessary to determine how the module is to be connected to the rest of the system. This depends, to a certain extent, on the size of the substrate; thus, preliminary estimates of area requirements should be made at this point and a desirable shape factor established. Before the circuit layout problem itself can be considered in detail, the types of terminations must be established (these, in themselves, may determine the overall size of the module) and a method of circuit packaging determined.

The first part of the design of a circuit layout is primarily a topological problem. The input is the circuit diagram, and the boundary conditions are the plane of the substrate surface and its defining edges. The constraints are numerous, including the location of external terminations, the minimization of crossovers, the shortness of a signal path, and the parasitics between critical conductors or components. These constraints are specific to the particular circuit being considered, but once determined they are often applicable to other circuits of the same system.

There is no unique solution to the topological problem. Components are normally moved around by the engineer until he finds an arrangement which seems to satisfy the requirements. In a complex circuit, there is seldom any real confidence that it is an optimum solution. Computer programs are available to help solve this problem, but in the present state of development, the computer requires a lot of interaction with the designing engineer.

Once a promising component arrangement is achieved, the next step is to give each component the approximate area it will require. It is clear that, at this point, film materials and film thicknesses must be chosen and, in fact, quantities such as resistor line widths must be designated. When the components are given their proportioned areas, the layout is adjusted to achieve the desired substrate size and shape. At this point, the conducting paths and land areas should be given suitable dimensions and margins between components established. The final step in the circuit layout problem is the detailing of each component. This step requires a complete knowledge of the process steps and process sequences to be followed, so that proper mask margins may be included in the design, as well as the various factors affecting electrical performance.

Once the patterns for the various components have been established in detail, the mask patterns for fabricating the circuit can be delineated. A single mask pattern may involve all types of components. For instance, the tantalum-pattern mask will normally include resistor lines, capacitor base electrodes, conducting paths, and land areas. Computer programs are being developed to aid in establishment of component areas, detailing of component dimensions, designation of conducting paths, and determination of mask patterns.

Examples of Film Circuit Layouts

The above description of the layout procedure is, of necessity, somewhat general. Some specific examples are now reviewed.

Twin-Tee Notch Filter. Consider the lumped-element, twin-tee, notch-filter network which was shown in Fig. 11-31. Separate filter networks of this type were fabricated to investigate procedures for tuning the thin film filters to provide the proper notch characteristics. Filters of this type were later included in the thin film data set shown in Figs. 11-1 through 11-4. The geometry was changed, however, to permit layout of other associated circuitry on the same substrate. There are 3 twin-tee filters on the data channel substrate shown in Fig. 11-3, and 28 filters were used in the entire data set.

The prototype filter was laid out with the following constraints. The terminations were all to be on one edge of the substrate and the two internal nodes of the circuit were to be brought out on the same edge for measurement purposes. The three resistors were to be physically close together so that their resistance values might track each other. This was particularly important for resistors R_1 and R_2.

The conventional circuit diagram for this filter is shown in Fig. 11-34(a); it has been redrawn in (b) to fulfill the above mentioned constraints. Figure 11-34(c) shows how capacitor C_2 can be used to eliminate the crossover that is inherent in this network. Figure 11-34(d) shows a rough layout after the components have been given proportioned areas and (e) shows the final layout drawing of the circuit. The heavy black lines are resistors, the hatched areas indicate capacitor dielectrics, and the white areas are conducting material. This same circuit layout was used for frequencies ranging from 700 to 2200 Hz. Since the impedance levels were to be the same at all frequencies, the different frequencies were achieved by changing only the capacitor counterelectrode pattern. The layout drawing shown in Fig. 11-34(e) is for a high-frequency filter and, hence, the capacitor counterelectrodes are very narrow. In the low-frequency version, the capacitor counterelectrodes cover nearly all the capacitor dielectric area.

The crossed-electrode capacitor pattern was used in this layout since capacitor precision was more important than series resistance effects at these low frequencies; resistors R_1 and R_2 were placed very close together to make it possible to anodize them together, rather than individually, when the filter was tuned. The requirement here was not so much that they have the design resistance ratio, but that they maintain a constant resistance ratio during resistor adjustment.

Figure 11-35 shows a picture of one of these filters and its notch characteristic after tuning. The notch frequency was adjusted to within ± 0.02 percent by the anodization of film resistors. The same tuning procedure was used to adjust precisely all 28 twin-tee filters of the thin film model of the data set. The ability to achieve a precise frequency response in a tuned RC circuit by resistor anodization is an important advantage of tantalum film resistors. When this frequency adjustment capability is coupled with inexpensive beam-lead active devices, it becomes economically favorable to replace conventional LC filters with active RC filters in many applications.

Impedance-Matching Pad. Another example of a thin film circuit layout is shown in Fig. 11-36. This circuit is an impedance-matching

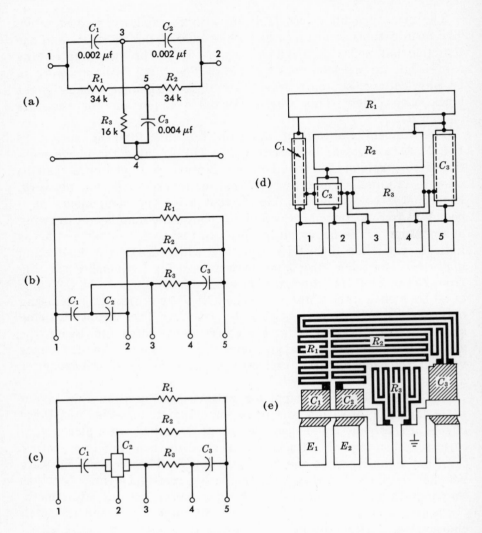

FIG. 11-34. Various steps in generating a thin film layout drawing for a twin-tee filter starting with the electrical circuit diagram [16].

pad in the A5 channel bank modem unit. One of the constraints in this layout was that terminals 6 through 12 be on one side of the substrate. The tantalum film resistors on this substrate were adjusted as a group while the series resistance of resistors R_3 and R_4 was monitored. Resistor tracking depended on film uniformity and pattern definition. It is seen that the dumbbell resistor design was

COMPONENT AND CIRCUIT DESIGN

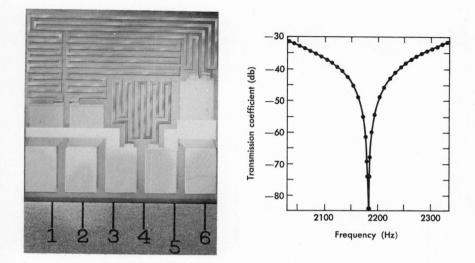

FIG. 11-35. Thin film twin-tee notch filter (scale in tenths of inches) and its transmission characteristic after tuning [16].

used to help achieve this goal and that the lowest value resistors were formed by putting four identical resistors in parallel, such that all resistors were made using identical line widths.

This circuit was used in a pilot trial and was the first thin film circuit manufactured by Western Electric Co., Inc. The resistor values averaged within 1 percent of nominal, with a standard deviation of 1 percent. The standard deviation of the capacitors was 3 percent.

Digital-to-Analog Decoder Network. Figure 11-37 shows an example of a thin film circuit where the control of parasitics was of prime importance. This is a high-speed PCM decoder network which operates at 12 MHz and converts nine parallel bits of digital information to an analog signal in about 80 nanoseconds. For proper operation, it was necessary that all 35 resistors in the circuit be adjusted and maintained within ±0.05 percent of their design value and that the series inductances and other parasitics be kept to a minimum. The reason for the single large substrate was to minimize differences in temperature and capacitive coupling between various parts of the circuit. When the film circuit model was tested, it came very close to providing theoretically perfect performance. The use of a thin film circuit of this type was the best way, if not the only way, to achieve the required performance.

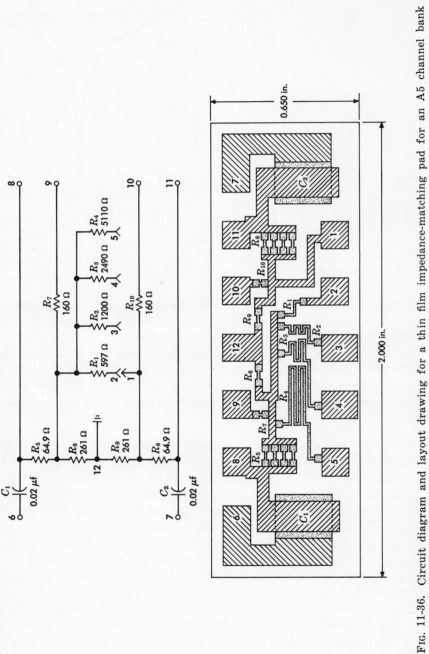

FIG. 11-36. Circuit diagram and layout drawing for a thin film impedance-matching pad for an A5 channel bank modem unit.

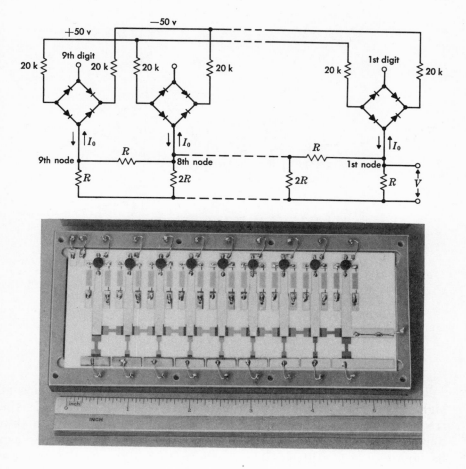

FIG. 11-37. Thin film experimental high-speed PCM decoder network [17].

100-MHz Preamplifier. Another example of a high-frequency circuit is shown in Fig. 11-38. This is a 100-MHz preamplifier circuit for a PCM repeater. The overriding design consideration in this case was the directness of the signal path, which has been indicated in the figure. Other techniques that were employed to minimize parasitics were the use of a minimum resistance pattern for the capacitors and the use of direct junctions for parts of the resistor network (see Fig. 11-38). It should be noted that some large-value capacitors and inductors have been appliquéd to the front surface along with the transistors.

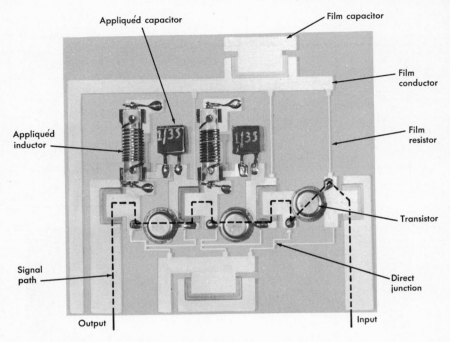

Appliquéd capacitor

Film capacitor

Film conductor

Appliquéd inductor

Film resistor

Signal path

Transistor

Direct junction

Output

Input

FIG. 11-38. Experimental thin film PCM preamplifier.

Thin films have provided a means for minimizing parasitics at high frequencies in the two PCM circuits described above. If the frequency is pushed even higher, the parasitics found, even in thin film designs, can no longer be minimized to the point that they can be ignored. It is still possible, however, to design circuits at these frequencies, if the presence of the parasitics is taken into account in the electrical design of the circuit. This approach is feasible only if the parasitics are reproducible and predictable, as they are in a thin film circuit.

UHF Amplifier. An example that illustrates a UHF amplifier is shown in Fig. 11-39. The circuit is a three-stage amplifier which operates from 0.4 to 1.4 GHz. At this high frequency, thin film inductors are feasible; they were used in the feedback loop of the first and third stages. The 100-MHz amplifier shown in Fig. 11-38 used appliquéd inductors to provide the same function. An equivalent circuit was prepared for a single stage of the 1-GHz amplifier. The circuit included all of the thin film and transistor parasitics that might affect the electrical performance. The magnitudes of the

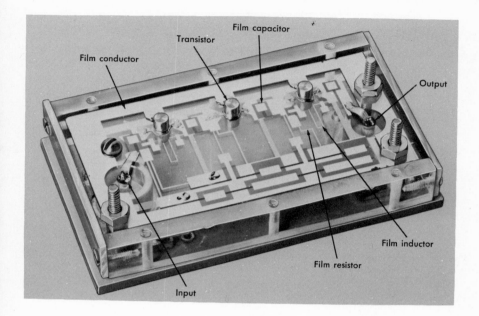

FIG. 11-39. Exploratory thin film UHF amplifier [18].

various parasitics were derived by calculations and independent measurements of components of the same type as those used in the design. A computer was then used to predict the performance of the amplifier based on this equivalent circuit. The variation of performance with variations in component parameters was also studied. Figure 11-40 shows how close the thin film models come to the computer-predicted performance. This result demonstrates the possibility of taking component parasitics into account when designing film circuits for these high frequencies. With the aid of a computer, the circuit performance can be optimized in terms of the component parameters and predictable parasitics before a circuit model is built.

A Frequency-Selective Oscillator. An example of the size and parts reduction that is achieved using thin film and beam-lead semiconductor devices is shown in Figs. 11-41 and 11-42. The thin film hybrid design represents an exploratory RC TOUCH-TONE® oscillator which is electrically equivalent to the conventional LC TOUCH-TONE® oscillator shown, with a 50 to 1 reduction in volume. Perhaps more startling is the fact that the component count of this RC TOUCH-TONE® circuit is about the same as for the thin film data channel circuit shown in Fig. 11-3, but it has about one-twentieth the volume.

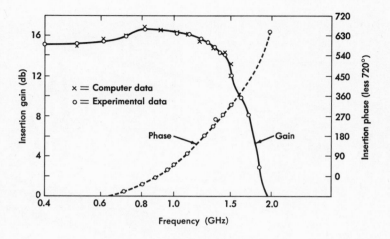

FIG. 11-40. Predicted and experimental performance curves for the amplifier shown in Fig. 11-39 [18].

FIG. 11-41. Size comparison of an LC TOUCH-TONE® oscillator and an exploratory RC TOUCH-TONE® oscillator that uses tantalum thin film passive components and unencapsulated beam-lead active devices.

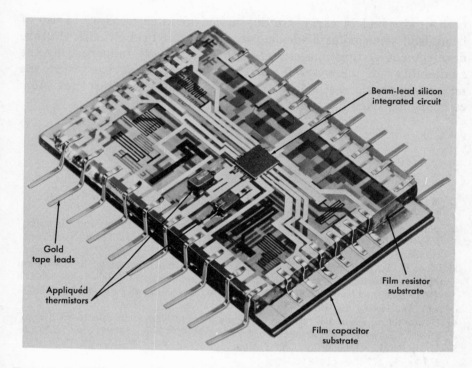

FIG. 11-42. Microcircuit for TOUCH-TONE® dialing which combines a silicon
integrated circuit with tantalum film resistors and capacitors [19].

In this microcircuit, the active devices have been integrated into a single chip of silicon that contains two complete amplifiers, one for each group of TOUCH-TONE® frequencies. The silicon chip also contains diodes for controlling the signal amplitude. The frequency-selective feedback loops are made up of tantalum film resistors and capacitors. The resistors and capacitors are on different substrates with gold leads providing the interconnections. Two thermistors, which provide temperature compensation for the signal amplitude, are appliquéd to the surface of the resistor substrate.

The microcircuit contains 10 transistors, 5 diodes, 27 resistors, 10 capacitors, and 2 thermistors for a total conventional component count of 54. In terms of the new devices, it contains 1 silicon circuit, 1 resistor substrate, 1 capacitor substrate, and 2 thermistors for a total count of 5. This vast reduction in the total number of component parts needed to form a circuit function is one of the keys to economies offered by the new device technologies. These savings, both in size and in parts, are due in considerable measure to use of beam-lead semiconductor devices in thin film hybrid circuits. Future development of thin film circuits will be markedly influenced by the availability of these small, inexpensive, active devices with the major emphasis being on higher packing densities for film components, finer lines, and finer spacings.

REFERENCES

1. McLean, D. A., N. Schwartz, and E. D. Tidd. "Tantalum Film Technology," *Proc. IEEE*, **52**, 1450-1460 (1964).
2. Schwartz, N., and R. W. Berry. "Thin Film Components and Circuits," *Physics of Thin Films*, vol. II, edited by G. Hass and R. E. Thun. New York: Academic Press, Inc., 1964. pp. 363-425.
3. Kuo, C. Y., J. S. Fisher, and J. C. King. "Thermal Processing of Tantalum Nitride Resistors," *Proc. Electronic Components Conference*, 123-128 (May, 1965); *idem, IEEE Trans. on Parts, Materials, and Packaging*, **PMP-1** (1), 123-128 (June, 1965).
4. Gerstenberg, D., and E. H. Mayer. "Properties of Tantalum Sputtered Films," *Proc. Electronic Components Conference*, 57-61 (May, 1962).
5. Hall, P. M. "Resistance Calculations for Thin Film Patterns," *Thin Solid Films*, **1**, 277-295 (1968).
6. Hagedorn, F. B., and P. M. Hall. "Right Angle Bends in Thin Strip Conductors," *J. of Appl. Phys.*, **34**, 128-133 (January, 1963).
7. Grover, F. W. *Inductance Calculations*. New York: D. Van Nostrand Co., 1946.
8. Happ, W. W. "Stray Capacitance in Thin Films," *Electronic Industries*, **21**, E4-E7 (June, 1962).
9. Compton, J. B., and W. W. Happ. "Capacitive Design Considerations for Film-Type and Integrated Circuits," *IEEE Trans. on Electron Devices*, 447-455 (October, 1964).
10. Wolfe, P. N. "Capacitance Calculations for Several Simple Two-Dimensional Geometries," *Proc. IEEE*, **50**, 2131-2132 (October, 1962).
11. Terman, F. E. *Radio Engineers' Handbook*. New York: McGraw-Hill Book Co., Inc., 1943.
12. Castro, P. S. "Microsystem Circuit Analysis," *Electrical Engineering*, 535-542 (July, 1961).
13. Wyndrum, R. W. "The Realization of Monomorphic Thin Film Distributed RC Networks," *IEEE International Convention Record*, **13** (10), 90-95 (March, 1965).
14. Dow, R. J. "The Conjugate Function Approach for Describing Current Density and Resistance in Odd-Shaped Conductors," *Proc. Electronics Components Conference*, 47-52 (May, 1962).
15. Happ, W. W., and P. S. Castro. "Distributed Parameter Circuit Design Techniques," *Proc. of National Electronics Conference*, **17**, 45-70 (1961).
16. Orr, W. H. "Precision Tuning of a Thin Film Notch Filter," *1964 International Solid-State Circuits Conference Digest*, **7**, 56-57 (February, 1964).

17. Jackson, W. H., and R. J. Moore. "A High-Accuracy Thin Film PCM Decoder Network Operating at 12 MC," *Proc. Electronic Components Conference*, 45-53 (May, 1965); *idem, IEEE Trans. on Parts, Materials, and Packaging*, **PMP-1** (1), 45-53 (June, 1965).
18. Wyndrum, R. W., Jr., and W. J. Pendergast. "A Tantalum-Film Gigacycle Amplifier," *1965 International Solid-State Circuits Conference Digest*, 8, 20-21 (February, 1965).
19. Morton, J. A. "The Microelectronics Dilemma," *International Science and Technology*, 35-44, July 1966.

PRINCIPAL SYMBOLS FOR CHAPTER 11

Symbol	Definition
A	Area
A_C	Active area of capacitor
A_R	Area of resistor
C	Capacitance
d	Film thickness
d_S	Substrate thickness
d_ϵ	Dielectric thickness
$\mathscr{E}_B$	Dielectric breakdown strength
f	Frequency
h	Heat transfer coefficient
K	Anodization constant
l_C	Active length of capacitor area
l_l	Line length
l_p	Pattern length
L	Inductance
L_m	Mutual inductance
n	Number of squares
P_R	Power of resistor
q	Area power density of resistor
R	Resistance
R_S	Sheet resistance of film
R'_S	Sheet resistance of substrate surface
R'	Resistance of electrodes and connections
t	Time
T	Temperature

SYMBOL	DEFINITION
T	Voltage transfer function
T_A	Ambient temperature
V	Electrical potential
V_B	Breakdown potential
V_R	Rated voltage
w_C	Active width of capacitor area
w_l	Line width
w_p	Pattern width
w_s	Space width
Z	Impedance
δ	Loss angle
δ'	Loss angle for dielectric
δ_S	Depth of skin effect
ϵ_r	Dielectric constant or relative permittivity
ϵ_0	Permittivity of free space
μ_r	Relative permeability
μ_0	Permeability of free space
ρ	Specific resistivity
ω	Angular frequency

Chapter 12

Lead Attachment

The number of electrical connections in a thin film circuit is decreased by the deposition of interconnections; however, appliquéd devices and components as well as subsystem or system interconnections still must be attached to the substrate. Early circuits utilized soldered connections, as shown in Fig. 12-1, primarily because of the ease of use. The development of silicon integrated circuits (with their associated microsize) and the temperature limitations of various hybria circuits stimulated extensive development of other thin film joining methods.

Out of these developments have come two methods that are currently in use for attaching to thin film terminations: solid-phase and alloy bonding. The most prevalent solid-phase bonds are formed by thermocompression and ultrasonic means. The term alloy bonding includes both soldering and brazing; however, brazing has not found application in thin film circuits. Techniques being used to solder include hand soldering, solder dipping, wave soldering, infrared soldering, and resistance soldering.

Another joining method, fusion welding, has been used with some success to attach gold leads to plated films; however, it has been less successful with evaporated films, which are considerably thinner. Lasers, electron beams, and resistance welders are the most common sources of heat in this method.

An important consideration in lead attachment is the metallurgical compatibility of the various metallic constituents of the joining system. The use of dissimilar metal bonds such as gold to aluminum, solder to gold, or solder to palladium-copper-Nichrome® may cause catastrophic circuit failures. These unfavorable results have motivated developers to restrict the connections to a one-metal system

(a) Glass substrate supported by a molded phenolic frame

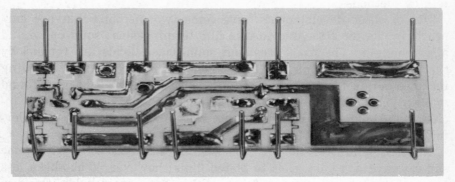

(b) Self-supporting ceramic substrate

FIG. 12-1. Circuit packaging method used for small thin film devices.

such as the aluminum-to-aluminum interconnections in transistors
and the gold-to-gold interconnections when attaching a beam-lead
device to a thin film. A one-metal system, although ideal from
a metallurgical standpoint, obviously restricts the designer or manufacturer in his choice of materials for the optimum circuit character-

istics, fabrication procedure, cost, or packaging configuration. A choice of metal system, compatible with the joining process parameters and the circuit operating environment, however, minimizes the probability of dissimilar metal connection failures.

With the promise of a factor of ten or greater improvement in cost, reliability, and performance of integrated electronics, the interconnection takes on greater importance than it has in the past. Efficient material control, process control, and an effective interconnection testing method will be critical requirements.

This chapter discusses fundamental atomic binding and the various factors and methods that play a part in achieving a metallurgical connection. Some of the considerations and methods used for testing connections are also discussed.

12.1   METALLURGICAL BACKGROUND

A brief discussion of certain topics in physical metallurgy is given in this section to permit a basic, unified approach to surface interaction and resultant joining between metals. For additional information, Refs. [1], [2], and [3] may be consulted.

Atomic Theory

In assemblages of atoms, the cohesive forces which bind* atoms together may be classified into four types: (1) ionic, (2) covalent, (3) van der Waals, and (4) metallic.

Ionic binding results from the actual transfer of electrons between different atoms and is thus electrostatic in nature. Ionic crystals, such as NaCl and LiF, are generally nonconductive and brittle.

Covalent binding results from the sharing of electrons between two or more atoms to achieve a stable electron energy configuration. Covalent solids like diamond and silicon usually possess high hardness and are poor electrical conductors.

A much weaker type of binding, *van der Waals*, always exists between all atoms. This force results from a polarization caused by the interaction of electron motion in adjacent atoms. In most materials, however, this mutual force is overshadowed by the stronger ionic,

---

*In this chapter, *binding* refers to interatomic forces. *Bonding* refers to the joining of metal parts without the formation of a fusion zone in the parts. *Welding* refers to the joining of metal parts by the fusion of the parts. The terms *joining* and *connecting* are used when referring to both bonding and welding.

covalent, or metallic forces. Where such strong forces do not occur, van der Waals binding becomes controlling.

The fourth principal type of binding is *metallic* and the rather unusual nature of this binding accounts for the distinctive properties of metals. In a metal atom, there are one or more electrons that are much less tightly bound to the atomic nuclei than the "inner electrons." These electrons (valence electrons) can be separated from their atoms rather easily, leaving the atom nucleus and the more tightly bound inner electrons intact. In a metal crystal, the valence electrons are relatively free to pass through the matrix structure, which consists of the positively charged ions. The strong binding force exhibited in metallic structures is due to the attractive forces between these "free" electrons and the positive ion matrix. These "free" electrons in the whole metal lattice constitute what is called the electron cloud or gas and are responsible for the high electrical conductivity of metals (see Chap. 6).

Crystal Structure Theory

As was shown above, the type of binding which exists between atom aggregates broadly determines the physical properties of materials. Material properties are also influenced by the particular arrangement of the atoms in relation to each other.

Periodicity and Ideal Crystalline Structures. The most striking feature of the solid state of materials is the three-dimensional periodicity observed on an atomic scale. Such a regular, repetitious structure is called *crystalline,* in which each like atom has identical atomic surroundings. Although there exist 14 different atom arrangements that satisfy the identical surroundings criterion, only a few of these are important in metals: body-centered cubic (bcc), face-centered cubic (fcc), and hexagonal close-packed (hcp) structures.

The atomic arrangement in body-centered cubic structures is shown in Fig. 12-2(a) where the atom nuclei and inner electrons are represented as fixed spherical balls. The atoms are arranged in orthogonal planes with identical interatomic spacing along the three axes; each atom is surrounded by eight adjacent atoms (nearest neighbor atoms) and the structure is said to possess a coordination number of 8. It can be demonstrated that such a structure is not "close-packed," i.e., the total volume occupied by the assembled atoms is not a minimum for their size. The relatively open nature of this structure is shown in Fig. 12-2(b). The bcc structure is exhibited by about 26 percent of the metallic elements.

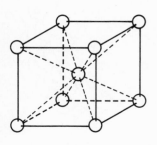

(a)

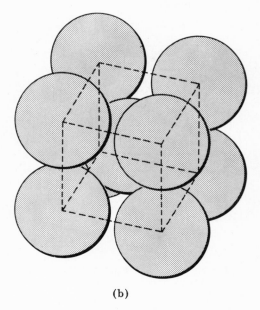

(b)

FIG. 12-2. Body-centered cubic structure.

A more prevalent cubic structure, fcc, is close-packed with the atoms arranged on orthogonal planes, Fig. 12-3(a). Each atom has 12 nearest neighbor atoms (a coordination number of 12). Although it is not apparent that this atomic arrangement is close-packed, Fig. 12-3(b) illustrates how each atomic layer is placed in the interstices of the underlying layer to achieve maximum packing density.

The other possible close-packed structure occurs in the hcp system, shown in Fig. 12-4(a). Such a structure is characterized by three coplanar axes spaced at 120° (the basal plane), and a fourth axis at 90° to this plane. The interatomic spacing, a, along the three axes is identical, and the perpendicular spacing, c, is ideally $\sqrt{8/3}\ a$.* Again, each atom has a coordination number of 12. The similarity between this structure and fcc is not immediately obvious, but can be shown in Figs. 12-3(b) and 12-4(b). Successive hcp atom layers are packed into the interstices of surrounding layers, with every

*By convention, crystals are classified as hcp even though the c/a ratio may be somewhat different.

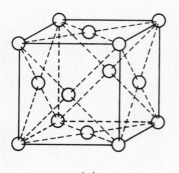

(a)

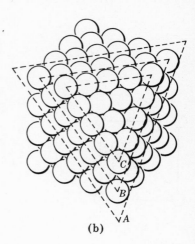

(b)

Fig. 12-3. Face-centered cubic structure.

other layer placed identically. In the fcc structure, the layers repeat by three's. About 66 percent of all metallic elements possess either fcc or hcp structures.

More complex structures, particularly oxides and other compounds, can have the same basic crystal patterns as described above, but with a group of perhaps two or even more atoms associated with each lattice point.

Imperfections and Real Crystalline Structures. Because of lattice irregularities, real crystals are somewhat different from the unvarying, periodic structures described above. If atoms of different elements are introduced into a lattice, solid solutions or second phases can be formed. Imperfections in atomic arrangements, quite apart from foreign atoms, are also present in real crystals. Line imperfections such as dislocations greatly influence strength properties of materials; point defects, such as vacancies (absent atoms), or interstitials (extra atoms), are important in diffusion.

Common metals are generally polycrystalline, unless special measures are taken to obtain a crystal of the same orientation throughout. Polycrystals consist of grains, with the atoms of each grain being of a single orientation (within the limitations of real crystals described above). The delineation between grains is termed the grain boundary, which is a transition region several atom layers thick between two

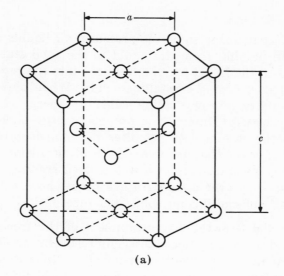

(a)

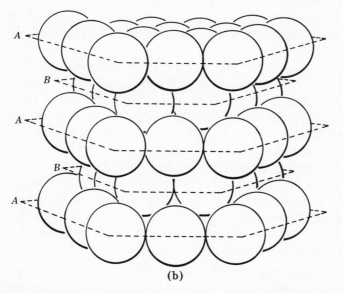

(b)

FIG. 12-4. Hexagonal close-packed structure. (After Azaroff and Brophy, *Electronics Processes in Materials.* Copyright 1963 by McGraw-Hill Book Co. Used with permission.)

differently oriented crystals. Depending upon the degree of misorientation between adjacent grains, the boundaries may be composed of simple dislocation arrays or more complex defect structures.

Metallic Solidification

If cooled or put under sufficient pressure, all liquids can be made to solidify. In passing from the liquid to the solid state, constantly moving atoms in the liquid may be captured into the regular periodic placements characterizing crystalline solids. When pure elements or compounds solidify into crystalline solids, an energy called the heat of fusion is released. During this process, energy added to or subtracted from the system does not alter the temperature but results in a change of state that involves the heat of fusion. Thus, pure elements and compounds exhibit a well defined freezing temperature. Alloys may solidify over a temperature range, however, and can be produced with nonequilibrium solid structures.

Nucleation and Growth. The attainment of the freezing temperature in a liquid phase does not necessarily result in solidification, even though this is favorable thermodynamically. Nucleation of critical-sized solid agglomerates (embryos) of atoms must first occur, either by chance meeting of a number of atoms within the melt itself (homogeneous nucleation) or by freeze-out on foreign solid surfaces present in the melt or on the walls of the container (heterogeneous nucleation).

The type of growth that occurs after nucleation is initiated depends upon the degree of supercooling below the melting point and the rate and direction of heat loss from the melt. The formation of dendrites (tree-like crystals) represents a type of growth not limited by the rate of heat transfer from the melt, and is most common for slowly cooled metals.

It is statistically probable that many embryos will nucleate simultaneously and produce a large number of crystals. Competition between these produces a multitude of grains during the course of solidification. Orientation differences between grains result in grain boundaries at their interfaces and lead to the formation of dislocations and other imperfections in the crystallized phase.

Alloy Solidification. Whenever a system of two or more components is caused to solidify from the melt, component concentration differences between various solidified areas may be produced. This is caused by a difference in solute concentration between the liquid and solid phases, since solute elements are often less soluble in the solid than in the molten alloy. Hence at the solidification interface, the freezing process continually rejects impurity elements into the melt. The first solid to freeze thus contains a lower concentration of solute than the last solidified fraction. Although this phenomenon is utilized to advantage in zone refining to purify materials, it often leads to coring and undesired segregation in welding applications.

Nonequilibrium Solidification. Effects due to rapid solidification in systems of any number of components can entrap gases, unsolidified liquid, or impurity atoms. Diffusion after freezing can then cause phase precipitation or even cracking due to gas pressure. Metallurgical structures with distinctly different mechanical properties are also easily formed by rapid removal of heat from the melt (quenching).

Alloying and Intermediate Phases

Alloying and the formation of compounds are important in certain types of joining, particularly where dissimilar metals are involved.

Atom Interaction. A metallic substance that is composed of two or more elements and is homogeneous on a macroscale is termed an alloy. In forming an alloy, as foreign (solute) atoms are added to an atom matrix (solvent), the interaction between the solvent atoms and the added atoms can be indifferent, attractive, or repulsive.

If solute and solvent atoms behave in a completely indifferent manner to one another, true random solid solutions are formed, and the mixture is homogeneous on a microscale. If the unlike atoms attract, the structure formed is an ordered solid solution, in which each atom is surrounded with unlike atoms as much as possible. If the elements differ greatly in electronegativity, intermetallic compound formation is possible with a tendency toward ionic-type binding. Repulsive forces between different atoms tend to produce separate and different atom groups known as phase mixtures. In metal joining, solid solutions and intermetallic compounds are more important than phase mixtures.

Solid Solutions. Solid solutions usually exist over a range of compositions. Binding in solid solutions closely approximates the binding found in pure metals and is not influenced greatly by exact atomic positioning or by concentration. Both *interstitial* and *substitutional* solid solutions exist. Interstitial solutions occur only where the added atoms are small enough to fit in the interstices between the solvent atoms. Substitutional solutions, in which solute atoms replace solvent atoms, may be either ordered, where the relationship between the two atom kinds is fixed and repetitious, or disordered, where the relationship is random.

Empirical rules, known as the Hume-Rothery rules of solid solubility, have been postulated for the relative extent of solid solutions and the stability of intermetallic compounds. It has been found that: (1) if the atoms differ in size by more than 13 to 15 percent, solid solubility is limited to a small range of compositions; (2) if stable intermetallic compounds are formed, the range of solid solubility will be restricted; and (3) for certain ratios of valence electrons to total number of atoms, stable intermediate phases can be formed, again restricting solid solubility.

Intermediate Phases. There are three main types of intermediate phases: electrochemical compounds, size-factor compounds, and electron compounds. Electrochemical compounds, formed between elements of different electronegativity, are limited in solubility range and have high melting points. Size-factor compounds can form when there is a difference of about 20 to 30 percent in size between the constituent atoms; they are referred to as Laves phases. The existence of these compounds is due to the higher-than-normal coordination number that is possible with a 20- to 30-percent size differential. All electron compounds are similar structures with identical valence-electron-to-atom ratios (3/2, 21/13, and 7/4); these compounds are particularly stable.

Diffusion

Many types of metallic joining depend upon the movement of atoms across the joint interface by diffusion. Thus, various aspects of mass transport in solids (diffusion) are discussed.

Energetics of Diffusion. Diffusion is the movement of atoms in relation to a fixed matrix of other atoms, from regions of high atom concentration into regions of lower concentration. The concentration gradient provides the driving force* for diffusion. Thus,

$$J = -D \frac{\partial C}{\partial X} \qquad (12\text{-}1)$$

*Strictly, the driving force is not the concentration gradient, but the free energy difference existing in a system. In multicomponent bodies, atom movements are more correctly attributed to driving forces such as thermal, electrical, and chemical potential gradients.

where J is the atom flux across an arbitrary plane, D is the diffusion coefficient, and $\partial C/\partial X$ is the concentration gradient perpendicular to the arbitrary plane. The flux is thus proportional to the gradient; when the structure is homogeneous, $\partial C/\partial X$ is 0 and there is no net flux across the plane. The diffusion coefficient, D, is not constant, but is a function of temperature and concentration.

Diffusion Mechanisms. The mechanism of diffusion is different in interstitial and substitutional alloys. Movement between adjacent interstitial positions is the predominant mechanism for interstitial alloys. The mechanism of atom movement in substitutional alloys is not as clear, and it is possible that any of the following may occur: (1) simultaneous movement of adjacent atoms (either by direct interchange or a ring mechanism), (2) movement via interstitial positions, or (3) movement into adjacent "empty" atom sites (vacancy diffusion). Vacancy diffusion is believed to be the main mechanism for diffusion within a metal (bulk diffusion).

Bulk diffusion is only one means for mass transfer of atoms in solids. Surface diffusion takes place at a relatively low energy level and results in the migration of atoms across a surface. Diffusion along grain boundaries occurs more readily than bulk diffusion because of the greater concentration of defect structures in such boundaries.

Although the above discussion of diffusion is stated in terms of solid materials diffusing into solid materials, diffusion into a solid material from a liquid phase is also common. This can lead to phase dissolution and grain boundary penetration by molten substances.

Wetting

In joining applications involving liquid phases, particularly in soldering, the function of wetting to form a metallurgical connection is basic. Wetting is the ability of a liquid to adhere more strongly to a solid upon which it is placed than to cohere to itself. The phenomenon of liquid-solid wetting is best explained with the aid of Fig. 12-5, which shows in cross section a liquid droplet in equilibrium on a solid surface in a gaseous atmosphere. At equilibrium, the angle in the liquid between the force vector at the liquid-gas interface and the vector at the liquid-solid interface is termed the contact angle, θ (dihedral angle). At point A in Fig. 12-5, all three interfaces

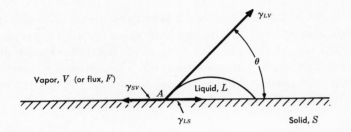

FIG. 12-5. Surface-tension equilibrium, showing contact angle, θ. (After H. H.
Manko, "Mechanism of Wetting During Solder Joint Formation,"
in *Papers on Soldering* (STP 319), 1963; courtesy American Society
for Testing and Materials.)

exist; the surface energy forces are in equilibrium, and

$$\gamma_{SV} = \gamma_{LS} + \gamma_{LV} \cos \theta \qquad (12\text{-}2)$$

where γ_{SV} is the force causing spreading or wetting of the solid by
the liquid. Since all of these surface energy forces are fixed for a
given system, the contact angle, θ, is used as a measure of wetting.
If ideal (complete) wetting occurs, $\theta = 0°$; for no wetting, $\theta = 180°$.
Partial wetting is indicated by $0° < \theta < 180°$.

12.2 METALLIC SURFACES

The approach to metal joining taken here begins with a consideration of the integral parts of a connection, namely the two (or more)
bulk masses which are joined together to form a cohesive structure.
For most of the joining applications discussed in this chapter, the
joint formation is primarily affected by the surface and a thin layer
of material beneath.

Ideal Surfaces

A surface is defined as the dividing entity between two phases.
This surface passes through all points which are similarly located
with respect to surrounding material, and usually coincides with the
physical boundary of the solid.

Surface Thermodynamics. All energy associated with a surface
can be defined as the excess energy over that which would be present
in the absence of the surface. Since the surface is an area of
discontinuity between two phases, the atoms in the vicinity of
the surface layer are not in their lowest (equilibrium) energy states.

This nonequilibrium arises because surface atoms have fewer neighbor atoms than interior atoms have, and hence possess extra energy since their binding forces are only partially satisfied [5]. The total energy of the system is composed of the contributions from the two bulk phases, E_1 and E_2, and from the boundary, E_s; that is,

$$E = E_1 + E_2 + E_s \qquad (12\text{-}3)$$

This extra energy develops an additional attractive force between adjacent atoms in the surface layer [6], the attraction being commonly referred to as "surface tension." The result of this increased energy force is represented schematically in Fig. 12-6.

If such a force exists, whether tension or compression, it is obvious that the metal system (surface plus bulk material) is not at the lowest possible energy state. One way for the system to tend to lower its energy is by a mechanical rearrangement in the underlying atom planes. The body thus becomes stressed in compression to balance the surface tensile stress [5]. If temperatures are sufficiently high to permit extensive diffusion to take place, the total energy can be lowered by atom migration from surface planes of high energy to those of lower [6]. It is also likely that such a stressed surface serves as a sink for imperfections, both vacancies and dislocations, to lower the body energy.

Physical Representation of Surfaces. An ideal surface may be pictured as the close-packed plane of an ideal crystal; however, on an atomic scale such a surface is neither flat nor smooth. Representations of a crystal system cut by various planes to produce a surface show that the less the number of nearest neighbor atoms a surface

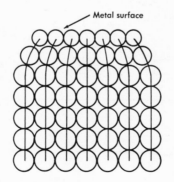

Fig. 12-6. Surface layer contraction caused by surface tension. (From H. H. Uhlig, "Metal Surface Phenomena," in *Metal Interfaces*, 1952; courtesy American Society for Metals.)

atom has, the rougher is the surface. In fact, certain cutting planes can produce "troughs" on the surface. Dense atomic planes in a structure form relatively smooth surfaces.

Real Surfaces

The ideal surfaces described above do not account for the way the metal bulk was originally formed as a body, the environment surrounding the body after formation, or the effects of any mechanical working.

Surfaces on a Microscale. A solid bulk phase, grown from a melt, possesses a roughened surface because of dendritic formation at the liquid-solid interface during solidification. Thin structures formed by condensation from the vapor are usually smoother. Surfaces formed by removal of already existing surface layers, such as by ion bombardment, are also smooth. But by whatever technique the bulk phase has been produced, its surface is not flat on an atomic scale, across even a small area [8].

In fact, departures from perfect order will still occur at surfaces even though the bulk material is free from imperfections. Such surface defects include additional metal atoms adsorbed on close-packed planes, vacant atom sites, kinks, incomplete atom rows, etc., as shown in Fig. 12-7. Of course, many additional surface defects arise from the intersection of interior dislocation lines and crystal boundaries [6]. Hence, it is virtually impossible to produce metal surfaces "smooth" on an atomic scale by any known method of solid growth or removal.

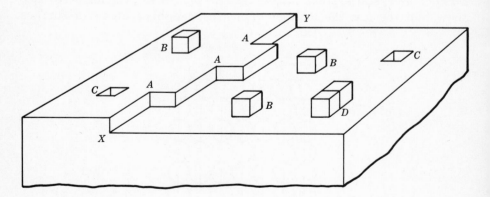

FIG. 12-7. Crystal surface containing step XY, kinks A, adsorbed molecules B, surface vacancies C, and pairs of adsorbed molecules D [8]. (From W. Boas, "The Nature of a Metal Surface;" courtesy *Journal of the Australian Institute of Metals*.)

Gaseous Sorption. As was explained previously, atoms at a surface have stronger binding ties to underlying atoms; this excess force produces a surface tension. The excess force can be reduced by the capture of atoms from the external environment; in gaseous atmospheres, physical adsorption and chemisorption of gas atoms on the surface are possible [6, 7] with the subsequent partial relief of surface tension, Fig. 12-8. Every surface, no matter how prepared, is either instantly or gradually contaminated (depending upon the gas pressure) by gaseous sorption [8]. Gas molecules are constantly bombarding surfaces, even at the highest attainable vacua; that portion of the total number of molecules "sticking" to the surface produces a higher concentration there than in the atmosphere. This phenomenon may be either physical or chemical in nature. Physical adsorption of gases is nonspecific, i.e., it occurs on all surfaces with all gases, and occurs with a van der Waals type of binding. Chemisorption is specific, depends on pressure and temperature, occurs with a much higher binding energy, and is irreversible. The irreversibility is due to the high binding energy and resultant chemical reactions at the surface. The most common example of chemisorption is the oxidation of metallic surfaces in oxygen atmospheres.

Oxide and Other Film Growths. All metals except gold react with oxygen to form oxide layers at room temperature [9]. Thin oxide films (thickness less than 10,000 A) are formed even at low temperatures. The rate of growth is usually controlled by the strength of an electric field resulting from the difference in chemical potential between adsorbed oxygen and the metal surface. Diffusion of metal cations and oxygen anions through films may also control the growth rate.

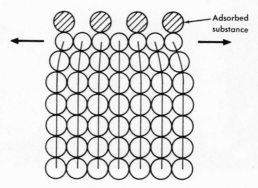

Adsorbed substance

FIG. 12-8. Metal surface showing how adsorbed gas atoms can partially relieve the surface tension [7]. (After H. H. Uhlig, "Metal Surface Phenomena," in *Metal Interfaces*, 1952; courtesy American Society for Metals.)

Oxide growth on an initially clean metal surface begins with the sudden formation of oxide nuclei, or isolated clumps of oxide. These nuclei probably form [10] on preferential crystal sites, Fig. 12-9, similar to the nucleation and growth phenomena discussed previously in this chapter and in Chap. 3. Growth occurs in three dimensions, i.e., along the surface and perpendicular to it, and finally leads to abutment of adjacent nuclei. Perfect joining of such oxide structures is no more probable than perfect joining between competing metal grain structures during solidification. Hence, the oxide pattern is imperfect. If the metal surface being oxidized is composed of poly-crystals, with each crystal at different orientation with respect to the surface and possessing different oxidation rates, it is likely that the oxide film covering the surface will be extremely irregular.

Oxide growth rates generally diminish as the film thickens, par-ticularly if the film is coherent, i.e., "protective" with respect to the metal. At temperatures below about 400°K, it has been observed that a coherent film tends to approach a limiting thickness on the order of 20 to 50 A [9].

It is also known that environments containing gases other than oxygen will produce contaminating layers on metal surfaces. Not only can adsorption take place, but chemical reaction and subsequent

FIG. 12-9. Model of a pyramidal oxide nucleus growing on a metal surface. (From Germer, Stern, and MacRae, "Beginning of the Oxidation of Metal Surfaces," in *Metal Surfaces*, 1963; courtesy American Society for Metals.)

film layer production can also occur. Adsorption of various gases and water vapor also occurs on oxide and other growth films. Residues of hydrocarbon products present in industrial atmospheres or used in the production and cleaning of surfaces are likewise present on metallic surfaces directly or on the overlying film surfaces [6].

Mechanical Surface Preparation. The structure of a metal body near its surface depends largely upon the method of forming. Deposition or dissolution techniques result in an essentially unworked surface and underlying layer; mechanical (abrasive) preparation, no matter how carefully conducted, leaves the surface and adjacent structures in a worked, plastically deformed condition.

12.3   REAL METAL COHESION

The observed physical properties of macroscopic metal specimens are determined by many factors. In particular, the strength of metals is influenced by interatomic forces, crystallographic defects, and chemical homogeneity. This section discusses the strength of real polycrystalline metals and introduces the grain boundary as the ultimate obtainable interface between two randomly oriented metal crystals. The grain boundary is developed as a model for macroscopic joining processes.

Electron Interaction

Metallic cohesive forces arise from the interaction between outer electrons and positively charged metal ions. The physical properties of a metal can be explained only if the outer electrons of each atom are assumed to belong to the whole metal aggregate. Thus, the concept of an electron gas, discussed in Chap. 6, offers a convenient model through which the conduction phenomena and plasticity of metals are explained. The forces and electron interactions which bind metal atoms together are not yet fully understood. In a real metal crystal, it is possible for different kinds of binding forces to coexist. Metallic binding is often closely related to both the covalent and to the van der Waals forces. Pauling [11] has suggested that metallic binding is a particular type of covalent binding wherein shared electrons are free to jump between adjacent atoms and thus produce the observed electron mobility. More recent observations [12] have shown that the incomplete inner electron shells of the transition metals also contribute to interatomic binding. No one theory is as yet available which will unequivocally predict metallic binding phenomena between either like or unlike metal atoms.

The Grain Boundary

Most real metals are composed of intimately packed small single crystal regions called grains. The strength and ductility of real metals for a given crystal structure are overwhelmingly influenced by crystallographic defects such as dislocations. A grain boundary is largely composed of such defects, and since it closely resembles a bond made by various joining processes, it is helpful to consider the properties of a grain boundary. In a homogeneous metal structure, each grain possesses the same atomic lattice symmetry (such as bcc, fcc, or hcp). In general, the crystal orientation of adjacent grains is dissimilar. The interface between two grains is thus composed of a region of relatively disordered material, typically several atom layers in width. This region of misfit is called a grain boundary and occurs throughout all polycrystalline metal structures. Figure 12-10 shows real polycrystalline metal specimens which illustrate the bounding facets of the individual grains for an unusually large-grained cast structure.

FIG. 12-10. Polycrystalline metal specimens, showing bounding facets of the individual grains. (From W. Rostoker and J. R. Dvorak, *Interpretation of Metallographic Structures*, 1965; courtesy Academic Press.)

Metallographic techniques of polishing and etching are used to reveal the grain structure of real metals. Figure 12-11 shows the

FIG. 12-11. Microstructure of rolled copper (magnification 200 ×).

grain boundary network of a typical polycrystalline structure. Figure 12-12 shows a copper microstructure which has a eutectic of Cu–Cu$_2$O at the grain boundaries. Metal grains can be surrounded by an envelope phase as illustrated in Fig. 12-13 for a tungsten-nickel-iron alloy. The central grains are almost pure tungsten, whereas the envelope phase is a binary alloy of nickel and iron. The metallic joining features shown in these micrographs are similar in many respects to interfaces produced by macroscopic joining processes such as thermocompression bonding, soldering, and brazing.

For a pure metal, the grain boundary represents a joined region where no second material is introduced, i.e., a grain boundary can be a connection in a one-component system. Since the mechanical properties of the two joined materials (the two grains) are usually identical and independent of direction, the properties of the boundary will not be adversely affected by moderate amounts of expansion introduced by heat or mechanical stress. Any material transfer (diffusion) between the boundary and the joined grains does not introduce new chemical phases or compounds which may degrade the joint strength. Even so, the properties of the boundary can be different from the parent metal. At high temperatures and under

FIG. 12-12. Microstructure of cast copper, with Cu-Cu$_2$O eutectic at the grain
boundaries (magnification 200 ×). (After E. P. Polushkin, *Structural Characteristics of Metals*, 1964; courtesy Elsevier Publishing
Co., Inc.)

the influence of prolonged stress, grains may undergo a slow relative
movement and give rise to "high-temperature creep." Grain boundaries are often more prone to corrosion because of the disordered
state of atom arrangement. This effect is utilized during certain
etching procedures used in metallographic preparation to delineate
structural details.

If a second phase is present in the grain boundary region, the
material properties may be altered, e.g., the brittle grain boundary
network of cementite (Fe$_3$C) in cast iron causes the whole structure
to be brittle even though the isolated grains are ductile. Diffusion of
mercury along aluminum or gold grain boundaries causes complete
loss of strength due to the weak amalgam coating on the grains.
A grain boundary network with a lower melting point will restrict
maximum service temperature. The principle illustrated is that the
very small (less than 1 percent) grain boundary volume can control
the entire material strength. The interfaces produced during macroscopic joining processes are subject to similar effects, since relatively
small amounts of interposed material can completely determine bond
or weld quality.

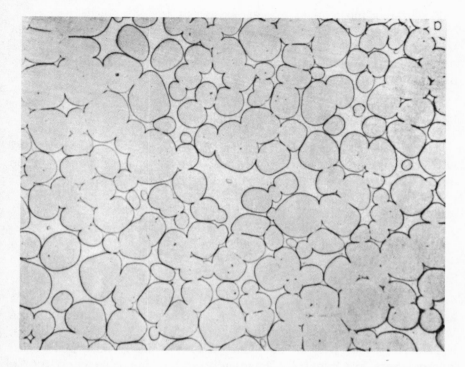

FIG. 12-13. Microstructure of a tungsten-nickel-iron alloy, showing grains of tungsten embedded in a nickel-iron matrix. (From W. Rostocker and J. R. Dvorak, *Interpretation of Metallographic Structures*, 1965; courtesy Academic Press.)

On an atomic scale, metals miscible in the solid state form solid solution alloys, wherein crystallographic lattice sites are arbitrarily occupied by metal A or metal B atoms. The metal grains have the composition of the alloy, and, in general, all the comments discussed for a pure metal apply to the solid solution alloy.

Metals which exhibit limited solid solubility may separate into grains whose composition is predominantly metal A and grains predominantly metal B. The grain boundary now represents a connection between dissimilar metals. Polycrystalline metals composed of two or more phases exhibit properties determined by the aggregate as a whole.

The interface produced by macroscopic joining processes can be interpreted in terms of grain boundary behavior discussed above. Based on this analogy, the ultimate obtainable macroscopic connection should have the following properties:

1. Complete metal-to-metal contact through a somewhat crystallographically misoriented transition zone.

2. Similar mechanical properties across the interface.
3. No degrading chemical interaction between the connecting area
   and the bulk parent metal.

A connection involving only one metal is one of the simplest ways
to achieve the above criteria.

12.4   FACTORS AFFECTING JOINING

There are many physical and chemical considerations associated
with macroscopic joining processes. These considerations can be
discussed in terms of the following:
1. Surface contamination.
2. Interface contact.
3. Binding activation energy.
4. Connection stability.

Surface Contamination

As described in Sec. 12.2, even well cleaned real metal surfaces
are covered with various layers of nonmetallic contaminants. These
layers can consist of corrosion products such as oxides, sulfides,
hydrides, etc., as well as chemisorbed or physically adsorbed gases.
These layers are generally chemically stable and thus constitute a
barrier to any joining process. The removal of these surface layers
and the prevention of their re-formation during joining is a primary
requisite in any bonding or welding process.

Fluxes are used to combat oxidation as well as to permit surface
wetting and spreading of molten metal during soldering and brazing.
A flux may react with and displace surface contaminant layers as
well as provide a physical atmospheric shield during bond formation.
Another important mechanism exhibited by some fluxes is the
deposition of surface films from the flux (by electrolytic or chemical
action) to modify the surface chemistry and to facilitate the spreading of the liquid metal.

Solid-phase bonding processes often depend on substantial plastic
deformation at the bond interface to mechanically break up and
disperse surface contamination. Thermocompression bonding has a
limited inherent ability to overcome intervening contaminants, and
for this reason reliable bonds have generally been restricted to the
use of small diameter gold or gold-plated wire [15].

Dispersion of a contaminant oxide layer may depend on the relative
mechanical properties of the oxide and the underlying metal. Thus,

the hard and brittle oxide coating on aluminum tends to crack and embed itself in the soft parent metal when subjected to sufficient stress. The success of aluminum ultrasonic bonding at room temperature is, in part, a result of this mechanical dispersion effect. As suggested, ultrasonic bonding has the ability to remove some interface contamination through tangential vibratory stresses and induced plastic deformation.

Wire brushing, filing, grinding, grease removal, and chemical etching are all common surface preparations used to reduce contamination effects in heavy metal joining applications. Most of these gross surface treatments, however, are not compatible with microelectronic fabrication. Use of the more noble metals renders surface contamination less of a problem.

If suitable temperatures are reached, surface contaminants such as oxides may become thermodynamically unstable and disassociate or dissolve into the parent metal. This reaction may be enhanced through the use of a protective gaseous atmosphere such as argon or forming gas.

Interface Contact

Intimate metal-to-metal contact is a necessary but *not sufficient* condition for joining. The influence of strong interatomic forces does not extend much beyond the atomic diameter of about 5 A, while weaker van der Waals forces are exerted over about 1000 A separation. Thus, the degree of interface contact for strong binding should approach that of a grain boundary.

In any joining process wherein one constituent is in the liquid state, a high degree of intimate metal-to-metal contact may be achieved, depending on the nature of the intervening layers as discussed above and also on the flow or wetting characteristics of the liquid. If the liquid is composed of the same material as the parent metal (as in fusion welding), wetting always occurs in the absence of contaminating layers. In the case of dissimilar metals (as encountered in soldering and brazing), the degree of wetting is determined by the cohesive energy of the liquid, as compared to the surface energy of the solid and energy of the liquid-solid interface. To review, we know that energy can be associated with any free surface. Simplified, this energy results from incomplete surface atom binding. From a thermodynamic viewpoint, any material system tends to proceed to the state of lowest free energy. Thus, solder will generally flow on a clean metal surface (reducing the high surface energy of the metal) and will not flow on an oxidized surface (already a low energy surface). It is still necessary, of course, to consider the particular interaction of the clean metals involved, for if the work of forming a solder-to-base-metal bond is greater than the reduction in surface energies, wetting will not occur despite the fact that the metal surfaces are clean. Subtle effects, such as the slight alloying of the solder with the base metal, may alter the "calculated" results considerably. Experimental observations provide the most complete evaluation of these effects.

Under normal forces, less than 1 percent of the apparent contact area between polished solid metal surfaces may be in real contact. This results from the roughness, on an atomic scale, of real metal surfaces, no matter how thoroughly polished. When joining occurs without liquid phase formation (i.e., solid-phase bonding), intimate interfacial contact is achieved through localized plastic deformation. It has been established [16] that tangential motion at the interface facilitates contaminant dispersion and intimate atomic contact. Tangential interface plastic flow can be induced through the action of large normal forces as in the case of thermocompression and roll bonding. Direct tangential stresses can be introduced during ultrasonic bonding if the vibratory coupling is parallel to the bond interface. Experiments [17] have shown that coefficients of adhesion between many metal couples are strongly increased due to the effect of tangential movement during the application of normal forces. Roll bonding of metals (commonly applied to sheets) often depends on severe deformation (up to 50 percent) to achieve bond formation. Reduced surface contaminants, however, will permit a considerable reduction in required deformation, indicating that contaminant dispersion, not surface conformity, is controlling. Increasing the temperature of the workpieces also lowers the required deformation because of increased metal plasticity.

Binding Activation Energy

The third consideration for joint formation—energy to activate atom-to-atom binding—is somewhat general, yet descriptive of many observations in joining. It has been found in solid-phase bonding that extended metal-to-metal contact is not usually sufficient to cause spontaneous joining between surfaces. Some additional energy is required to bring the atoms sufficiently close to achieve metallic binding.

Connection Stability

The final criterion for successful metal joining is the connection stability, as reflected by mechanical strength and electrical conduction, which is retained during intended service. *Mechanical equilibrium* should be achieved such that potential failure due to microcracks or locked-in stress is not· a controlling factor. High stresses can be generated by resolidification of molten metal as well as by solid-phase cold working mechanisms. *Chemical equilibrium* is also an important consideration for connection evaluation. Thermocompression bonds between gold and aluminum, used extensively throughout the microelectronics industry, may fail prematurely because of the formation of a brittle intermetallic compound (Au_2Al). The characteristic color of Au_2Al has prompted this effect to be called the "white plague" [18]. Connections between two dissimilar metals which form brittle intermetallic compounds may remain quite ductile and satisfactory as long as intradiffusion is suppressed. Subsequent heat cycles, however, may cause the formation of the brittle compound and result in complete loss of connection strength. Metals that form brittle intermetallic compounds may give quite satisfactory service only when the joining operation and subsequent environmental stresses do not activate the formation of the compounds.

12.5   THIN FILM JOINING

Special precautions must be taken when joining leads to a thin film electronic circuit. The rather unique problem of connecting a bulk member to an essentially surface layer introduces complexities such as the effect of surface damage (through mechanical abrasion or chemical dissolution) on the performance of the interconnection. The connection region is initially composed of a diversified material combination consisting of the substrate (usually glass or ceramic), metal film, and metal conductor. The overall bond or weld strength and electrical integrity may be dependent on a multiplicity of interfaces that are present before joining and produced during joining. This section discusses the adhesion aspects of thin films as well as some of the "boundary conditions" which the thin film structure imposes on the selection of interconnection bonding or welding methods.

Thin Film Adhesion

Metal thin films exhibit varying degrees of adhesion to either glass or ceramic substrates as well as to other metal films. The macroscopic strength of the interface between a thin film layer and adjoining surface depends on many factors such as:

1. Influence of interposed contaminant layer.
2. Inherent affinity between materials (solid solubility and chemical reactivity).
3. Energy supplied to interface.
4. Defect structure of original surface.
5. Mechanical stress developed at interface.
6. Composition gradient at interface.
7. Aging effects which alter interface structure.

The above factors are influenced by cleaning procedures and deposition process variables, both of which affect thin film nucleation and growth behavior. The disciplines of vacuum technology, surface physics, and chemical thermodynamics all must be applied to predict thin film adhesion phenomena. The chemistry and physics associated with thin films may deviate considerably from their behavior in bulk material, and this must be taken into account.

The interface between a thin film and adjoining surface (conductor or nonconductor) can be classified according to some typical characteristics as stated by Mattox [19]. A summary of this classification is presented below:

1. The mechanical interface: characterized by interpenetration of the thin film into the pores and crevices of the substrate, producing a mechanical interlocking interface.
2. The monolayer-to-monolayer interface: characterized by an abrupt change from the film material to the substrate material in a distance comparable to the separation between atoms, with no diffusion or chemical reaction between the thin film and substrate.
3. The compound interface: characterized by the formation of a compound of constant composition between the thin film and the substrate due to chemical interaction. Depending on the substrate surface, this compound may be an intermetallic or other chemical compound such as an oxide.
4. The diffusion interface: characterized by a gradual change in composition between the film and substrate. Some degree of mutual solubility is required as well as the presence of sufficient energy to overcome barriers to diffusion.

5. "Pseudo-diffusion" interface: characterized by an interface which exhibits a compositional gradient that is not due to normal diffusion reactions. This can be caused by the physical interpenetration of energetic bombarding atoms into the substrate (as observed in the case of sputtering). No mutual solubility or diffusional activation energy (in the thermodynamic sense) is necessary.

The energies associated with pertinent deposition processes and material interactions are helpful in understanding film adhesion, and a list of these energies is given in Fig. 12-14.

Atom interactions	Energy (ev)
Physically adsorbed gases	0.1 to 0.5
Chemisorbed gases	1 to 10
Compound layers on surfaces	1 to 10
Thermally evaporated atoms	{ Mean 0.2 to 0.3 { Maximum 1
Sputtered Cu atoms using 900 ev Hg^+ ions	{ Mean 10 { Maximum 100
Binding energy in solids	2 to 5

FIG. 12-14. Approximate energy associated with atom interactions [19].

The adherence of a thin film to a substrate is both material dependent and process dependent in that a wide range of depositing atom energy may be encountered, e.g., between evaporation and sputtering. Differences in impinging atom energy can alter the role of contaminants (which would be disrupted by high energy atoms) as well as change the nature of the interface. For example, an evaporated metal thin film deposited onto a mutually insoluble metal substrate would exhibit the monolayer-to-monolayer type interface. The same film sputtered at high energy levels may well exhibit the "pseudo-diffusion" interface, along with increased adhesion derived from interpenetration.

In general, the more oxygen-active metals, such as titanium and chromium, tend to form quite adherent films when evaporated onto oxygen-bearing substrates such as glass or ceramic. Substrate temperature plays an important role in this process since chemical interaction between the depositing metal and the substrate requires sufficient activation energy to become operative. If multilayered films are produced in two separate deposition cycles such that the film sees atmospheric contamination between steps, an oxygen-active metal is often used as an intermediate layer to act as a "glue" for the second metal thin film deposition. To form contacts in thin film manufacture, Nichrome® or titanium is deposited on tantalum nitride resistive film to provide adhesion for the subsequently evaporated contact layers. It is probable that successive deposition of multi-layered films without intermediate atmospheric exposure would elim-inate the necessity for Nichrome® (or similar) films.

Thin films can exhibit a wide variety of aging effects as reflected by the change in adhesion with time. Three such aging phenomena, typical of three classes of film-interface interaction, are shown in Fig. 12-15. As can be seen from the data, adhesion of evaporated films on glass substrates can increase, remain constant, or decrease with time. Generally, an increase in adhesion is attributed to the diffusion of oxygen to the interface and subsequent chemical reaction to form a transition interface boundary. The oxygen-active metals are most subject to this mechanism. The relatively low adhesion of oxygen-inactive metals such as gold is probably caused by a lack of chemical interaction with the substrate; here, weak van der Waals forces are responsible for any observed adhesion. The adhesion of the lower melting point metals appears to be influenced by a mobility consideration wherein microscopic physical separation from the sub-strate (or at least a tendency to separate) is favored (because the

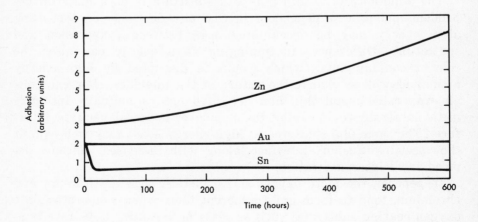

FIG. 12-15. The variation in adhesion with time for typical films of Zn, Au, and Sn [20].

cohesive forces are greater than the adhesive forces). This factor is believed to be responsible for the *decrease* in adhesion of tin with time, the decrease reflecting less film-substrate contact. It should be remembered that the above discussion on adhesion and aging phenomena was for films evaporated at a particular set of deposition conditions. It has also been observed that increased reactive gas partial pressures during deposition increase initial adhesion for reactive metals [20], as well as for gold films [21]; sputtered films generally exhibit greater initial adhesion because the higher energy impinging atoms cause contamination dispersion and possible physical penetration into the substrate.

Phase changes induced in thin film layers can cause loss of adhesion due to volume changes and the resulting high stress. For example, an initially adherent copper film over a tantalum film subjected to a palladium sputtering cycle may show either of the following: (1) at low sputtering potentials, the palladium film merely adheres to the copper; or (2) at high sputtering potentials, the energetic palladium penetrates the copper film (causing an ordered Cu-Pd alloy to form) and results in complete loss of adhesion between the alloy and the underlying tantalum film.

In conclusion, the adhesion of thin films is a complex dynamic phenomenon which often is extremely difficult to measure quantitatively. In thin film joining, the bonding or welding process itself may be the most severe adhesion test.

Thin Film Joining Constraints

The two-dimensional nature of a thin film renders it incapable of effectively dissipating energy. Thus, during any joining process, both the mechanical and thermal input energies are transmitted to the *substrate* material. The resulting stresses developed in both the thin film layer and the underlying substrate must be considered as potential limitations on the connection integrity. It has been observed that the use of a standard tin-lead solder for the interconnection of leads to thin films on glass may result in substrate crazing and failure. The substrate failure is caused by the stresses induced by soldering. Soldering with an indium-base solder alleviates this problem by reducing these stresses. Another example of substrate damage during joining may be experienced when leads are ultrasonically bonded to a thin film on glass. The stresses transmitted to the substrate may be reduced by increasing the plasticity of the metal conductor through the application of heat. The above examples illustrate that the controlling factor in connection quality can be the effect of bonding or welding process energies on the substrate.

The extremely small volume associated with a thin film imposes further restrictions on any joining process which tends to degrade the film through abrasive action, dissolution, or evaporation. Changes in the chemical composition in the joining zone as well as geometrical changes (such as film coalescence) may give rise to potential electrolytic corrosion cells, electrical hot spots, and highly stressed regions. Observations of ultrasonic bonds of aluminum leads to tantalum films deposited on glass have shown that a wide range of film disturbance can occur. The film damage (observed through the underside of the substrate) varies with the bonding parameters. The use of soldering or brazing techniques can cause the complete removal of the initial thin film layer by dissolution into the molten metal. This film removal can destroy the ability to remelt and resolder the interconnection. The possibility of depleting protective film layers adjacent to the solder bond region may also cause potential failure sites. Fusion welding processes, such as laser or electron beam welding, must be applied with the utmost caution since relatively small amounts of energy can cause the film to melt or evaporate, resulting in electrical discontinuity.

The desire to achieve a small bond or weld zone consistent with high density electronic circuitry imposes yet another restriction on any potential joining process. The luxury of using a low strength joining material (such as soft solder) and increasing the joint area to achieve high overall strength may not be practical. Solid-phase bonding and fusion welding processes are, at least theoretically, able to achieve interface strengths equal to that of the lead or the film.

The strength of any interconnection is dependent on its weakest link. Consequently, the initial adhesion of the thin film(s) to the substrate, as discussed above, is always a prime factor which must be evaluated. Not only must the initial adhesion be sufficiently high to withstand the bonding or welding cycle and provide strength to the interconnection, but it must also be maintained during the joining cycle. Joining processes that do not appreciably disturb the various interfaces in the connection region can be expected to retain initial adhesion qualities. Joining processes that alter these interfacial properties (through fusion, dissolution, or diffusion) must provide the required adhesion to form an acceptable connection.

12.6   ALLOY BONDING

Alloy bonding (brazing and soldering) is "joining metals by flow-ing a thin layer, capillary thickness, of nonferrous filler metal into the space between them. Bonding results from the intimate contact produced by the dissolution of a small amount of *base metal* into the molten *filler metal,* without fusion of the base metal . . . The term *brazing* is used where the temperature exceeds some arbitrary value, such as 800°F; the term *soldering* is used for temperatures lower than the arbitrary value" [22]. In both brazing and soldering, the wetting properties of the solid-liquid interface and capillary action of the molten metal control bond formation and subsequent physical properties. Once strong interface bonds are achieved, the properties of the joint are controlled by the properties of the filler materials in much the same way as was described for grain boundary regions which contain a second phase. Thus, the strength of the filler metal often determines the strength of the bonded materials.

Because brazing is not generally used in joining to thin films, this section is limited to a discussion of soldering [23].

Fundamentals of Soldering

As with all the other metallurgical joining processes, the formation of a solder joint is not spontaneous, but is dependent on overcoming the always present metal surface barriers. Both the solder material and the base metals may have an oxide or other nonmetallic coating, in addition to contaminants such as grease and dirt deposited from environmental exposure. In solid-phase and fusion joining, pressure, interfacial movement, or gross melting is used to disperse coatings and contaminants; in soldering, they must be removed by the addition of fluxes as a preliminary and concurrent step with solder material melting and wetting.

As was shown in Sec. 12.1 under Wetting, the equilibrium estab-lished during wetting can be represented by an equation relating the component interfacial tensions [see Eq. (12-2) and Fig. 12-5]. If, however, the "atmosphere" explicit in Fig. 12-5 is replaced by a layer of flux, the equilibrium becomes

$$\gamma_{SF} = \gamma_{LS} + \gamma_{LF} \cos \theta \qquad (12\text{-}4)$$

where γ_{SF} is now the interfacial tension between the workpiece and the flux, and γ_{LF} is the tension between the solder material and the flux. The added flux promotes wetting if the replacement of the atmosphere by the flux results in a decreased dihedral angle.

The ability of a solder material to wet a surface is represented by the particular dihedral angle, θ, exhibited in a given system. With an angle of $\theta = 180°$, total nonwetting has taken place. If the solder comes to equilibrium before solidifying and θ is between 90° and 180°, the wettability of the system is poor and good solder bonds cannot be obtained. Such an angle may also result from the solder "dewetting," i.e., after the initial wetting and spreading of the solder over the surface, the force balance [indicated in Eq. 12-4)] changes, with the solder tending to revert into a ball. A satisfactory joint cannot be made under this circumstance either. With θ between 75°* and 90°, partial wetting has taken place. Joints made at these angles are usually marginal. At angles of less than 75°, good wetting action has taken place.

The soldering operation is thus made successful by: (1) removal of contaminants by fluxes, and (2) wetting of all joint surfaces by the solder. Unless both of these requirements are met, the solder joint, if formed at all, will be of poor mechanical and electrical properties.

Fluxes

The fluxing action depends upon chemical activity, thermal stability, temperatures of activation and deactivation, and wetting ability. A flux must have sufficient chemical activity to combine with the inorganic contaminants and dissolve or reduce these layers to provide a clean surface for the solder. The chemical cleaning action of a flux must be maintained throughout alloy bonding to prevent the recontamination of the surfaces. A flux must not evaporate or decompose at soldering temperatures. The fluxing action must be initiated by the soldering temperature, yet not deactivated by accidental overheating. The flux must also be able to displace the adsorbed gaseous surface layers and wet the surfaces.

Fluxes can be classified by their chemical nature as inorganic or organic, with the latter including both nonrosin base and the common rosin fluxes. Inorganic acid fluxes such as HCl promote fast oxide removal and are stable, but are very corrosive to the base metals. Inorganic salts (for example, ZnCl) possess similar properties of fast oxide removal and stability, but are only slightly less corrosive. Because they are corrosive and difficult to remove, inorganic acid and

---

*An approximate value.

salt fluxes do not find wide use for electrical connections, where long connection life is paramount. Inorganic gases, such as HCl, are suitable as fluxes only at high temperatures and must be used on relatively clean surfaces. Such fluxes find some use in semiconductor manufacture.

Organic nonrosin base fluxes, such as lactic or citric acids, aniline hydrochlorides, or ethylenediamine, are corrosive and temperature-unstable and, similarly, do not find use in the electronic components industry. Rosin fluxes are derived from pine sap and are a mixture of abietic and pimaric acids. Although these fluxes are used extensively for electrical connections because of their noncorrosive nature, rosin is not, by itself, a first-rate oxide remover. Many claims have been laid to "activated" rosin fluxes, i.e., rosin types with the addition of more active fluxing agents, to overcome the relatively poor oxide removal ability of rosin. Controversy exists concerning the degree of "noncorrosiveness" of such fluxes and their use on long-life electrical equipment.

Solders

Alloys of tin and lead are the most commonly used solders, probably because of their good wettability relative to their cost. Figure 12-16 shows the binary tin-lead equilibrium phase diagram with several common solders indicated. The eutectic alloy (approximately 63-percent tin) melts at 361°F, the lowest melting temperature for any tin-lead solder composition. All other compositions liquefy completely at higher temperatures and pass through a liquid-solid (pasty) region when cooled. Use of solders other than the eutectic composition can increase the probability of a "cold" solder bond, which occurs if the joint surfaces are moved when the solder alloy is in the pasty range. Cold bonds appear frosty and are lacking in strength.

With the increased use of soldering on other than bulk materials, the range and diversity of solder materials have increased. Basically, any alloy or element that is able to wet a surface and solidify into a bond could be considered as a solder material. Figure 12-17 is a diagrammatical representation of some of the more commonly used alloys; in addition to the tin-lead system, silver-containing solders (for high temperature use), indium solders (for intermediate temperature use), and bismuth solders (for low temperature use) have been employed extensively. Silver-containing solders generally require more heat reactive fluxes and are not as workable as tin-lead solders.

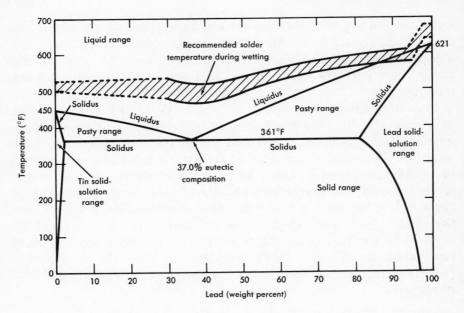

FIG. 12-16. An engineering version of the tin-lead phase diagram. (From H. H. Manko, *Solders and Soldering*. Copyright 1964 by McGraw-Hill Book Co. Used with permission.)

Lower melting silver alloys are especially useful on silver or silver-plated parts, since soldering with tin-lead alloys usually results in appreciable quantities of silver being dissolved into the solder and changing its characteristics. Intermediate temperature indium-base compositions are used extensively as solders to give better wetting, higher electrical conductivity, lower liquidus temperature, and greater ductility. Indium solders have been used successfully to solder to various thin films where the use of tin-lead solders is detrimental because of glass substrate microcracking and where gold dissolution is a problem. Bismuth solders have the property of expanding during solidification; these solders find application in printed wire board component "through-hole" connections, where mechanical strength and freedom from shrinkage cracks are important; see Fig. 12-18.

In addition to the basic consideration of melting temperature, the mechanical properties of the solder must be considered. It is important to realize that the solder itself is usually the weakest part of a joint, and undue stressing can lead to quick failure. For instance, in soldering wires to terminals, it is common to impart mechanical strength to the connection by twisting or looping the wire through the terminal, with the solder being used for electrical continuity only. It

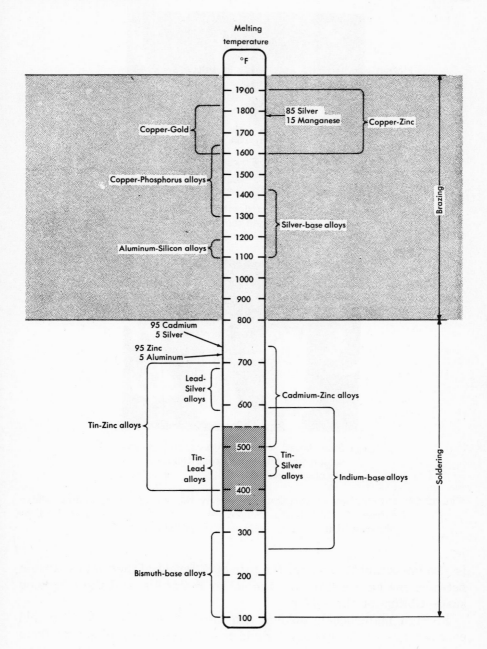

F<small>IG.</small> 12-17. Common alloys used for soldering and brazing. (From H. H. Manko,
        Solders and Soldering. Copyright 1964 by McGraw-Hill Book Co.
        Used with permission.)

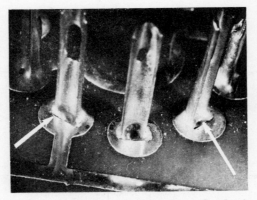

(a) Printed wire board with tube-socket leads.
    The cracks around the pins are
    attributed to solder shrinkage.

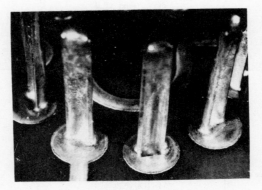

(b) Same board soldered with expanding-type
    solder, under the same conditions.
    No cracks were found.

FIG. 12-18. Elimination of shrinkage cracks by the use of an expanding solder. (From H. H. Manko, *Solders and Soldering.* Copyright 1964 by McGraw-Hill Book Co. Used with permission.)

is also important to consider the rate of differential thermal expansion between the base metals and the solder materials, and the corrosion susceptibility of the solder.

Foreign materials in solders can also affect bond reliability and consistency. Contaminants include metallic elements dissolved from solder holding pots, soldering tools, or the base metals themselves. Such elements as zinc, aluminum, and gold can radically alter the solder resistivity, spreading properties (Fig. 12-19), and liquidus temperature. Nonmetallic contaminants such as sulfur and phos-

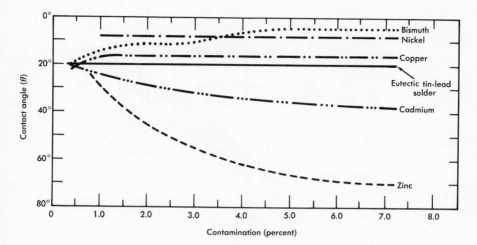

FIG. 12-19. Effect of contamination on the spread of eutectic solder. (From H. H. Manko, *Solders and Soldering*. Copyright 1964 by McGraw-Hill Book Co. Used with permission.)

phorus cause dewetting problems; tin and lead oxides can form inclusions in the solidified solder, resulting in poorer wetting and surface roughness.

Soldering Equipment and Methods

One basic requirement for soldering is the transfer of sufficient heat to the components being joined to permit proper solder melting and spreading. Conduction, convection, and radiation are all employed to transfer heat for soldering; the result is a large number of different soldering tools [23, 24].

Heat conduction is used in soldering irons, wave soldering machines, and parallel-gap and dual electrode resistance soldering units. The soldering iron, Fig. 12-20, is a common and familiar soldering tool. The copper or iron-plated copper tip is electrically heated and transfers heat to the joint by conduction. For consistent bonds, the tip temperature and rate of heat abstraction during soldering must be closely controlled. A method that is used in current thin film and printed wire board production is called fountain or wave soldering, in which a steady stream of solder is flowed onto the joint from a molten bath reservoir as shown in Fig. 12-21. This technique achieves rapid soldering rates, which result in less heat and distortion being imparted to the workpieces. For consistent joints, however, close control of the molten solder material is important, since it is rela-

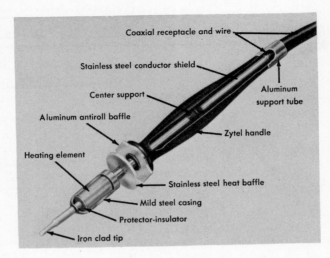

FIG. 12-20. Cutaway view of small soldering iron. (Courtesy American Electrical Heater Co.)

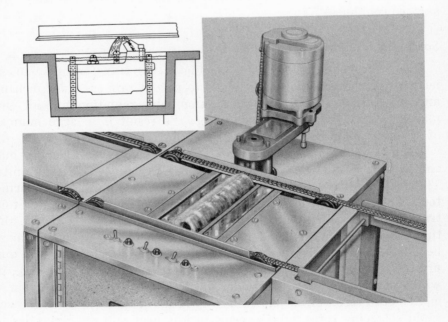

FIG. 12-21. Wave soldering machine. (Courtesy Leesona Corp.)

tively easy to dissolve contaminating materials from the joined parts into the bath, thus changing the bath composition and hence soldering quality. Another form of soldering by conduction of heat is the use of a parallel-gap welder to generate heat by interfacial resistance to electric current. Dual electrode soldering units, Fig. 12-22, which operate on a similar principle, provide an easily automated source of instantaneous, localized heat.

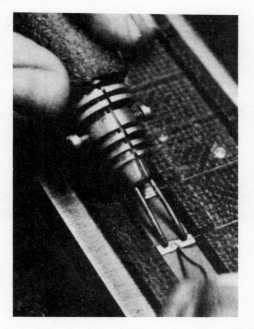

Fig. 12-22. Dual electrode resistance soldering unit. (From H. H. Manko, *Solders and Soldering*. Copyright 1964 by McGraw-Hill Book Co. Used with permission.)

Microtorch soldering is an example of a soldering process which operates by the convection of heat to the workpieces. Another convection heating method is employed in hot gas soldering equipment, in which a heated stream of gas (usually inert) is applied to a joint.

Infrared energy, both focused and unfocused, is used to radiantly heat joints for soldering, as in Fig. 12-23. This technique is a noncontact method which produces a rapid, precisely positioned heating pulse [25].

In many applications, it is desirable to add an extremely small or precise amount of solder to the joint at the time the workpieces are at the soldering temperature. Also, the joint may be inaccessible

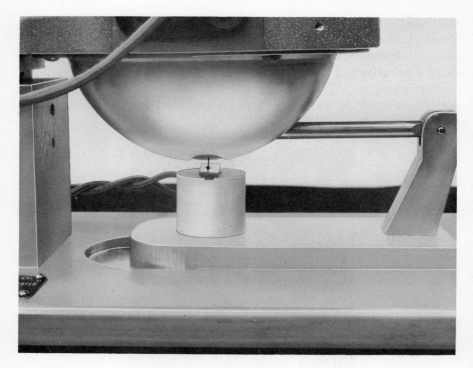

(a)

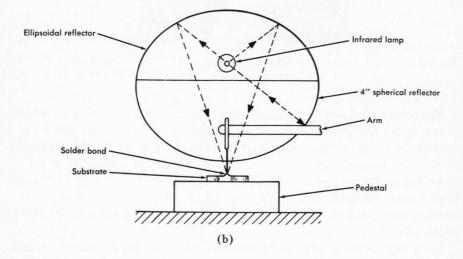

(b)

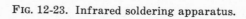

FIG. 12-23. Infrared soldering apparatus.

for the addition of solder when heat is applied. In these cases, a solder *preform* may be advantageous. A preform is a specially shaped quantity of solidified solder, usually flux coated, which can be pre-positioned in the joint so that it melts and flows to make the connection when the soldering temperature is reached. Where the parts to be joined are extremely small or fragile, a process known as *pretinning* is sometimes employed. In pretinning, the workpieces are coated separately with solder, assembled, and heated to effect the joint. This is particularly common in the electronics industry where the addition of solder is inconvenient at the time of heating. In this application, the solder coating is used as a protective layer to minimize surface oxidation during storage.

Soldering irons and baths can be made more effective by the addition of ultrasonic energy. This energy, acting in the molten solder pool or bath, is helpful in oxide or other contaminant breakdown and removal from surfaces to be soldered. Such tools can be efficient enough that the use of fluxes becomes unnecessary.

Soldering to Thin Films

Soldering is the most commonly used process in joining leads to thin films. This is a result of its advantages of bonding to the widest range of materials with a variety of tools, proved bond reliability, and low cost. To utilize these advantages, however, it is important to recognize the special requirements necessary to effect a strong and conductive solder bond on thin films.

With the increased use of microminiature solid-state electrical devices in electronic assemblies, connecting points of smaller size and mass are required. Beam-lead integrated circuits, for instance, have many connecting leads on the order of 7 mils long, 2 mils wide, and 0.5 mil thick. Since the strength of the bond between the solder material and the base metals is much lower than the strength of the base metals alone, it has been common practice in soldering wires to terminals to increase the solder contact area (anchor area) so that the overall bond strength is increased. Bond configurations, shown in Fig. 12-24, have been developed to augment the small bond area to give an increased strength to the connection. With the advent of miniaturization, connection points are spaced too closely to permit an increased anchorage area, and the overall bond strength is lower. This is partially offset by the smaller mass of the joined parts, which limits the stressing imposed on the bond.

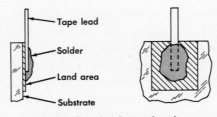

(a) Surface lead termination

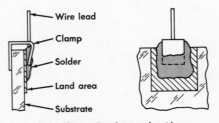

(b) Clamp lead termination

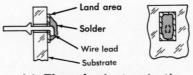

(c) Through-wire termination

FIG. 12-24. Three types of external lead configurations used in soldering to thin film land areas.

An additional problem with device bonding on films is the difficulty in using conventional equipment on ultrasmall connector points. Conventional soldering tools present difficulties in localizing the soldering heat where desired without unsoldering adjacent bonds. These problems have given impetus to such techniques as the use of a conventional parallel-gap welder for a heat source to make bonds on pretinned small leads as in Fig. 12-25. Parallel-gap welders have also been used for experimental beam-lead device soldering on pretinned contact areas [26].

The temperature required to make solder contacts on thin films with conventional tin-lead solders can cause damage such as loss of film adhesion, warpage of the substrate, or microcracking [27]. The use of a lower melting solder can reduce these effects.

0.050

FIG. 12-25. Small pretinned leads solder bonded with a parallel-gap welder.

The use of soldering techniques on tantalum thin films has made it desirable or even necessary to use a multilayer thin film to present a solderable film surface. In tantalum film circuits, for instance, copper is used (over a Nichrome® layer) as a high conductivity bonding land area. Since copper oxidizes readily during storage, however, such thin films cannot be reliably soldered with rosin flux solders after storage, and it is necessary to use a protective thin film layer of gold or palladium for oxidation reduction. These gold or palladium surfaces are then solderable. Tin-containing solders have been used extensively on gold-plated parts and on gold thin films, but many unrepairable solder joint failures were observed. An extensive investigation of the gold-tin-lead system [28] revealed such defects as joint roughness, porosity, and brittleness, particularly where the amount of solder present in a connection was relatively small on a heavy gold film. Metallurgical examination showed the presence of long, acicular (needle-like) crystals in the solder, Fig. 12-26. These were shown to be a gold-tin compound formed during solder solidification after the tin-lead solder had dissolved approximately 10- to 15-percent gold. Such dissolution and compound formation resulted in a very brittle joint structure which was prone to easy failure under slight bending stresses or vibration. This com-

Fig. 12-26.   Acicular gold-tin crystals in solder. (After F. G. Foster, "Embrittlement of Solder by Gold from Plated Surfaces," in *Papers on Soldering* (STP 319), 1963; courtesy American Society for Testing and
Materials.)

pound formation is particularly troublesome for wave soldering since
the entire solder bath becomes contaminated from the gold parts.

Recent observations show that whiskers may be found near soldered
connections on thin film resistors, as shown in Fig. 12-27. On test
samples that are soldered with tin-lead, tin-indium, and lead-indium
alloys, this effect is noted after aging at high power levels. The
whiskers consist of tin, lead, or alloys of tin and lead which migrate
over nickel, Nichrome®, or chromium films, but not over tantalum,
titanium, or copper films. One hypothesis [29] is that at high operating temperatures electromigration of molten tin or lead causes penetration of the nickel or chromium oxide, interaction with the underlying film, and production of whiskers. It has been observed that,
if the copper film is removed during the soldering of contacts and the
Nichrome® layer beneath is exposed, whiskers are readily formed.
Thus, use of a thicker copper layer to prevent depletion and substitution of titanium for Nichrome® alleviate this problem.

12.7   SOLID-PHASE BONDING

Solid-phase bonding [30] implies the absence of any gross melting
during bond formation. The concepts of temperature and melting
usually are associated with a large aggregate of atoms. These terms
become less clear in the limiting case of monolayer atomic arrange-

FIG. 12-27. Tin-indium whiskers.

ments. Thus, the terms "no melting" and "below the melting point" are used to describe observable amounts of material, that is, many hundreds of atom layers. The temperature reached on surfaces in some solid-phase bonding processes may be considerably higher than macro observations indicate.

Solid-phase bonding is achieved by the formation of regions of intimate metal-to-metal contact by the application of energy. The dispersion of interface contamination through plastic deformation and interface tangential movement constitutes a major factor in solid-phase bonding. Plastic behavior of metals is determined by both material properties (such as crystal structure, purity, and microstructure) and external parameters such as temperature and pressure. In general, the following all contribute to increased ductility and hence aid solid-phase bonding: face-centered cubic crystal system, high purity metals, large strain-free grains, and elevated temperatures and pressures.

Attempts have been made, with mixed success, to predict the solid-phase bonding characteristics of various metal combinations using adhesion and friction data. Ductile materials, as a general rule, are easier to bond than hard materials. The more noble metals are particularly suited to solid-phase bonding because of their more nearly oxide-free surfaces. Adhesion data indicate that metal pairs of different crystal systems, especially a cubic and a hexagonal combination, exhibit low coefficients of adhesion, and hence indicate the existence of barriers to bond formation.

Figure 12-28 presents a correlation between the coefficient of adhesion (ratio of the breaking load to the force applied during a twist-friction test) and hardness for metals of various crystal systems. The soft fcc and tetragonal metals exhibit the highest coefficient of adhesion. Figure 12-29 lists the coefficient of adhesion for five metal combinations. The coefficients of adhesion for the unlike metal combinations are smaller than the coefficients of adhesion obtained for the metals to themselves. Additional data [31] suggest that miscible metals exhibit adhesion, while immiscible metal systems show little adhesion. Results of the data are:

Adhesion	No Adhesion
Fe-Al	Cu-Mo
Cu-Ag	Ag-Mo
Ni-Cu	Ag-Fe
Ni-Mo	Ag-Ni

Thermocompression Bonding

Thermocompression bonding [32] is the solid-phase bonding process most commonly used for the connections of integrated circuits and transistors. The bond is formed by inducing a suitable amount of material flow in one or both members of the bond by the application of pressure and heat so that adhesion takes place in the absence of a liquid phase. The development of this process was inspired by the important problem of attaching metal conductors directly to semiconductor surfaces and metallized substrates.

Fundamentals of Thermocompression Bonding. In forming thermocompression bonds, several barriers have to be overcome to some degree. The bringing together and formation of actual contact area between two mating surfaces is one of these barriers. Since even the most carefully prepared metal surface is rough by microscopic

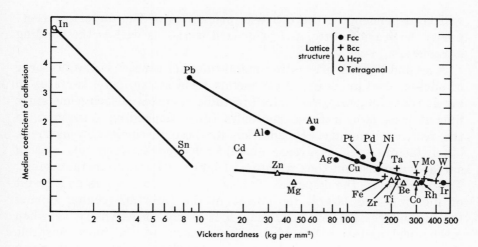

FIG. 12-28. Correlation between coefficient of adhesion and hardness for metals of various lattice structures [17].

Metal couple	Crystal structure	Coefficient of adhesion
Al-Cu	fcc to fcc	0.60
Ag-Fe	fcc to bcc	0.002
Pt-Fe	fcc to bcc	0.16
Rh-Fe	fcc to bcc	0.03
V-Fe	bcc to bcc	0.07

FIG. 12-29. Coefficients of adhesion for various metal combinations [17].

standards, two surfaces in contact touch at only a few junctions. The extent of the area in contact and its behavior under load have been the subject of considerable research [33]. It was concluded that due to strain hardening at the points of contact, these regions are difficult to deform compared with the adjacent bulk material, and complete contact across the interface is difficult to achieve. An increase in load will result in deformation and plastic flow near the junctions, with the result that other asperities will flow to further increase the contact area. As plastic flow in the regions of the asperities occurs, strain hardening is developed which increases the resistance to form more contact area. The amount of strain hardening will be dependent on the intrinsic properties of the materials (e.g., purity, lattice structure, and prior cold work) as well as the bonding parameters.

A second barrier preventing metal-to-metal contact is the presence of surface contaminants. Most metals form stable oxide surfaces in air at room temperature, and all metallic surfaces adsorb a moisture film at least two or three molecules thick, depending somewhat on the relative humidity. In addition to these intrinsic film barriers, other factors such as grease and dust lodged between two mating surfaces can prevent the formation of intimate metal-to-metal contact. To some degree, the dispersal of these surface films occurs as a result of piercing and lateral movement in the interface region during thermocompression bonding, though it is not as extensive as when mechanical agitation is added. The removal of the more unstable oxides prior to mating may also occur by disassociation or chemical reduction.

With the formation of added contact area of nascent metal regions, interdiffusion of the two materials can occur. Though there is evidence that the transfer of atoms from one material to another is not essential in forming a solid-phase bond [32], diffusion can enhance the bond properties. Exceptions to this occur when brittle intermetallic compounds such as Au_2Al ("white plague") form. The formation of this phase is dependent on time and temperature. Intermetallic compounds such as the white plague can be avoided by bonding similar metals, e.g., gold to gold, or tolerated by accepting limitations on the operational and storage temperatures of the bonds.

Thus, the formation of a thermocompression bond includes the formation of extended contact area and a limited disruption of surface films, followed by possible diffusion. The extent to which these occur is dependent upon the chemical and physical properties of the materials being joined as well as the bonding parameters.

Techniques of Thermocompression Bonding. The basic equipment, Fig. 12-30, required to form a thermocompression bond consists of a probe to press the mating surfaces together with a suitable pressure, a technique to deliver thermal energy to the bond region, and an anvil to support the workpieces during the bonding operation. Depending upon the application, heat may be applied either to the substrate or to the bonding probe, or to both (Fig. 12-31). The bond region must be maintained at the desired bonding temperature in spite of the heat sinking effects of the probe and metal conductor. This is particularly troublesome when substrate heating is employed. Substrate

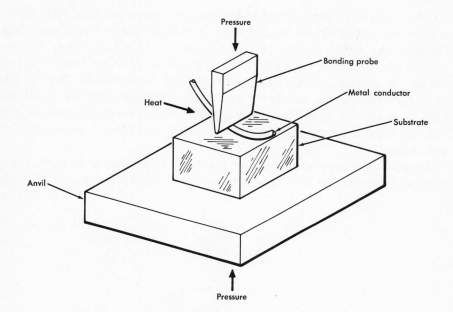

FIG. 12-30. Basic conditions and equipment required to form a thermocompression bond.

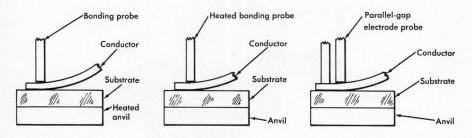

(a) Substrate heating   (b) Bonding probe heating   (c) Resistance heating

FIG. 12-31. Methods of thermocompression bonding.

heating is generally accomplished by a heating element attached to
the anvil. Thermocompression bonding by this method is widely used
where the components on the heated substrate can withstand the
required temperature.

A thermocompression bond may also be accomplished by a heated
probe which compresses the metal conductor onto an unheated sub-

strate. A process called mechanical thermal pulse bonding [34] over-
comes the problem associated with a heated probe, i.e., heat loss
through the metal conductor and substrate. The heat input is controlled by the heat capacity and application time of the bonding
probe. A mechanical thermal pulse bonder is shown in Fig. 12-32.

A combination of substrate and probe heating may be employed
and is generally most desirable since higher bonding temperatures
can be attained with a minimum amount of thermal shock in the
substrate.

Parallel-gap resistance welders may also be employed to form ther-
mocompression-type solid-phase bonds. In this design, Fig. 12-31(c),

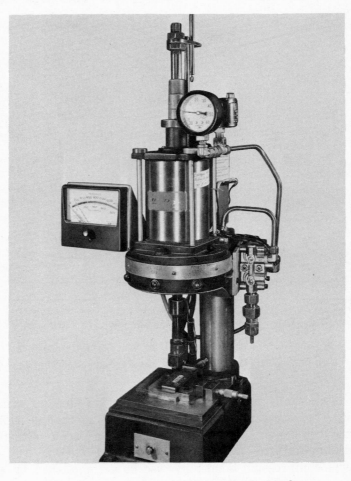

FIG. 12-32. A mechanical thermal pulse bonder.

two electrodes are placed beside each other and separated by an insulator. In some cases, the insulator also serves to bond the two electrodes together. Air may also be used as the insulating medium and, in this case, the electrodes can be made to move relative to each other and thus comply to irregular surfaces. The bond is formed by joule heating resulting from the passage of current from one electrode through the bond region to the other electrode. In this manner, some of the current flows in parallel paths across the conductor-film interface. Since the amount of heat developed during the process depends on the electrical properties of a number of interfaces, the physical and chemical properties of the surfaces involved must be controlled for reproducibility. For example, the probe surface condition, weld geometry, material composition, and contamination have to be fixed within limits.

In all of the above techniques, the bonding probe must be physically and chemically stable during the bonding operation to insure a consistent transmission of energy to the bond region. The probe must be a material that does not adhere to the metal conductor during bonding so that removal of the probe does not unduly stress the bond region. Refractory metals such as molybdenum and tungsten as well as alumina and beryllia ceramics are acceptable materials for this purpose. When a heated probe is used, as in mechanical thermal pulse techniques, it must have sufficient heat capacity to transfer large quantities of heat to the bond region.

Thermocompression Bonding to Thin Films. Experience throughout the electronics industry has indicated that thermocompression bonding can produce a reliable mechanical and electrical connection. The range of lead materials used for thermocompression bonding is limited because some metals require more deformation and lateral movement at the interface than are available with this process. Fortunately, the materials compatible with thermocompression bonding (e.g., gold and gold-plated copper) possess the desirable electrical and mechanical properties. Metals, such as copper, that form relatively unstable oxides also become suitable for thermocompression bonding if surface preparations and bonding atmospheres are employed to remove the oxide films.

The size as well as the material of the metal conductor affect the bond properties. The force per unit area required to break a thermocompression bond decreases as the conductor diameter or thickness increases, as shown in Fig. 12-33. Thermal losses along the con-

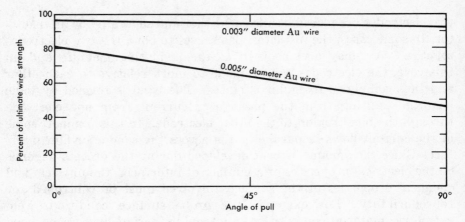

F IG. 12-33. Effect of metal conductor size and angle of pull testing on
bond strength [15].

ductor, as well as the increased forces required to adequately deform
larger leads, contribute to this loss in strength.

The reliability of a thermocompression bond is dependent on maintaining the integrity of the thin film and substrate during bonding.
This requisite is generally realized since a thermocompression bond
is formed at temperatures low enough that the substrate and film are
not damaged.

Depending upon the application, a horizontal or vertical bond
geometry may be used in attaching a metal conductor to a thin film.
In the horizontal method (Fig. 12-30) , the substrate is positioned
on an anvil directly under the bonding probe, and the metal conductor
(ribbon or wire) is horizontally positioned on the bonding surface.
The probe is lowered onto the lead with a predetermined force, deforming it in the region of the probe. The method of heating may
vary, as previously indicated. In the vertical method, Fig. 12-34, a
wire is fed vertically through a capillary hole in the bonding probe.
In the case of bonding gold, the wire is severed by a small hydrogen
flame which forms a sphere (due to surface tension) greater in
diameter than the metal conductor. The probe applies the force to
deform the ball on the surface to be bonded. After the probe is
raised, the hydrogen flame severs the wire, producing another ball
for the next bonding cycle. The vertical bond is mechanically superior
to the horizontal configuration because there is little possibility of
peel-type failure. In the vertical bond configuration, a constant

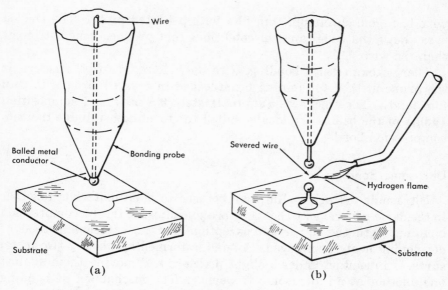

Fig. 12-34. Thermocompression bonding using a vertical geometry.

strength is observed at most angles of pull since the entire bond area is always being acted upon when stressed.

The three parameters for thermocompression bonding are force, temperature, and time; fortunately, reliable bonds can be produced over a wide range of these parameters. For example, Fig. 12-35 shows the conditions of temperature and pressure for bonding gold

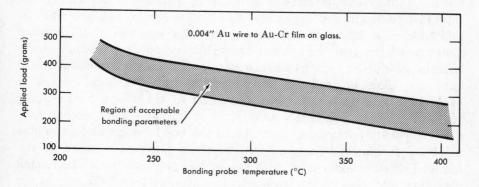

Fig. 12-35. Range of temperature and pressure (for a given application time) in which reliable horizontal thermocompression bonds are produced [35].

wire to a gold-chromium thin film with a heated probe. The shaded area shows the wide range of conditions that produce bonds stronger than the wire.

Thermocompression bonding, with its wide range of parameters, is the prime method for joining beam-leaded integrated circuits to thin film lands. The thickness (approximately 0.5 mil) and composition (gold) of the beams are ideally suited for forming a reliable thermocompression bond.

Ultrasonic Bonding

Ultrasonic vibrations [36] can be used to bond materials together in the same general way that they are joined using thermocompression bonding, with the important exception that no external method of developing heat is employed. A bond is formed by placing two metal surfaces in contact under a slight pressure and moving them against one another at an ultrasonic frequency. The mechanism of bonding is not fully understood, but it seems closely akin to friction seizure, where bonding occurs between two metal surfaces moving in contact in the absence of a lubricant. As a result of ultrasonically induced mechanical vibrations, together with the resulting frictional heat (both internal and interfacial), plastic flow, which aids in obtaining increased contact area, occurs in the bond region. This material flow permits obtaining a satisfactory contact area with lower mechanical loads and temperatures.

Fundamentals of Ultrasonic Bonding. Ultrasonic bonding is believed to lower the barriers to bonding by plastically deforming the interface region of the workpieces to disperse the oxide films, adhered moisture, and other contaminants. In this manner, the irregular surfaces of the two workpieces are made to conform to each other, thereby producing regions of intimate contact between nascent metal surfaces. This desired plasticity developed in the bond region is associated with several factors. The first of these is a temperature rise at the bonding interface as a result of friction. This temperature rise temporarily increases ductility in the bond region and promotes plastic deformation. Secondly, the material flows as a result of the induced stresses associated with the ultrasonic vibrations. Increased plasticity may also derive from selective attenuation of the ultrasonic vibration at imperfections such as dislocations [37]. As a result of these factors, deformation occurs without the gross heating techniques employed with thermocompression bonding.

The ability of ultrasonic vibration to disperse oxides and obtain metal-to-metal contact appears to be substantial with metals such as aluminum, but somewhat limited in others. Greater abrasiveness or hardness of specific oxides such as alumina and the higher hardness ratio of metal oxide to parent metal are believed to be influencing factors in oxide dispersion during ultrasonic bonding. Also, the adhesion between the oxide and its parent metal influences the ability to obtain metal-to-metal contact. Because effects at the interface are highly material dependent, ultrasonic bonding cannot be used successfully with all material combinations.

Techniques of Ultrasonic Bonding. Two types of ultrasonic bonding systems are shown in Fig. 12-36. The equipment for either system includes an electronic oscillator, a transducer, a coupling assembly (including a velocity transformer), a bonding tip, a support for the workpieces, and a method for applying a clamping force to the workpieces.

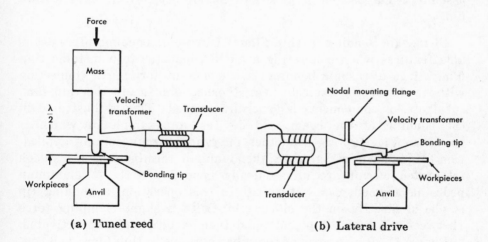

(a) Tuned reed   (b) Lateral drive

FIG. 12-36. Ultrasonic bonding systems [38].

A magnetostrictive- or piezoelectric-type transducer is employed to convert electrical energy into vibratory motion. This vibratory motion travels along the coupler system, a portion of which is tapered to serve as the velocity transformer. The velocity transformer amplifies the oscillatory motion and delivers it to the bonding tip. The stepped, Fourier, and exponential transformers are among those generally employed [39, 40]. The design of a velocity transformer for a specific bonding application involves its shape, magnification, and physical properties.

The bonding tip transmits the amplified motion to the workpieces under a clamping force. The tip generally moves in a plane parallel to the bond surface, though other modes (e.g., perpendicular to the surface) are also employed. In the case of bonding leads to thin films, the substrate is clamped to the anvil, while the lead is clamped between the substrate and bonding tip. In the tuned-reed system, Fig. 12-36(a), the velocity transformer (or horn) is connected to the workpieces by a resonant flexing bar designed to provide a vibration antinode at the end of the bonding tip. This system is generally employed for bonding larger workpieces where the clamping force on the material is usually applied directly through the bonding tip. In the lateral-drive system, Fig. 12-36(b), a nonresonant bonding tip is attached directly to an antinode of the velocity transformer and the clamping force is applied by the transducer-coupling assembly. This system is generally employed to attach leads to thin films.

Ultrasonic Bonding to Thin Films. Ultrasonic bonding offers desirable features when connecting a metal conductor to a metallic thin film. Since ultrasonic bonding is a means of forming a connection without the use of external heat, thermal damage to the thin film, substrate, or components is generally avoided. This is illustrated in the bonding of the beam lead devices, an example of which is shown in Fig. 12-37. In another integrated circuit joining application [41], shown in Fig. 12-38, the aluminum terminations of a silicon integrated circuit are ultrasonically bonded directly to aluminum pedestals formed on a substrate. In this operation, a bonding tip is placed directly on the silicon chip with a slight clamping force (between 0.5 and 1.0 pound) and then is ultrasonically activated.

Figure 12-39 lists some of the substrates, metal thin films, and conductors which have been bonded with the use of pressure and ultrasonic energy. This wide range of materials that are compatible with ultrasonic bonding offers flexibility in choosing appropriate land materials for thin film circuits.

Hot Work Ultrasonic Bonding. During conventional ultrasonic bonding, seizure and compressive deformation of the metal conductor occur. As the conductor undergoes increasing strain, the force required to produce additional deformation increases as a result of

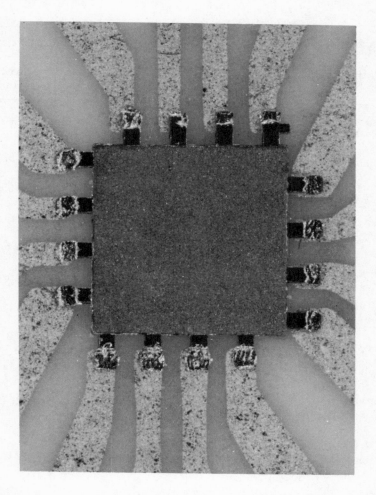

FIG. 12-37. Ultrasonically bonded beam lead device. All the leads were bonded simultaneously.

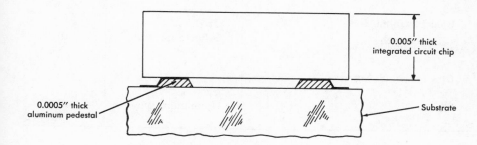

0.005″ thick
integrated circuit chip

0.0005″ thick
aluminum pedestal

Substrate

FIG. 12-38. Cross section through bonded chip and substrate.

Film	Conductor	
	Material	Thickness (mil)
On Glass Substrate		
Aluminum	Aluminum wire	2 to 10
	Gold wire	3
Nickel	Aluminum wire	2 to 20
	Gold wire	2 to 3
Copper	Aluminum wire	2 to 10
Gold	Aluminum wire	2 to 10
	Gold wire	3
Tantalum	Aluminum wire	2 to 20
Chromel®	Aluminum wire	2 to 10
	Gold wire	3
Nichrome®	Aluminum wire	2 to 20
On Alumina Substrate		
Molybdenum	Aluminum ribbon	3 to 5
Gold-Platinum 7-microinch	Aluminum wire	10
On Ceramic Substrate		
Silver	Aluminum wire	10

FIG. 12-39. Combinations of electrical conductors, thin films, and substrates which have been ultrasonically bonded [42].

strain hardening. This strain hardening prevents the desired deformation and increases the stress levels in the substrate during bonding. The use of higher ultrasonic power to increase deformation has
been found to produce substrate cracking, especially for thick leads,
thus limiting the lead size that can be reliably bonded to a thin
film [43].

A combination of thermocompression and ultrasonic bonding techniques referred to as hot work ultrasonic bonding has been found to
dramatically decrease the deleterious effects caused by strain hardening [43]. In this type of bonding, the portion of the lead about to be
deformed is maintained in the hot work temperature range [22]
during application of the ultrasonic energy. In this temperature range,
strain hardening does not occur. This is because recovery processes
occur simultaneously with deformation by the nucleation and recrystallization of strain-free grains during bonding. Under these conditions, the slope of the plastic portion of the stress-strain curve
approaches zero, Fig. 12-40, which indicates the large amount of
deformation obtained for a given stress. The lower limit of the hot
work ultrasonic temperature range depends on such variables as
composition and prior cold work of the metal.

Hot work ultrasonic bonding reduces the stresses in the substrate
by increasing the contact area, as well as by eliminating lead strain
hardening during bonding. Since the bonded lead is in a strain-free
condition, the physical stability of the bond is enhanced. Hot work
ultrasonic bonding is capable of bonding a wide range of materials

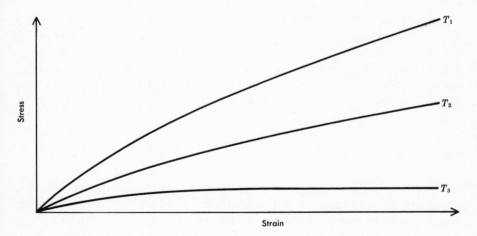

FIG. 12-40. Effect of temperature $(T_3 > T_2 > T_1)$ on the stress-strain characteristics of a metal. T_3 is in the hot working range.

and minimizing the deleterious effects of strain hardening associated with conventional ultrasonic bonding. In addition, it is capable of forming the necessary contact area at lower temperatures and pressures than thermocompression bonding, and at lower power levels than conventional ultrasonic bonding. This increases the lead size capability and reduces the probability of damage in the bond region and nearby components. Copper (0.020-inch) [44] and aluminum (0.035-inch) [43] wires have been reliably bonded with this process, resulting in bond strengths greater than the strength of the wire. A bonder with the capability of ultrasonic, thermocompression, and hot work ultrasonic bonding offers the flexibility often required in the integrated circuit field.

12.8   FUSION WELDING

Fusion welding involves melting of the base metals at the weld interface. A fusion weld consists of the two base metals joined by a resolidified interfacial structure. The weld can be divided into a resolidified zone, a heat-affected zone, and the base metals. High stresses (with possible microcracks caused by rapid resolidification) and a weakened heat-affected zone are potential weaknesses in fusion welding.

If dissimilar metals are welded, the resolidified zone, as well as some layers in the unmelted base metals, may consist of alloys of varying composition. The binary phase diagram for the metals may be used to estimate the various compositions and phases formed. If the two metals form complete solid solutions, behavior similar to single-metal welding may be expected. If the two metals form intermediate phases, the properties of the weld depend on the phases that form and their distribution in the weld zone.

Laser Welding

Laser welding employs a high-intensity monochromatic beam of coherent light energy to join materials. When this form of energy impinges on a metal surface, a quantity of heat is generated which generally causes the temperature to exceed the melting point of one or both materials being welded. The increased potential of fusion welding in the integrated circuit field is a result of the precise control of the intensity, duration, and placement of the laser beam. This control has not been offered by other fusion processes such as arc and flame welding.

An optically pumped solid-state laser is generally employed for welding. Though a large number of materials have exhibited laser action, a rod-shaped ruby single crystal is generally used for this purpose. The crystal contains about 0.05-percent chromium ions. A xenon flash lamp is employed to excite the chromium ions in the crystal to a higher energy level. Unless the exciting radiation is sufficiently intense, the only energy emitted from the ruby will be fluorescent radiation. When the exciting radiation exceeds a certain threshold, however, the excited chromium ions are induced to give up their excitation energy simultaneously. This produces radiant energy which is transmitted through the end of the ruby rod. This avalanche of radiant energy is characteristically monochromatic, coherent, collimated, and very intense.

When this beam of energy interacts with a metal surface, a portion is reflected and the remainder is absorbed with a resultant release of heat. The amount of heat generated for welding is largely dependent on the reflectivity of the material. This thermal energy generally produces fusion in the weld region. Thus, the fundamental considerations previously described for producing a solid-phase bond are not applicable in laser welding where the formation of the weld is facilitated by rapid interdiffusion in the liquid state. As a result of this mechanism, laser welding is capable of joining materials having a wide range of physical and chemical properties.

Laser Welding to Thin Films. Laser welding has had only limited success in welding to thin films on glass and glazed ceramic substrates. Such substrates are usually damaged by thermal shock or glass softening. Localized melting in the thin film portion of the weld may also result in loss of electrical continuity to the circuit. If the laser is to find a place in welding to thin films on glass or glazed ceramic substrates, further refinements in the equipment and control of the energy input will be required.

With the increased use of unglazed alumina substrates which are more stable, laser welding is showing promise. Laser welding has formed acceptable connections of 0.001-inch gold ribbon and 0.002-inch nickel ribbon to an evaporated aluminum thin film on an alumina substrate [45].

Recent studies [46] have shown the feasibility of the simultaneous laser welding of 16 gold beam leads of an integrated circuit to a gold-titanium thin film pattern on alumina, Fig. 12-41. The multiple welding operation was accomplished by use of a cylindrical lens system which generated a four-line pattern from a single laser source. All the leads formed a fusion weld with their respective thin film lands, while the adjacent thin film areas subjected to the laser beam showed no signs of having melted. This was attributed to the higher surface reflectivity of the gold-titanium thin film relative to that of the titanium-platinum layer of the beam lead.

FIG. 12-41. Beam-lead device laser welded to gold-titanium thin film.

Because of the inherent characteristics of laser welding, it has gained increasing interest in the field of integrated circuits. For example, laser welding the beam-lead integrated circuit does not require intimate physical contact of a bonding probe with the leads, whereas contact is essential in most other joining processes. Therefore, damaging stresses caused by bonding of the workpieces to the bonding probe, bond contamination, and the problem of transmitting sufficient energy from a bonding probe to the fine beam leads are eliminated with laser welding. Other advantages of laser welding such as precise control of the intensity, duration, and placement of the weld energy, welding in inaccessible areas, welding a wide range of materials, and transmitting the weld energy through transparent materials (e.g., glass substrates and windows) may also increase interest in its use as a process for making electrical interconnections in the integrated circuit field.

Electron Beam Welding

Electron beam welding employs a high-density stream of accelerated electrons to join materials. As the electrons bombard the workpieces, most of their kinetic energy is converted to thermal energy to provide the heat required for a fusion weld. The characteristics of electron beam welding to integrated circuits are similar to those of laser welding.

The electron source (gun) is generally a triode system. The electrons are emitted from a hot cathode, prefocused by the shape of a bias cup, and accelerated through a hole in an anode. The electrons are accelerated to approximately two-thirds the speed of light and maintain this velocity, in a vacuum environment, until they strike the workpieces. The electron stream is electromagnetically aligned and focused to provide a highly concentrated beam. A stigmator serves to round the beam, which is normally oval in cross section. An electromagnetic deflection system provides the means of deflecting the beam in a predetermined pattern and is adaptable to automatic control. Some focusing systems are capable of concentrating the electrons to a beam a few microns in diameter.

Electron Beam Welding to Thin Films. Electron beam welding to thin films on glass or glazed ceramic substrates is expected to present the same problems as experienced in laser welding, i.e., thermal shock and glass softening. Experimental studies [47] have indicated, however, that metal conductors can be electron-beam welded to gold thin films deposited on more stable substrates such as quartz.

Since the electron beam is limited to a vacuum environment, special precautions have to be taken with materials that degas or have high vapor pressures. Pulse techniques [48] are commonly used to overcome this difficulty. Though the initial cost and required vacuum environment of the electron beam welder may delay its use, certain advantages (no requirement of intimate physical contact with the weld region, precise control of the weld energy, ability to weld in inaccessible areas, and ability to weld a wide range of materials) may warrant its future use as a welding process. The use of electron beam welding may be further justified when employed in additional capacities (separately or in conjunction) allied to the integrated circuit field such as evaporating metal and dielectric materials, and scribing discrete lines for thin film patterns.

Parallel-Gap Resistance Welding

The parallel-gap resistance welder, discussed under Thermocompression Bonding, is also capable of fusion welding. With this type of welding, the joining of bulk metal conductors to thin films on glass substrates has produced limited success. For example, from a mechanical testing evaluation, 0.003-inch by 0.010-inch gold ribbons and 0.004-inch diameter gold wires produced acceptable results [49]. In contrast, metallographic examination of similar welds revealed cracking in the borosilicate glass substrates. Indications of forming some acceptable welds with glass substrates were also reported by the same investigator. As in the other fusion welding processes, the use of the more stable alumina substrates is expected to improve weld quality.

As a fusion process, parallel-gap resistance welding is generally limited by the properties of the thin film and substrate. When employed to make thermocompression bonds and solder connections, its short bond time and efficient control of bond energy lead to numerous integrated circuit applications.

12.9 CONNECTION EVALUATION

The field of connection evaluation is wide and complex; a connection that is entirely satisfactory or stable in a given situation may fail in short order under slightly different conditions. At the time of manufacture, the exact operating conditions each unit will experience in service are usually unknown, and the expected time to failure and the identification of the failure mode are virtually impossible to postulate. The most error-proof method of establishing these facts is to monitor all units throughout the lifetime of each and examine each failure. This may be impossible when large numbers of a given unit are involved, and economically unfeasible even when small quantities are to be employed. In either case, the time delay is too great to significantly influence manufacture.

Testing is utilized to evaluate the uniformity of product and the suitability of product for expected service environmental conditions. Any individual test can be considered on the basis of whether it is destructive or nondestructive to the product tested. Obviously, destructive testing may be employed only on a sampling basis; nondestructive testing can be used in a similar manner or to monitor the entire production.

This section describes several destructive tests employed to evaluate thin film bond or weld quality, some potential nondestructive tests which may have application to bonding or welding process characterization, and a few environmental testing techniques used to simulate service conditions.

Destructive Testing

Most of the destructive tests applied to thin film lead attachment may be classified as either mechanical or metallographic. Mechanical testing includes those test techniques which apply a stress to the joint interface in order to evaluate bond or weld quality. Tensile, centrifuge, torsional, and fatigue tests are all mechanical in nature. Although the joint interface is of major interest in mechanical testing, in all of the above tests the entire lead-connection-film-substrate *system* must be considered. It would be erroneous to evaluate, for instance, bond or weld quality from data on samples all of which failed by lead wire breakage. Of what use is a "perfectly" joined interface if the strength of the lead wire has been decreased by a reduction in cross-sectional area due to joining deformation, or if the thin film adhesion has been destroyed as a result of joining? Destructive metallographic testing is applied to the lead-connection-film-substrate system to permit examination by light or electron microscopy, or electron microprobe analysis.

Tensile Testing. The tensile test is probably the most widely used test in connection evaluation because of its speed and simplicity; in many cases, it is the only test employed. The test consists of simply stressing the joint interface to failure under the action of some mechanical tensile force [50]. The load may be applied by hand, by a spring gauge, or by a dead-weight system. Testing equipment of a more sophisticated nature, employing strain gauge transducers, is shown in Fig. 12-42. To insure meaningful data, the angle of pull (defined as the angle between the substrate and wire axis) and loading rate have to be controlled.

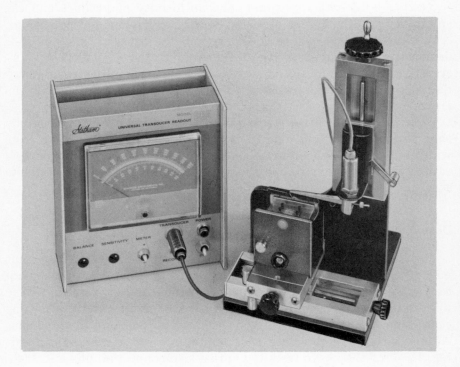

FIG. 12-42. Microtensile tester.

Since the loading force is usually applied through the lead wire, the wire also undergoes tensile testing along with the joint interface. The stressing at the interface differs, depending upon the particular type of joint (horizontal or vertical) and the testing angle. Thus, as joining techniques have been improved, various test angles have been used, with 0°, 45°, and 90° being the most widely employed. For horizontal-type joints, testing at 0° results in almost pure shear forces acting on the joint interface; as the angle is increased, a tensile-peel pull is generated. At angles other than 0° for these connections, the bending properties of the lead must be considered, for the lead may be so bend-resistant that it is stressed appreciably even before loading has taken place. While connection "strengths" are generally reported as load at failure, it is obvious that a variation in results need not entirely be due to a difference in the true bond or weld strength, defined as load per unit joined area.

Centrifuge Testing. Centrifugal action is employed in tensile testing where the size of the leads and other factors make it impractical to stress the joint area by direct contact. Centrifuges operating at 10 to 30 × 10³ G's are used in the semiconductor industry for fine wire bonds, and instruments capable of up to 200 × 10³ G's have been used for special testing. Various loadings can be impressed on the connections, depending upon the details of the mounting in the centrifuge. Although centrifuge testing is employed routinely in semiconductor manufacture (probably due to military requirements), it usually suffers the disadvantage of being performed unseen.

Fatigue Testing. The study of fatigue as a potential failure mechanism depends on the fact that metallic structures are often observed to fail, if placed in a cyclic type of stressing, at strengths lower than the uncycled tensile strength limits. One form of a fatigue tester of leads is shown in Fig. 12-43, where the sample substrate is rotated through an angle (variable from 0° to 180°) at the same time a load is placed on the lead. The number of cycles to failure is taken as a measure of the fatigue capability of the lead-connection combination. An important variable in fatigue testing is the radius about which the lead is bent during the test; the smaller the radius, the lower the fatigue limit.

Metallographic Testing. Metallographic examination of connection areas by both light and electron microscopy can be applied on a broad scale for bond or weld interface analysis. With this technique, visual evidence of the connection quality can be obtained. Such effects as completeness of point interface, presence of voids and deleterious interfacial layers, extent of fusion and recrystallization, and damage to lead-film areas can all be found directly.

FIG. 12-43. Lead fatigue tester.

Standard metallographic practice employs cross sectioning of the lead-connection-film-substrate areas. Sectioning techniques permit not only light microscope examination, but also electron microscope and microprobe analyses. The procedures necessary to obtain acceptable micrographs, however, require time and specialized training. One technique, which obviates this problem, that has been used to study the film-connection area directly is lead removal by electrolytic or chemical etching.

The lead etching technique was developed [27] specifically to study aluminum bonded to tantalum films, although the procedure is not limited to these materials. Etching techniques can be found which dissolve the lead without damage to the film. The entire interface

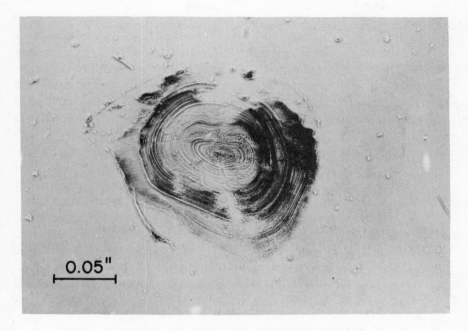

FIG. 12-44. Ultrasonic bond interface revealed by lead-etching technique.

area then becomes available for microscopic examination, as shown in Fig. 12-44.

Nondestructive Testing

The importance of nondestructive test methods has increased as device assemblies have become miniaturized [51]. While the number of interconnections has been reduced drastically by the use of integrated circuits, the potential effect of a single interconnection failure is vastly augmented. Particularly with complex systems where destructive testing techniques cannot be employed for economic reasons, nondestructive testing of connections can be the only way of evaluation before service application.

The several nondestructive test methods discussed briefly include visual examination, electrical resistance and noise, infrared (thermographic), ultrasonic, and radiographic. Although all have been employed in testing joints between large metal parts, the application of these methods to small bond or weld interface areas or to connections made on thin films has been severely limited by the present state-of-the-art.

Visual Testing. Visual examination is perhaps the oldest and most used nondestructive testing method. When used in a nondestructive manner, such techniques are essentially limited to surface examinations. Either the examination is carried out unaided or by image magnification with a low-powered microscope. In either case, the amount of illumination reflected from the surface to the eye must be adequate to allow sufficient definition of defects. Dye penetrant techniques are used to better delineate defects.

Direct microscopic examination of connections made on transparent substrates is possible [27] with a commercially available optical accessory, the Nomarski Interference Contrast microscope attachment. This unit permits definition of height variations down to 35 A on relatively flat samples at the usual levels of linear magnification. By sighting through the back of the substrate onto the connection area, disturbances in the film due to the joining operation can be readily detected. Figure 12-45 shows an aluminum wire ultrasonically bonded to a tantalum thin film as seen through the underside

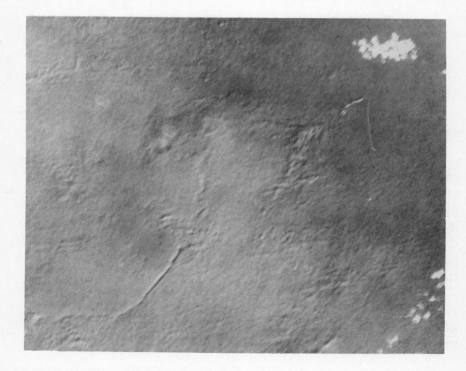

FIG. 12-45. Film disturbances in an ultrasonic bond as revealed by interference contrast technique (magnification 100 ×).

of the glass substrate. The aluminum can be seen in the areas where the film has been removed (white spots). The other visible features indicate film and substrate roughness resulting from bonding.

Electrical Resistance Testing. Resistance measurements of connections have been proposed as a nondestructive test technique, probably due to their speed and closeness to expected service conditions. The electrical resistance of the connection is generally measured with a Wheatstone or Kelvin double bridge, but the measurements are difficult to interpret. Although there are some bases for correlation of interfacial resistance to connection physical properties, the correlation has been questioned experimentally [52]. A somewhat different technique of measuring transient connection resistance may have better application, however.

Electrical Noise Testing. The passage of current through a metal generates spurious and undesirable electrical fluctuations called noise. Some types of noise are functions of geometry and material properties, and hence it may be possible to detect connections of different quality by noise measurements. Experimental measurements of the noise of lead connections made to tantalum thin films [53] showed little relation to connection quality, although transient noise measurements may permit more significant results.

Infrared Testing. Infrared energy detectors to measure small temperature differences on small-area surfaces have been developed recently. Units capable of 0.3-mil linear resolution and 0.5°C thermal resolution are presently available. Since a connection possesses an inherent interfacial resistance, the heating effect due to this resistance can be measured by an infrared device when current is passed across the interface. Abnormal interface conditions can be evaluated by this technique. For meaningful results, the emissivity of the infrared emitting areas must be known or controlled.

Although not strictly a test technique, at least one application has been developed to use an infrared detector to monitor a fusion welding process (parallel-gap welding) and in this way sort out poor welds [54]. A detector mounted between the two welding electrodes measures the time response of the heat generated at the interface during welding. Differences are observed in the response between good and poor (i.e., high and low strength) welds.

Ultrasonic Testing. Vibrational waves in the ultrasonic range (greater than 20,000 Hz) can be used to detect discontinuous features in solid materials. The ultrasonic energy is coupled into the material and a detector is so placed that attenuation or delay in wave transmission due to defects can be observed as discontinuities in the reception pattern. Resolution of present instruments has been reduced to 250μ, which, while sufficient for large inhomogeneities in large castings or welds, is of questionable value for detection in small bonded or welded areas. In addition, variations in body surface finish and the need to employ liquid transmission media complicate unduly the potential use of this technique. Furthermore, interpretation of instrument response is not easy for unskilled personnel.

Radiographic Testing. X-ray or gamma radiation can also be used in defect detection. Although radiographic techniques have been used with considerable success in testing of large samples, their applicability to microjoining does not appear promising. Since such radiation responds primarily to differences in body density, defects such as porosity or inclusions can be readily detected but interfaces in intimate, but not bonded or welded, contact would probably not be detected.

Environmental Testing

The response of a device to potentially degrading environments is of interest to evaluate device design and manufacturing process quality. Usually these test environments accelerate failure mechanisms so that years of normal service may be compressed into a short time span. It is also possible to initiate failure modes which are not normally observed in service. Temperature cycling, abnormally high or low humidity, high temperature aging, thermal or mechanical shock, and corrosion have all been used to promote premature failure. The tests can be conducted to failure, or the devices can be destructively or nondestructively tested at various time intervals before failure.

REFERENCES

1. Guy, A. G. *Elements of Physical Metallurgy*. Reading: Addison-Wesley Publishing Company, 1959.
2. Cahn, R. W., editor. *Physical Metallurgy*. New York: John Wiley and Sons, Inc., 1965.
3. Azaroff, L. V., and J. J. Brophy. *Electronic Processes in Materials*. New York: McGraw-Hill Book Company, Inc., 1963.
4. Manko, H. H. "Mechanism of Wetting During Solder Joint Formation," *Papers on Soldering* (STP 319). Philadelphia: American Society for Testing and Materials, 1963.
5. Gomer, R., and C. S. Smith, editors. *Structure and Properties of Solid Surfaces*. Chicago: University of Chicago Press, 1953.

6. *Symposium on Properties of Surfaces* (STP 340). Philadelphia: American Society for Testing and Materials, 1963.
7. Uhlig, H. H. "Metal Surface Phenomena," *Metal Interfaces*. Cleveland: American Society for Metals, 1952.
8. Boas, W. "The Nature of a Metal Surface," *Journal of the Australian Institute of Metals*, 4, 79-93 (1959).
9. Kubaschewski, O., and B. E. Hopkins. *Oxidation of Metals and Alloys*. London: Butterworths, 1962.
10. Germer, L. H., R. M. Stern, and A. V. MacRae. "Beginning of the Oxidation of Metal Surfaces," *Metal Surfaces*. Cleveland: American Society for Metals, 1963.
11. Pauling, L. "A Resonating Valence Bond Theory of Metals and Intermetallic Compounds," *Proceedings of the Royal Society*, 196A, 343-363 (1948).
12. Kiessling, R. "Bonding in Metals," *Metallurgical Reviews*, 2, 77-106 (1957).
13. Rostoker, W., and J. R. Dvorak. *Interpretation of Metallographic Structures*. New York: Academic Press, 1965.
14. Polushkin, E. P. *Structural Characteristics of Metals*. New York: Elsevier, 1964.
15. Conti, R. J. "Thermocompression Joining Techniques for Electronic Devices and Interconnects," *Metals Engineering Quarterly*, 6, 29-35 (1966).
16. Anderson, O. L. "Adhesion of Solids: Principles and Applications," *Bell Laboratories Record*, 35, 441-445 (1957).
17. Sikorski, M. E. "Correlations of the Coefficient of Adhesion with Various Physical and Mechanical Properties of Metals," *Journal of Basic Engineering—Transactions ASME*, 85D, 279-285 (1963).
18. Warner, R. M., and J. N. Fordemwalt. *Integrated Circuits*. New York: McGraw-Hill Book Company, Inc., 1965.
19. Mattox, D. M. "Interface Formation and the Adhesion of Deposited Thin Films." Sandia Corporation Monograph SC-R-65-852 (1965).
20. Benjamin, P., and C. Weaver. "Adhesion of Evaporated Metal Films on Glass," *Proceedings of the Royal Society*, 261A, 516-531 (1961).
21. Mattox, D. M. "Influence of Oxygen on the Adherence of Gold Films to Oxide Substrates," *J. Appl. Phys.*, 37, 3613-3615 (1966).
22. *Metals Handbook*, vol. I. Cleveland: American Society for Metals, 1961.
23. Manko, H. H. *Solders and Soldering*. New York: McGraw-Hill Book Company, Inc., 1964.
24. Powell, R. "What's New in Soldering," *Electronic Products*, 8, 28-45 (December, 1965).
25. Costello, B. J. "Soldering with Infrared Heating," *Western Electric Engineer*, 8, 11-16 (July, 1964).
26. Heightley, J. D., and P. Mallery. "Exploratory Study of Bonding Methods for Leads on 2.5 to 50 mil Centers," *Proceedings 1966 Electronic Components Conference*, 168-177 (1966).
27. Avila, A. J., and G. E. Kleinedler. "Lead Attachment to Thin Film Circuits," *Advances in Electronic Circuit Packaging*, 6, 10.1-10.23 (1965).
28. Foster, F. G. "Embrittlement of Solder by Gold from Plated Surfaces," *Papers on Soldering* (STP 319). Philadelphia: American Society for Testing and Materials, 1963.
29. Berry, R. W., G. M. Bouton, W. C. Ellis, and D. E. Engling. "Growth of Whisker Crystals and Related Morphologies by Electrotransport," *Applied Physics Letters*, 9, 263-265 (1966).

30. Milner, R. D., and G. W. Rowe. "Fundamentals of Solid-Phase Welding," *Metallurgical Reviews*, 7, 433-480 (1962).
31. Keller, D. V. "Adhesion Between Atomically Pure Metallic Surfaces, Part IV," Syracuse University Research Institute Report, Met. 1100-653SA, March, 1965.
32. Anderson, O. L., H. Christensen, and P. Andreatch. "Technique for Connecting Electrical Leads to Semiconductors," *J. Appl. Phys.*, 28, 923 (1957).
33. Bowden, F. P., and D. Tabor. *Friction and Lubrication.* London: Methuen and Company, 1956.
34. Cushman, R. H. "Mechanical Thermal Pulse Metal Joining," *7th International Electronic Circuit Packaging Symposium*, 4/12.1-4/12.15 (1966).
35. Phillips, L. S. "Microbonding Techniques," *Microelectronics and Reliability*, 5, 197-201 (1966).
36. Jones, J. B., F. J. Turbett, and J. V. Kane. "Exploratory Research on the Application of Ultrasonics to Spotwelding," Aeroprojects, Inc., Research Report 50-5 (1950).
37. Kearns, T. F. "MPEP—The Metalworking Processes and Equipment Program," *Metals Engineering Quarterly*, 4, 1-11 (1964).
38. Brown, B. *High Intensity Ultrasonics.* Princeton: D. Van Nostrand Co., Inc., 1965.
39. Carlin, B. *Ultrasonics.* New York: McGraw-Hill Book Company, Inc., 1960.
40. Eisner, E. "Design of Sonic Amplitude Transformers for High Magnification," *Journal of the Acoustical Society of America*, 35, 1367-1377 (1963).
41. Cubert, J. S., S. Markoe, T. J. Matcovich, and R. P. Moore. "Face-Down Bonding of Monolithic Integrated Circuit Logic Arrays," *Proceedings 1966 Electronic Components Conference*, pp. 156-167 (1966).
42. Peterson, J. M., H. L. McKaig, and C. F. DePrisco. "Ultrasonic Welding in Electronic Devices," *IRE International Convention Record* (Part 6), pp. 3-12 (1962).
43. Coucoulas, A. "Ultrasonic Welding of Aluminum Leads to Tantalum Thin Films," *Transactions AIME*, 236, 587-589 (1966).
44. Coucoulas, A. Private communication.
45. Rischall, H. "Laser Welding for Microelectronic Interconnections," *IEEE Trans. Comp. Parts*, CP-11 (2), 145-156 (June, 1964).
46. Epperson, J. P., and F. P. Gagliano. Private communication.
47. Meier, J. W., and F. R. Schollhammer. "Application of Electron Beam Techniques for Electronics." Paper delivered at National Electronics Conference, Chicago, October, 1962.
48. Stohr, J. H., and J. Briola. "Vacuum Welding of Metals," *Welding and Metal Fabrication*, 26, 366-370 (1958).
49. Clarke, D. J. "Welded and Bonded Connections for Microelectronics," *Microelectronics and Reliability*, 5, 73-83 (1966).
50. Wason, R. D. "Thermocompression Bond Tester," *Proceedings of the IEEE*, 53, 1736-1737 (1965).
51. McGonnagle, W. J. *Nondestructive Testing.* New York: Gordon and Breach, 1961.
52. Sawyer, H. F. "3-D Welded Module Design and Manufacturing Control Parameters," *Advances in Electronic Circuit Packaging*, 4, 325-346 (1964).
53. Cook, H. C. Private communication.
54. Crowley, A. H. "Weld Quality as Seen by IR." Paper delivered at SNT Session, ASM National Metal Congress, Chicago, November 3, 1966.

Chapter 13

Techniques and Facilities for
Large Scale Manufacture

Previous chapters have covered the fundamentals of thin film processing. This chapter discusses the procedures, facilities, and problems associated with large scale manufacture of thin film components and circuits. Special emphasis is placed on manufacturing variables such as yield, output, reliability, and cost. Manufacturing techniques and facilities are illustrated by specific examples drawn mainly from the experience of the Western Electric Company at Allentown, Pa.

13.1 SUBSTRATES

This section considers the economic factors in the use of substrate materials for the production of thin film circuits. The technical aspects were discussed in Chap. 9 and are not considered in this section.

For reasons of economy, several individual circuits are fabricated on one large substrate. The effect on cost of increasing the number of circuits on a substrate can be seen by referring to Figs. 13-1 and 13-2. For simplicity, a circuit size requiring an area of 1 square inch per circuit was assumed. Thus, the abscissa indicates total substrate size in square inches as well as circuits per substrate. In addition, it was assumed that the maximum substrate length available was 5 inches.

The curve of Fig. 13-1 shows that the cost per circuit increases when going from one to two circuits per substrate. This increase

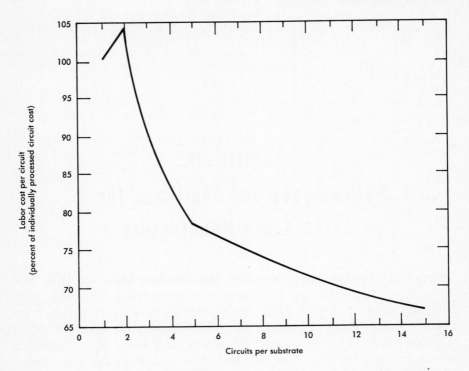

FIG. 13-1. Labor cost per circuit versus number of circuits per substrate [1]. (After Paulsen and Steiner, "Economic Substrate Size for Thin Film Circuits;" courtesy *Ceramic Age*.)

occurs because the cost of substrate separation is greater than the saving due to multiple processing. Beyond this point, the cost goes down rapidly until the number of circuits per substrate goes from 5 to 6. The change in slope at this point occurs because separation is required in two dimensions, as can be seen from Fig. 13-2. Thus, the size should be as large as practical to afford maximum economy, but small enough to be readily handled with production facilities.

The primary limitation on size is likely to be the sputtering machine. The sputtering facility at the Western Electric Company in Allentown limited the substrate dimensions to 3-3/4 by 4-1/2 inches. Once this size was determined, it was established as the standard size for thin film manufacture and was used for all subsequent codes. As new circuits were designed, they were laid out to produce the maximum number of circuits per substrate. Often, slight adjustments in circuit size could significantly increase the number of circuits per substrate.

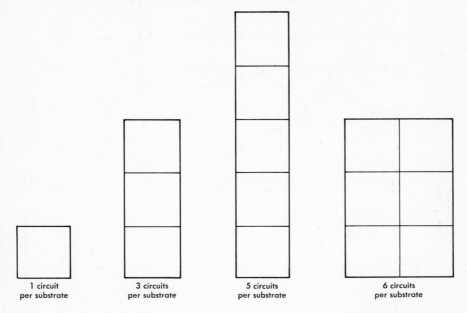

FIG. 13-2. Circuit arrangement [1]. (After Paulsen and Steiner, "Economic
Substrate Size for Thin Film Circuits;" courtesy *Ceramic Age*.)

Using the standardized size of 3-3/4 by 4-1/2 inches, relative costs
can be shown for the various substrate materials available. To
establish a relative cost scale, a base value of 1.0 has been assigned
to unglazed 99.5-percent alumina ceramic substrates. On this basis,
the relative costs are as shown in the table below.

Material	Relative Cost Scale
Soda lime glass	0.04
Corning 7059 glass	0.65
Unglazed 99.5% alumina	1.00
Unglazed 99.5% alumina with grooves for hand breaking	1.40
Glazed 95% alumina	1.66
Sapphire	(400.)

Sapphire is not commercially available in the 3-3/4 by 4-1/2 inch
sizes. It is used in a 1- by 1-inch size, and the 400 figure is the
relative cost of 16 square inches of sapphire in 1- by 1-inch square
pieces.

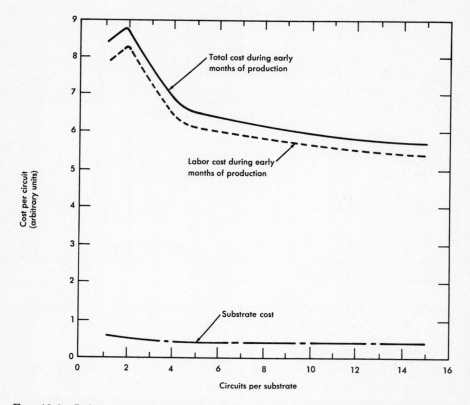

FIG. 13-3. Labor and material cost curve for early production [1]. (After Paulsen and Steiner, "Economic Substrate Size for Thin Film Circuits;"
courtesy *Ceramic Age*.)

The choice of substrate material, of course, depends on many
factors in addition to cost, as discussed in Chap. 9. All of the listed
substrate materials are useful for certain applications.

The relative effect of the substrate cost on the total circuit cost
varies with the period of manufacture. During the early production
period, the substrate contributes only a small percentage to the total
cost, as shown in Fig. 13-3. This is because labor costs in this period
are relatively high. During the "objective" cost period, however, the
percentage contribution made by the substrate takes a substantial
jump, as can be seen in Fig. 13-4. This is because the amount of
labor decreases with experience in manufacture.

Recently, prescored ceramic substrates have been developed which
can be broken by hand, thus eliminating the need for diamond
sawing. The economic significance of this development is shown in
Fig. 13-5. The two curves show the effect on the total labor cost of

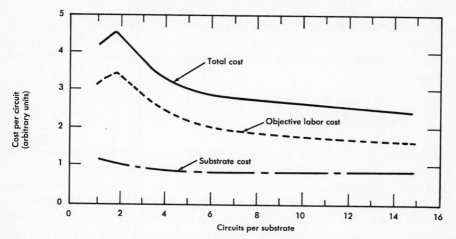

FIG. 13-4. Labor and material cost curve objective [1]. (After Paulsen and Steiner, "Economic Substrate Size for Thin Film Circuits;" courtesy *Ceramic Age.*)

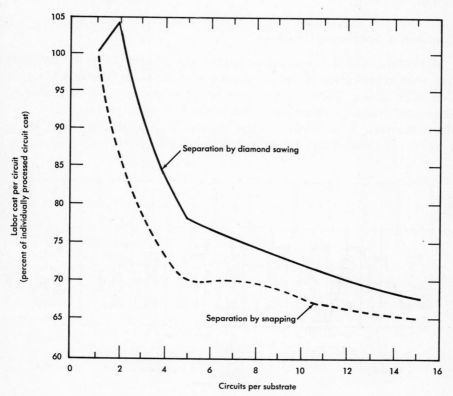

FIG. 13-5. Labor cost per circuit versus number of circuits per substrate [1]. (After Paulsen and Steiner, "Economic Substrate Size for Thin Film Circuits;" courtesy *Ceramic Age.*)

manufacturing thin film circuits when changing from diamond sawing to prescored ceramic substrates. The initial rise in the labor curve is removed when separation is performed by breaking prescored substrates.

13.2   SPUTTERING

In Chap. 4, the basic principles of sputtering were discussed. It was shown that film properties such as sheet resistance, thickness, electrical resistivity, and temperature coefficient of resistance (TCR) are controlled by the sputtering voltage, current, time, and nitrogen-doping level. It was also pointed out that sputtering can be accomplished by several different methods, including both the batch-type bell jar and the continuous (in-line) sputtering machine. These approaches are examined here from the standpoint of uniformity, output, and cost.

Continuous Sputtering Machine

A machine for the continuous sputtering of nitrided tantalum film has been in operation at the Allentown Works of the Western Electric Company since 1964. The machine consists of 11 individually exhausted vacuum chambers interconnected by high-impedance restrictions (see Fig. 13-6). This results in the development of a pressure of about 4 torr in the end chambers, progressively improving

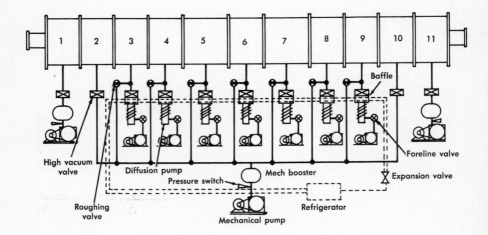

FIG. 13-6. Schematic of a continuous, open-ended sputtering machine [2].

in the chambers approaching the center of the machine until in chambers 5, 6, and 7 a vacuum of about 5×10^{-6} torr is attained.

The first four chambers are used for stepping the pressure down and outgassing the substrates by preheating. Chambers 5 and 7 are used for sputtering, with chamber 6 acting as an isolation chamber. Substrate carriers move through the machine on tracks and restrict the openings between chambers, minimizing the leakage between chambers.

Chambers. The heliarc-welded chambers are made from 304-type stainless steel. They are sealed to one another using special molded Viton gaskets. Chambers 5 and 7, the sputtering chambers, measure 12 by 12 by 30 inches and the others measure 13 by 12 by 15 inches. Each chamber has three removable sides to facilitate access for the installation and maintenance of processing fixtures. The back face of each chamber, an integral part of the machine, has a 4-inch port for high vacuum pumping and a 1-1/2 inch roughing port.

Substrate Carriers. The carrier shown in Fig. 13-7 is 7 inches long, 5 inches high, and 0.438 inch thick. It is recessed to accommodate the substrates, and grooved along the top and bottom to guide it as it slides along the track. The carriers, which were designed for minimum usable clearance in the throats between vacuum chambers, were hardened and ground to correct size to minimize wear and galling. A one-piece design was used to minimize difficulties in maintaining alignment.

Track. The carriers ride along a track consisting of two hardened-steel guide rails located in the same vertical plane. The bottom rail serves to support and guide the carrier along its bottom surface. The upper rail acts as a guide for the top surface of the carrier. The track system, made up in short sections, is supported at the interchamber walls and extends through the entire length of the machine. The loaded carriers are pushed end-to-end through the successive vacuum chambers along this track.

Main Feeder Drive. The carriers are pushed on the track alternately by two cam-operated fingers. As one finger engages a recess in the rear of one carrier, it is pushed into the machine. Meanwhile, a second finger moves back to pick up a second carrier, moving it forward until it contacts the first carrier. The forward finger then releases and moves back to pick up a third carrier, etc. Thus, one of the two fingers is always engaged in pushing to effect continuous motion through the sputtering chambers and thus aid in uniformity of deposition.

FIG. 13-7. Substrate carrier for sputtering, containing a standard 3-3/4 by 4-1/2 inch ceramic substrate.

The presently attainable range of feeder speeds is 1.7 inches/minute to 18.2 inches/minute (15 to 156 boats per hour).

Pumping Systems. Two types of pumping systems are used. One is a mechanical booster-pump system consisting of a 150 cubic feet per minute (cfm) mechanical Roots-Connerville blower backed up by an 80-cfm Beach-Russ mechanical pump. Both the blower and mechanical pump are water-cooled to allow continuous operation at relatively high pressure without overheating. This system is designed to pump against high pressure and is used to pump the chambers nearest the end of the machine (Fig. 13-6). A total of three such mechanical booster-pump systems is used. One is connected individually to each of the end chambers, and the third is connected to both chambers 2 and 10 through a 4-inch manifold. From this manifold, connections can also be made through roughing valves, when desired, into the seven middle chambers to provide them with a high-speed roughing system during the initial pumpdown.

The other type of pumping system consists of a 1500-cfm diffusion pump (4-inch) backed up by a 15-cfm mechanical pump. Each diffusion pump is provided with a water-cooled cold cap to reduce the backstreaming of oil. Further trapping is provided by an optically dense chevron baffle located between the diffusion pump and the associated vacuum chamber. The baffle is refrigerated to $-40°C$ using Freon 12 as the refrigerant. This pumping system is capable of taking the pressure down to the high vacuum range. Each of the seven middle chambers is pumped by such a system.

Cathode Design. The cathodes fabricated from rectangular tantalum sheet (Fig. 4-12) are installed in a vertical plane parallel to the incoming substrates. The overall size is 26 inches by 9-1/2 inches by 1/8 inch thick. The distance from the cathode to the substrate carrier is adjusted to 2.6 inches.

As discussed in Chap. 4, a strong field is set up around the periphery of the cathode during sputtering. This causes selective erosion along the periphery, which is much more rapid than the normal thinning of the balance of the cathode. To extend the life of the cathode, tantalum straps 1/2 inch wide by 1/8 inch thick are riveted to the periphery. With this design, the erosion occurs on the straps rather than on the periphery of the cathode sheet. As the straps wear out, they are replaced by new straps, thereby greatly extending the life of the cathode.

Power Supplies. The power supplies use 3-phase, full-wave, bridge-type solid-state rectifiers with saturable core reactors for voltage regulation. The output ratings are 750 to 7500 volts and from 0 to 750 ma of current. The maximum power rating is 5 kw.

Heater. Substrates are heated by infrared heating lamps, which are controlled by thermocouples imbedded in the track in each chamber. Chambers 1 and 2 are maintained at 250°C, while chambers 3 and 4 are maintained at 200°C. The estimated substrate temperature during sputtering is 400° to 450°C.

Operation. Substrates are loaded into carriers in a controlled-atmosphere room located at the input end of the machine. As each carrier is loaded with substrates, it is positioned in a vertical plane in an indexing mechanism which feeds it into position in line with chamber 1. Here, the main feeder drive previously described pushes the carrier into the machine. The carrier being driven pushes the entire line of carriers already in the machine. Thus, the line of carriers moves through the machine continuously.

Inside the machine, the carriers first enter the preheating chambers where the substrates and boats are outgassed by heating. They then pass through the sputtering chambers. As mentioned earlier, the machine has two sputtering chambers (numbers 5 and 7). The use of the two chambers doubles the sputtering output as follows: Each carrier is loaded with two substrates (one on each side). The sputtering chambers have their cathodes oriented vertically and located on opposite sides of the track. Thus, as a carrier passes through chamber 5, one substrate is coated with tantalum, and as it passes through chamber 7 the other substrate is coated with tantalum. After leaving the sputtering chambers, the carriers containing sputtered substrates pass through the remaining chambers through increasing steps of pressure. Then, through several transfer mechanisms, they are returned to the original loading position.

Cleaning

In order to assure good film adherence and consistent film properties, it is important to properly clean the substrates just prior to sputtering. The cleaning procedure described below is both effective and economical. Thirty substrates (3-3/4 by 4-1/2 inches) are loaded into an open-type cleaning basket, as shown in Fig. 13-8. They are then ultrasonically degreased for 5 minutes in trichloroethylene and blown dry with high-pressure filtered air. This is followed by 5 minutes in boiling, overflowing, deionized water containing approximately 0.25-percent detergent. The detergent is removed by spraying with hot, deionized water for 5 minutes and rinsing in boiling, overflowing, deionized water. Parts are not removed from the rinse until the conductivity of the efflux is 0.2 micromho per centimeter. The rinse requires a minimum of 10 minutes for glass and 20 minutes for ceramic. The substrates are finally dried in filtered nitrogen at 180°C for a minimum of 10 minutes with glass and 20 minutes with ceramic.

Cleanness Evaluation

To assure adequate substrate cleanness, a suitable evaluation method should be used. Each of three methods has been used at

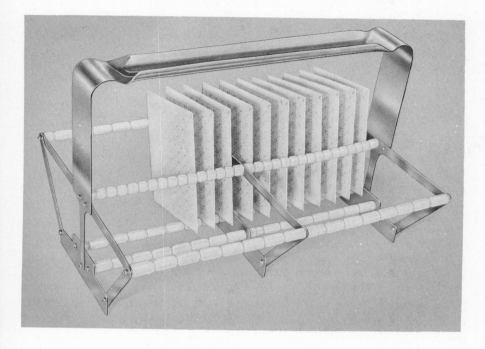

FIG. 13-8. Cleaning basket for standard 3-3/4 inch by 4-1/2 inch substrates.

various times with varying degrees of success. These methods are the water-break test, the atomizer test, and the "goniometer" angle-measuring test. All are a measure of hydrophobic contaminants.

Water-Break Test. In this method, the cleaned substrate is immersed completely in deionized water and then removed. The cleanness is determined from the degree to which the substrate surface is coated with a film of water. A highly clean substrate surface will show a continuous film of water; a contaminated surface will show droplet formation. This method is adequate for determining gross differences in cleanness but cannot be used effectively for finer differentiation often necessary in thin film work. It also relies heavily on operator interpretation.

Atomizer Test. The atomizer test may be considered a refinement of the water-break test. A fine, controlled spray of deionized water is directed onto the substrate surface. The completeness of water-film adherence is again used to measure the degree of cleanness. Ultraclean substrates will show a continuous film of water, and contamination is detected by the degree to which the water pulls away into droplets. Interpretation of the results again relies on operator judgment.

"Goniometer" Angle-Measuring Test. A more quantitative measure of cleanness may be made by measuring the wetting angle made by a water droplet on the substrate surface. A drop of deionized water is placed on top of the cleaned substrate surface. A narrow beam of light is then directed onto the top of the droplet, which is observed through an eyepiece with a line of sight parallel to the beam of light. The light source and the eyepiece, which are connected together mechanically, are then rotated in unison until the reflection of the light beam from the droplet disappears from the field of sight. The angle of rotation, observed on a protractor built into the instrument, is equal to the wetting angle and thus is a quantitative measurement which is directly related to the cleanness of the substrate.

The goniometer method is preferred since it is easy to perform and does not depend on operator judgment. One disadvantage is that its measurement is restricted to a very small area and many readings are required to determine any variation in cleanness across the substrate surface.

Environment

It does little good to clean substrates if they are to become recontaminated from contact with lint fibers, particles, vapors, and other foreign matter that will be present in the work area. For this reason, the room air should be passed through high-grade filters before it enters the room. The exhaust from the vacuum pumps should be fed to the outside to prevent oil vapors from entering the work area. In addition, the work area itself should receive a thorough daily cleaning.

Control of Film Properties

Operational control of the continuous sputtering machine to produce the desired film properties involves several interacting variables affecting the film properties. The task can be simplified by the use of the technique of "piecewise linear approximation" [3]. This technique relies on the fact that a small piece of any curve can always be approximated by a straight line. As long as this approximation is used only for small adjustments of machine variables near the operating point, the error is not great and the results are usable.

The sputtering machine can be visualized as a "black box" with four measurable input variables, namely, sputtering voltage, sputtering current, feeder speed, and nitrogen level.* These can be related to the four output variables: sheet resistance, film thickness, resistivity, and temperature coefficient of resistance.† These output variables are, of course, the film properties to be controlled. A rough guide showing the relationships between the measurable input variables and the output variables (or film properties) is shown in Fig. 13-9. This guide indicates the effect on each of the output variables of varying one of the input variables while holding the other input variables constant. Another output variable is cost per substrate. This variable is inversely related to feeder speed and is almost independent of the other input variables.

Using the rough guide, it is possible to select economical combinations of input settings. Thus, if economy and sheet resistance were the only considerations, it would be desirable to operate at high voltage and current levels and at low nitrogen levels to permit high feeder speeds. The above method of operation produces a less negative TCR. There are, of course, limitations imposed by the equipment on the maximum voltage and current. Also, too little nitrogen can result in unstable film, with the "negative dip" problem as discussed in Chap. 7.

One may be concerned at this point about the effect of the nonmeasurable input variables, such as the temperature and outgassing of the carrier and substrates, the outgassing of the chamber, etc. To compensate for variations in these nonmeasurable variables, the rough guide previously described may be used to prescribe adjustments when any of the output variables is found to be out of tolerance. This process is generally necessary at the start of each run as the machine becomes stabilized.

Bell Jar Versus In-Line Machine

Bell jar, or batch-type, sputtering methods are discussed in detail in Chap. 4. A comparison of the bell jar with the in-line machine reveals major differences in sputtering time, adjustment procedures, output rate, equilibration time, initial investment, and operating cost per circuit.

---

*Specifying the current and voltage specifies the argon pressure as well, so the argon pressure is not independently controllable.

†Recall that $\rho = R_s d$. Thus, there are really only three independent output variables.

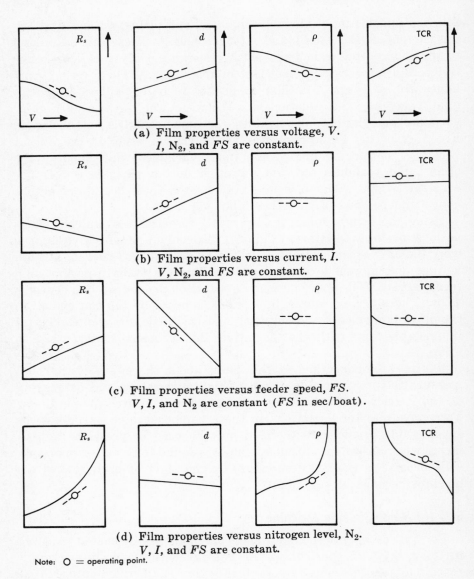

(a) Film properties versus voltage, V.
I, N_2, and FS are constant.

(b) Film properties versus current, I.
V, N_2, and FS are constant.

(c) Film properties versus feeder speed, FS.
V, I, and N_2 are constant (FS in sec/boat).

(d) Film properties versus nitrogen level, N_2.
V, I, and FS are constant.

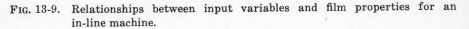

Note: O = operating point.

FIG. 13-9. Relationships between input variables and film properties for an in-line machine.

In a bell jar system, sputtering times may be as short as a few minutes, and it is not very practical to try to vary the voltage, current, or nitrogen-doping level during sputtering. In an in-line machine, however, sputtering may continue for 24 hours a day over several days. Thus, it is not only practical but often necessary to adjust voltage, current, feeder speed, and doping level during a run.

When sputtering is performed in a bell jar, the voltage, current, and doping level are usually preset based on past experience, and only the sputtering time is varied. When sputtering is performed in an in-line machine, it is common practice to run a few samples before starting a production run and to readjust the machine settings if necessary, based on the sheet resistance, thickness, and TCR of these samples. The result is that the in-line machine can produce more reproducible film than a bell jar system.

Further advantages of the in-line machine lie in cost and capacity considerations. It is conservatively estimated that the output of one continuous machine is equal to that of about 70 bell jars. Thus, by this comparison, the continuous machine is economical from the standpoint of both investment and space savings. In addition, a labor saving of approximately an order of magnitude can be achieved in operating the continuous machine compared to the equivalent in bell jars.

On the other hand, the continuous machine has some drawbacks. The time required to establish equilibrium conditions when starting is long. Moving parts cannot be lubricated, and this results in considerable friction. Parts must be held to critical tolerances to allow for the expansion and contraction of the track due to temperature changes. Monitoring the deposition is impractical because the substrates are in continuous motion. Lastly, the initial investment for one machine is high and would normally require a sizable production program for justification.

In summary, the advantages of the in-line machine based on cost, product quality, and product uniformity far outweigh the disadvantages, and for high-volume production the in-line machine is definitely to be preferred over bell jars.

13.3 EVAPORATION

The previous section considered the deposition of tantalum nitride by sputtering techniques. This section considers the deposition of thin films by evaporation techniques.

Vacuum evaporation is the most common approach for depositing conductive and dielectric layers for thin film microcircuits. These methods are readily adaptable to high-level manufacture, although significant capital expenditures are required to provide suitable facilities.

Conductive films are used in thin film devices for interconnections within a circuit and for termination pads to which leads and appliquéd components may be attached. Dielectric films can be used either singly or in conjunction with other dielectrics in manufacturing thin film capacitors. They can also be used for crossovers.

The conductive films are normally multilayered since a single material cannot provide all the characteristics required. An ideal material would have the following properties:

1. High electrical conductivity to assure the low resistance requirement of circuit interconnections.
2. High resistance to chemical attack, especially to environmental conditions encountered during processing and operation.
3. Tenacious adherence to the substrate and associated films.
4. Joinability for reliable lead connections.

For the most part, present thin film circuits are fabricated with a conductive multilayer consisting of Nichrome®, copper, and palladium. Each metal serves a purpose in attempting to meet the ideal requirements of a conductive film.

Nichrome® serves as the bonding agent of the multilayer by providing the necessary adherence to the tantalum substrate surface. It is an alloy of 80-percent nickel and 20-percent chromium in which the two metals evaporate fractionally (as described in Chap. 3). Chromium is the principal constituent of the initial deposit. This is followed by a mixture of the two materials which increases in nickel content until finally the alloy is primarily nickel with very little chromium. This fractionation is an advantage for the adherence, since chromium is more adherent than nickel. The chromium probably combines with oxygen to form an oxide which produces the tenacious bond. Although chromium is the constituent necessary for interfilm adherence, Nichrome® is used because of its availability in wire form, which is most desirable for resistance evaporation. Because it lacks ductility, pure chromium is available only in pellet form, and only boats or conical basket filaments could be used for evaporation. Titanium is now being used increasingly in place of Nichrome® since it provides at least as good adherence without allowing for the growth of tin whiskers from solder, which can occur when Nichrome® is used.

Copper is the second metal of the conventional multilayer. Copper is chosen because of its solderability, low electrical resistivity (1.673 micro-ohm centimeters), and relatively low cost.

The deposition of palladium on the copper completes the multilayer and provides a wettable, nontarnishing surface to which reliable lead connections can be made. The primary purpose of the palladium is to prevent excessive oxidation of the copper during subsequent processing. Thus, the chemical stability of the palladium is its prime asset.

To achieve high-volume batch processing at minimum cost, large vacuum systems must be used. A standard high vacuum system used for high-volume production is shown in Fig. 13-10. It consists of a 32-inch diameter stainless steel bell jar, a mechanical pump, an oil diffusion pump, a liquid nitrogen cold trap, two thermocouple gauges, and an ionization gauge. The system has automatically operated roughing, fore-line, high vacuum, and vent valves to perform the proper valving sequence and to prevent damage to the vacuum components during a possible malfunction of the machine. The pumpdown is initiated by pushing a button, and the venting sequence is initiated by pushing a second button.

One drawback of such systems is the complexity of maintenance. Since the valves are electromechanically operated, maintenance operations require both vacuum and electrical specialists, and any required repairs are more time-consuming.

To further minimize the cost of batch processing, the fixturing must be designed to allow a maximum number of substrates per batch. Figure 13-11 shows a fixture which provides acceptable uniformity of the deposited films on 3-3/4 by 4-1/2 inch substrates and minimizes the material build-up on the inside wall of the bell jar to reduce maintenance efforts. This fixture can accommodate forty-five 3-3/4 by 4-1/2 inch substrates. With a slight loss in uniformity of deposition, it can be modified to accommodate 60 substrates, provided the bell jar is tall enough to accept another row of substrates. Evaporation sources are located on a rotary turret capable of holding three to six tungsten filaments. The shielding necessary to prevent cross-contamination of the filaments is provided by a triangular divider as shown in Fig. 13-11. The substrates are held vertically to minimize pinholes in the film due to any particles dropping during loading of the bell jar.

For evaporation on a large scale, tungsten filaments of a 3-strand coil configuration find widest use for two reasons. First, a maximum surface area is provided which allows most evaporant to be held in the molten state by surface tension. Second, the coil configuration accommodates large quantities of evaporant to permit the deposition of relatively heavy coatings on large areas.

FIG. 13-10. Large automatic vacuum system for evaporation.

To assure purity of the deposited films, the tungsten filaments must be cleaned to remove the high-vapor-pressure contaminants which are generally present on the surface. This cleaning can be

FIG. 13-11. Evaporation fixture.

done either thermally or chemically. The thermal method includes a vapor degreasing followed by heating to about 1700°C in a vacuum of 5×10^{-6} torr. The temperature is chosen to be greater than any temperature experienced during the evaporation process. The chemical treatment consists of a degreasing operation followed by a 5-minute etch in a mixture of H_2O_2 and NH_4OH.

Two general approaches are used to form the conductive pattern. The first is to evaporate the multilayer conductive film over the entire substrate and to follow this by a photolithographic etching procedure, covered in detail later in this chapter. The second approach is to evaporate through a mechanical mask to give the final configuration desired.

A common mechanical mask (Fig. 13-12) uses 30-mil beryllium copper with a 1-mil electroplated nickel finish, as described in Chap. 10. The pattern image is defined by the nickel, and the copper provides mechanical rigidity. These masks have a number of disadvantages. The bulky fixture required to assure intimate contact limits

FIG. 13-12. Combination mechanical mask with holder and substrate.

the number of substrates that will fit in the bell jar, and significantly increases load time. After repeated use, contact material builds up on the mask and reduces the precision of the defined pattern. Etch-cleaning of the masks may be impractical, especially if it is made of one of the evaporant materials. Mechanical stripping of the deposited film, even with predeposited parting layers, is difficult to perform without causing mechanical damage to the edge of the nickel coating which defines the image.

The disadvantages of the combination mask may be avoided by the use of a 5-mil molybdenum mask. The back of this mask may be coated with a silicone-based adhesive to permit the mask to be attached to the substrate. This feature provides the intimate contact between the mask and substrate required to achieve well delineated patterns, even for very intricate geometries. In addition, since the materials evaporated do not include molybdenum, etching techniques may be used to remove material build-up. The flexibility of the thin molybdenum sheet is an asset in providing conformity to uneven substrate surfaces. Finally, and perhaps most important, this adherent-masking approach does not decrease the evaporation capacity since bulky fixturing is not necessary.

One serious drawback of the adherent mask is that the silicone adhesive loses effectiveness when subjected to high temperatures such as may be encountered during evaporation. Despite this drawback, the adherent molybdenum mask provides the four basic desirable properties of an evaporation mask, which are precise pattern definition, reasonable life expectancy, minimum maintenance, and a minimum of associated tooling.

To evaluate adherence of the evaporated film, a Scotch® tape test may be performed on every substrate. If this test is made immediately after the evaporation process, it may not be sufficiently reliable since film that passes the test at this point occasionally shows poor adherence after soldering operations. It has been found that the best time to make this test is after the 250°C thermal stabilization. Although the Scotch® tape pull test is not a sophisticated one, it has proven to be the best practical adherence test presently known within the industry.

Although batch processing is currently considered more economical than continuous processing, it is probable that this condition will be reversed at some future date, possibly by integration of the conductor deposition with the sputtering of the tantalum film.

13.4   PATTERN GENERATION

The first four major process steps in the manufacture of thin film circuits comprise the deposition of the tantalum nitride (Ta-N) film, the deposition of conductive film, the formation of the Ta-N pattern, and the formation of the conductive film pattern. These steps can be performed in various sequence combinations. The most common, to be discussed in further detail, are shown in Fig. 13-13.

As can be seen in Fig. 13-13, two distinct patterns must be defined to make a thin film circuit, namely, the Ta-N and the conductive film patterns. In all methods, the Ta-N pattern is generated by photolithographic techniques which will be covered in detail later in this section. The conductive pattern, on the other hand, may be generated by photolithographic methods as in A, B, and D or defined by evaporating directly to the desired pattern through a mask as in method C.

In method A, the Ta-N and conductive films are deposited over the entire substrate. This is followed by the sequential generation of the patterns for the conductive film and Ta-N by photolithographic methods. This method is readily adaptable to production and, since it requires a minimum of handling between sputtering and evapora-

Method A	Method B	Method C	Method D
1. Sputter Ta-N	1. Sputter Ta-N	1. Sputter Ta-N	1. Sputter Ta-N
2. Evaporate conductive film	2. Generate Ta-N pattern	2. Evaporate conductive film through mask	2. Evaporate conductive film
3. Generate conductive film pattern	3. Evaporate conductive film	3. Generate Ta-N pattern	3. Generate patterns of conductive film and Ta-N by double-etch techniques
4. Generate Ta-N pattern	4. Generate conductive film pattern		

FIG. 13-13. Thin film processing sequences.

tion, it reduces the probability of contamination between the resistive and conductive films. It is the method most frequently used at present.

In method B, the Ta-N is sputtered over the entire substrate. The Ta-N pattern is then generated prior to the evaporation of the conductive film. After evaporation, the conductive film pattern is generated. This method is used in preference to method A when the coating of photoresist is applied by spinning techniques. Spinning is not used in method A since undercutting would occur in generating the Ta-N pattern due to poor resist coverage. This occurs because the photoresist does not hug the corners of the previously etched conductive film. Method B is also used when resistors must be partially anodized prior to the deposition of conductive film. Method B is not recommended where extremely low noise levels are required since there is a possibility of producing contact resistance.

In method C, the Ta-N is deposited over the entire substrate. The conductive film is then evaporated as in method A, except that mechanical masks are used to provide the pattern directly. This is followed by the generation of the Ta-N pattern by photographic methods. One of the photoresist sequences is eliminated in this process, which may reduce the cost of the process.

The final method, D, is similar to method A in that Ta-N and conductive films are deposited over the entire substrate prior to further processing. In generating the subsequent patterns, however, a single photoresist application is used and the patterns of both Ta-N and conductive film are etched out from this single application. Thus, this method may reduce cost, but is limited in its application because it is difficult to avoid undercutting during the etching operations.

Photolithography

Photolithography is important in most pattern generation methods and was discussed in Chap. 10. The application of this process to manufacture is covered in detail here. The basic process sequence is:
1. Application of the photosensitive emulsion (photoresist) to the substrate.
2. A prebake to harden the resist and increase its etch resistance.
3. Exposure of the desired pattern using ultraviolet light.
4. Development of the resist image.
5. A postbake to remove any developer and further harden the resist.
6. Removal of the unwanted metal film by etching.
7. Removal of the residual resist from the substrate.

These operations are performed twice in methods A and B to define both the Ta-N and the conductive patterns, and once in method C to define the Ta-N pattern. A modification is made to effect the double-etch approach of method D (only "positive" working resists may be used in this method). The revised sequence is:
1. Application of the photoresist to the substrate.
2. A prebake to harden the resist.
3. Exposure of the Ta-N pattern using ultraviolet light.
4. Development of the Ta-N pattern.
5. A postbake to remove any developer and further harden the resist.
6. Etching of the unwanted conductive and Ta-N areas from the substrate (both through the previously formed Ta-N pattern).
7. Reexposure of the resist to ultraviolet light, but this time using the conductor pattern.
8. Development of this second image.
9. Etching of the remaining unwanted portions of conductive film which at this point are still covering the resistive elements.
10. Removal of the residual resist from the substrate.

Although this latter sequence reduces cost by using a single application of resist to define two patterns, undercutting is inherent with the process and this limits its application.

Negative Versus Positive Resists

Two general types of photoresist are commercially available. One classification is the "negative" working type marketed by the Eastman Kodak Company as KMER (Kodak Metal Etch Resist), KPR (Kodak Photo Resist), and KTFR (Kodak Thin Film Resist). These resists are polymers which become crosslinked when exposed to ultraviolet light, making them highly resistant to chemical attack.

The Azoplate Company manufactures a series of "positive" working resists, designated as AZ-17, AZ-340, and AZ-1350, that provide the same purpose as the Kodak products but function differently (see Chap. 10). These resists are organic ester solutions that change chemically upon exposure to light to become soluble in an aqueous alkaline solution. The unexposed material remains resistant to acid attack.

Although the results are comparable, the Azoplate resists are used for thin film production since they are more easily removed after processing. AZ-340 is presently used for the majority of thin film applications. It possesses a higher viscosity and higher solids content than the other positive resists. These properties result in improved acid resistance at the expense of some loss in resolution.

Resist Coating Methods

The application of a uniform coating of photoresist to a substrate may be accomplished by spinning, dipping, or spraying techniques. Figure 13-14 depicts an installation in which photoresist is applied by spinning onto glass substrates. A substrate is placed on the turntable, and a premeasured amount of resist is automatically dispensed (through a filter) onto the substrate from a reservoir. The spinning is initially done at a slow rate to assure complete coverage of the substrate. This is followed by fast spinning for a longer duration. The fast spinning minimizes the build-up of a heavy layer of resist along the periphery of the substrate and produces a more uniform coating. The equipment is enclosed to minimize environmental contamination.

For ceramic substrates with holes, the spinning process is not suitable since the holes prevent uniform coverage. Hand dipping is

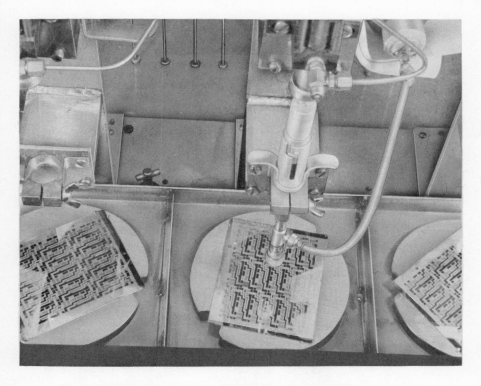

FIG. 13-14. Machine for spinning photoresist onto glass substrates.

possible but is expensive because of the labor required. Spray techniques give good results and can be inexpensive.

Machine spraying (performed inside a laminar flow station) is the preferred method, and is illustrated in Fig. 13-15. The labor involved is reduced to the loading and unloading of the substrates and the operation of a single switch. For the machine illustrated, each cycle coats 15 standard-size substrates (3-3/4 by 4-1/2 inches) uniformly and reproducibly.

Exposure

After the substrates are coated with photoresist, they are baked and then exposed to ultraviolet light through a photomask. Carbon-arc light sources have long been used for this purpose. The printing is accomplished by positioning the substrate onto a fixture which holds the glass photomask. The substrate and photomask are held in intimate contact by means of a vacuum frame built into the carbon-arc printer, shown in Fig. 13-16.

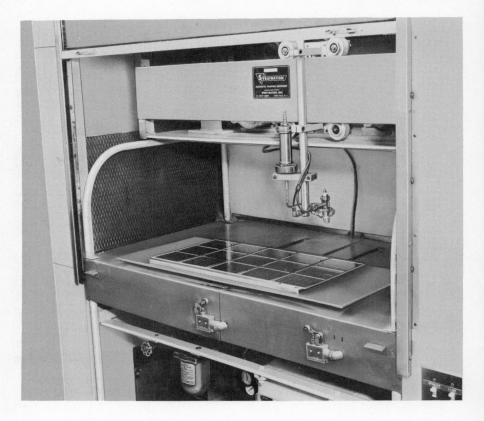

FIG. 13-15. Machine for spraying photoresist.

The carbon-arc light source and the glass photomasks are commonly used with good results throughout the industry. The use of ceramic substrates, however, introduces problems because the ceramic is not as flat as glass. Warpage in ceramic substrates can be sufficient to result in breakage because of stresses during exposure. The warpage can also cause excessive reduction in resistor line widths because there is not intimate contact between the photomask and the substrate. To overcome these difficulties, flexible film photomasks are used in place of the rigid glass photographic plates, and a mercury light source is used in place of the carbon arc. Also, the reflectors behind the light source may be darkened so that the source more closely approximates a point source.

The flexible photographic film has a Mylar® base coated with the same high-resolution emulsion used on the glass plates. Although

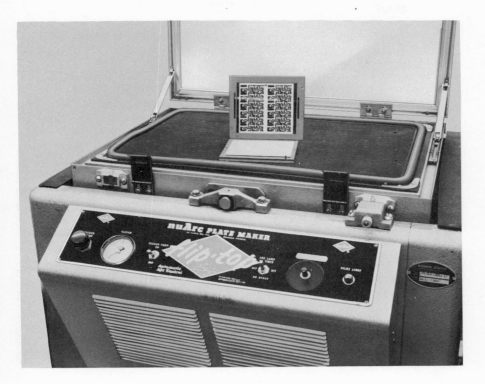

FIG. 13-16. Carbon-arc printer.

the dimensional stability of Mylar® is not as great as glass, it is within acceptable limits for practical use. The light transmission of the Mylar® is 10 to 20 percent lower than that of glass in the important wavelength range (3400 to 4500 angstroms), but the high intensity of the mercury vapor source more than compensates for this loss.

A special printing fixture, shown in Fig. 13-17, is used to achieve intimate contact between the Mylar® and the substrate. The stainless steel bases contain vacuum ducts, milled grooves, and rubber gaskets necessary to accomplish vacuum seals.

In operation, the substrate is oriented by two alignment pins in the base which fit through location holes in the substrate. The Mylar® photomask is then placed over the substrate with three hardened tool steel stops used for location. The mask is held in intimate contact with the substrate by evacuating the fixture. A mercury vapor light is then moved horizontally over the assembly to expose the image on

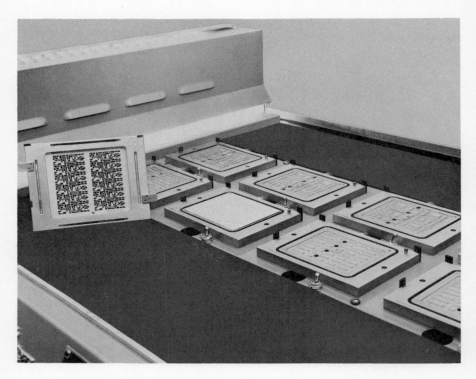

FIG. 13-17. Mercury light printer for use with Mylar® photomasks.

the substrate. Sixteen substrates may be mounted on the tabletop and exposed in one pass of the moving mercury vapor light. The time required for the light to traverse the table is 25 seconds. Since about 2-1/2 minutes are required to expose a maximum of four substrates using the carbon-arc light source, this is a significant improvement.

The use of the Mylar® essentially eliminates breakage because the substrate is not compressed between rigid bodies, as it is with the glass masks. The Mylar® follows the deviation of the substrate from flatness when the vacuum is drawn and is in intimate contact across the entire substrate surface. Thus, both of the problems previously mentioned have been made negligible.

After exposure, the Azoplate photoresist is developed in a solution of 1-percent sodium hydroxide and the film is then ready for etching. The etchant used depends on the metal to be removed. Tantalum nitride, for example, is removed by a solution of hydrofluoric and nitric acids. The conductive film of Nichrome®, copper, and palladium is first etched with a potassium iodide-iodine solution to remove the palladium and the copper. Nichrome® is difficult to remove, and a three-step process is used. The three etchants are a dilute hydrochloric acid, cupric chloride, and finally ammonium hydroxide to stop the action of the cupric chloride.

As the last step to the pattern generation sequence, the substrates are rinsed in a series of acetone baths to remove the photoresist.

Photolithography Versus Mechanical Masks

Although the major discussion in this section has been on the photolithographic method of pattern generation, mechanical masks held in intimate contact with the substrate during the film deposition process are also useful. The use of mechanical masks may be more economical than photolithography and may be preferred where the pattern accuracy allows; photolithography is more precise and allows more intricate geometries to be defined, so it is used where these factors are important.

Thermal Stabilization

After the patterns have been generated, the substrates are given a thermal treatment at 250°C for 2 to 5 hours in air. This accelerated aging is found to stabilize the resistances of Ta-N resistors (see Chap. 7 for a discussion of preaging effects). In this process, the substrates are placed in an open basket, shown in Fig. 13-18, and into a box-type oven. Air is recirculated continuously throughout the oven to ensure uniform temperature distribution.

13.5   TRIM ANODIZATION

As mentioned previously, tantalum film resistors may be trimmed to their final value by anodization. Analytical treatments of this process are found in Chaps. 5 and 7. The essential elements of an anodizing facility are: a power supply to provide the anodizing current; controls to turn off the anodizing current when the desired resistance value is reached; and a fixture to contain the electrolyte, resistor circuit, electrodes, and contacts.

Trim-Anodization Sets

The power supply to provide anodizing current and a measuring device that automatically terminates the anodization when the desired

Fig. 13-18. Open basket for aging.

resistance value is reached are generally incorporated within a single trim-anodization set. For economical manufacture, the set must be capable of trimming a large number of resistors simultaneously. In practice, this may be accomplished by constructing trim sets in modules, with each module containing a resistance-measuring circuit and a power supply.

A photograph of a 12-module trim set is shown in Fig. 13-19. In this set, the anodizing current is obtained by half-wave rectification of 60-Hz alternating current. The maximum anodizing current is limited by a resistor in series with each cathode terminal, and the maximum voltage is limited by a zener diode. The anodizing voltage, therefore, appears as half a sine wave until the limiting voltage is reached, at which time the peak of the wave begins to clip. During the interval

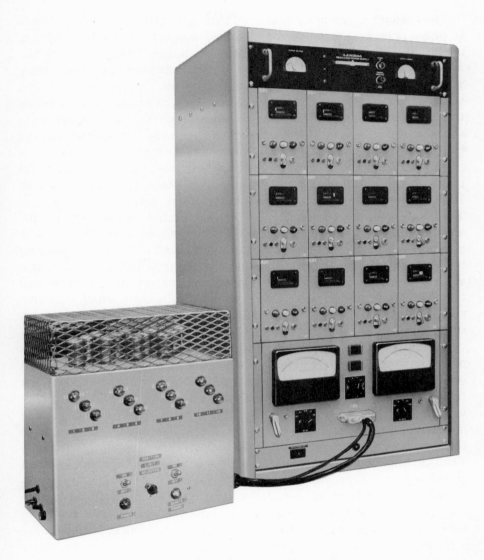

FIG. 13-19. 12-module trim-anodization set.

when the anodizing current is zero (once every cycle), the thin film resistor is switched into the bridge circuit for measurement.

The measuring voltage applied to the bridge is also 60-Hz alternating current. The bridge circuit is a double Wheatstone bridge, i.e., it has two standard arms, one adjusted to the nominal or resistance value to be trimmed to and the other set to the upper limit.

The outputs of the bridges are fed through individual a.c. amplifiers to phase detectors that operate indicator lamps to show the condition of R_x as "high," "low," or "OK." To automatically control the anodization, the output of one of the phase detectors operates a relay which switches the anodizing current on or off. A milliammeter and voltmeter on the front panel of the set can be switched to monitor the anodizing progress of any module. Such faults as electrolyte shorting to a resistor terminal, or the cathode not contacting the electrolyte, can be diagnosed by observing the meter readings.

In operation, a substrate with the resistors to be trimmed is placed into a suitable fixture. Contact is made to a maximum of 12 resistors simultaneously, and these 12 are anodized as a group. The next group of 12 resistors is then connected to the trim set for anodization by means of an external switching network called a converter. The converter has a maximum of 12 steps, corresponding to a maximum of 144 resistors, that can be trimmed on a substrate at one setting. Changing the value of resistance that a module will trim to is accomplished by replacing a plug-in card which contains the precision resistors for the bridge and the limiting resistor which controls the anodizing current.

An improvement over the 12-module set, which makes simultaneous measurements on all resistors, is a 24-module set which performs sequential measurements (see Fig. 13-20). In this model, the resistance-measuring circuits are commercial units which operate by scanning the trim resistors sequentially at an adjustable rate. The measurement time is of the order of 50 milliseconds, and this fixes the time for one complete cycle of the set at approximately 1.2 seconds. Trim anodization of the resistor occurs during the interval between measurements. The value to which a resistor is trimmed is determined by the setting of decade switches on the front panel of the set. A unique feature, incorporated into the set, allows the trim rate to be reduced as a resistor approaches its final value. In operation, a resistor is initially trimmed at some fast rate (e.g., 1000 volts per minute) until it reaches a set percentage below the decade setting on the front panel. At this point, the trim rate is automatically reduced to some fraction of the original value. In this way, a high machine output is achieved without sacrificing precision. After 24 resistors have been anodized to the desired value, the test set automatically sequences to the next group of 24 resistors and continues the cycle until six steps are completed, thus trimming a maximum of 144 resistors at one time in one fixture.

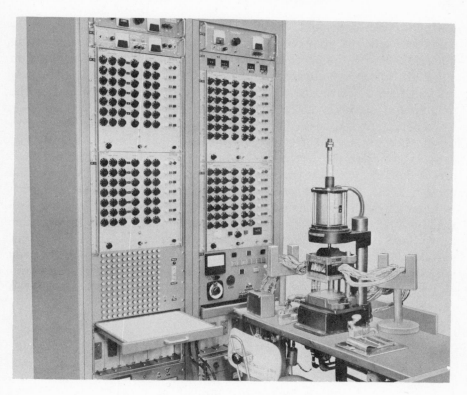

FIG. 13-20. 24-module trim-anodization set.

Fixturing

In early thin film work, beakers were used to hold the substrates, electrolyte, and electrodes. In this approach, the resistor contact areas were coated with an insulating material, such as Apiezon grease, using a screening process. Anodization was performed by immersing the substrate in the beaker containing an aqueous (0.1-percent citric acid) electrolyte. The resistor film was the anode, and the cathode was a separate strip of metal immersed in the electrolyte. Voltage was applied to all resistors, simultaneously anodizing all the resistors on the substrate while a single resistor was monitored. The accuracy of the final resistors was determined by the uniformity of the sputtered film and the pattern precision.

This "beaker" approach has a number of disadvantages which preclude its use for high-volume manufacture. First, resistors cannot be mass-anodized to a high degree of accuracy since the film is not

uniform when deposited. Second, considerable difficulty may be encountered due to spurious leakage through the screened masking
material. Third, the process is not suitable for economical production
since low anodizing currents (long trim times) are required to provide uniform anodization. Consequently, a new approach based on the
principle of capillary attraction has been developed.

Whereas in the beaker approach the contact areas are masked to
avoid shorting to the electrolyte, the capillary fixture, Fig. 13-21,
draws the electrolyte vertically through narrow slots and brings it into
direct contact with the resistor areas only, at the exclusion of the
contact areas. This approach minimizes the problem associated with
leakage through the masking material. It also results in improved
resistor tolerances and lower cost, although it still shows undesirable
limitations in these areas. One serious disadvantage is that it is un-
suitable for use with ceramic substrates. The warpage associated
with glazed ceramic, though small, is sufficient to prevent good contact with the capillary fixture; on unglazed ceramic, electrolyte creep
causes leakage paths to the contact areas.

13.6   CIRCUIT SEPARATION

In order to achieve low-cost objectives in thin film manufacture,
it is important that a maximum number of circuits be processed
simultaneously on a single substrate through as many processing
steps as possible. Normally, lead attachment is the first process
which requires that circuits be available individually in separated
form. Thus, circuits are separated just prior to this lead attachment
operation.

The method of circuit separation depends on the type of substrate
material used. Glass substrates, for example, are most economically
separated by diamond scribing and then hand breaking. Ceramic
substrates, however, cannot be economically separated by this technique, and either of two other methods may be used. The usual method
is by sawing with a circular wheel with diamond chips imbedded in
the cutting edge. Another method is the simple hand breaking of
substrates pre-scribed by the ceramic manufacturer.

Glass Scribing Machines

A scribing machine with multiple diamond scribes is shown in
Fig. 13-22 and has been used in production with consistently good
results (i.e., yields greater than 99 percent). This multiscriber has

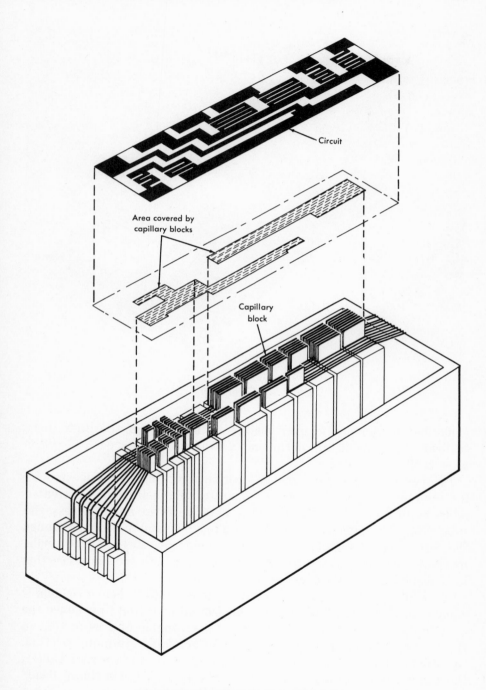

Circuit

Area covered by
capillary blocks

Capillary
block

FIG. 13-21. Capillary fixture schematic.

FIG. 13-22. Multiscriber.

a number of diamond-tipped heads, one for each of the lines to be scribed. These heads are arranged in two sets, one for each direction to be scribed.

In operation, the substrate is positioned against stops located on the table of the machine. An air-operated plunger is then activated to move one set of diamond heads at a uniform speed across the substrate. The substrate is then rotated 90° and positioned under the second set of diamonds, and the plunger reactivates. The scribe movement is mechanized to get a uniform smooth motion since this is indispensable to good yields.

Each diamond tool is accurately fastened into a holder which is capable of accurate location in the horizontal direction to produce the exact circuit dimensions desired. It is also capable of free motion in the vertical direction with minimum effort and minimum friction. To produce the proper scribe depth (in spite of surface variations), the heads are individually weighted. The play in the scribing heads is kept to a minimum to achieve tight scribing tolerances. As scribing begins, a cam arrangement lowers the diamonds gently onto the substrate surface to prevent the diamond from being damaged by riding up over the substrates. As scribing terminates, the diamond is allowed to drop off the glass surface but is prevented from striking the holding platform. Setup and maintenance are simple, and the movements of the diamond scribes are accurate and smooth.

Scribe Line and Diamond Details

In order to obtain optimum breaking results, the scribing diamond must penetrate the glass sufficiently to generate a uniform scribe line which is free from excessive chipping. A good scribe line stresses the glass enough to cause easy breaking. To achieve this, the scribe must create a vertical fissure perpendicular to the glass surface.

The cutting point of the diamond consists of the trihedral junction of three of its faces (Fig. 13-23). The leading edge (called the ridge edge) of the point should be aligned in the direction of travel. The deviation from this condition is called the axial angle shown in Fig. 13-23(a). If the axial angle is not close to zero, the scribe may vibrate or chatter during the scribing. From a side view of the scribe, the angle that the ridge edge makes with the direction of travel is called the cutting angle, shown in Fig. 13-23(b). By adjusting the cutting angle, the depth of the scribe line can be changed.

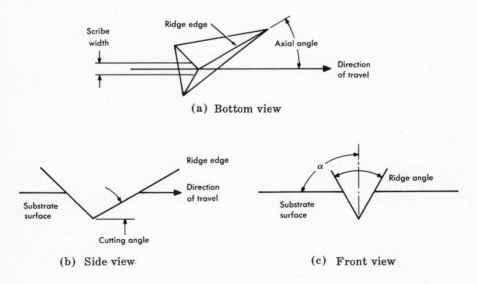

(a) Bottom view

(b) Side view.   (c) Front view

FIG. 13-23. Diamond scribe angles.

A small cutting angle produces a shallow cut, and a larger cutting angle produces a deeper cut. The optimum cutting angle must be determined for best scribing. The ridge angle, shown in Fig. 13-23 (c), is the apparent angle between the two trailing edges, as viewed from the front. This angle determines the width-to-depth ratio. The last angle to be considered is the alpha angle, or the apparent angle between the ridge edge and the surface of the glass as viewed from the front. This angle should be set at 90° to provide uniform and direct pressure at the cutting point.

As mentioned previously, the weight or pressure applied to the diamond point is important since it also affects the cutting depth. Insufficient pressure produces a light scribe line without the necessary vertical fissure for good breaking; excessive pressure produces excessive stress which causes the glass to chip and results in poor breaking. The weight for any particular diamond is determined by a trial and error method and is an important part of setting up a diamond scribing machine. The optimum weight for different situations can vary greatly. This weight depends on such things as the diamond point configuration, the surface conditions of the glass, the glass thickness, and the possible variations in the functioning of the diamond head assembly.

Although there are many types of diamond tools and diamond point geometries, the one that has worked best to date and eliminates some of the variables mentioned is a conical point with a cutting radius of 0.0005 to 0.0009 inch. The required weight increases as the radius increases. The smaller radius point is presently used for production, and it requires weights ranging from 1 to 5 ounces. The light weight necessary with these points has an added advantage when cross-scribing since excessive chipping is minimized where the scribe lines cross one another.

Breaking

Breaking can be accomplished by either mechanical or hand methods. Highest yields have been consistently achieved through hand breaking rather than mechanical breaking. Since multiscribing generates all the scribe lines before any separation is attempted, hand breaking allows a large number of different breaking sequences to be tried if problems are encountered. Normally, breaking is performed so that the least number of breaks is necessary to separate a substrate. When using hand breaking, however, the operator can "feel" if excessive pressure is being applied and can then change the sequence to suit. For example, if one of the long scribe lines appears to require excessive pressure for breaking, the operator can quickly change sequence and make the shorter breaks first. Thus, after the substrate is broken into small sections, the "difficult" scribe line will have been divided into small lengths, allowing breaking of the shorter sections to become more feasible. Another advantage is that hand breaking allows the operator to observe each scribe line during breaking. Thus, poor lines or excessive healing can be readily detected.

The type of scribe line that is generated by a conical diamond should be very light, with no indication of frosting or healing. Extreme stress in the scribing area tends to relieve itself by releasing many small chips of glass. Depending upon the amount of stress, this healing can occur either during the generation of the scribe line, shortly after scribing, or hours later. As the glass relieves itself, thousands of tiny fractures and cracks are generated, producing a telltale frosty appearance. The result is that, when breaking, the crack may begin to propagate along the scribe line but may detour from this line by following any one of the spurious fractures. Thus, a minimum of time should elapse between scribing and breaking.

Ceramic Substrate Separation

A high alumina ceramic, too hard for normal scribing and breaking, is presently used for thin film substrates. Ceramic substrates are separated by diamond sawing, which consists of four basic steps: mounting and orienting the substrate, sawing, demounting, and cleaning.

Mounting and Orienting. The substrates are mounted onto metal blocks which hold them in position during the sawing operation. This is done by cementing the substrates to the blocks with glycol phthalate under heat. Alignment is maintained by locating the substrates accurately with respect to a pin in the block and aligning the potential saw cuts in the substrate parallel to the squared edges of the block. The assembly is then mounted against a backstop on the magnetic chuck in the saw. Slots are provided in the block(s) to allow cutting through the substrates without cutting into the block(s).

Sawing. Precision sawing machines equipped with magnetic chucks are used for the sawing operation. Diamond-tipped blades are multistacked on arbors to produce the number of parallel cuts required. The distance between blades is accurately adjusted by the use of metal spacers and plastic shims. The initial alignment of the block to the sawing blade is made by adjusting the table crossfeed. There-after, it can be assumed that all blocks will be in alignment so long as they are securely and accurately positioned against the backstop on the magnetic chuck. To minimize setup time, saws are used in pairs, with one saw set up to cut in one direction and the other to make the crosscut at right angles to the first.

Normally, the highest speed is desirable for maximum output, but the choice of speed is limited by a number of factors. These consist of the number of blades per arbor, the size and concentration of the diamond grit, the substrate material, the required edge finish, and the desired blade life. A small-size, high-concentration diamond grit used at slow speeds will produce a fine, polished-type edge. Increasing the grit size and increasing the cutting speed produces a rough, more coarse cut and, if carried to extremes, can lead to chipping and cracking. Since the overall circuit dimensions must be closely con-trolled and the edges must be reasonably smooth with a minimum amount of chips or cracks, a premium-quality saw blade should be selected for production use, with cutting speeds of 4 to 8 inches per minute.

Demounting and Cleaning. After sawing in both directions, the block assemblies are removed from the magnetic chuck and placed in a solvent. When the bonding cement softens, the circuits are physically removed from the blocks and residual bonding cement is removed from the circuits by further soaking in a solvent such as acetone.

Prescored Snappable Substrates. Since diamond sawing facilities are expensive and separation by sawing is slow, other methods of circuit separation for ceramic substrates have been sought. Scribing with diamonds and glass cutters has been tried and found impractical since the breaks were poor and the cutters wore out rapidly because of hardness of the ceramic. Studies have also been made in collabora-tion with ceramic manufacturers on substrates that were prescored prior to firing. With glazed ceramic, the glaze causes considerable difficulty in obtaining good fractures, but with unglazed ceramics prescoring is more successful.

The best method developed to date involves fine grooves embossed in the ceramic. The grooves may be typically 0.005 inch deep and 0.0025 inch wide. The depth is most important in achieving good breaking yields. Depressions that are too shallow require excessive pressure to break and often produce poor breaks. Excessively deep depressions produce premature breaking during processing.

Good manufacturing control by the ceramic supplier can result in very high separation yields. The edges are quite uniform, although some flaring may occur. The breaking process is much faster and cheaper than the sawing process. It also eliminates the need for costly sawing facilities, tooling, and supplies. This cost saving, however, is somewhat offset by a 40-percent increase in cost of the raw substrate with the snappable feature. Again, as with glass, higher yields and more consistent results are achieved by hand breaking as compared to a mechanical means of breaking because of the amount of control that hand breaking affords.

13.7 LEAD AND COMPONENT ATTACHMENT

It has often been said that the success of thin film technology rests on economically connecting circuits to the "outside world." This statement carries much truth. A study of the cost of thin film manufacturing shows that a disproportionately high percentage of the cost is involved in lead and component attachment. Significant improvements in the overall cost can be realized by improvements in this area of operation. Furthermore, special emphasis must be applied to quality in this area since it is essential that the connections be highly reliable.

A number of attachment approaches have been tried with varying degrees of success, both from cost and reliability considerations. This section covers primarily the soldering approach, since this is the most common technique used in production. New methods of bonding are required to attach beam lead transistors and other more compatible semiconductor structures, and to take full advantage of miniaturization possibilities. These methods are discussed in Chap. 12.

Soldering to Glass Substrates

Leads may be attached to glass substrates by soldering, as in Fig. 13-24. The leads, made of nickel ribbon, are soldered with an alloy consisting of 50-percent indium and 50-percent tin. This alloy is chosen for its high plasticity, which minimizes the thermal strains caused by temperature cycling. If regular 60-40 tin-lead solder is used on the glass, cracking may occur, resulting in unreliable joints. The leads are attached by a production machine, in which the circuits are moved around a turntable to each of two soldering positions. In the first position, leads are attached to one side of the circuit, and in

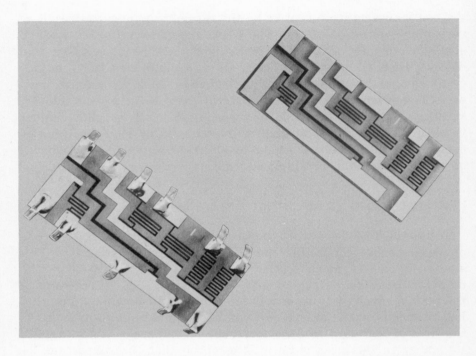

FIG. 13-24. Film circuit on glass substrate with and without soldered leads.

the second position, to the other side of the circuit. A number of functions take place at each of these soldering positions: the solder and nickel ribbon are fed to the contact pads from reels, the nickel ribbon is formed and cut to size and, finally, heated soldering-iron tips come down into position and melt the solder onto the ribbon and pads.

An alternative to this approach is wave soldering, Fig. 13-25. In this technique, the soldering is done by passing the circuits (with their leads) upside down over the apex of a wave of solder. The wave is generated by pumping molten solder up through a slot-like orifice.

Wave soldering may require redesign of the lead configuration to provide a suitable method for holding the leads in position during the soldering operation. The lead configuration shown in Fig. 13-26 has been designed for the circuit shown in Fig. 13-24. The clip leads are held together in groups by a common strap. After the soldering operation, the common strap is cut off. This clip-lead design results in a better joint because of the mechanical clamping action.

FIG. 13-25. Wave-soldering machine.

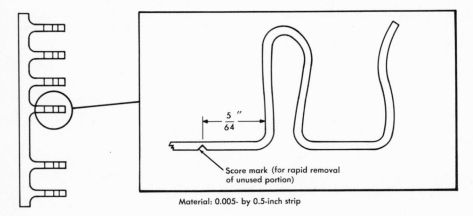

Material: 0.005- by 0.5-inch strip

(a) Bottom view of   (b) Side view of an individual clip
     rack and clips

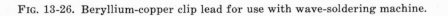

FIG. 13-26. Beryllium-copper clip lead for use with wave-soldering machine.

Soldering to Ceramic Substrates

Ceramic substrates have the advantage of being available with prepunched holes. Conventional round leads may be riveted into these holes prior to soldering, thus adding mechanical strength to the solder connection. The main advantage, however, is that the prepunched holes permit conventional components (transistors, diodes, capacitors, etc.) to be attached, thus providing completely functional circuits. Other advantages of the ceramic over the glass include its mechanical strength and higher thermal conductivity (see Chap. 9).

Some leads designed for ceramic substrates are made with shoulders on them (Fig. 13-27). The leads are inserted through the holes in the ceramic until the shoulder makes contact with the substrate. The top of each lead is then compressed to hold it in place prior to soldering, as in Fig. 13-27(a). In practice, the mechanical strength of the lead connection may not be adequate. Because of the normal variations in thickness of the ceramic substrate, the compressed leads vary in their fit. Substrates on the low side of the thickness tolerance have their leads held too loosely to assure good mechanical strength. Substrates that are too thick may crack as a result of the pressure to which they are subjected during the compressing operation. Another difficulty is that the solder joints cannot be inspected since the lead head is buried by the solder.

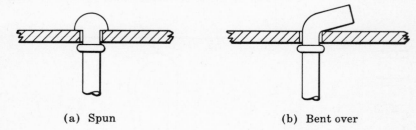

(a) Spun   (b) Bent over

FIG. 13-27. Leads for a ceramic circuit.

Instead of compressing the lead head, it may be bent over (essentially at right angles), tending toward contact with the substrate as shown in Fig. 13-27(b). This design results in a decided increase in torque strength. It also provides visual proof of wetting since the outline of the bent-over portion of the lead can be identified through the solder. Unfortunately, this design does not provide mechanical strength independent from soldering strength, as originally hoped for when ceramic was first introduced. The strength, as in the case of the glass substrates, depends primarily on the reliability of the solder joint, and the quality of the connections on the ceramic substrates is comparable with those on glass.

The use of 60-40 tin-lead solder on ceramic substrates does not seem to result in the cracking problem experienced with glass, provided that conditions are properly controlled. Cracking of glaze can occur when conditions are not well controlled; e.g., when repairing circuits by hand soldering, severe glaze cracking may occur.

Wave Soldering. Ceramic substrates may be soldered using the same wave-soldering machine mentioned previously with respect to glass substrates. The circuits, leads, and appliquéd components are assembled in position in a fixture prior to the soldering operation. To prepare the assembly, the appliquéd units are first individually attached to the circuits by inserting their leads through holes in the ceramic and bending them over. By means of a vibratory feeder, the separate leads are loaded and oriented into a magazine with many small cylindrical chambers. Each chamber is designed to line up with one lead hole in the final lead-holding fixture. The leads are then transferred to the lead-holding fixture by shaking.

The circuits with their appliquéd components are then assembled in the lead holder and fed through a roller that bends the leads over against the circuits. The assembly passes sequentially over a fluxing station, preheater, soldering station, and finally a solvent-cleaning station. The lead holder itself is coated with Teflon® in order to prevent any adherence of solder to the fixture.

Soldering Problems

Many factors enter into the achievement of good soldered connections. These can be grouped into two areas, namely, the condition of the contact material and the combination of temperatures of the preheat and soldering stations.

It has been found that a heavy amount of copper is perhaps the most critical requirement for good contact material. Palladium aids in soldering by minimizing the oxidation of copper during the 5-hour,

250°C thermal aging treatment, but the degree of protection depends on the thickness of the palladium. Since copper always oxidizes to some degree in this aging treatment, the oxide must be removed by an etch to assure that the surface is adequately clean prior to soldering. Since there is some loss of copper, this must be accounted for in the copper deposition to assure that an adequate amount remains after the etching operation. If the deposit of either the palladium or the copper is insufficient, the solder will wet poorly, or perhaps not at all.

If there are a few marginally acceptable solder bonds after wave soldering, the parts may be touched up with a hand soldering iron. This touch-up, however, can be disastrous if not done properly. It is easy to crack the substrate since excessive localized heat can be generated with hand touch-up.

In cases where contact material is exceptionally poor, the contact areas will look black after passing through wave soldering. These circuits must be discarded since the black color indicates the lack of any significant amount of contact material.

The soldering quality is also affected by the combination of the temperatures in the preheat and soldering stations. Proper control of the preheater temperature minimizes the shock when the circuits hit the soldering bath. It also aids in more uniform soldering conditions and the avoidance of cracked substrates. In addition, it preheats the flux, increasing its effectiveness.

If the solder is too cold, solder icicles will form; if it is too hot, excessive contact material will go into solution, resulting in poor wetting. Excessive solder temperatures may also crack the ceramic or damage some of the appliquéd components.

Soldering has been the primary lead-attachment method used to date in manufacture because a better, more economical approach has not yet been developed. Soldering is expensive since it cannot be highly mechanized. It also limits the subsequent application of heat in accelerated-aging tests since it can remelt. In addition, it requires tight control in both the actual soldering operation and the many processes that precede it. The problems become more difficult as the size of the circuit is reduced, a growing trend in the microelectronics field. It is for these reasons that investigations into other areas of lead attachment are under continuous study. Several approaches which show promise are parallel gap welding, ultrasonic welding, and thermal pulse bonding. These approaches are discussed in Chap. 12.

13.8 TESTING

To ensure process control and thin film circuit performance, a number of tests are made at various stages of processing. Tantalum nitride film is evaluated after sputtering to qualify it for further processing. Resistors are measured after trim anodizing to be sure they are within specified tolerances. The completed thin film circuits with their appliquéd transistors and other components must meet input and output functional requirements.

Tantalum Nitride Film Evaluation

Before sputtered tantalum nitride films are processed, a number of properties are evaluated. These include sheet resistance, thickness, resistivity, and temperature coefficient of resistance. It is also advantageous to make an accelerated-aging test to qualify the stability of the film.

Sheet Resistance. The first property measured is sheet resistance, R_s (see Chap. 7). Sheet resistance is measured by the equipment shown in Fig. 13-28. This test set consists of a four-point probe, a constant-current power supply, a digital voltmeter, and a printer. The measurement is performed by passing a constant current (I) between the two outside probes of a four-point probe, while measuring the voltage (V) between the inner two probes. The spacing between each of the probes is 0.050 inch. The sheet resistance, R_s, in ohms per square is given by

$$R_s = \frac{V}{I} C$$

where C is a constant which depends upon the geometry of the substrate and the position of the four-point probe on the substrate. (Values for various geometries and probe placements are given in Figs. 7-1 and 7-2.) If the current is set equal to the value of the constant, the reading on the digital voltmeter is numerically equal to the sheet resistance.

Thickness and Resistivity. To evaluate resistivity, the film thickness is measured, using beta back-scattering (see Chap. 3). A typical instrument of the type used for thickness measurement is shown in Fig. 13-29. The rod at the top of the test platform is a weight to hold the substrate in position. In the base of the platform are a

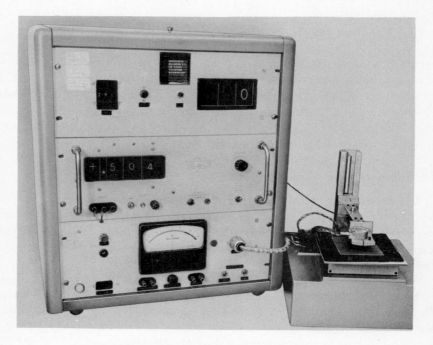

FIG. 13-28. Sheet resistance test set.

FIG. 13-29. Beta back-scatter test set for thickness measurement.

Promethium 147 beta ray source and a Geiger-Muller tube to monitor the back-scattered particles. The detector is connected to an electronic counter which integrates the counts over a fixed period of time. To determine the actual thickness, the following equation is employed:

$$t = \frac{F - S}{KS}$$

where t = thickness in angstroms, F = number of counts for a substrate with film, S = number of counts for a blank substrate, and K = calibration constant determined from a standard of known thickness. The resistivity, ρ, is then calculated as

$$\rho = R_s \times t$$

Temperature Coefficient of Resistance. The temperature coefficient of resistance is also used as an indication of process control. A rapid evaluation of the TCR is desirable, and this is accomplished by using a special test pattern defined by a screened photoresist. The substrate is etched, and aluminum leads are ultrasonically welded to the large contact pads as shown in Fig. 13-30. Resistance is then measured at

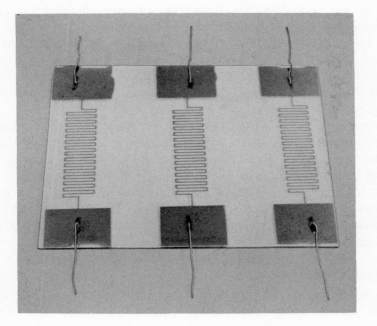

Fig. 13-30. Ultrasonically welded aluminum leads.

both room temperature and liquid nitrogen temperature ($-196°C$). A direct-reading TCR test is used to make this measurement. The test sample is placed in one leg of a Wheatstone bridge, which is then balanced at room temperature. When the test sample is placed in liquid nitrogen, the bridge becomes unbalanced. The null detector is calibrated to read the unbalance directly as the TCR. Using this system, TCR information needed to control sputtering can be made available less than 10 minutes after a substrate leaves the sputtering machine.

Resistance Measurements

A determination of resistance value is perhaps the most basic measurement used in circuit manufacture. Such a measurement may be made at various stages of manufacture, but the principles involved are similar.

A resistance-measuring test set has been designed with enough flexibility to be useful for a variety of circuits. This test set, shown in Fig. 13-31, contains 14 testing channels. It is a high-speed, dual-limit, go/no-go resistance bridge. For each test position, the control unit selects the proper resistor nominal value and tolerance limits. The test set is a universal type which is programmed for specific circuits by the use of punched cards.

To test a circuit, it is placed into a fixture and a start button is pressed, after which the test set sequentially tests the resistance of resistors and any critical conducting paths. Measurement time is important, and this set allows 20 resistances to be measured per second.

Fixturing also presents a problem in resistance-testing the thin film circuits. Contact areas may be quite small, causing difficulty in maintaining proper circuit location and good electrical contact to the contact pins in the fixture.

The ideal time to make final resistance measurements is at resistor trim anodizing. Before this can be done, however, all subsequent processing must be under such control that no change in resistance value can occur for any circuit.

Overload Pulse Test

In order to assure high reliability of the resistors, some method is needed for detecting resistors with flaws that might degrade reliability. One such method is to inspect each resistor under a microscope and arbitrarily reject resistors with pattern defects (such as

FIG. 13-31. High-speed resistance measuring bridge.

scratches or pinholes) larger than 50 percent of the resistor line width. This is a costly procedure. In addition, there is some question as to whether all poor resistors are rejected and all good circuits are accepted.

A better method involves the application of an overload pulse to each resistor. The amount of overload is tailored for each resistor on a circuit, and is chosen to be the maximum power which does not change a good unit by more than about 0.05 percent during the time of the test. Resistors with scratches, pinholes, or other defects of a serious nature develop "hot spots" and burn out during this pulse test.

The test set built for this operation contains four timers and five power supplies (see Fig. 13-32). The use of four timers instead of one provides the flexibility of pulsing different resistors on the same circuit for a different length of time. The five power supplies pro-

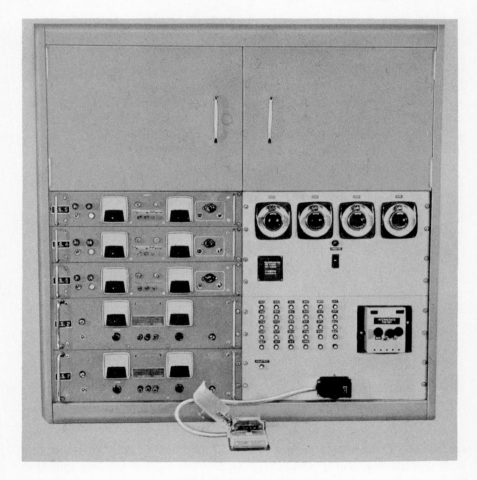

FIG. 13-32. Overload pulse test set.

vide a range of 0 to 575 volts d.c. This range of voltages permits the pulsing of a wide variety of circuits.

In operation, the overload information is contained in a "program plug" for the circuit code under test. One program plug encompasses the voltages and pulse times used for each resistor on any one code. A change to a different code is made by simply replacing the program plug.

Circuit Performance Test

Once a thin film circuit has been completed and its active devices (and any other appliquéd components) attached, it must be tested for overall circuit performance. The required test specifications must be supplied by the circuit designer and are likely to be unique to the circuit. Test equipment, therefore, is usually specific to a particular circuit or to a family of circuit types. Complex circuits may require many functional tests, and the test equipment may thus also be quite complex.

Accelerated-Aging Test

In order to assure that the film has the desired stability, sample lots of circuits are taken weekly from the production line and are given an accelerated-aging test. The test consists of applying an overload voltage to produce a hot-spot temperature on the test resistor of approximately 215°C. Resistance measurements are then made at logarithmic intervals of time from 0 to 1000 hours to determine the change in resistance, and a plot is made of the log percent change in resistance against log time, Fig. 13-33. To qualify the film for stability, the plotted line must have both an acceptable slope and intercept.

To make a test, a sample (24 to 30 circuits) is taken at random from the weekly production. These are mounted on modules which may hold 8 to 10 circuits each. The modules, in turn, are placed on an aging rack, shown in Fig. 13-34, which supplies the required d.c. voltage to the resistors.

When modules are removed from the rack for resistance measurements, they are placed in a temperature-controlled oven at 70°C. The oven is used to ensure constant temperature for all measurements, thus eliminating calculations to compensate for the TCR. After one-half hour of stabilization in the oven, the resistor values are measured by a five-place digital ohmmeter coupled with an IBM keypunch which

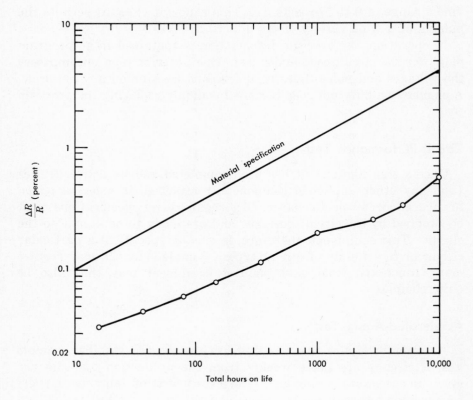

$\frac{\Delta R}{R}$ (percent)

Total hours on life

FIG. 13-33. Accelerated-aging test.

provides punched cards identifying each resistor and its present resistance value. These cards are processed through a computer for calculation of percent change in resistance over each measurement period. This information is manually plotted and compared to the specification slope line limit.

This method of life testing does have some problems associated with it. For example, since a constant test voltage is used, the power dissipated in the resistor depends on where the resistance value lies within its tolerance limits. The temperature reached also depends on the test fixtures and, unless adequate precautions are taken, it may deviate from that desired.

A high-temperature oven test (as described in Chap. 7) may be used to eliminate some of these difficulties. The major problem with this approach is the requirement of special contact material to with-

FIG. 13-34. Accelerated-aging facility.

stand the high temperature. This need not be a limitation, however, if a special resistor pattern such as that used for the TCR test were used for this test. Another advantage would be that film qualification could be achieved before any further processing is done.

REFERENCES

1. Paulsen, J. F., and E. F. Steiner. "Economic Substrate Size for Thin Film Circuits," *Ceramic Age*, 82, 36-40 (1966).
2. Charschan, S. S., R. W. Glenn, and H. Westgaard. "A Continuous Vacuum Processing Machine," *Western Electric Engineer*, 7 (2), 9-17 (April, 1963).
3. Hanfmann, A. M. "Simplified Operations Analysis in Continuous Sputtering of Thin Films," *Western Electric Engineer*, 10, 11-17 (1966).

Index

A/volt conversion factors, 279, 386

Aberration, spherical, 427

Abnormal glow, 199

Accelerated aging
  capacitors, 373ff.
  resistors, 359ff.
  thin film circuits, 685ff.

Acid plating baths, 261

Acids for etching thin films, 451

Adherence (*see* Adhesion)

Adhesion
  changes with time, 586
  evaporated films, 133ff., 653
  sputtered films, 213ff.
  thin film conductors, 365ff.
  thin films, 584ff.

Adhesion coefficients, and hardness, 605

Adjacency effect, 443

Adsorption, 124ff.
  chemical, 573
  physical, 573

Aging of capacitors, accelerated, 373ff.

Aging of resistors, 349ff., 476
  d.c. load, 352ff., 477
  distribution of along resistor, 355
  effect of ambient, 479
  equivalence of temperature and
    power, 356
  function of nitrogen content, 354
  generalized aging curves, 363
  oven, 349ff.
  preaging, 359ff.
  temperature *vs* power, 354
  vs sheet resistance, 480

Aging temperature, d.c. power, 354, 479

Aging test, 685

Air, properties of, 22

Alkaline plating baths, 260

Alloy bonding, 559, 589ff.

Alloy deposition, by sputtering, 212ff.

Alloy evaporation, 149ff.

Alloy solidification, 566

Alloying, 567ff.

Alloys
  definition of, 567
  resistivity of, 304, 306ff.

Aluminum, etchant for, 451

Aluminum evaporation, 146

Amplifier, UHF, 550

Angle of contact, 569
  cleanness test, 416, 644
  wetting angle in soldering, 589

Anode glow, 201

Anodic oxide films
  capacitor dielectrics, 380ff.
  crystallization of, 277
  dielectric constants, 272
  dielectric properties, 283
  growth rate, 276
  interference colors in, 284ff.
  leakage current, 282
  optical properties, 284
  properties of, 281ff.
  thickness *vs* time, 279

Anodic Ta_2O_5
  conversion factor (A/volt), 279
  density of, 276
  equivalent weight, 276

Anodization, 271ff.
  capacitor dielectrics, 380
  capillary fixture, 666
  constant current, 276
  constant voltage, 276

Anodization *(cont)*
    current density, 274ff.
    differential field strength, 276
    film growth rate, 276
    fixturing, 665
    general description, 9
    kinetics of, 274ff.
    mass, 493ff.
    mechanism of, 272ff.
    resistor adjustment, 344ff.,
        492ff., 661ff.
    technique, 277
Anodization constant (A/volt)
    anodized tantalum nitride, 386
    tantalum oxide, 279
Anodization electrodes, 281
Anodization electrolytes, 279
Anodization parameters
    cell voltage, 278
    current density, 278
    electrodes, 281
    electrolyte, 279ff.
Anodized tantalum nitride,
        conversion factor (A/volt),
        279, 386
Aperture masks, 462
Appliquéd components, 673
Area efficiency factor, 374
    of various dielectrics, 392
Area required
    for a capacitor, 512
    for a resistor, 474, 485
Area resistor networks, 533ff.
Arrival rate of molecules, 24
Art master, 422, 455
    generation of, 455
Astigmatism, in camera optics, 427
Atomic interactions, energies of,
        585
Atomic theory, 561
Atomizer test, 417, 643
Attachment of leads, 559ff.
Avogadro's law, 20
AZ photoresists, 448
A5 channel bank circuit, 545

Back-etching, 381
Backstreaming, diffusion pumps,
        50ff.
Bayard-Alpert gauge, 87
Beam leads, 599, 612, 614, 620
Bell jar system
    leak detection, 71ff.
    maintenance, 71ff.
    operation of, 69ff.
Beta back-scattering, 175ff., 679
    effective atomic numbers for, 178
    film-substrate combination, 177
Beta tantalum, 221, 342
Binding
    atomic, 561
    metallic, 575
Binding activation energy, 582
Body-centered cubic structure, 562
Bond
    factors affecting, 580ff.
    ultimate properties of, 579
Bonding
    alloy, 559
    definition, 561
    mechanical thermal pulse, 608
    solid phase, 559, 602ff.
    testing of, 622ff.
    thermocompression, 604ff.
    ultrasonic, 612ff.
Bourdon tube, vacuum gauges, 78
Brazing, definition, 589
Breakdown strength, 373, 512
Breakdown voltage
    counterelectrodes and, 383
    tantalum oxide, 382
Breaking after scribing, 670
Cadmium evaporation, 149
Cameras, 426ff.
    fixed reduction, 427, 458
    multiple array photorepeat, 461
    optics of, 427ff.
    photorepeat, 461
    variable reduction, 427, 460

Capacitance, shunting, in film
    resistors, 500ff.
Capacitance density, 512
    of variance dielectrics, 392
Capacitor dielectrics, 376ff.
    duplex, 388
    organic polymer, 389ff.
    parylene, 390
    properties, 392
    silicon monoxide, 378ff.
    tantalum oxide, 380ff.
Capacitor materials, 377ff.
Capacitor pattern
    adjustable capacitance, 523
    crossed electrode, 519
    minimum resistance, 521
    nonpolar, 522
Capacitor properties, 371ff., 392
Capacitors, 371ff., 510ff.
    adjustable capacitor pattern, 523
    allowable stress, 392
    area efficiency factor, 376, 392
    capacitance of, 371
    charge storage factor, 376, 392,
        514
    crossed electrode pattern, 519ff.
    design of, 510ff.
    dielectric loss in, 376ff.
    dielectric thickness, 511ff.
    dissipation factor, 376, 515ff.
    duplex dielectric, 388ff.
    energy stored in, 372
    equivalent circuit, 377, 518
    failure in, 374
    fundamental properties of, 371ff.
    geometrical patterns, 519ff.
    leakage current, 282, 372, 525
    losses in, 515ff.
    made from doped tantalum,
        384ff.
    made from tantalum nitride, 384
    manganese-oxide tantalum-oxide,
        387ff.
    material properties, 377, 510
    maximum working stress, 373ff.
    minimum resistance pattern, 521
    nonpolar pattern, 522
    organic polymer, 389
    parasitics, 523
    parylene, 390

Capacitors (cont)
    rated voltage, 513
    series inductance, 523
    series resistance, 377, 516
    silicon monoxide, 378
    silicon-monoxide tantalum-oxide,
        388ff.
    special patterns, 519
    step-stress testing, 374
    surface roughness and, 281
    tantalum oxide, 380ff.
    thin films, 15
    working stress, 392
    yield, 514
Carbon films, 13
Cathode
    electroplating, 261
    sputtering, 238
Cathode glow, 201
Centrifuge testing, 625
Ceramic manufacture, 397
Ceramic substrates, 397
    comparison of, 413
Cermet resistors, 14, 339
Characteristic curve, of photo-
        graphic emulsions, 440
Charge storage factor, 376, 514
    tantalum oxide, 381
    various dielectrics, 392
Chemical methods of film
        formation, 255ff.
Chemical reduction, deposition by,
        5
Chemical stability, substrates,
        406ff.
Chemical vapor plating, 266ff.
    deposition parameters, 267
    deposition techniques, 268
    dielectrics, 270
    film properties, 268
    general description, 6
    mechanism, 267
    silicon dioxide, 270
    silicon films, 269
    single crystal germanium, 268
Chemisorption, 573
Chromium, film resistors, 14, 337
Chromium evaporation, 146
Chromium silicon monoxide,
        resistors, 339

692

THIN FILM TECHNOLOGY

Circuit design, 467ff.
  DATA-PHONE® data set, 468ff.

Circuit layout, 543ff.
  digital to analog decoder, 547
  examples of, 544

Circuit separation
  glass breaking, 670
  sawing, 671
  snapping, 672

Circuits
  accelerated-aging test, 685
  design of, 467ff.
  distributed components, 533ff.
  electroplating on, 262
  layout, 543ff.
  manufacture, 633ff.
  performance test, 685
  processing sequences, 654
  separation of multiple, 666ff.
  testing, 679ff.

Clapeyron-Clausius equation, 118

Cleaning
  acid, 259
  alkaline, 259
  by firing, 415
  emulsion, 258
  prior to electroplating, 258
  solvent, 258, 415
  substrates, 413ff., 642
  testing of, 416ff.
  typical procedure, 415
  ultrasonic, 258, 414

Cleanness
  atomizer test for, 643
  contact angle measuring, 644
  water-break test for, 643

Cleanness testing, 642ff.

Closed-end sputtering, 249

Cohesion, real metals, 575ff.

Coma, camera optics, 427

Combination mask, 465, 651

Component design, 467ff.

Compressibility, metals, 305

Computer-aided masks, 425

Conductance for gas flow, 32ff.
  function of temperature and
    molecular weight, 38
  intermediate, 37
  molecular flow, 36
  viscous flow, 36

Conducting paths, thin film
    circuits, 364ff., 526ff., 648

Conduction, anodic oxide films, 282

Conduction in metals, 289ff.

Conduction in real films, 323ff.
  effect of anodization, 324
  effect of crystallite size, 324
  tantalum nitride, 325

Conduction in thin films, 310ff.
  Sondheimer theory, 310ff.

Conductivity, of thin films, 3

Conductor films, 364ff., 525ff., 648
  adherence of, 365ff.
  joinability, 366, 583ff.
  manufacture, 647ff.
  plated, 528
  series inductance in, 495ff.
  sheet resistance, 366, 527
  skin effect, 509ff.
  thermal stability of, 367
  uses and materials, 525ff.

Conformal masks, 452

Connection evaluation, 622ff.

Connection stability, 583

Contact angle, 569
  cleanness test, 416, 644
  wetting angle in soldering, 589

Contaminants
  solder, 594
  thermocompression bonding and,
    606

Contamination
  effect of on sputtering, 208ff.
  sources of in sputtering, 208ff.

Continuous-feed vacuum system,
  74ff.

Conversion factors (A/volt), 279, 386

Coordinatograph, 423, 455

Copper, etchant for, 451

Copper evaporation, 147

Corner corrections, resistance, 487

Corrections for ends and corners, 487ff.

Cost per circuit, 633

Counterelectrodes, breakdown voltage and, 383

Covalent binding, 561

Crossovers, 530

Crucibles, for vacuum evaporation, 139

Crystal imperfections, 564
  in surfaces, 572

Crystal structure, 562ff.
  body-centered cubic, 562
  face-centered cubic, 563
  hexagonal close-packed, 563
  imperfections in, 564
  polycrystalline, 564
  sputtered films, 232
  tantalum films, 342
  thin films, 2

Crystallite size, 3

Crystals, grain boundaries of, 564

Current noise, 335

Curvature of field, camera optics, 427

Dark spaces, 200ff.

DATA-PHONE® data set, 468

D.c. load aging
  equivalence with oven aging, 357
  tantalum film resistors, 353

Debye temperature, 297
  for common metals, 298

Decay of latent images, 443

Decoder, digital to analog, 547

Density gradients, optical, 440

Density of tantalum films, vs
  sputtering voltage, 229

Deposition methods, 3ff.

Depth of field, 426

Depth of focus, 426, 435

Design
  thin film capacitors, 510ff.
  thin film circuits, 467ff.
  thin film resistors, 473ff.

Desorption of gases, 116

Destructive testing, 623

Development
  chemical, 437
  photographic, 437
  physical, 437

Diameter, molecular, 24

Diamond scriber, 666

Dielectric absorption, 373

Dielectric constant
  anodic films, 272
  various dielectrics, 392

Dielectric films, 648
  organic polymer, 15
  tantalum oxide, 15

Dielectric loss, 376ff., 515ff.

Dielectric properties, anodic oxide films, 283

Dielectric thickness, film capacitors, 511ff.

Dielectrics
  breakdown strength, 373, 512
  chemical vapor plating of, 270
  sputtering of, 217

Diffraction, 428

Diffusion, 568ff.

Diffusion pumps, 44ff.
  backstreaming in, 50ff.
  fluids, 46
  fractionating, 45
  operation of, 52
  pump performance, 47
  pumping speed of, 49
  speed factor, 49

Digital to analog decoder, circuit layout, 547

Dissipation factor, 376ff.
  tantalum oxide, 382

Distortion, in camera optics, 427

Distributed electrode resistance, 516

Distributed film components, 533ff.

Distributed RC structure, 535ff.
  adjustment of, 542
  analysis of, 538ff.
  bandpass function, 540
  tapered, 536
  three-terminal, 536
  transmission coefficient of, 538
  uniform, 536

Distribution
  of sputtered atoms, 194
  of thickness, 158ff.

Doctor blading, 397

Drift velocity, electrons, 290

Drumheller evaporation source, 155

Dumbbell pattern, 490ff., 492

Duplex dielectric capacitors, 388ff.

Elastomers, in vacuum systems, 103ff.

Electrical breakdown field, 373

Electrical conduction
  bulk metals, 289ff.
  electron energy distribution in, 291ff.
  electron mean free path, 294ff.
  gases, 30
  ideal thin films, 310ff.
  phonon scattering, 296ff.
  real films, 323ff.
  substrates, 411

Electrochemical compounds, 568

Electroless plating, 263ff.
  deposition parameters, 265
  film properties, 265ff.
  general description, 5
  mechanism, 263ff.

Electrolytic oxidation, 9

Electron beam evaporation, 139ff.
  glow discharge in, 141
  ionization during, 141
  schematic drawings, 142

Electron beam welding, 621ff.

Electron compounds, 568

Electron interaction, metal cohesion and, 575

Electron mean free path, in metals, 294ff., 302

Electron scattering, by impurities and defects, 303ff.

Electroplating, 4, 255ff.
  baths, 259ff.
    copper, acid, 261
    copper, alkaline, 260
  cathode, 261
  cleaning for, 258
  conductor films, 528
  deposition parameters, 257ff.
  technology, 258ff.
  thin film circuits, 262

Embossed ceramic resistor, 348

Emulsions
  characteristic curve, 440
  development, 437
  light absorption in, 434
  light scattering in, 434
  photographic, 431
  refraction at, 433
  thickness, 434

End corrections, resistors, 487ff.

Energy of atom interactions, 585

Envelope phase, 577

Etch rate, glass substrates, 407

Etchants, for various metal films, 451

Etching, 451ff.

Etching procedures, 655

Evaporated film adhesion, 133ff., 653

Evaporated film growth
  factors affecting, 130ff.
  initial, 125ff.

Evaporated films, 113ff.
  conductor, 648ff.
  properties of, 7
  thickness distribution of, 158ff.

Evaporated tantalum, resistivity, 341

Evaporation
  charge required, 162
  collisions enroute, 115
  electron beam, 139ff.
  environment, 113ff.
  filaments, 137ff., 649
  fixturing, 649, 651
  general description, 7
  glow discharge used in, 156ff.
  insulators, 151ff.
  manufacturing facility for, 647ff.
  metals, 146ff.
  molecular collision in, 115ff.
  RF heated, 142ff.
  rate, 119
  residual gas, 113
  sources, 135ff.
    crucibles, 139
    electron bombarded, 139ff.
    emissive characteristics, 161
    filaments, 137ff., 649
    foil sources, 138
    induction heated, 142ff.
    plane source, 159
    point source, 158
    production, 649
    radiation heated, 145
    resistance heated, 135ff.
  specific materials, 145ff.
  system, 649
  vaporization in, 117ff.

Exhaust speed, 35

Exposure of photoresists, 657ff.

f-number of a lens, 428

Face-centered cubic structure, 563

Faraday's law, 276

Fermi distribution, 292

Field curvature, in camera optics, 427

Filaments, vacuum evaporation, 137ff., 649

Film growth, 123ff.
  factors affecting, 130ff.

Film properties
  dielectric conductivity, 3
  mechanical, 2
  metal resistivity, 3

Film resistors (see Resistors)

Film thickness (see Thickness)

Films (see Thin films)

Fixing bath, 438

Fizeau fringes, of equal thickness, 171

Flatness, substrate, 400

Fluxes, soldering, 580, 589ff.

Focus, in camera optics, 430

Foreline valves, 96

Fore-pumps, 38ff.

Four-point probe, 330ff.
  measurement near holes, 333
  off-center measurements, 332

Freezing, 566

Fusion welding, 559, 618ff.

Gamma, photographic emulsion, 440

Gas ballasting, 40

Gas conductance, 32ff.

Gas discharge, 199ff.

Gas flow, 31ff.
  intermediate range correction, 37
  viscous and molecular, 35ff.

Gas flow rate, 32
  units of, 32

Gas manifold for sputtering, 241

Gas quantity, 31
  units of, 31

Gases
  electrical conductivity of, 30
  mean free path in, 23
  molecular collision in, 23, 115ff.
  molecular diameters, 24
  molecular motion, 22ff.
  pressure in, 25ff.
  pressure units, 29
  rate of desorption of, 116
  thermal conductivity, 30

Gauges, vacuum (see also Vacuum gauges), 78ff.

Germanium, sputtering of, 234

Getter pumps, 53ff.

Getter sputtering, 209

Glass breaking, circuit separation, 670

Glass manufacture, 397

Glass scribing
  machines, 666ff.
  tool, 669

Glass substrates, 397
  comparison of, 412
  etch rate of, 407
  ion migration, 408ff.
  sodium oxide content, 408ff.

Glaze resistors, 340

Glow discharge, 199ff.
  appearance of, 199ff., 201
  conduction in, 200
  current-voltage characteristics, 199
  electron beam evaporation, 141
  evaporation in the presence of, 156
  film deposition in, 204ff.
  normal and abnormal, 202
  relation of dark space to pressure, 202

Gold, etchant for, 451

Gold evaporation, 147

Goniometer test for cleanness, 644

Grain boundary, 564, 576ff.
  similarity to bond region, 577

"Green" sheet ceramics, 397

Growth of thin films, 123ff.

Gruneisen relation, 297

Halation, 435

Heat of vaporization, 118

Hexagonal close-packed structure, 563

High-frequency resistors, 336, 494

High vacuum pumps, 44ff.

High vacuum valves, 94ff.

Ho coefficient, 49

Hot work temperature range, 617

Hot work ultrasonic bonding, 614ff.

Hume-Rothery rules, 568

Humidity
  effect on resistors, 336
  effect on SiO capacitors, 379
  effect on Ta_2O_5 capacitors, 382

Ideal gas laws, 20ff.
  derivation, 26ff.
  molecules per cc, 21

Illumination of art master, 431

Image development, in photographic emulsions, 437

Impedance-matching pad (A5), circuit layout, 545ff.

Imperfections, in crystals, 564, 572

In-line sputtering, 248, 638ff.
  vs bell jar, 645

In-line sputtering system, 74ff.
  pressures in, 75ff.

Inductance, 531ff.
  capacitors, 523ff.
  conductors and resistors, 495ff.
  meander pattern, 498
  thin film resistors, 499

Inductors, thin film, 531ff.

Infrared soldering, 597

Interface classification, 584

Interface contact, effect on joining, 581ff.

Interference colors, anodic oxide films, 284ff.

Interference contrast technique, 628

Interferometry, 170ff.
  Fizeau fringes, 171
  fringes of equal order, 173

Ion migration, glasses, 408

Ion plating, 156

Ion pumps, 53ff.

Ionic binding, 561

Ionization gauges, 85ff.
  cold-cathode, 91ff.
  limitations of, 89ff.
  sensitivity of, 87ff.

Ionization monitors, 185

Joining
  effect of interface contact, 581ff.
  factors affecting, 580ff., 587ff.
  to thin films, 583ff.

Joints for vacuum systems, 108

Kinetic theory of gases, 19ff.

KMER, 445, 447

Knudsen number, 35

Knudsen's cosine law, 25, 26

Kodak photoresists, 445ff., 656

KOR, 445

KPR, 445

KPR-2, 445

KTFR, 445, 447

Land areas, in thin film circuits, 526ff.

Laser welding, 618ff.

Latent image decay, 443

Lattice pattern, for resistors, 489

Laves phases, 568

Layout of film circuits, 543ff.

Lead attachment, 559ff.
  ceramic substrates, 676
  glass substrates, 673
  manufacturing techniques, 673ff.
  testing, 622ff.
    centrifuge, 625
    electrical noise, 629
    electrical resistance, 629
    environmental, 630
    fatigue, 625
    infrared, 629
    mechanical, 623
    metallographic, 623, 625ff.
    nondestructive, 627ff.
    radiographic, 630
    tensile, 624
    ultrasonic, 629
    visual, 628

Lead configurations, soldering to thin films and, 600

Lead-etching technique, 627

Leak detection, in vacuum systems, 72ff.

Leak valves, 99

Leakage current
  capacitors, 372, 525
  resistors, 507
  tantalum oxide, 382, 525

Lenses, photoshaping, 427

Levitation evaporation, 145

Line width
  minimum, 484
  resistors, 482

Low-density tantalum, 343
  capacitors from, 387
  properties of, 227
  sputtering of, 226

Low-pressure sputtering, 214ff.

Magnetic applications, 12

Manganese oxide
  etchant for, 451
  in thin film capacitors, 387

Manufacturing thin films, 633ff.

Masks
  adherent, 652
  aperture, 426
  by anodization, 453
  combination metal, 452, 465, 651
  comparison of, 465
  conformal, 452
  contact printing of, 424, 458
  flexible photomasks, 658
  machined metal, 423, 452
  mechanical, 451, 462ff.
  metal, 464
  metal-on-glass, 444
  molybdenum foil, 464
  photoetching, 451
  photo *vs* mechanical, 661
  processing, 456ff.
  production, 457
  semiconductor device, 460
  silk screen, 463
  tantalum oxide, 454
  tolerances, 465
  types, 422

Mass anodization, 493ff.

Master mask, 422, 460

Matthiessen's rule, 303, 306

Maxwell-Boltzmann distribution, 22, 26

McLeod gauge, 80ff.

Mean free path
electrons, 294, 302
molecules in a gas, 23
residual, 316
tantalum nitride film, 325

Meandering-line pattern, 475

Mechanical masking, 451, 462ff.

Mechanical pumps, 38ff.

Mechanical strength
structure and, 2
substrates, 402

Mechanical thermal pulse bonding, 608

Mechanical trimming, resistors, 347

Melting points, metals, 120ff.

Metal cohesion
electron interaction and, 575ff.

Metal-to-metal contact, joining and, 581

Metallic binding, 562, 575

Metallic solidification, 566ff.

Metallic surfaces, 570ff.
film growth on, 573
oxide growth on, 573

Metallurgy, 561ff.

Metals
adhesion coefficients of, 604
electrical conduction in, 289ff.
melting points of, 120ff.
vapor pressure curves, 120ff.

Molecular arrival rate, 24

Molecular flow conductance, 36

Molecular flow of gases, 35

Molybdenum evaporation, 148

Monolithic resistor networks, 533

Negative glow, 201

Nichrome®
etchant for, 451
glue for thin films, 365, 586
thin film resistors, 14, 337ff.

Nichrome®-copper-gold, films, 526

Nichrome®-copper-palladium, films, 648ff.

Nichrome® evaporation, 150

Nickel evaporation, 147

Noise, electrical, 629

Noise coefficient, 335

Nondestructive testing, 627

Normal glow, 199

Notch filter
circuits, 538
twin-tee, 544

Nucleation, 126ff.

Nucleation and growth, in solidification, 566

Number of squares, 486

O-ring seals and materials, 103ff.

Ohms per square, 330

Opacity, 439

Optical applications, thin films, 11

Optical density, 439
edge gradient, 440

Optics, camera, 427ff.

Oscillator, RC, 551

Oven aging, 476
equivalence with load aging, 357
tantalum film resistors, 349ff.

Overload pulse test, 358, 682

Oxide film growth, on metallic surfaces, 573ff.

Oxide underlay, 408

Palladium, etchant for, 451

Palladium evaporation, 148

Palladium-silver glaze, resistors, 340

Parallel-gap resistance welders, 608

Parallel-gap resistance welding, 622
Parasitics
  capacitors, 523ff.
  resistors, 494ff.
Parylene, capacitor dielectric, 390
Parylene films (poly para-
      xylylene), 11
Pattern generation, 419ff.
  aluminum resist, 454
  anodic resist, 454
  conformal masks, 452
  indirect methods, 452ff.
  manufacturing facilities, 455,
      653ff.
  photographic techniques, 425ff.
  sequences, 419, 653
  silk screened resist, 454, 463
PCM decoder, 547
PCM preamplifier circuit, 419
PCM repeater, 549
Phonon resistivity
  temperature coefficient, 299
  temperature dependence, 300
Phonon scattering, 296ff.
Photoetching, 445ff., 655ff.
  etchants for, 451
  manufacturing process, 655
  resists, 445ff., 656
Photographic emulsions, 431ff.
  characteristic curve, 440
  chemical development in, 437
  density and density gradient in,
      440ff.
  image development in, 437ff.
  physical development in, 437ff.
Photographic plates, 431ff.
  spectral transmission of, 436
  types, 437
Photographic technique, 425ff.
Photolithography, manufacturing
      process, 421ff., 655
Photomasks, 422ff., 661
  density gradient, 440
Photoresists, 445ff., 656
  AZ, 11, 448
  AZ 1350, 448
  coating methods, 656
  exposure of, 449

KMER, 447
KOR, 447
KPR, 447
KPR-2, 447
KTFR, 447
  light exposure, 657ff.
  negative, 445, 656
  positive, 448, 656
  prebake, 449
  processes, 449
  spectral sensitivities, 450
  use of, 655
Photoshaping tools, manufacturing
      facility for, 455ff.
Physical adsorption, 124, 573
Physical properties, thin films, 2
Pirani gauge, 84
Plane source, 159
Plastic flow, 603, 605, 612
Plates, photographic, 431ff., 437
Plating
  electroless, 263ff.
  vapor, 266ff.
Plating baths, 259ff.
Platinum films, 13
Point source, 158
Polycrystals, 2, 564, 576
Polymer films, 10
  capacitor dielectrics, 15, 389
Polymerization of photoresists,
      446ff.
Poly(vinyl cinnamate), 445
Porosity
  low-density tantalum films, 226
  substrates, 401
Pourbaix diagrams, 273
Power density
  resistor, 474
  upper limit, 479
Preaging effect on tantalum film
      resistors, 359ff.
Preamplifier, 100 MHz, 549
Prescored substrates, 636, 672
Pressure, 25ff.
  measurement in vacuum systems,
      78ff.

Pressure units, 29

Process sequences, 654

Pump speed, 34

Pumpdown dynamics, 65ff.

Q (quality factor)
  of a capacitor, 376
  of an inductor, 531

Quartz, crystal thickness monitors,
    184, 252

Raoult's law, 150

Reactive sputtering, 210ff., 220ff.
  reaction rate, variables, 211

Relaxation time, electrons, 291, 296

Residual mean free path, 316

Residual resistance, 316

Residual resistivity, 303

Resist coating methods, 656

Resistance (see also Resistivity)
  effect of stress on, 304ff.
  measurement of, 682

Resistance monitors, 186

Resistance welding, 622

Resistivity, 679
  alloys, 306ff.
  bulk and film, various metals, 527
  definition, 291
  effect of stress on, 304ff., 317ff.
  function of thickness, 311
  heterogeneous mixtures, 307
  impurity, 303
  nitrogen-containing tantalum
      films, 221
  phonon, 296ff.
  solid solutions, 306
  substrates, 411
  tantalum films, 341, 342
  temperature coefficient of, 299
  thin films, 310ff., 313
  very thin films, 323

Resistivity of alloys, 306ff.
  function of concentration, 304
  magnetic alloys, 308ff.
  typical binary alloys, 309

Resistivity of tantalum films
  vs sputtering current, 227
  vs sputtering voltage, 228
  vs substrate bias, 231

Resistivity of thin films, 310ff.
  temperature dependence, 314

Resistor adjustment, 344ff.
  anodization, 344ff., 492ff., 661ff.
  mechanical trimming, 347
  thermal trimming, 347

Resistor pads, 533

Resistor patterns, end corrections,
    487, 490

Resistor testing, overload pulse,
    359, 682

Resistors, 473ff.
  adjustment, 344ff.
  aging (see also Aging of resis-
      tors), 349ff., 476, 685
  carbons, 13
  cermet, 14, 339
  chromium, 14, 337
  design, 473ff.
  discrete ceramic, 348
  dumbbell pattern, 492ff.
  early development, 13ff.
  effect of humidity on, 336
  film, 473ff.
  high- and low-valued, 486
  high-frequency performance, 336
  inductance of, 499
  line width, 482ff.
  materials, 13, 329ff., 337
  Nichrome®, 14, 337ff.
  overload pulse test, 358, 682
  palladium-silver glaze, 340
  parameters, 329ff.
  parasitic resistances, 507
  parasitics, 494ff.
  platinum, 13
  power density, 336
  preaging, 359ff., 661
  properties, 474
  series inductance in, 495ff.
  shunting capacitance, 500ff.
  stability of, 336
  stability, power, and size, 474ff.
  stabilization, 349, 661
  tantalum, 341ff.

Resistors *(cont)*
  tantalum nitride, 343
    temperature rise in, 404
    temperature distribution along, 352
  testing, 359
  thermal stabilization of, 661
  tin dioxide, 339
Resolving power, optics, 428
RF heated evaporation, 142ff.
RF sputtering, 216
Roll bonding, 582
Roots blower pumps, 40
  pumping speed curves, 42
Rotary oil pumps, 38ff.
Roughing pumps, 38ff.
Roughing valves, 96ff.
Roughness of substrates, 398

Sapphire substrates, 398
Scotch tape test, 134, 653
Scribe-and-break, 666
Scribing angles, 669
Scribing parameters, 669
Separation of circuits, 666ff.
  sawing, 671
  snapping, 672
Sequences, for thin film circuit processing, 654
Sheet resistance, 480ff.
  definition, 329
  measurement, 330, 679
  near holes, 333
  of thin film conductors, 366, 527
  off-center measurement, 332
  practical limits, 481ff.
  uniformity of, 484ff.
Shipley photoresists, 448, 656
Silicon, chemical vapor plating of, 269
Silicon dioxide, chemical vapor plating of, 270
Silicon monoxide
  capacitor dielectric, 378ff.
  vapor pressure *vs* temperature curve, 153

Silicon monoxide evaporation, 153ff.
  Drumheller source for, 155
Silicon-monoxide tantalum-oxide, duplex dielectric capacitors, 388
Silk screen stencils, 454, 463
Silver halide emulsion, 432
SiO capacitors, dielectric properties, 379
SiO-Ta_2O_5 capacitors, 388
Size-factor compounds, 568
Skin effect, 529
  film conductors, 509ff.
Smoothness, substrate surfaces, 398
Softening points, 406
Solder joints, quality of, 678
Soldering, 591, 673ff.
  contact area and strength, 599
  damage to thin films, 600
  definition, 589
  dissolving of gold in, 601
  effect of foreign materials, 594
  equipment, 595ff.
  infrared heating in, 597
  lead configurations and, 600
  methods, 595ff.
  microtorch, 597
  performing in, 599
  pretinning in, 599
  problems in, 677
  to ceramics, 676
  to glass, 673
  to thin films, 599ff., 673ff.
  whisker growth and, 602
Solders, 591ff.
  expanding, 592
  melting temperatures, 593
Solid-phase bonding, 559, 580, 602ff.
Solid solubility, Hume-Rothery rules, 568
Solid solutions, 564, 567, 579
  resistivity of, 306
Solidification, 566

Sondheimer relation, 310ff.
  approximations, 312
  conductance as function of thick-
      ness, 325
  plot of, 311
  residual mean free path, 316
Sorption, gaseous, 124, 573
Sorption pumps, 41ff.
  pumpdown curves, 43
Spherical aberration, camera
      optics, 427
Sputter evaporation, 156ff.
Sputter-ion pumps, 53ff.
Sputtered atoms, energy of, 198
Sputtered films
  adherence of, 213ff.
  geometric influences, 233ff.
  properties, 219ff.
  substrate influences, 235
Sputtered tantalum films, 219ff.,
      341
  asymmetric, a.c., 228
  deposition parameters, 225ff., 644
  deposition rate, 226, 645
  effect of nitrogen, 220ff., 645
  effect of oxygen, 225
  film thickness monitors, 252
  low density, 226
  resistivity, 341, 645
  sputtering current, 226, 645
  sputtering environment, 220
  sputtering pressure, 226
  sputtering voltage, 226, 645
  substrate bias, 230
  TCR, 645
Sputtering, 191ff., 638ff.
  alloys, 212ff.
  angular distribution, 194
  artificially supported discharge,
      214ff.
  atmosphere, 220ff.
  cathode-substrate distance, 206
  continuous, 638ff.
  deposition parameters, 211ff.
  dielectrics, 217
  discovery of, 191
  distribution of deposition, 207
  effect of contamination, 208ff.
  effect of extraneous structure,
      206

Sputtering (cont)
  effect of substrate bias potential,
      211, 212, 230ff.
  energy distribution, 198
  film deposition by, 204ff.
  film deposition rate, 205ff.
  gas supply system, 241
  general description, 8
  geometric influences, 233ff.
  glow discharge, 199ff.
  history of, 191
  low-density tantalum, 226
  low energy, 215
  low pressure, 202, 214ff.
  manufacturing facilities, 638ff.
  operational variables, 644
  reactive, 210ff.
  sources of contamination, 208ff.
  substrate influences, 235
  theories of, 193ff.
  thermionic electron emission, 214
  using RF fields, 216
Sputtering conditions
  asymmetric a.c., 228ff.
  atmosphere, 220ff.
  deposition rate and, 205
  deposition rate and current, 226
  effect of substrate temperature,
      232
  effect on film properties, 219ff.
  low pressure, 214ff.
  voltage and pressure, 226
Sputtering facility
  bell jar, 236
  in-line, 248
  operation of, 246
  special designs, 248ff.
Sputtering hardware, 237ff., 638ff.
  film thickness monitors, 252
Sputtering theories, 193
  evaporation, 193
  focusing, 194
  momentum transfer, 193
Sputtering yield
  atoms per ion, 195ff.
  definition, 193
  vs ion energy, 196
Square counting, 487ff.
Squares, 330, 486

Stability, thin film resistors, 336, 474ff.

Stabilization of resistors, 349, 661

Step and repeat printer, 458

Step-stress testing, capacitors, 374

Sticking coefficient, 125

Strain effects on tantalum nitride, 320

Strain gauge factor, 319

Strain hardening, 605, 617

Stress effects on thin films, 317ff.

Stress-strain curve, 617

Stresses due to joining, 587, 600, 621

Structures, crystal, 562ff.

Substrate materials
    comparison of, 412ff.
    properties of, 398ff.

Substrates, 395ff.
    chemical stability, 406ff.
    choice of, 395ff.
    circuit separation, 666ff.
    cleaning, 413ff., 642
    composition, 406
    cost, 633ff.
    density, 412
    dielectric constant, 412
    effect of on resistor aging, 408
    effect of Na$_2$O, 408ff.
    electrical conductivity, 411
    etch rate, 407
    flatness, 400
    ideal, 396
    lead attachment, 583, 673ff.
    loss tangent, 412
    manufacture, 396ff.
    mechanical properties, 402
    porosity, 401
    power dissipation, 404
    prescored, 636, 672
    properties, 412
    refractive index, 412
    relative cost, 411, 635
    sawing, 671ff.
    smoothness, 398
    softening points, 406
    surface temperature, 404
    tantalum oxide underlay on, 408

Substrates (cont)
    thermal conductivity, 403
    thermal expansion, 402
    thermal shock resistance, 405
    thermal stability, 406
    water evolution, 401

Surface energy, 570, 581

Surface tension, 571

Surfaces
    defects in, 572
    gas sorption at, 124, 573
    mechanical preparation of, 575
    metallic, 570ff.
    real, 572ff.

Tafel, slope, 275

Talysurf®, 165ff.

Tan delta, 377, 516

Tantalum
    etchant for, 451
    evaporation, 148

Tantalum films
    advantages of, 16
    beta, bcc, and nitride, 221, 222
    carbon doped, 224
    density vs voltage, 229
    effect of sputtering atmosphere on, 220ff.
    low density, 226, 343
    nitrogen doped
        resistivity of, 221
        TCR of, 221
    oxygen doped, 224
        resistivity of, 225
    resistivity vs sputtering current, 227
    resistivity vs sputtering voltage, 228
    resistivity vs substrate bias, 231
    stability of, vs nitrogen content, 223
    structure of, 221, 342
    TCR vs sputtering current, 227
    TCR vs sputtering voltage, 228

Tantalum nitride films
    resistivity, 221
    resistors, 343
    stability, 223, 349ff.
    temperature coefficient of resistance, 221

Tantalum oxide
    anodization constant, 381
    anodization reaction, 273
    capacitor dielectric, 380ff.
    density, 276, 381
    dielectric films, 15
    equivalent weight, 276
    etchant for, 451
    growth rate, 276
    temperature coefficient of capaci-
        tance, 381
Tantalum oxide capacitors, 380
    breakdown voltage, 382
    capacitance change with fre-
        quency, 381
    dielectric properties, 381ff., 392
    dissipation factor, 382
    effect of tantalum type on, 383
    formation voltage, 381
    leakage current, 382
    nitrogen-containing, 384ff.
    nonpolar, 386
    properties of, 381ff., 392
Tantalum oxide films, resistivity,
    225
Tantalum resistivity, function of
    temperature, 341
TCR (see Temperature coefficient
    of resistance)
Temperature coefficient of capaci-
    tance tantalum oxide, 381
Temperature coefficient of resis-
    tance, 333ff.
    and expansion mismatch, 321
    definition, 333
    effect of expansion mismatch on,
        321ff.
    measurement, 681
    of tantalum vs nitrogen pres-
        sure, 221
    of tantalum vs sputtering cur-
        rent, 227
    of tantalum vs sputtering volt-
        age, 228
    tantalum films, 343
    used as a thermometer, 334
Temperature coefficient of resis-
    tivity, 299
    function of temperature, 301
    thin films, 314

Temperature dependence of resis-
        tivity, typical binary alloys,
        309
Temperature distribution, film
        resistors, 352
Termination material, 364, 526,
        648ff.
Testing, 679ff.
    completed circuit, 685
    connections, 622
    overload pulse, 682
    resistance, 682
Thermal conductivity
    gases, 30
    substrates, 402ff.
Thermal expansion, substrates and
        metals, 402
Thermal shock resistance, sub-
        strates, 405
Thermal stability, substrates, 406
Thermal trimming, resistors, 347
Thermocompression bonding, 580,
        582, 604ff.
    failure of bonds, 583
    methods, 606
    parameters in, 611
    to thin films, 609ff.
Thermocouple gauge, 83
Thickness, 480, 679
    maximum and minimum, 481
Thickness distribution, 484
    evaporated films, 158ff.
    on a plane surface, 161
Thickness measurement, 164ff.
    beta back-scattering, 175ff., 679
    in-process monitors, 183ff.
    interferometry, 170ff.
    ionization monitors, 185
    quartz crystal monitors, 184
    resistance monitors, 186
    stylus instrument (Talysurf®),
        165ff.
    weighing, 164
    X-ray fluorescence, 181ff.
Thickness monitors, 183ff.
    during sputtering, 252
    ion gauge, 185
    quartz crystal, 184
    resistance, 186

Thickness nomograph, charge
    needed for evaporation, 162
Thickness uniformity, ways of
    increasing, 162ff.
Thin film adhesion
    changes with time, 586
    conductor films, 365ff.
    evaporated films, 133ff., 653
    factors affecting, 584ff.
    sputtered films, 213ff.
Thin film capacitor materials, 377
Thin film capacitors (see Capaci-
    tors)
Thin film circuits (see Circuits)
Thin film conductors (see Con-
    ductor films)
Thin film deposition, 3ff.
Thin film inductors, 531ff.
Thin film resistors (see Resistors)
Thin film transistors, 15
Throughput, 31ff.
    units of, 32
Tin dioxide, resistors, 339
Tin-lead, phase diagram, 592
Tin-lead solders, 591
Titanium
    etchant for, 451
    glue for thin films, 586
Titanium-copper-palladium, films,
    648
Titanium-palladium-gold, films, 367
Titanium evaporation, 148
TOUCH-TONE® oscillator, circuit
    layout, 551ff.
Transistors, thin film, 15
Transmittance, 439
Trim anodization, 492ff.
    fixturing, 665ff.
    manufacturing facilities, 661ff.
    of several resistors at once, 493ff.
Trimming of resistors (see Resistor
    adjustment)
Tungsten evaporation, 148
Turbidity, photographic emulsions,
    434

Turbomolecular pumps, 56ff.
    pumping speed curves, 58
Twin-tee notch filter, circuit layout,
    544ff.

UHF amplifier, 550

Ultrasonic bonding, 612ff., 681
    examples of, 616
    hot work, 614ff.
    techniques of, 613
    thin films, 614
Ultrasonic cleaning, 258, 414

Vacuum evaporation (see
    Evaporation)
Vacuum gauges, 78ff.
    Bayard-Alpert, 87
    Bourdon tube, 78
    cold-cathode ionization, 91ff.
    diaphragm, 78
    hydrostatic, 79
    ionization, 85ff.
    ionization gauge sensitivity, 87ff.
    location of in vacuum system, 93
    McLeod's 80ff.
    Pirani, 84
    thermocouple, 83
    types, ranges of, 79
Vacuum pumps
    fore- and roughing, 38ff.
    high vacuum, 44ff.
    selection, 67ff.
Vacuum systems, 60ff.
    analysis of, 61ff.
    bell jar system, 60
    design, 101ff.
    elastomers in, 103ff.
    gauge locations, 93
    in-line, 74ff., 638
    leak detection, 72ff.
    maintenance, 71ff.
    materials, 101ff.
    metal joining techniques in, 108
    operating curves, 62
    operation, 69ff.

Vacuum technology, 19ff.

Vacuum valves, 94ff.
  disk, 97
  foreline, 96
  gate, 95
  leak, 99
  needle, 99
  roughing, 96
  ultrahigh vacuum, 97ff.

"Valve metals," 271

van der Waals binding, 124, 561, 573

Vapor degreasing, 415

Vapor plating (*see also* Chemical vapor plating), general description, 6

Vapor pressure, 118ff.

Vapor pressure curves, of various metals, 120ff.

Vaporization, 117ff.
  heat of, 118

Viscous flow conductance, 36

Viscous flow of gases, 35

Voltage coefficient of resistance, 335

Water-break test, substrate cleanness and, 417, 643

Wave soldering, 595, 674, 677

Waviness of substrates, 400

Weathering of substrates, 406

Welders, parallel-gap resistance, 608

Welding, 561
  electron beam, 621ff.
  fusion, 618ff.
  laser, 618ff.
  parallel-gap resistance, 622

Wetting, 569ff., 581
  in soldering, 589

Wetting angle, 589

Wetting angle test, for substrate cleanness, 416, 644

Whisker growth, and soldering, 366, 602

White plague, 583, 606

X-ray fluorescence, 181ff.

Zinc evaporation, 149